Wireless IP and Building the Mobile Internet

Wireless IP and Building the Mobile Internet

Sudhir Dixit
Ramjee Prasad

Editors

Artech House
Boston • London
www.artechhouse.com

Library of Congress Cataloging-in-Publication Data
Wireless IP and building the mobile Internet / Sudhir Dixit, Ramjee Prasad, editors.
    p.   cm. — (Artech House universal personal communications series)
  Includes bibliographical references and index.
  ISBN 1-58053-354-X (alk. paper)
  1. Wireless Internet.   2. TCP/IP (Computer network protocol).   3. Wireless
communication systems.   I. Dixit, Sudhir.   II. Prasad, Ramjee.   III. Series.
TK5103.4885 .W572   2002
621.382'12—dc21

                                        2002027778

British Library Cataloguing in Publication Data
Wireless IP and building the mobile Internet. — (Artech House universal personal
  communications series)
  1. Wireless Internet—Congresses   2. Mobile communication systems—Congresses
  3. TCP/IP (Computer network protocol)—Congresses
  I. Dixit, Sudhir   II. Prasad, Ramjee   III. International Symposium on Wireless
  Personal Multimedia Communications (4th : 2001 : Aalborg, Denmark)
  621.3'845
  ISBN   1-58053-354-X

Cover design by Igor Valdman. Text design by Darrell Judd

International Standard Book Number: 1-58053-354-X
Library of Congress Catalog Card Number: 2002027778

10 9 8 7 6 5 4 3 2 1

To my wife Asha, daughter Sapna, and son Amar, who have been endless sources of strength, encouragement, and purpose in my life

—Sudhir Dixit

To my wife Jyoti, to our daughter Neeli, to our sons Anand and Rajeev, and to our granddaughters Sneha and Ruchika

—Ramjee Prasad

Contents

Preface

इन्द्रियाणां हि चरतां यन्मनोऽनुविधीयते ।
तदस्य हरति प्रज्ञां वायुर्नावमिवाम्भसि ॥६ ७

indriyanam hi charatam
yan mano 'nuvidhiyate
tad asya harati prajnam
vayur navam ivambhasi

As a boat on the water is swept away by a strong wind, even one of the roaming senses on which the mind focuses can carry away a man's intelligence.
  —*The Bhagvad Gita* (2.67)

The Technical Program Committee of the Fourth International Symposium on Wireless Personal Multimedia Communications (WPMC'01), held September 9–12, 2001, in Aalborg, Denmark, decided to organize special invited sessions to highlight the trend of the fusion of the packet IP wireless networks. This task was assigned to us. We realized the importance of the area, as well as the shortage of technical material in a single place in the field of wireless IP and closely related technologies that form the critical success factors. Therefore, we decided to invite the experts who are truly active in the field: the equipment manufacturers, mobile operators, and those working in research laboratories and universities.

Wireless IP and Building the Mobile Internet is the first book to take a comprehensive look at the convergence of wireless and Internet technologies giving rise to the mobile wireless Internet as we know it. In short, the book endeavors to provide an overview of all the elements required to understand and develop the future IP-based wireless multimedia communications and services.

The primary audience of this book is practicing engineers and designers, as well as engineering managers. The book is organized, however, in a format that makes it easily adaptable for a graduate-level textbook. We believe that this book provides sufficient exposure and knowledge in multiple technologies with a good mix of theory and practice to understand the internal working of the wireless Internet. This book attempts to bridge the gap between research in wireless and IP communications by including chapters from experts who have hitherto confirmed their work in their respective specialties with their peers.

The major objective of this book is to focus not only on the latest developments in mobility, wireless, and Internet technologies, but also to integrate these to provide workable end-to-end solutions. We have encouraged the authors to be concerned about the adjoining layers and technologies even though their primary interest is to focus deeply on only one aspect of the technology spectrum. To meet this objective of seamless interworking, we felt that it was good to have some overlap in the subject materials since the context is often different from one focus area to another and the perception of a certain issue or problem can vary widely.

This book covers a broad range of topics and has been organized into several sections: wireless IP evolution, quality of service (QoS) and resource management, TCP/IP in wireless IP networks, handoff, mobility, signaling, and services and applications. We illustrate the coverage of this book with the help of Figure P.1.

We have tried our best to make each chapter complete in itself. This book is a single authoritative source of information both for the industry professional and the academic in the combined field of wireless and Internet protocols. Any feedback that would improve the book or correct any errors is greatly appreciated.

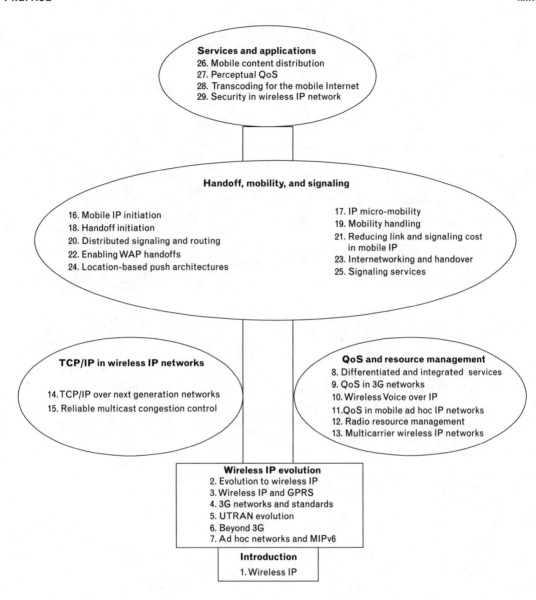

FIGURE P.1   *Illustration of the topics covered in the book and their organization into sections.*

Acknowledgments

The material in this book originates from the *Proceedings of the Fourth International Symposium on Wireless Personal Multimedia Communications* (WPMC'01), held September 9–12, 2001, in Aalborg, Denmark. We would like to thank all our colleagues involved in the Organizing Committee and the Technical Program Committee for their support and cooperation, which made this project not only possible but a huge success as well. Also, the authors and reviewers of each chapter are deeply acknowledged for their diligent and expert work.

WPMC'01 was organized by the Center of Personkomunikation (CPK), the Yokosuka Research Park (YRP) R&D Committee, and the Communication Research Laboratory (CRL). Their support is greatly appreciated.

Junko Prasad helped to prepare the manuscript, freeing us from the enormous editorial burden, for which we are immensely grateful. We thank the wonderful editorial staff of Artech House for their continuous support and advice during the course of writing this book. We especially thank Dr. Julie Lancashire and Ms. Ruth Harris for keeping us on track, and Ms. Rebecca Allendorf for producing this book in record time.

Sudhir Dixit
Burlington, Massachusetts

Ramjee Prasad
Aalborg, Denmark
November 2002

CHAPTER 1

Wireless IP

Sudhir Dixit and Ramjee Prasad

1.1 Introduction

The Internet era started in 1969. A family tree of the Internet is shown in Figure 1.1. The key driver for the *Internet protocol* (IP)-based network is a common application and service environment across multiple types of networks. In fact, the IP has created an open platform for innovative, flexible, and fast service creation, has enabled existing services to be supported, and has provided IP-based mobility for all types of wired and wireless transport in both the access network and the core network.

The target setting is to (1) create a world class all-IP system with rapid time-to-market and future-proof design, (2) enable flexibility for providing new, revolutionary services while ensuring smooth network evolution and service continuity, (3) provide access independent design for globally seamless services, and (4) enable growth of revenue-generating systems now by leveraging the newly emerged wireless data market.

1.2 Wireless IP

The basic concept of wireless IP is shown in Figure 1.2. It is basically a powerful confluence of the network interworking layer and the tetherless connectivity with or without mobility in a heterogeneous networking environment with the promise of seamless connectivity across network subdomains. Combining the best of both wireless and IP technologies has brought us into the era of wireless IP. Wireless IP will enable cost-effective, high-quality IP-based wireless multimedia services, including voice over IP, in large volumes [1–8].

In most developed markets the volume of data traffic has already surpassed that of voice traffic, and this trend will only continue to accelerate.

FIGURE 1.1
*Family tree of the
Internet. Branches and
leaves of the tree are
not shown in chrono-
logical order.*

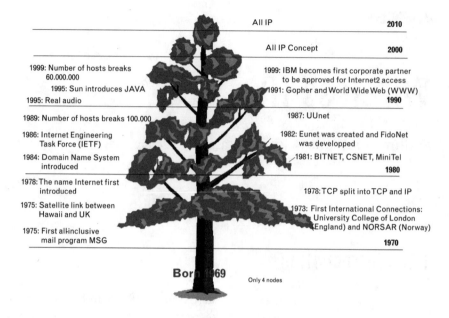

1999: Number of hosts breaks
60.000.000

1995: Sun introduces JAVA

1995: Real audio

1989: Number of hosts breaks 100.000

1986: Internet Engineering
Task Force (IETF)

1984: Domain Name System
introduced

1978: The name Internet first
introduced

1975: Satellite link between
Hawaii and UK

1975: First all-inclusive
mail program MSG

All IP   2010

All IP Concept   2000

1999: IBM becomes first corporate partner
to be approved for Internet2 access

1991: Gopher and World Wide Web (WWW)
1990

1987: UUnet

1982: Eunet was created and FidoNet
was developped

1981: BITNET, CSNET, MiniTel
1980

1978: TCP split into TCP and IP

1973: First International Connections:
University College of London
(England) and NORSAR (Norway)
1970

Born 1969   Only 4 nodes

FIGURE 1.2
What is wireless IP?

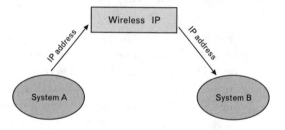

Wireless IP

IP address   IP address

System A   System B

Consequently, operators and vendors have realized that there are major economic advantages to multiplexing all types of traffic over packet switched
networks rather than over circuit switched networks. No doubt, the same
trend is being observed in the wireless mobile world. This evolution is being
supported in the core network and the radio access network by retrofitting
the various *second-generation* (2G) standards and networks and by building
new infrastructure based on *third-generation* (3G) standards.

The vision of enabling end-to-end connectivity has propelled IP to be
adopted as a unifying layer to support a multitude of link layer standards and
technologies. This end-to-end "all-IP" vision has caused everyone to begin
looking beyond 3G systems, commonly referred to as *fourth-generation systems*. Though the 3G systems are primarily limited to cellular/GSM wireless
access limited to terminal mobility, the next generation all-IP systems of the
future will enable both terminal and user mobility across a range of wireless

access networks [e.g., wireless *local area network* (LAN) in hot spots, fixed access, ad hoc networks].

Figure 1.3 shows an all-IP network architecture where access is through a variety of wireless technologies—with the intelligence residing in the access and the backbone primarily providing the packet transport. The various functions, features, and capabilities of the network and services will be supported by specialized servers potentially attached anywhere in the IP network. The new architecture will need to be based on the paradigm that every user is potentially mobile, and the network is to be able to carry all types of traffic, some requiring strict *quality of service* (QoS) and others requiring only best effort service. Some key areas that are being studied aggressively are (1) transporting heterogeneous traffic in a heterogeneous access network using Internet protocols, (2) seamless mobility, (3) seamless QoS and resource management, (4) ubiquitous networking for dynamic ad hoc networks, (5) security, and (6) network services and end user applications.

The IP has emerged as a unifying network layer protocol that transparently works over heterogeneous link layer and physical layer protocols. Efforts are underway to ensure that QoS, signaling, routing, resource management, mobility, and security functions and features are provided at the IP layer and above and are mapped suitably to the lower layers so as to be consistent and meaningful end-to-end. In this book we have focused on the different aspects of wireless IP. Keeping this objective in mind, the book is divided into five different sections: (1) wireless IP evolution, (2) QoS and resource management, (3) TCP/IP in wireless IP networks, (4) handoff, mobility, and signaling, and (5) services and applications.

FIGURE 1.3
All-IP network architecture.

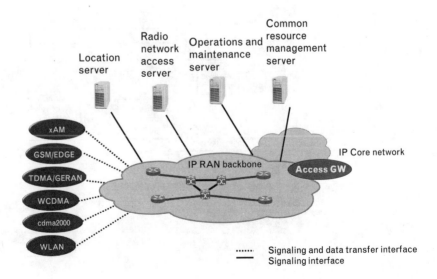

This chapter provides a high-level overview of the challenges that wireless IP poses for the mobile Internet to become a reality for the masses.

1.3   Challenges for the Heterogeneous Environment

Figure 1.4 illustrates an example of a hierarchical, heterogeneous network.

The present-day terrestrial mobile systems operate in the licensed radio spectrum. It is anticipated, however, that networks whose spectrum allocation is not regulated in the ISM band will be abundant, requiring terminals to support multiple air interfaces and corresponding *medium access control* (MAC) layer and radio standards. The problem of incompatibility with different frequency bands and standards in different parts of the world has led to the development of self-configuring multimode phones and terminals capable of adapting to the spectrum and standard available in a particular operating region. Some key technologies operating in the ISM band are Bluetooth, IEEE 802.11a and b, ETSI HIPERLAN I and II, and broadband wireless [1]—all of which are quite suitable for "hot-spot" locations where there is a large concentration of users carrying mobile devices (PDAs, phones, and laptops) and who could benefit from locally available low-cost high bandwidth. Examples of such hot spots are major transportation centers, conference/exhibition halls, museums, and shopping malls.

Most wireless access networks today are point-to-point, and the core and the backbone networks are mostly mesh based. Point-to-point access topology suffers from unreliable connectivity and unbalanced traffic

FIGURE 1.4
*Hierarchical,
heterogenous integrated
networks.*

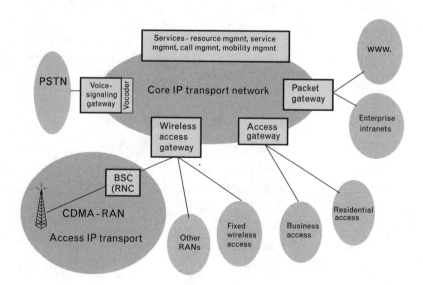

distribution resulting in poor utilization of network resources. It is clear that longer-term mesh topology will be the preferred topology, and the artificial boundaries of *radio access network* (RAN) and *core network* (CN) will disappear. The flat distributed IP network, as it exists today in the Internet backbone, will extend all the way to the end user connected via wireless access. Increasingly the access networks will be heterogeneous, and the IP layer, which will be the integrating common layer across the networks, will need to deal with different access topologies, from full mesh to point-to-point, from dedicated bandwidth to shared bandwidth, and from best effort service to guaranteed QoS, across different link layer technologies. Since different access networks will offer varying bandwidth capabilities, the resulting traffic profiles will be varying as well. Adaptation of (and to) the appropriate radio interface, power control, radio resource, and mobility-enabled access and handover are some of the key requirements stipulated by the technology developers and the operators alike.

Although in the foreseeable future a vast number of applications will be transaction-oriented of short duration, in the long-term the multimedia flows will be of longer duration requiring stringent QoS. The content of the *World Wide Web* (WWW) is already sourced from different locations during the same session; the same will happen in the wireless Internet much sooner than anyone can imagine. The model based on setting up connections prior to data transfer is clearly not workable because of the long latencies involved, especially for short flows. Therefore, an always-connected connectionless model will most likely be the dominant approach. Traffic management and optimum use of radio resources are certainly going to be major challenges since the same infrastructure will need to deal with a variety of flows of different lengths, durations, QoS requirements, and subscription agreements. Wireless spectrum, a finite resource, is closely regulated and licensed to the operators. For example, we illustrate in Figure 1.5 the frequency spectrum allocated to second- and third-generation and beyond networks. We strongly advise the reader to peruse the chapters in Part I of this book for an overview of the evolution of the wireless IP technologies.

1.4 QoS and Resource Management

Future wireless Internet architecture will be based on IP and will utilize IPv6 mobility. Although the legacy IPv4 is by far the dominant and accepted protocol and will probably remain as such for a long time, it has many weaknesses, such as limited address space, lack of mobility support, and poor or unproven support for guaranteed QoS over both wireless and wired links [2]. Many of these limitations will be overcome when IPv6 is universally deployed. From a QoS perspective, the various services and

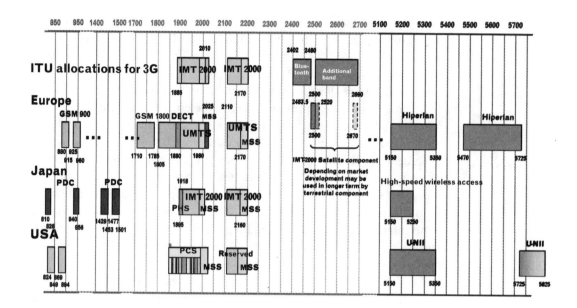

FIGURE 1.5   *Frequency ranges of second- and third-generation networks and beyond.*

applications can be categorized in real-time and nonreal-time classes with different QoS requirements. The real-time applications (and consequently the resulting traffic) can be symmetrical or asymmetrical, putting stringent requirements on delay, delay jitter, loss, and maintenance of QoS in inter-domain/intertechnology handovers. The IP was originally designed as a connectionless best-effort network layer transport protocol without any QoS guarantees, with the vision of keeping the protocol simple, resilient, distributed, self-configuring, and plug-and-play. This has resulted in nondeterministic performance guarantees, and any packet- or bit-level reliable transfer is assured at the transport (TCP) layer. Because of IP's proliferation and the embedded base of deployed IP-enabled network and end user terminals, it is only natural to add QoS support at the IP layer to meet the varying requirements of the users and applications. This would enable the operators and the service providers to start charging for those value-added, guaranteed-quality services depending upon the users' willingness and ability to pay. Technology developers, however, are faced with formidable difficulties in meeting this challenge in a heterogeneous network environment. When and if the IP QoS standards are in place and implemented, the mapping of the QoS mechanisms and parameters to the link layers and the physical layers will be important. These will need to be coordinated with the radio resource and connection admission mechanisms. Uniform mapping of IP QoS over the layer 2 QoS classes (at radio layer and core network) will enable the terminals to roam globally and operate across heterogeneous networks. Parts II and III of this book address many of the above issues in QoS and resource management, and TCP/IP in wireless IP networks, respectively.

Currently proposed QoS techniques, such as *integrated service* (IntServ), *differentiated service* (DiffServ), and *multiprotocol label switching* (MPLS), are either nonscalable, too immature, or both, rendering them unable to enforce and manage end-to-end QoS throughout an IP-based heterogeneous network. Commercial-grade IP telephony requires linkage between call setup, end-to-end QoS setup, interdomain authorization, and accounting. This section describes the IP network model for interdomain QoS for access and the backbone network necessary to support QoS-aware application services in both public and business contexts. In particular, this section provides a survey of exciting protocols and addresses how to effectively combine them across IP-based networks.

There is widespread consumer expectation that commercial IP-based 3G devices and fixed IP devices will need to provide QoS equal to that of cellular digital circuit switched telephony, such as the *Global System for Mobile Communications* (GSM). In order to achieve this, a mechanism is needed to incorporate end-to-end QoS in IP networks [3]. The support of QoS in any network requires the use of network resources, and the primary objective is to allocate and manage all dedicated bandwidth, control jittering, and bound latency (required for real-time and interactive traffic), as well as to meet data rate and reliability commitments. QoS support enables premium services to prioritize the delivery of certain IP packets at the expense of packets carrying best-effort traffic. Thus, best-effort packets will suffer from degraded performance (e.g., delay) when traffic is heavy. Consequently, service providers must either over-dimension the network to ensure adequate capacity, or limit the admission of best-effort users, in order to ensure an acceptable QoS for applications that are not real-time sensitive, such as ftp, e-mail, and Web browsing. QoS delivery in access networks such as *Universal Mobile Telecommunications System* (UMTS) and *wireless LANs* (WLANs) is determined by local network usage policy. Implementing *Session Initiation Protocol* (SIP) servers in such networks can help to enforce policy for all SIP calls. SIP call parameters, such as endpoint addresses, call ID, time, authorization requests, and tokens, must be exchanged with policy servers, trusted authorization and accounting servers, in order to install and tear down QoS policy in *Resource Reservation Protocol* (RSVP)-enabled routers. These parameters may apply to either mobile or fixed end access points. Mobile IP has been standardized by the Internet Engineering Task Force (IETF) to support mobile users and Internet devices.

1.4.1 QoS Network Model

Figure 1.6 illustrates a simple QoS reference model for various real-time IP communications; this covers call setup, QoS reservation, policy, authorization, and payments. The approach used here applies standard end-to-end RSVP in access networks and DiffServ with MPLS in the backbone networks. Key elements in the hierarchical model include the access network, backbone network, and clearinghouse (centralized QoS management unit).

Strict priority for real-time traffic (such as voice) facilitates simple and robust QoS implementation in a private IP network where policy control and individual specific accounting are not required. Extension, however, of the strict priority to real-time traffic between the different domains across the Internet would prevent service providers from exercising policy control and accounting, which would make the services very costly and reduce the incentive for service providers to deploy QoS mechanisms in their networks. Furthermore, the model of strict priority service without policy and accounting may be under pressure when applied to high-bandwidth applications, such as video on demand. Economic trade-offs between the subscription charge and the guaranteed QoS level will be required.

- *Access network*: An IP network to which users directly connect their hosts/clients for IP connectivity. The access network is part of a single administrative domain, such as those operated by *Internet service provid-*

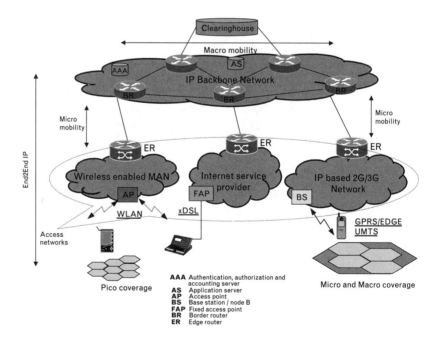

FIGURE 1.6
A reference model for QoS support for IP telephony in an IP-based heterogeneous network.

ers (ISPs), corporate networks, the government, and educational organizations.

• *Backbone network*: One or more backbone networks may be between two or more access networks. The backbone network in our model has no knowledge of individual microflows, such as phone calls between parties connected to access networks.

• *Clearinghouse*: Given the large number of access networks belonging to different administrative domains, it is not possible to have *service-level specifications* (SLSs) between all domains on the Internet. Clearinghouses can facilitate the authorization and logging or accounting between domains for premium services, such as QoS. Current SLSs are static in nature, although there is interest in signaling for dynamic delivery of QoS between service providers, such as in the case of bandwidth-broker-mediated services.

1.4.2   Resource Management Problems

Spectrum resource management is an important topic and will continue to be in the near and distant future. Resource management takes on new dimensions and can no longer be restricted to being a matter of spectrum utilization only. Other important components are mobile equipment management and infrastructure deployment and cost structure.

Future systems are expected to require much higher data rates than current systems, since most of the current resource management systems are not directly tied to any specific development of new methods. Data rates in personal communication systems, however, will in many cases be limited by propagation conditions such as distance loss and multipath. The primary constraining factor is the link budget. Since the required transmitted power increases linearly with the bandwidth, high-speed wireless access will have but a limited range. This will increase the complexity of the resource management schemes.

If the bandwidth as such is not important to the design and performance of *radio resource management* (RRM) algorithms, the traffic characteristics are. The key resource management problems in multimedia type systems are related to the data rates, and delay constraints of traffic in small cell environments will exhibit very large peak-to-average capacity demands. Video users with absolute delay requirements may require considerable portions of the spectrum that they share with e-mail message traffic with no such absolute constraints. Dynamic channel allocation (i.e., statistical multiplexing) will provide even larger capacity gains in these situations than in today's mobile phone scenarios.

1.5 Seamless Mobility and IP

Fast-forward to a few years in the future, and we will be living in an era where seamless mobility across heterogeneous wireless networks will be taken for granted [2, 4]. In today's networks, though, mobility is supported at layer 2 in WLANs and 2G/3G networks, which prohibits roaming across heterogeneous access networks and routing domains. In contrast, mobility support at the network layer (IP) will allow Internet-wide (global) mobility where the layer 2 and physical layers will be completely transparent, albeit at the cost of increased complexity and longer propagation delays. Any issues of complexity, cost, and performance will eventually become irrelevant with the everyday advances in the technologies involved and the volumes that are expected. There are three key reference models for mobility under study in the IETF: *Mobile IP* (MIP), *Handoff Aware Wireless Access Internet Infrastructure* (HAWAII), and *Cellular IP* (CIP). Each reference model has its own pros and cons. Supporting them will require the mobile terminal to be mobility-aware, and the legacy IP protocol stacks that have already been implemented in them will have to be replaced with the new reference implementations. The reader will learn more about seamless mobility in many of the chapters devoted to this topic in Part IV of this book.

Most work until now has focused on terminal mobility, which has already been successfully proven in commercial networks. Terminal mobility allows the network to route calls or packets to a mobile device regardless of the type of network to which it is attached [2]. If we add to this the capability of user or personal mobility (i.e., the user is not tied to a personal terminal), an additional set of requirements and complexities have to be dealt with. Personal mobility enables the users to access their services regardless of their point of attachment or the type of terminal they are using [5], where the device they use may not belong to them. It is akin to the very successful use of the e-mail alias in the Internet today (e.g., Yahoo!, MSN, Hotmail, and AOL) where one can access his or her e-mail from any terminal, any time, and from anywhere in the world as long as he or she is connected to the Internet. Personal mobility allows a user's calls and environment to be forwarded from one terminal to another. The user should be contactable with only one identity that should be mappable to an address where his packets can be routed, and this mapping should not be tied to a single terminal or a single operator since a user may roam from one operator's network to another and his packets may follow paths that may cross many different operators' networks of one or more types. The control of the identity mapping must remain with the user who should ultimately decide who can reach him or her on which network during what times and with what level of security/confidentiality.

A vision of a seamless network of complementary access systems is depicted in Figure 1.7.

1.6   Ubiquity and Dynamic Ad Hoc Networks

Ubiquitous computing or networking has received a great deal of attention recently [9]. Ubiquitous networking refers to the dynamic ad hoc formation of collaborating entities (people and devices), which adapt to network conditions and network types, and are basically access–network agnostic. They self-configure autonomously when nodes and services appear, negotiate, migrate, and disappear [2, 3, 9, 10]. Although there are many different directions being pursued toward a ubiquitous networking infrastructure, there are some key fundamental characteristics that are common to them all. Architecturally, these fundamental characteristics may be classified into the *collaboration-level* infrastructure and *communication-level* infrastructure [10]. The former enables devices to automatically discover each other, form collaborative regions, and interact with each other at the application or service level; the latter enables the exchange of collaboration-level messages and of application data itself by means of the networking infrastructure.

Both the collaboration-level and the communication-level infrastructures will have major implications on all the layers of the protocol stack (including the wireless IP) and vice versa, depending on how soon and to what extent the protocols can be enhanced. In the meantime, the industry is making every effort to use whatever standards are available today to provide the necessary enhancements to deliver dynamic ad hoc networks using the

FIGURE 1.7
Seamless network of complementary access systems.

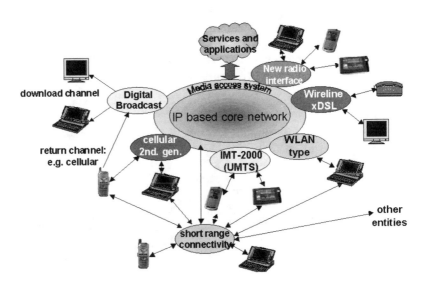

concepts of ubiquitous networking. For example, work is already in progress to interconnect multiple wireless devices on one's person by a *personal* (or *body*) *area network* (PAN or BAN). PAN is a person-centered network concept that enhances our personal experience by connecting all known and future personal devices and equipment within a limited range (of, say, 2m) using wireless techniques (e.g., Bluetooth). PAN will cover the personal space surrounding the person within the distance to which the voice reaches. It will have a capacity in the range of 10 bps to 10 Mbps. Figure 1.8 illustrates the position of PAN with respect to other systems (B-PAN stands for broadband-PAN).

The following are the challenges and open issues for PAN:

- Low-power, low-cost radio integration;

- Definition of possible physical layers and access techniques;

- Ad hoc networking;

- Middleware architecture;

- Security (different security techniques, gatekeeping functionalities);

- Overall system concept;

- Human aspects.

1.7   Security Considerations

Authentication, data security, and privacy are of paramount interest in the wireless networks [2]. Security features in 2G and 3G systems focus on mainly two aspects: one to authenticate the user with a billing system and the other to encrypt the data so that it cannot be eavesdropped. If the call is prepaid, then there is no association of the billing system with the identity of the caller. In future IP networks, this dependence on the operator (and the

FIGURE 1.8
PAN with respect to other systems.

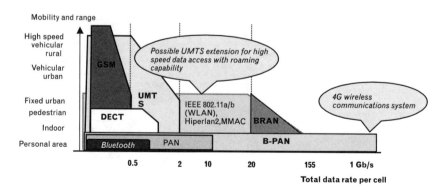

associated billing system) would need to be minimized as much as possible, and any negotiation with respect to authentication and identity disclosure should rest with the end users. The security mechanisms present in today's wireless networks are very different from those present in the fixed networks. There are typically four levels of security applicable in both fixed and wireless networks: (1) authentication at login, (2) end-to-end security at the application level, (3) network level security, and (4) link level security.

In a heterogeneous environment some of the major issues are: (1) how to get two access routers that may have never known each other to trust each other, (2) the level and type of security support may be different at different access routers connected to different types of access networks, (3) the provision of security and trust features for access routers at the client level and ramifications of this at the IP layer, (4) the enabling of different levels of identity authentication and data security (either on demand or depending on the context) without reference to operator, and (5) the optimization of end-to-end security while removing security at certain layers to minimize duplication while the data path crosses multiple technologies and operator domains. The implications on the IP or vice versa may be significant. Chapter 29 examines the various aspects of security in wireless and IP networks.

1.8 Concluding Remarks

An open IP-enabled wireless network will provide vast opportunities for a myriad of new services and applications to be developed, upon which one cannot begin to speculate. The younger generation is already comfortable using mobile devices and PCs while continuously challenging established business models and ethical boundaries. On the data front, *short message services* (SMS) have already been a huge success, and similar efforts are underway now to develop and offer mobile multimedia messaging with a big push for *multimedia messaging services* (MMS) [11]. The MMS can use up to 57.6 Kbps for sending and retrieving data. MMS-compatible handsets are already beginning to ship. It will not be long before location-aware services begin to appear as well. Part V of this book examines the issues of wireless IP from the standpoint of some of the key services and applications that will need to be supported in the near future. The real challenge for the future can be explained by the following equation to achieve IP-based wireless multimedia communications:

$$E \propto m.c^4 \tag{1.1}$$

where E is evolution of wireless communications, m is multimedia communications, and c is consumer electronics, computer technology, communication technology, and contents. Figure 1.9 illustrates the clue to the (e)-/(re)-volution of wireless IP-based multimedia communications. The rest of the book is devoted to deliberating the variables in (1.1).

FIGURE 1.9
*(E)-/(Re)-volution
of wireless IP–based
multimedia
communications.*

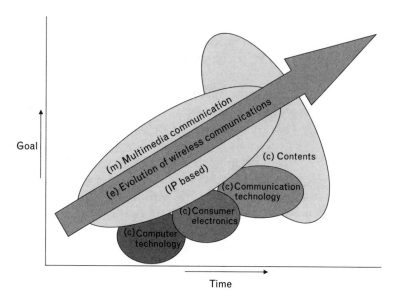

REFERENCES

[1]   Leeper, D. G., "A Long-Term View of Short-Range Wireless," *IEEE Computer*, Vol. 34, No. 6, June 2001, pp. 39–44.

[2]   Dixit, S., "Wireless IP and Its Challenges for the Heterogeneous Environment," *International Journal of Wireless Personal Communications*, August 2002.

[3]   Snoeren, A. C., and H. Balakrishnan, "An End-to-End Approach to Host Mobility," *Proceedings of the 6th Annual International Conference on Mobile Computing and Networking (MOBICOM 2000)*, Boston, MA, August 2000.

[4]   Prasad, R., and M. Ruggieri, (Group Editors), Special Issue on Unpredictable Future of Wireless Communications, *International Journal of Wireless Personal Communications*, August 2002.

[5]   Akyldiz, I., et al., "Mobility Management in Current and Future Communication Networks," *IEEE Network Magazine*, Vol. 12, No. 8, July/August 1998, pp. 39–40.

[6]   Alam, M., R. Prasad, and J. Farserotu, "Quality of Service Among IP-Based Heterogeneous Networks," *IEEE Personal Communications*, December 2001, pp. 18–24.

[7]   Ojanpera, T., and R. Prasad, *WCDMA: Towards IP Mobility and Mobile Internet*, Norwood, MA: Artech House, 2001.

[8]   Prasad, R., W. Mohr, and W. Konhauser, *Third Generation Mobile Communication Systems*, Norwood, MA: Artech House, 2000.

[9]   http://oxygen.lcs.mit.edu/.

[10] Kalafonos, D., internal Nokia memo, 2001.

[11] http://www.3gpp.org.

Part I:
Wireless IP Evolution

CHAPTER 2

Evolution to Wireless IP

Mark Epstein

2.1   Introduction

The press is full of horror stories about the evolution to 3G and wireless IP. The press has reported that equipment is late, that licenses have cost service providers more than they can hope to get paid back, and that consumers may not really want the services that 3G wireless will provide. In May 2001, NTT DoCoMo reported that they would not cover all of Japan with full 3G *wideband code division multiple access* (WCDMA) mobile phone service for at least 3 years. In its May 28, 2001, issue *Newsweek International* reported of the potentially devastating collapse of the race to roll out 3G mobile phones in Europe because of greed and government policy. The article noted that British Telecom was crumbling under a debt of £27.9 billion, had to spin off its wireless division in order to survive, had accepted the resignation of its longtime chairman, and postponed its first trial run of 3G technology. AFX News, on May 24, 2001, reported that Sonera will delay investments in and the rollout of UMTS services in Finland and elsewhere because of a shortage of compatible handsets. This article asserted that Sonera would not likely set another launch date until handset makers could guarantee mass produced phones free of glitches. They noted that some of the handset makers are dealing with phones that consume too much power, leaving the batteries drained after only a few minutes. On the more positive side, initial commercial deployments of 3G cdma2000 phone service have occurred in Korea and in Japan.

The negative press observations neglect the fact that complicated systems take time to develop, test, produce, and deploy. This necessity results in a reevaluation of initial estimates of deployment strategies and timing for the 3G revolution. This chapter addresses alternate wireless evolution strategies for transitioning from 2G to 3G capabilities. Discussed are evolution paths, time to market of the various alternatives, economies of scale, ease of international roaming, and relative performance. This discussion is followed by a presentation of key applications that will drive the momentum to wireless IP, including mobile text, mobile chat, avatars, position location, entertainment, and games. A discussion on providing access to wireless IP users for these applications will focus on platforms that can provide downloadable applications on demand. Also discussed are the supporting infrastructure and middleware to successfully tie third-party developers of these applications and their products with the carriers and subscribers.

2.2   Motivation for High Data Rates and IP

Consumer demand for higher rate wireless connectivity is illustrated in Figure 2.1. Listed in this figure are services, which form the basis for the motivation for 3G and IP connectivity. Many of these services are already available through the wired Internet. Generally, it is the success of the wired Internet and wireless telephony that form the basis for projections of success of wireless IP services. Shown in Figure 2.1 are examples of mobile text communications, mobile chat, avatars, position location services, entertainment, and games. NTT DoCoMo launched some of these services in Japan using a relatively narrowband wireless channel. The terrific success of these i-mode services in Japan is believed to be indicative of potential public demand for these capabilities.

Going beyond the classic wired Internet services by adding services based on position location capabilities will also be of high interest to wireless consumers for safety, security, and commercial applications. Consumer interest in various wireless IP services is illustrated in Figure 2.2, which

FIGURE 2.1
*Motivation for higher
data rates.*

FIGURE 2.2
Consumer preferences.
(Source: Answers
Research Inc.)

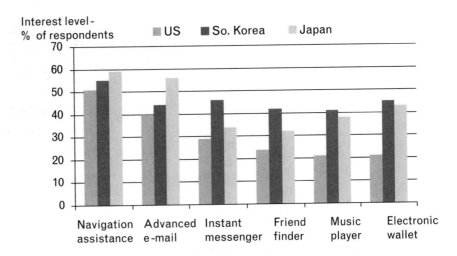

shows the results of surveys by geographic area of consumers' desire for various types of wireless services. A key area here is the provision of e-mail and file download in an anytime, anyplace mode. These needs, in turn, drive the marketplace, as shown in Figure 2.3.

Figure 2.3 shows potential revenue estimates for provision of wireless services. Most noticeable in this figure is the predominance of the Asian marketplace. Early 3G commercial deployments in Korea and Japan and the expected commercialization in China in 2002 have strongly influenced these results.

FIGURE 2.3
Expected revenues.
(Source: Morgan
Stanley Dean Witter,
International Data
Corp., and
Dataquest.)

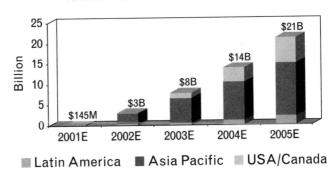

FIGURE 2.4
Technologies for higher data rates.

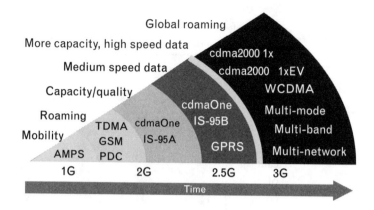

2.3   Radio Interface Technologies

The evolution of wireless technologies for providing higher data rates are shown in Figure 2.4. Shown here for all the current commonly used technologies is the progression from the analog technologies of *first generation* (1G), to the digital technologies that form the basis of 2G, 2.5G, and 3G. Also presented on this chart is the possibility of further progression to technologies providing multimode, multiband, and multinetwork capabilities for the facilitization of worldwide roaming. Possible evolution paths for these technologies as a function of time are shown in Figure 2.5. The other two *International Telecommunication Union* (ITU)-approved 3G technologies

FIGURE 2.5
Evolution paths.

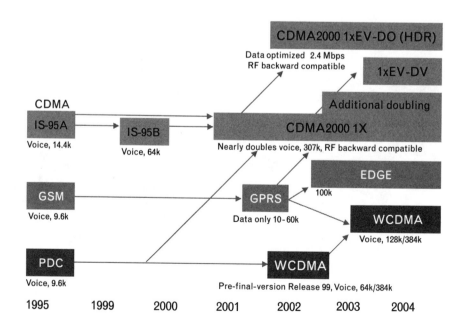

of *Digital European Cordless Telecommunications* (DECT) and *time division synchronous code multiple access* (TD-SCDMA) are not shown because there has been no announced interest by service providers in these radio interface approaches. The designation "additional doubling" refers to novel approaches using two rather than one antenna on subscriber equipment as a method to provide diversity.

It is expected that the process of improving and advancing radio interfaces will continue as the industry moves from 3G toward what may be called *fourth-generation* (4G) technology. The ITU is now in the process of incorporating into its relevant recommendations advanced forms of cdma2000 1x-EV, called DO and DV, and an advanced form of WCDMA, called *HSDPA*.

The actual evolution path chosen by a particular service provider depends on many factors. These include regulatory limitations, such as those applied in Europe (where in many cases the specific technology that may be employed in a particular radio frequency band is specified by the nation's regulatory authority or by the EC); existing technology that may provide an easier path forward owing to corporate and personal experience; financing or other incentives provided by manufacturers; and trade politics that might favor certain regions or domestic industry.

Much of the press reports noted above give the impression that 3G commercial rollouts have been significantly delayed. This is not true. As shown in Figure 2.5, cdma2000 1x was the first commercially deployed, ITU-recommended 3G technology. It was commercially deployed in 2001 by LG Telecom and SK Telecom in Korea. The cdma2000 launches are continuing with deployments by Telesp Celular, Bell Mobility, KDDI, Pegaso PCS, ALLTEL Communications, Verizon Wireless, and Sprint PCS. NTT DoCoMo in Japan launched the pre-final version of WCDMA in late 2001. Further upgrades to this version will be needed for it to be fully compatible with the full-up WCDMA version to be launched in Europe.

Not shown in Figure 2.5 is the 2G technology called TDMA. This radio interface approach does not have an easy migration path to 3G. Accordingly, service providers who are now using TDMA are in the process of deciding whether to migrate on a path followed by GSM or IS-95 adherents.

The time-to-market of the various 3G technologies varies because of the varying maturity of the underlying technology. The generic steps in the deployment process for any new wireless system involve the building and testing of several prototype systems, setting a firm specification for systems and handsets, optimizing system and handset performance, and adding rich feature sets for multimedia. In proceeding through these steps, cdma2000 is ahead of WCDMA in time-to-launch. Thus, cdma2000 is deployed, while WCDMA is still finalizing its standard and optimizing its system and handset performance. This is because cdma2000 is closely based on IS-95A/B and uses its synchronization approach. The initial attempts by the designers of WCDMA to choose other than the basic approaches proven in the IS-95A/B systems have resulted in some additional work, risk, and time delay during the specification and optimization process.

Owing to the earlier commercial launch of cdma2000 and the desire by some GSM carriers to first deploy 2.5G GPRS and perhaps EDGE technology, before migrating to WCDMA, subscriber volume for cdma2000 will initially exceed that of WCDMA. This circumstance is enhanced by the ease of in-band deployment of cdma2000 in 2G spectrum. This may give the cdma2000 approach an earlier learning curve and the related advantage of having less costly subscriber equipment, as a function of time. This is shown in Figure 2.6.

Although some press reports have commented on the difficulties of global roaming among the various ITU-recommended standards, there are technical approaches in development to solve the roaming and interoperability issues. One approach is illustrated in Figure 2.7. This figure shows multimode chip sets now in development, which will bridge the technology gap between cdma2000 and WCDMA. This is a major technological paradigm shift. Thus far, only one company has announced this solution to the roaming problem engendered initially by the inability of the wireless industry to agree on a common chip rate for cdma2000 and WCDMA. This particular approach solves serious problems for roaming within commonly owned or allied networks such as that of Vodafone in Europe and Verizon in the United States.

2.4   Cost Advantages of 3G Wireless IP

Some press articles have been critical of the future of 3G technologies; however, there is a significant advantage to be gained through the use of 3G versus 2.5G approaches. This is illustrated in Figure 2.8, which shows the

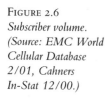

Figure 2.6
Subscriber volume.
(Source: EMC World
Cellular Database
2/01, Cahners
In-Stat 12/00.)

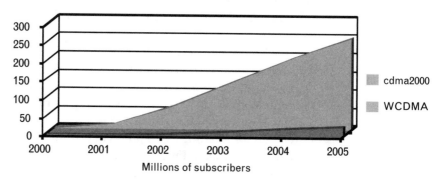

EMC world cellular database 2/01, Cahners in-stat 12/00

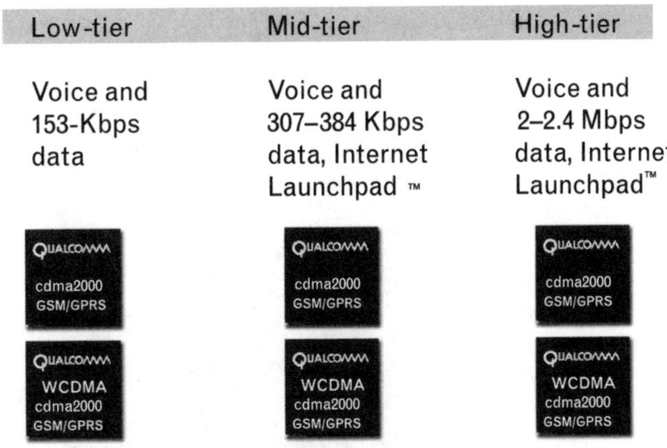

FIGURE 2.7
Multimode chips.

Low-tier	Mid-tier	High-tier
Voice and 153-Kbps data	Voice and 307–384 Kbps data, Internet Launchpad ™	Voice and 2–2.4 Mbps data, Internet Launchpad™

QUALCOMM
cdma2000
GSM/GPRS

QUALCOMM
cdma2000
GSM/GPRS

QUALCOMM
cdma2000
GSM/GPRS

QUALCOMM
WCDMA
cdma2000
GSM/GPRS

QUALCOMM
WCDMA
cdma2000
GSM/GPRS

QUALCOMM
WCDMA
cdma2000
GSM/GPRS

FIGURE 2.8
Data transfer time.

Air interface	Data rate	Download time	
GSM	9.6 Kbps	41 minutes	
IS-95A CDMA	14.4 Kbps	31 minutes	
GPRS	45 Kbps	9 minutes	
IS-95B CDMA	64 Kbps	6 minutes	
cdma2000 1x	307 Kbps*	1.5 minutes	- 1.25 MHz
WCDMA	384 Kbps	61 seconds	- 5 MHz
cdma2000 1xEV	2.4 Mbps	11 seconds	- 1.25 MHz

* Peak data rate for first commercial release of 1x terminals will be 153.8 Kbps

approximate download times for a 3-minute MP3 song using various 2G, 2.5G, and 3G systems. The differences are significant and are further quantified when the download times are translated to costs. The network cost per megabyte is roughly $0.42 for GPRS, $0.07 for WCDMA, $0.06 for cdma2000 1x, and $0.02 for cdma2000 1x-EV. Of special note is the cost associated with downloading data files using the i-mode network, which is $23.44 per megabyte. This approach, initiated by NTT DoCoMo with significant success, is now being planned for use by KPN in the Netherlands. Note, however, i-mode's very high cost compared to the other technologies. The success of i-mode indicates a good future for much less costly 3G approaches. Furthermore, this is indicative that the marketplace will not stop at 2.5G solutions, since 3G provides a much more economically efficient and timely service.

2.5 Technology Trade-Offs for 3G Voice and Data

The capabilities of the current GSM GPRS, EDGE, cdma2000, and WCDMA systems to support voice and data are compared in Figure 2.9.

FIGURE 2.9
Technology comparison.

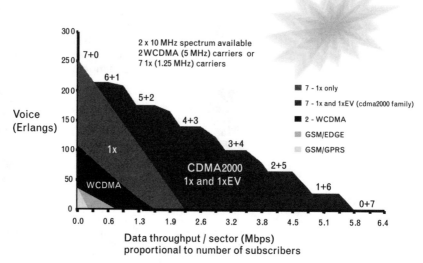

Data is presented assuming 2×10 MHz spectrum is available. The chart displays voice channels in Erlangs versus the data throughput per sector in megabits per second. Similar to the cost advantage figures, Figure 2.9 shows the strong reasons to move to 3G as soon as possible. The bumpy curve at the top of the cdma2000 1x and 1x-EV part of the chart illustrates the capabilities offered when a carrier chooses different mixes of 1x and 1xEV(DO) channels within the available bandwidth. Note that these results are not static. It is likely that all the capabilities shown will improve as revisions to the various standards are implemented. ITU-R Working Party 8F, called IMT-2000 and Systems Beyond IMT-2000, is a key venue for these advances.

2.6   Other Market Segments

There are complementary data solutions in market segments other than wide area cellular. Illustrated in Figure 2.10 are the IEEE 802.11 LAN and Bluetooth approaches. It is likely that all of these will be applied in subscriber equipment offerings to enhance user services whether for telephony, PDAs, or home networks. WCDMA and cdma2000 *wide area network* (WAN) approaches (here called 3G CDMA) may be broadened in devices that support the complementary technologies. Chipsets may soon include all of these capabilities. These would then provide in-home LAN solutions, both house-wide and with Bluetooth to minimize or eliminate interdevice wiring. IEEE 802.11 could also provide subscribers with wideband access at specialized locations in stores, airports, and so forth, as a complement to the

FIGURE 2.10
3G and WLAN.

- 3G CDMA will provide ubiquitous coverage tying WAN and public spaces and providing seamless data roaming

- 802.11
    - High peak data rates
    - Low cost equipment
    - Unlicensed spectrum
    - Technology for home and enterprise

Bluetooth

Peripherals

Wireless connections

WAN cellular connectivity. This arrangement will markedly broaden the user experience.

2.7   Open Application Platforms for Wireless Devices

One of the concerns regularly expressed in the press deals with pessimism concerning the wide and profitable use of services with 3G subscriber equipment. There have been some historic issues associated with the necessity to customize services/applications for each phone. One way to eliminate this problem is to provide an open platform for the offering and download of software packages to the subscriber. Thus, the true "killer application" may be the service itself that supplies multiple downloadable services. NTT DoCoMo is successfully employing this idea today for their i-mode service in Japan. Figure 2.11 illustrates one implementation of this approach, called *Binary Runtime Environment for Wireless* (BREW). This is just for illustration purposes. Similar approaches are being developed: J2ME, for Java 2 Platform Micro Edition; PCA, for Personal Internet Client Architecture; and Club Nokia.

In these approaches, the open platform is made available to any third-party developer through the use of a *software developers kit* (SDK). It is offered free to any developer. The product that is developed by the third party is certified to assure that the software works, has no viruses, and will not harm subscriber equipment to which it is downloaded. The open platforms can support native applications written in C/C++, and the system supports applications written for other environments, including all kinds of browser-based applications as illustrated in Figure 2.11. Middleware services are critical elements to the success of this approach. These consist of billing

FIGURE 2.11
Platform architecture.

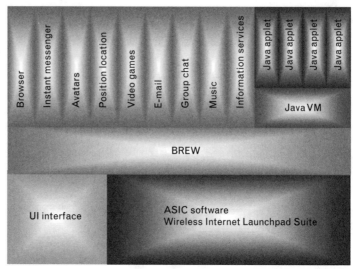

with multiple pricing models, authentication of certified applications, download management for end user purchases, device configuration management, application version control, and administration. Generally, the developer can specify the price he wants for his product, on either a per use, per month, or other method. The revenue is split among the third-party developer, the service provider, and the open platform provider. Interest in this kind of approach has been very high, as indicated by the very large number of third-party developers and service providers who are now participating in these programs.

2.8 Concluding Remarks

This chapter indicates that the evolution toward 3G and wireless IP capabilities is continuing aggressively. Systems are being developed to support these capabilities, and they are in varying states of final development, testing, and commercial deployment. 3G and wireless IP will provide markedly less expensive and more highly capable services than their predecessors. Owing to delays and differences in approaches, service providers have choices to make in the radio interfaces they use and the evolution path they employ to get there. New ideas are being implemented to provide open platforms for the provision of software applications on demand to wireless subscribers that have the potential of revolutionizing services available to the user and to assure the commercial success of 3G and wireless IP systems.

CHAPTER 3

Wide-Area Wireless IP Connectivity with the General Packet Radio Service

Apostolis K. Salkintzis

3.1 Introduction

This chapter investigates how wireless IP connectivity can be provided with the *General Packet Radio Service* (GPRS). The chapter thoroughly discusses the fundamental GPRS concepts, protocols, and procedures and demonstrates the main functionality provided by the GPRS network. The key procedures discussed and explained are the registration procedure, the routing/tunneling procedure, and the mobility management procedure, which enable mobile IP sessions.

The GPRS is a bearer service of GSM, which offers packet data capabilities. The key characteristic of the data service provided by GPRS is that it operates in end-to-end packet mode. This means that no communication resources are exclusively reserved for supporting the communication needs of every individual mobile user. On the contrary, the communication resources are utilized on a demand basis and are statistically multiplexed between several mobile users. This characteristic renders GPRS ideal for applications with irregular traffic properties, because, with this type of traffic, the benefits of statistical multiplexing are exploited (i.e., we obtain high utilization efficiency of the communication resources). A direct effect of this property is the drastically increased capacity of the system, in the sense that we can support a large number of mobile users with only a limited amount of communication resources. The increased capacity offered by GPRS, combined with the end-to-end packet transfer capabilities, constitute the main drive factors for using GPRS in providing wide-area wireless Internet access.

This chapter investigates the key operational and conceptual aspects of GPRS, and we demonstrate how it is used to provide wide-area wireless IP connectivity. We start the discussion with an introduction to GPRS technology and the necessary terminology. Further tutorial material that explains several GPRS aspects can be found in [1–5].

3.2   GPRS Overview

In general, a GPRS network can be viewed as a special IP network, which offers IP connectivity to IP terminals on the go. To provide such a mobile connectivity service, the GPRS network must feature additional functionality compared with standard IP networks. From a high level point of view, however, the GPRS network resembles a typical IP network in the sense that it provides typical IP routing and interfaces to the external world through one or more IP routers.

Figure 3.1 captures schematically this high level conceptual view of a GPRS network. By using shared radio resources, the mobile users gain access to remote *packet data networks* (PDN) through a remote access router (in GPRS terminology this is designated as GGSN). The access to a remote PDN can be envisioned as being similar to a typical dial-up connection. Indeed, as discussed in Section 3.4, a user establishes a virtual connection to the remote PDN. With GPRS, however, a user may "dial-up" to many remote PDNs simultaneously and can be charged by the volume of the transferred data, not by the duration of a connection.

GPRS can offer both *transparent* and *nontransparent* access to a PDN. With transparent access the user is not authenticated by the remote PDN

FIGURE 3.1
High-level conceptual view of a GPRS network.

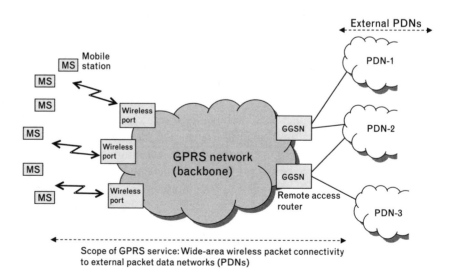

Scope of GPRS service: Wide-area wireless packet connectivity to external packet data networks (PDNs)

and is assigned an IP address from the address space of the GPRS network. With nontransparent access, the user credentials are sent to the remote PDN and the user is permitted to access this PDN only if he or she is successfully authenticated. In this case, the user is assigned an IP address from the address space of the remote PDN. Note that, irrespectively of the access type to a PDN, a user is always authenticated by the GPRS network before he or she is permitted access to GPRS services (this is further discussed is Section 3.3). The nontransparent access is particularly useful for accessing secure intranets (e.g., corporate networks) or ISPs, whereas the transparent access is most appropriate for users who do not maintain subscriptions to third-party ISPs or intranets. As illustrated in Figure 3.1, the GPRS network forms an individual subnet, which contains all users who use transparent access to remote PDNs. External PDNs perceive this subnet as being a typical IP network.

Figure 3.2 illustrates some more detailed aspects of a GPRS network. A *mobile station* (MS) is shown on the left, and the *gateway GPRS serving node* (GGSN) is shown on the right. Among other things, the GGSN offers IP routing functionality and it is used for interfacing with external IP networks. From the MS point of view, the GGSN can be thought of as a remote access router. It must be noted that, in general, the GGSN may interface not only with IP networks but also with several other types of PDNs (e.g., with X.25 networks) [6]. In this chapter we mainly focus on IP and, unless otherwise indicated, it is assumed that GPRS interfaces with IP PDNs only.

FIGURE 3.2
*The GPRS bearer
service.*

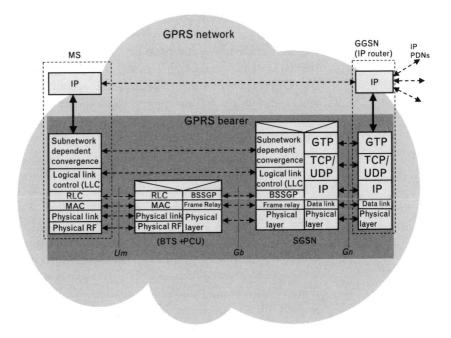

3.2.1   GPRS Bearers

As illustrated in Figure 3.2, the GPRS network effectively provides a GPRS bearer; that is, it provides a communication channel with specific attributes between the MS (the terminal) and the GGSN (the router). Over the GPRS bearer, the MS may send IP packets to the GGSN and it may receive IP packets from the GGSN. As explained below, the GPRS bearer is dynamically set up at the beginning of an IP session (when the user "dials" to a specific PDN) and it can be tailored to match the specific requirements of an application. For example, it can be set up with specific QoS attributes, such as delay, throughput, precedence, and reliability [6].

Figure 3.2 also illustrates the internal structure of a GPRS bearer, which includes the protocols and the GPRS nodes involved in the provisioning of this bearer. A brief explanation follows.

The MS communicates through the radio interface (the so-called Um reference point) with a *base transceiver station* (BTS), which provides mainly physical-layer functionality. In GPRS, the BTS handles the transmission and the reception of packet data on the GPRS physical channels. Data received by the BTS is processed (e.g., decoded and deinterleaved) and then relayed to the next hierarchical node in the GPRS architecture, that is, to the *packet control unit* (PCU). The PCU offers radio resource management and is responsible for allocating uplink and downlink resources to the various MSs on a demand basis. As discussed later, the radio resource allocation is implemented with a packet scheduling function that takes into account the QoS committed to each active MS.

The PCU communicates with the *serving GPRS support node* (SGSN) over a Frame Relay interface (Gb). As discussed below, the SGSN provides mobility management functionality, session management, packet scheduling on the downlink, and packet routing/tunneling. The interface between the SGSN and the GGSN (Gn) is entirely based on IP, typically on IPv4. The GGSN provides mainly routing and optionally screening functionality, and can be considered as a remote access router interfacing with the external PDNs. The fact that we have two IP layers within the GGSN implies that some sort of IP-to-IP tunneling is applied across the Gn interface. This is discussed in more detail in Section 3.3.1.

Not all GPRS bearers feature the same attributes. The particular attributes of a GPRS bearer are specified mainly by the operational mode of each protocol and by the level of precedence applied in the scheduling procedures. For example (see Figure 3.2), in one GPRS bearer, the *logical link control* (LLC) protocol may operate in acknowledged mode, whereas in another GPRS bearer, it may operate in unacknowledged mode. By definition, the acknowledged mode of operation offers increased reliability compared with the unacknowledged mode of operation. Similar distinctions between different GPRS bearers may apply to the *Radio Link Control* (RLC) Protocol and to the *GPRS Tunneling Protocol* (GTP). In addition, one GPRS bearer, which is given high precedence in the scheduling procedures, would typically feature lower delay compared with another GPRS bearer, which is given lower precedence in the scheduling procedures.

3.2.2 GPRS Protocols

The *Subnetwork Dependent Convergence Protocol* (SNDCP) runs between the MS and the SGSN, and it is specified in [7]. It is the first layer that receives the user IP datagrams for transmission. SNDCP basically provides (1) acknowledged and unacknowledged transport services, (2) compression of TCP/IP headers (conformant to RFC 1144 [8]), (3) compression of user data (conformant to either V.42bis or V.44), (4) datagram segmentation/reassembly, and (5) PDP context multiplexing (see Section 3.4). The segmentation/reassembly function ensures that the length of data units sent to LLC layer does not exceed a maximum prenegotiated value. For example, when this maximum value is 500 octets, then IP datagrams of 1,500 octets will be segmented into three SNDCP data units. Each one will be transmitted separately and reassembled by the receiving SNDCP layer.

As discussed in Section 3.4, a *Packet Data Protocol* (PDP) Context essentially represents a virtual connection between an MS and an external PDN. The PDP Context multiplexing is a function that (1) routes each data unit received on a particular PDP Context to the appropriate upper layer and (2) routes each data unit arrived from an upper layer to the appropriate PDP Context. For example, assume a situation where the MS has set up two PDP Contexts, both with type IP but with different IP addresses. One PDP Context could be linked to a remote ISP, and the other could be linked to a remote corporate network. In this case, there are two different logical interfaces at the bottom of IP layer, one for each PDP Context. The SNDCP layer is the entity that multiplexes data to and from those two logical interfaces.

The LLC protocol also runs between the MS and the SGSN, and it is specified in [9]. LLC basically provides data link services. In particular, LLC provides one or more separate *logical links* (LLs) between the MS and the SGSN, which are distinguished into *user*-LLs (used to carry user data) and *control*-LLs (used to carry signaling). There can be up to four user-LLs, while there are basically three control-LLs: one for exchanging GPRS mobility management and session management signaling, another to support SMS [10], and a third to support *location services* (LCS) [11]. The user-LLs are established dynamically, in the context of the PDP Context Activation procedure (see Section 3.4), and their properties are negotiated between the MS and the SGSN during the establishment phase. Negotiated properties typically include (1) the data transfer mode (acknowledged versus unacknowledged), (2) the maximum length of transmission units, (3) timer values, and (4) flow control parameters. On the other hand, the control-LLs have predefined properties that are automatically set up right after the MS registers to the GPRS network (see Section 3.3). It should be noted that each user-LL carries data for one or more PDP Contexts, all sharing the same QoS.

Control-LLs operate only in unacknowledged mode, which basically provides an unreliable transport service. User-LLs operate either in unacknowledged mode or in acknowledged mode, depending on the reliability requirements. The latter mode provides reliable data transport by (1) detecting and retransmitting erroneous data units, (2) maintaining the sequential order of data units, and (3) providing flow control.

Another service provided by the LLC layer is ciphering. This service can be provided in both acknowledged and unacknowledged mode of operation, and therefore, all LLs can be secured and protected from eavesdropping.

The RLC and MAC protocols run between the MS and the PCU, and they are specified in [12]. The RLC provides the procedures for unacknowledged or acknowledged operation over the radio interface. It also provides segmentation and reassembly of LLC data units into *fixed-size* RLC/MAC blocks. In RLC acknowledged mode of operation, RLC also provides the error correction procedures that enable the selective retransmission of unsuccessfully delivered RLC/MAC blocks. Additionally, in this mode of operation, the RLC layer preserves the order of higher layer data units provided to it. Note that, while LLC provides transport services between the MS and the SGSN, the RLC provides similar transport services between the MS and the PCU.

The MAC layer implements the procedures that enable multiple mobile stations to share a common radio resource, which may consist of several physical channels. In the uplink direction (MS to network) in particular, the MAC layer provides the procedures for the arbitration between multiple mobile stations, which simultaneously attempt to access the shared transmission medium. In the downlink direction (network to MS), the MAC layer provides the procedures for queuing and scheduling of access attempts.

The MAC function in the network maintains a list of active MSs, which are mobile stations with pending uplink transmissions. These MSs have previously requested permission to content for uplink resources and the network has responded positively to their requests. Each active MS is associated with a set of committed QoS attributes, such as delay and throughput. These QoS attributes were negotiated when the MS requested uplink resources.

The main function of the MAC layer in the network is to implement a scheduling function (in the uplink direction), which successively assigns the common uplink resource to active MSs in a way that guarantees that each

MS receives its committed QoS. A similar scheduling function is also implemented in the downlink direction.

From the above, it is obvious that every GPRS cell features a central authority, which (1) arbitrates the access to common uplink resources (by providing an uplink scheduling function) and (2) controls the transmission on the downlink resources (by providing a downlink scheduling function). These scheduling functions are part of the functions required to guarantee the provisioning of QoS on the radio interface, and are implementation dependent.

The *Base Station Subsystem GPRS Protocol* (BSSGP) runs across the Gb interface and it is specified in [13]. BSSGP basically provides (1) unreliable transport of LLC data units between the PCU and the SGSN and (2) flow control in the downlink direction. The flow control aims to prevent the flooding of buffers in the PCU and to match the transmission rate on Gb (from SGSN to PCU) to the transmission rate on the radio interface (from PCU to MS). Flow control in the uplink direction is not provided because it is assumed that uplink resources on Gb are suitably dimensioned and are significantly greater than the corresponding uplink resources on the radio interface. BSSGP provides unreliable transport because the reliability of the underlying frame relay network is considered sufficient enough to meet the required reliability level on Gb.

BSSGP also provides addressing services, which are used to identify a given MS in uplink and downlink directions, and a particular cell. In the downlink direction, each BSSGP data unit typically carries an LLC data unit, the identity of the target MS, a set of radio-related parameters (identifying the radio capabilities of the target MS), and a set of QoS attributes needed by the MAC downlink scheduling function. The identity of the target cell is specified by means of a *BSSGP virtual channel identifier* (BVCI), which eventually maps to a frame relay virtual channel. In the uplink direction, each BSSGP data unit typically carries an LLC data unit, the identity of the source MS, the identity of the source cell, and a corresponding set of QoS attributes. The mobility management function in the SGSN uses the source cell identity to identify the cell wherein the source MS is located.

As shown in Figure 3.2, the GTP runs between the SGSN and the GGSN. In general, however, GTP also runs between two SGSNs. GTP provides an unreliable transport function (usually runs on top of UDP) and a set of signaling functions primarily used for tunnel management and mobility management. The transport service of GTP is used to carry user-originated IP datagrams (or any other supported packet unit) into GTP tunnels. GTP tunnels are necessary between the SGSN and the GGSN for routing purposes. This is further explained in Section 3.4.1. They are also necessary for correlating user-originated IP datagrams to PDP Contexts. By means of this correlation, a GGSN knows how to treat an IP datagram received from an SGSN (e.g., to which external PDN to forward this datagram), and an SGSN knows how to treat an IP datagram received from a GGSN (or another SGSN) (e.g., what QoS mechanisms to apply to this datagram and to which cell to forward this datagram).

The following sections investigate the typical procedures carried out to enable wireless IP sessions over GPRS. In particular, we discuss the registration procedure, the routing/tunneling procedures, and the mobility management procedures.

3.3   Attach Procedure

Before an MS can start a wireless IP session, or any other packet data session over the GPRS network, it has to perform the registration procedure. In the GPRS specifications [6], the registration procedure is formally referred to as the *attach procedure*. During this procedure, the MS is actually informing an SGSN that it wants to have access to the GPRS network, and, at the same time, it identifies its comprehensive set of capabilities. In response, the SGSN authenticates the mobile, retrieves its subscription data, and checks if it is authorized to have access to the GPRS network from its current routing area (one or more cells served by the same SGSN [6]). If none of the checks fails, the SGSN accepts the attach request of the mobile and it returns an accept message. After that, the SGSN becomes the *serving* SGSN of that particular mobile.

The entire attach procedure is schematically illustrated in Figure 3.3 in the form of a message sequence diagram. In step 1, the MS sends an *Attach Request* message to the SGSN (labeled as *new SGSN*), which serves the routing area wherein the mobile is located. In the *Attach Request* message, the MS typically includes a temporary identifier, called *packet temporary MS identity* (P-TMSI). This P-TMSI has previously been allocated, possibly by another SGSN (e.g., the *old SGSN* shown in Figure 3.3) and, possibly, in another routing area. However, the P-TMSI is stored in the nonvolatile memory of the MS and, as long as it is valid, it is used as an MS identity. The use of a temporary identity instead of the permanent MS identity [i.e., the *international MS identity* (IMSI)] provides user identity confidentiality. As explained below, the GPRS network allocates a new P-TMSI value to the MS whenever appropriate. Along with the P-TMSI, the *Attach Request* includes the identity of the routing area wherein this P-TMSI was allocated, as well as information related to the MS capabilities (e.g., supported frequency bands, multislot capabilities, ciphering capabilities).

In step 2, the new SGSN tries to acquire the permanent MS identity (i.e., its IMSI). If the P-TMSI included in the *Attach Request* message has previously been allocated by the new SGSN, then the new SGSN also

FIGURE 3.3
*GPRS attach
procedure.*

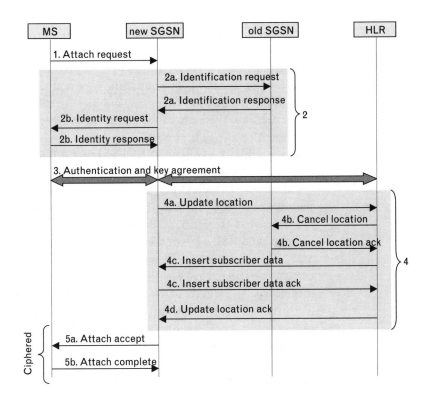

knows the IMSI of the MS. In the example shown in Figure 3.3, however, we assume that the P-TMSI has previously been allocated by the old SGSN. Therefore, the new SGSN may try to contact the old SGSN and request the IMSI value that corresponds to the P-TMSI reported by the MS. This is accomplished in step 2a with the *Identification* messages exchanged between the two SGSNs. We have to note that the address of the old SGSN is derived by the new SGSN with the aid of the *routing area identity* (RAI) included in the *Attach Request*. The exact mapping between an RAI and an SGSN IP address is implementation specific and can typically be based on preconfigured mapping tables or DNS queries.

If the new SGSN cannot acquire the MS's IMSI value in step 2a (e.g., because the old SGSN has deleted the relevant information, or because the IP address of the old SGSN can not be resolved), then the new SGSN requests from the MS to send its permanent identity. This is accomplished in step 2b. The obvious drawback of this step is that it introduces additional signaling over the radio interface and it compromises the user identity confidentiality, since the IMSI is transmitted unciphered on the radio interface.

In step 3, the authentication and key agreement procedure is executed. During this procedure, the new SGSN contacts a *home location register* (HLR), which maintains the subscription data of the identified IMSI, and requests from this HLR the authentication data to authenticate the MS. The address of the appropriate HLR is derived by translating the routing information contained in the IMSI value. An HLR is typically accessible over the international SS7 network and either MTP transport or IP transport can be used for SS7 signaling. The authentication and key agreement procedure is identical to the one used in GSM, and more details can be found in [14, 15]. Typically, after the authentication and key agreement procedure, ciphering is enabled on the radio interface, and therefore, additional messages transmitted on this interface are enciphered. In GPRS, the ciphering function is performed at the LLC layer [9].

In step 4 of Figure 3.3, the new SGSN tries to update the HLR database with the new location of the MS. For this purpose, it sends an *Update Location* message to the HLR containing its own IP address, its own SS7 address, and also the IMSI value of the MS. Subsequently, the HLR informs the old SGSN that it can now release any information stored for this MS. This is done with the *Cancel Location* message. Typically, when the old SGSN receives this message, it will release the previously allocated P-TMSI for this MS (and make it available for reallocation), it will delete any other information possibly stored for this MS, and it will respond with a *Cancel Location Ack* message. In step 4c, the HLR sends to the new SGSN the GPRS subscription data of the MS. At this point, the new SGSN may perform several inspections (e.g., it may check if the MS is allowed to roam in its current routing area). If none of the checks fails, then the new SGSN builds up a *GPRS Mobility Management* (GMM) Context for this MS and returns a positive acknowledgment to the HLR. On the other hand, if an inspection routine fails (e.g., because of roaming restrictions), the new SGSN sends a negative response to the HLR and subsequently it sends an *Attach Reject* message to the MS, including the specific reason for rejecting the attach request. The GMM Context can be considered as a database record that holds GPRS mobility management information pertaining to a specific MS. Such information includes the IMSI value of the MS, the routing area and the cell where the MS is currently located, the P-TMSI allocated to the MS, the ciphering algorithm used to encipher packets for this MS, the GPRS capabilities of the MS, and the authentication data that can be used to authenticate the MS in the future. In step 4d, the HLR acknowledges the location update that was previously requested in step 4a.

In step 5a, the new SGSN sends an *Attach Accept* message to the MS to indicate that the MS has successfully been registered for GPRS services. Typically, with the *Attach Accept* message, the new SGSN assigns a new P-TMSI value to the MS. At the final step (5b), the MS responds with an *Attach Complete* message, which acknowledges the correct reception of the new P-TMSI value. Note that the messages transmitted in steps 5a and 5b are typically enciphered, therefore, the new P-TMSI value cannot be eavesdropped.

3.4   Setting Up PDP Contexts

After a successful GPRS attach procedure, the MS is permitted to use the mobile GPRS services in a secure fashion (a security context is established between the mobile and the network). Further actions, however, are needed for accessing an external PDN. In particular, a virtual connection has to be set up with that PDN. This is accomplished with the (formally referred to) *PDP Context Activation* procedure.[1] Roughly speaking, this procedure can be conceptually associated to the well-known dial-up procedure used over the *Public Switched Telephone Network* (PSTN) to establish connectivity, for example, with ISPs. However, GPRS PDP Contexts (virtual connections) operate in connectionless mode, as opposed to the connection-oriented mode of the PSTN dial-up connections. In this section we discuss the concepts behind GPRS PDP Contexts and the PDP Context Activation procedure.

As mentioned before, the GPRS network can be considered an access network that offers connectivity between a number of MSs and a number of external PDNs. For this purpose, the GPRS network offers access ports where the MSs can be connected and access ports whereto the external PDNs can be connected. This concept is schematically illustrated in Figure 3.4. This figure also depicts two established PDP Contexts, one for MS A and one for MS B.

Each connection between the GPRS network and an external PDN features a unique official name, similar to a domain name used in the Internet. This unique official name is formally called the *access point name* (APN) and it is represented as: PDN_name.PLMN_name.gprs. The PDN_name is a sequence of labels in the form label1.label2. … and identifies an external PDN. The PLMN_name identifies the *Public Land Mobile Network* (in this case, the GPRS network) that is used to provide access to the external PDN. The encoding of PLMN_name depends on the *mobile country code* (MCC) and the *mobile network code* (MNC) allocated to the given PLMN. For instance, for a PLMN with MCC = 10 and MNC = 202, the PLMN_name is mnc202.mcc010. To simplify the illustration, however, the PLMN_name in Figure 3.4 is shown either as PLMNA or PLMNB. The APN names shown in this figure are typical examples, used to explain the APN structure.

Each PDP Context is characterized by (1) a specific PDP type (e.g., IPv4, IPv6, X.25, or PPP), which specifies the type of the payload transferred on the PDP Context, (2) a specific APN, which represents an external PDN, and (3) by a specific GPRS bearer (i.e., by specific transmission properties). The GPRS bearer is a key characteristic of a PDP Context

---

1   The term "PDP Context," used instead of IP Context, emphasizes the fact that GPRS supports not only IP contexts but also other types of packet contexts, such as X.25 and PPP.

FIGURE 3.4
*Establishment of PDP
Contexts for accessing
external PDNs.*

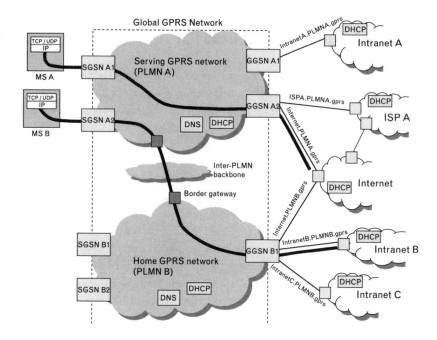

because it specifies QoS properties such as reliability, delay, throughput, and precedence.

It is important to note that a GPRS mobile may have one or more simultaneously active PDP Contexts. This means that one GPRS mobile may simultaneously exchange data with one or more external PDNs, for example, with one that provides Internet access and with another one that provides access to a corporate intranet. Of course, this is not possible with a single PSTN dial-up connection.

The message flow sequence for establishing a new PDP Context is illustrated in Figure 3.5. When an MS wants to establish a new PDP Context, it sends a specific signaling message (*Activate PDP Context Request*) to its serving SGSN. This message specifies all the previously mentioned characteristics of the requested PDP Context. The SGSN checks the requested APN

FIGURE 3.5
*PDP Context Estab-
lishment—message
sequence diagram.*

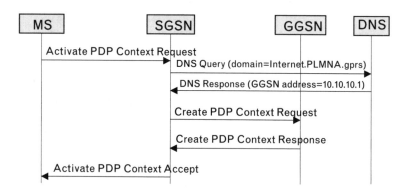

and it identifies (e.g., by using the DNS system) the IP address of the GGSN that provides access to that APN. This GGSN may be located either in the serving GPRS network or in the home GPRS network (see Figure 3.4). If the MS specifies only the PDN_name (e.g., Internet), instead of the full APN name, the SGSN will first try to use a GGSN in the serving GPRS network. If that fails, it will then try to locate a GGSN in the home GPRS network. If, on the other hand, the MS specifies the full APN name in the *Activate PDP Context Request* (e.g., Internet.PLMNA.gprs), a GGSN in PLMN A may only be used to offer connectivity to the Internet. It is evident that for establishing a PDP Context like the one shown in Figure 3.4 for MS B, specific inter-PLMN connectivity means must exist and the operators of PLMN A and PLMN B must have established a roaming agreement.

After identifying a GGSN, the SGSN sends a GTP signaling message (*Create PDP Context Request*) to that GGSN to request the activation of the requested PDP Context. Typically, the GGSN checks if the MS is authorized to access the requested APN, and, if so, it allocates a new IPv4 address to this PDP Context (assuming that the requested PDP type is IPv4). It must be pointed out that the GGSN may request a new IPv4 address either from an internal *Dynamic Host Configuration Protocol* (DHCP) server of from an external DHCP server located in the requested PDN. In the first case, the MS is allocated an IPv4 address from the address space of the serving or home GPRS network, and the MS becomes a new IPv4 node within this network. In the latter case, however, the MS is allocated an IPv4 address from the address space of the external PDN, and it effectively becomes a new IPv4 node inside this PDN. This is equivalent to the case where the access to PDN is accomplished through a dial-up connection. It is typically used when the external PDN is an intranet, which may use private (rather than public) IP addresses.

Under normal conditions, the GGSN accepts the request to create the new PDP Context and returns a positive GTP response (*Create PDP Context Response*) to the SGSN. Subsequently, the SGSN returns an accept message (*Activate PDP Context Accept*) to the MS, which includes the IPv4 address allocated to the new PDP Context.

At this point, a new PDP Context (i.e., a new virtual connection) has been established. One aspect to be highlighted here is that the establishment of a PDP Context does not involve the reservation of dedicated communication resources in the GPRS network. This applies to both the radio interface and the wireline part of the GPRS network. The establishment of a PDP Context involves only the storage of new information in the GPRS nodes (i.e., the creation of new PDP Context records in the SGSN and the GGSN). This new information is subsequently used to route the packets correlated with that PDP Context. This routing procedure is discussed in the next section.

3.4.1 Routing and Tunneling

After having established a PDP Context, a tunneling procedure is used to transfer PDP packets from the MS to the GGSN. Assume, for instance, that a PDP Context, of type IPv4, has been established between the MS and the GGSN shown in Figure 3.6.

In this case, each IP packet transmitted from the MS is put into an envelope that carries two important addressing identifiers: the *traffic flow identity* (TFI; see [12]) and the *network service access point identity* (NSAPI). The TFI effectively identifies an active GPRS MS in a certain cell, and the NSAPI identifies one of the PDP Contexts that has been activated by that GPRS MS. The PCU that receives this packet translates the TFI into a *temporary logical link identifier* (TLLI) and forwards the packet to the SGSN. The TLLI is another MS identifier, which, as opposed to TFI, is decoupled from the cell wherein the MS is located. In particular, the TLLI is a unique identifier in the SGSN and is used to identify a specific MS served by that SGSN. It is essentially derived from P-TMSI, which is another identifier for the MS. The difference between TLLI and P-TMSI is their range of applicability: the first is applied as an identifier in the LLC protocol, whereas the second is applied as an identifier in the GMM protocol (this is a special signaling protocol that handles GPRS mobility management issues; see [6]). Since TLLI is derived from P-TMSI, a unique TLLI is also assigned to every MS when it is registered with an SGSN.

The SGSN that receives the packet from the PCU tries to correlate this packet with a preestablished PDP Context. For this purpose, the SGSN searches its PDP Context database and identifies the PDP Context that has stored TLLI and NSAPI values matching the TLLI and NSAPI values contained in the envelope of the received packet. From the information

FIGURE 3.6
*Tunneling IP packets
from the MS to the
GGSN.*

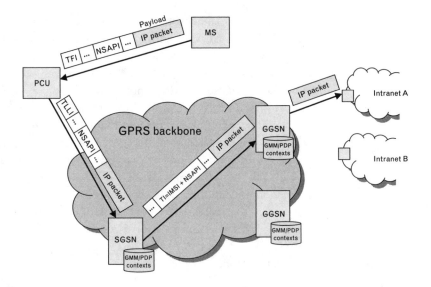

contained in the identified PDP Context, the SGSN finds out the IP address of the GGSN associated with this PDP Context. Subsequently, it makes up a new IP packet, it addresses this packet to the identified GGSN and encapsulates in it the original IP packet transmitted from the MS (this is called IP-IP encapsulation or tunneling). Afterwards, this new packet is transported through the GPRS IP backbone to the addressed GGSN. Note that, in general, the new IP packet may have encapsulated other types of payload, such as X.25, PPP, and IPv6. The type of the payload will match the type of the PDP Context.

The envelope of the IP packet transmitted by the SGSN contains a *tunnel identifier* (TI), which is the concatenation of the MS's IMSI and the NSAPI. The TI is used by the receiving GGSN to correlate this packet with the correct PDP Context. When the GGSN identifies the PDP Context that has a stored TI, which matches the TI in the envelope of the received packet, it discovers the APN associated with this packet and effectively knows the external PDN to which the payload (i.e., the original IP packet transmitted by the MS) should be forwarded.

In the downlink direction, the routing procedures carried out in the GGSN and SGSN are similar. In this case, the GGSN identifies from the destination address of an inbound packet the PDP Context associated with this packet. It then identifies the SGSN address associated with this PDP Context and forwards this packet to that SGSN, after including in the header the correct TI value. In a sense, the packet is sent to the SGSN over a particular tunnel, identified by the TI value. The SGSN uses the TI to identify the associated PDP Context record in its database. From the contents of the identified PDP Context record, the SGSN finds out the TLLI of the target MS and finally it forwards the packet to that MS through the correct PCU. The correct PCU is the one wherefrom the last uplink packet from that MS was received.

3.5 Mobility Handling

Throughout this section, we consider an MS, which is communicating with an Internet host—let us say that it is downloading a file from that host. Our main effort is to illustrate how the file transfer can be sustained when the MS is on the move and roams between different radio access points (i.e., base stations). In such situations, the GPRS network needs to dynamically cope with the location changes of the MS and carry out procedures to modify the associated PDP Contexts according to the identified location changes. All these procedures are typically termed as mobility management procedures and are the basis for all mobile networks. In this section, we focus only on the mobility management procedures executed when an MS is in the packet transfer mode. The mobility management procedures executed when an MS is in the idle mode (i.e., involved with no packet transfer) are not discussed.

Figure 3.7 is a network schematic diagram that will be used throughout the discussion. In this figure, the lines connecting the various network elements are used merely to illustrate the connectivity between these network elements. They do not imply that the network elements are physically interconnected by point-to-point links. For efficiency and cost reasons, it is common in practice to deploy sophisticated transport means to interconnect the network elements. For instance, between an SGSN and a PCU there is typically a frame relay network. In this case, the line connecting an SGSN with a PCU corresponds to a permanent virtual circuit of the frame relay network.

At this point, we assume that the MS has established an appropriate PDP Context and is currently within the coverage area of BTS1 receiving a series of downlink packets, each one belonging to the ongoing file transfer session. From Figure 3.7 we note that every downlink packet traverses a series of network nodes in order to be delivered to the MS (i.e., from the GGSN to SGSN1, to PCU1, and finally to the BTS1). The routing procedures used to transfer the packets between successive nodes were explained in Section 3.4.1. The series of network nodes traversed by the packets belonging to the same PDP Context define the *transmission path* of the PDP Context. As we will see below, the transmission path of a PDP Context changes dynamically (e.g., from one SGSN to another) in order to facilitate the location changes

FIGURE 3.7
*Cell change—new cell
in the same routing
area.*

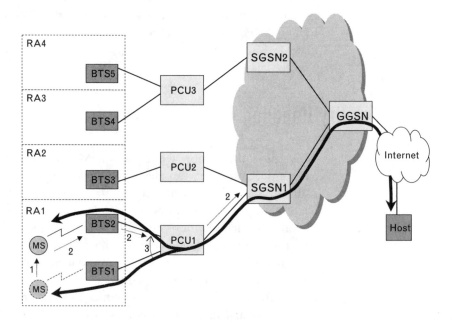

of the MS. However, the GGSN in the transmission path of a PDP Context can never change, and therefore, it serves as an anchor point. This anchor point effectively hides the mobility of the MSs and makes possible for an external PDN to reach a specific MS through the same GGSN no matter where the MS is located.

3.5.1 Cell Change

Let us now assume that the MS starts moving towards the BTS2 (see arrow 1 in Figure 3.7). At some instant, the *radio resource* (RR) layer in the MS will identify that BTS2 can provide better communication quality and will camp on a *radio frequency* (RF) channel controlled by BTS2. This will happen by suddenly switching RF channels and camping on a new one. This procedure is known as *mobile originated handover* since the handover from one cell to another is decided and performed by the MS alone. In GPRS, this procedure is also referred to as *cell reselection*.

Note now that both PCU1 and SGSN1 will not know that the MS has moved to another cell until the MS makes an uplink transmission in the new cell. For some time, therefore, the connection with the MS is inevitably lost, and consequently, downlink packets that may be sent by PCU1 are not received by the MS. This means that during a handover, the packets transmitted by SGSN1 in unacknowledged LLC mode will be lost. On the other hand, packets transmitted by SGSN1 in acknowledged LLC mode (remember that the LLC mode is specified during the establishment of a PDP context) will not be lost but will stay unacknowledged and will be retransmitted later, when the communication with the MS is made feasible. These recovery procedures are handled by the LLC layer, which copes with the occasional blackouts that may happen due to the MS mobility. Here we observe that, even when all single-hop links between the MS and the SGSN1 are perfectly reliable (can transfer data with no errors), the link between the MS and the SGSN can still be unreliable. This observation explains the need for the LLC Protocol—a reliable data link protocol between the MS and the SGSN.

After the handover procedure, the RR layer monitors the broadcast control channel of the new cell. Over this channel, BTS2 transmits the cell ID of the new cell and a RAI, which identifies the *routing area* (RA) wherein this cell belongs. The RR layer will inform the GMM layer that the cell ID has changed but the RAI is still the same (according to Figure 3.7, BTS1 and BTS2 belong to the same RA). In response, the GMM layer will command the LLC layer to transmit a *NULL* frame on the uplink (arrow 2). This is a special LLC frame, which aims to notify the network about the cell change. When this *NULL* frame is received by the LLC layer in SGSN1, the cell change is recorded and subsequent downlink packets are forwarded via BTS2 (arrow 3). This procedure is illustrated in Figure 3.7. Any downlink packets that were sent from SGSN1 to PCU1 during the blackout period are transmitted in the old cell and they are never acknowledged by the RLC layer. Typically, these packets are discarded as soon as their lifetime expires.

3.5.2   Intra-SGSN RA Change

Suppose now the MS moves further and, suddenly, the RR layer makes another handover, this time from BTS2 to BTS3 (see Figure 3.8). This again is a cell change and what was mentioned in the previous section applies here too. In this case, however, the routing area changes too and the RR layer in the MS informs the GMM layer that the mobile has entered into a new RA. In response, the GMM layer does not send a *NULL* frame, but rather it sends an *RA Update (RAU) Request* message (arrow 2). This is done because the MS does not know if the new RA is handled by the same SGSN (SGSN1), and therefore, it has to include additional information in its uplink transmission. This additional information is included in the *RAU Request* message and it can be used by a new SGSN to retrieve subscription and other mobility-related information about that MS. This is explained in more detail in the next section. In the example shown in Figure 3.8, the *RAU Request* will reach SGSN1 and will be treated merely as a cell change. That is, it will simply notify the SGSN1 that the MS can now be reached in a new RA (i.e., through PCU2 and BTS3). The SGSN1 will confirm that the MS is eligible to roam to the new RA and will accept the *RAU Request* by replying with an *RAU Accept* message. Subsequently, it will change the

FIGURE 3.8
Cell change—new cell in another RA handled by the same SGSN.

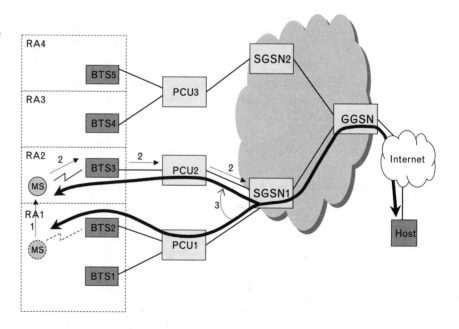

transmission path of the PDP context from PCU1 to PCU2 (arrow 3). This means that further downlink packets related to that PDP Context are sent via PCU2.

3.5.3   Inter-SGSN RA Change

When the MS performs a handover from BTS3 to BTS4, as shown in Figure 3.9, it again transmits a *RAU Request* message (arrow 2). This time, however, the new routing area is controlled by another SGSN, namely, the SGSN2. At this point, SGSN2 needs to acquire some information about the MS. For this purpose, it sends an *SGSN Context Request* message to SGSN1, asking for the needed information (arrow 3). The address of SGSN1 is effectively derived from the RAI parameter included in the *RAU Request* message. Only now, the SGSN1 identifies that the MS has moved to another routing area and it stops sending downlink packets to the MS (3a).

Note that, between the instant where the handover took place and the instant where the SGSN1 received the *SGSN Context Request* message, another blackout period exists. During that period, the SGSN1 could have been transmitting downlink packets to the MS in the context of the ongoing file transfer. These packets would be unacknowledged and would need to remain buffered at the SGSN1. It is important to note also that, the GGSN still assumes that the MS is reachable through SGSN1 and could be transmitting new downlink packets to the SGSN1. The latter would need to buffer these packets too. Observe that, if the transport protocol in the GPRS backbone is based on *User Datagram Protocol* (UDP), which does not support

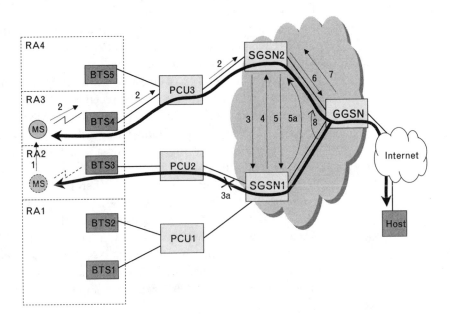

FIGURE 3.9
Cell change—new cell in another RA handled by another SGSN.

any flow control, when SGSN1 runs out of buffering resources, it has no means to signal that to the GGSN, and therefore, downlink packets can actually be lost. That observation justifies that UDP transport is unreliable, even when no transmission errors occur in the backbone links. If, during the PDP establishment, no reliable transport in the GPRS backbone is requested, then the applications running at the MS and the host are responsible for correcting the packet drops that may happen in the GPRS backbone. In our example, the MS and the host could deal with such potential drops by carrying out the file transfer over a reliable transport protocol, such as the *Transport Control Protocol* (TCP).

Again, any potential downlink packets that were transmitted to PCU2 before the SGSN1 was informed about the RA change, will be discarded later on.

An interesting situation arises when the *RAU Request* message sent by the MS (arrow 2) is lost (e.g., due to bad radio conditions). (Typically, another *RAU Request* would be transmitted after 15 seconds.) In this case, it would take quite a long time for the SGSN1 to identify that the MS has changed routing area. During this period, the SGSN1 (1) would keep buffering any new packets sent by the GGSN and (2) would periodically retransmit the buffered downlink packets that remain unacknowledged. To tackle such situations, a careful dimensioning of the buffering resources of the SGSN is required. In addition, if the SGSN1 attempts the maximum number of LLC retransmissions before receiving the *SGSN Context Request* message, the LLC connection would be released and any activated PDP Contexts would be deactivated. In this situation, the file transfer would effectively be dropped. When setting up the LLC parameters, these situations must be taken into account.

Let us now continue with the normal message flow. As soon as the SGSN1 receives the *SGSN Context Request* message (arrow 3), it will reply with an *SGSN Context Response* message (arrow 4) passing the requested information to SGSN2. The latter sends an *SGSN Context Ack* message (arrow 5), which verifies that it has received the requested information and it is ready to receive any buffered packets for the MS that are still unacknowledged. At this point, the SGSN1 forwards all the buffered packets for the MS to the SGSN2 (arrow 5a) within a new tunnel. At the same time, the SGSN2 sends an *Update PDP Context Request* (arrow 6) to the GGSN to inform it that any further downlink packets for the MS should be forwarded to SGSN2. In response, the GGSN replies with an *Update PDP Context Response* (arrow 7). Now the transmission path of the PDP context changes (arrow 8) to accommodate the location change of the MS.

Note that in the case of inter-SGSN routing area change, all the LLC connections in the MS are released and new LLC connections are established with the SGSN2. After the short interruption required for modifying the PDP Context and for reestablishing the new LLC connections, the ongoing file transfer is resumed.

3.6 Summary

In this chapter, we discussed the key GPRS concepts and procedures, and we demonstrated how these procedures enable the provision of wireless packet data connectivity services, including wide-area wireless IP connectivity. In particular, we described the registration procedure, the activation of virtual connections (PDP Contexts), the routing and the tunneling of the data in the GPRS backbone, and, finally, we discussed how wireless IP connectivity is sustained while a user roams between different areas.

The discussion made it apparent that the GPRS system is a versatile and cost-effective solution for the provision of wireless packet data connectivity. It provides increased capacity and efficient utilization of the communication resources, and it can support different types of packet data protocols, such as X.25, IPv4, and IPv6. In addition, it supports access to both intranets and extranets, and it can offer roaming capabilities, which would ultimately provide for ubiquitous access to data facilities.

REFERENCES

[1] Salkintzis, A. K., "Mobile Packet Data Technology," in F. E. Froehlich and A. Kent (Eds.), *Encyclopedia of Telecommunications*, Vol. 18, Article 346, New York: Marcel Dekker, 1999, pp. 65–103.

[2] Priggouris, G., et al., "Supporting IP QoS in the General Packet Radio Service," *IEEE Network*, September/October 2000, pp. 8–17.

[3] Kalden, R., et al., "Wireless Internet Access Based on GPRS," *IEEE Personal Commun. Magazine*, April 2000, pp. 8–18.

[4] Cai, J., and D. J. Goodman, "General Packet Radio Service in GSM," *IEEE Commun. Magazine*, October 1997, pp. 122–131.

[5] Brasche, G., and B. Walke, "Concepts, Services and Protocols of the New GSM Phase 2+ General Packet Radio Service," *IEEE Commun. Magazine*, August 1997, pp. 94–104.

[6] 3GPP, "General Packet Radio Service (GPRS); Service Description; Stage 2," Technical specification *3GPP TS 03.60*, version 6.10.0, Release 1997, January 2002.

[7] ETSI, "General Packet Radio Service (GPRS); Mobile Station (MS)—Serving GPRS Support Node (SGSN); Subnetwork Dependent Convergence Protocol (SNDCP)," Technical specification *ETSI TS 101 297 (GSM 04.65)*, version 6.7.0, Release 1997, March 2000.

[8] Jacobson, V., "Compressing TCP/IP Headers for Low-Speed Serial Links," *IETF RFC 1144*, February 1990.

[9]   3GPP, "General Packet Radio Service (GPRS); Mobile Station—Serving GPRS Support Node (MS-SGSN); Logical Link Control (LLC) Layer Specification," Technical specification *3GPP TS 04.64*, version 6.10.0, Release 1997, December 2001.

[10]   ETSI, "Technical Realization of the Short Message Service (SMS); Point-to-Point (PP)," Technical specification *ETSI TS 100 901 (GSM 03.40)*, version 6.1.0, Release 1997, July 1998.

[11]   3GPP, "Functional Stage 2 Description of Location Services (LCS) in GERAN," Technical specification *3GPP TS 43.059*, version 5.1.0, Release 5, November 2001.

[12]   3GPP, "General Packet Radio Service (GPRS); Mobile Station (MS)—Base Station System (BSS) Interface; Radio Link Control/ Medium Access Control (RLC/MAC) Protocol," Technical specification *3GPP TS 04.60*, version 6.13.0, Release 1997, April 2001.

[13]   3GPP, "General Packet Radio Service (GPRS); Base Station System (BSS)—Serving GPRS Support Node (SGSN); BSS GPRS Protocol (BSSGP)," Technical specification *3GPP TS 08.18*, version 6.8.0, Release 1997, June 2001.

[14]   Mehrotra, A., *GSM System Engineering*, Norwood, MA: Artech House, 1997.

[15]   Salkintzis, A. K., "Network Architecture," in W. Willie (Ed.), *Broadband Wireless Mobile—3G Wireless and Beyond*, New York: John Wiley & Sons, Ltd., Chapter 3, 2002.

CHAPTER 4

3G Networks and Standards

Neeli R. Prasad and Kim K. Larsen

4.1 Introduction

The *Third-Generation Partnership Project* (3GPP) has already specified the Release 99 standards, which focus on the *asynchronous transfer mode* (ATM) as the backbone network. Recent developments are focusing on an all IP-based network to be standardized for Release 2000 (split into release 4 and release 5) of 3GPP. The all-IP network is evolving from the packet switched mobile core network of Release 99.

It has been forecast that there will be more than 1 billion mobile users by the end of 2002, and packet based multimedia services, including IP telephony, will account for more than 50% of all wireless traffic. There is a momentum in the industry to evolve the current infrastructure, network services, and end user applications toward an end-to-end IP solution capable of supporting QoS that meets the needs of the dominant data traffic. At present, there are three types of 2G networks: GSM [1, 2], *time division multiple access* (TDMA), and *code division multiple access* (CDMA).

There are several 2.5G data transport standards, which are being implemented by many operators. Decisions are based on user demand, spectrum availability, equipment and spectrum license costs, backward compatibility, and the assessment of which will be the dominant 3G worldwide standard.

Evolution toward 3G is described in Section 4.2. Section 4.3 gives the full details of 3G and its releases (Release 99 and Release 2000). The 3G deployment scenario is discussed in Section 4.4. The chapter concludes with the impact of this deployment on the existing network in Section 4.5.

....

4.2   Evolution from 2G to 3G

There are several 2G-to-3G evolution scenarios for the operators, and some would be content with using 2.5G [2–5] technologies to make their networks reach 3G characteristics and features. Figure 4.1 shows the different evolution scenarios. Although WCDMA, also known as IMT-2000 or UMTS [6], has emerged as the dominant worldwide standard, other 3G standards (e.g., cdma2000 [3, 7]) are still being considered by some operators and countries.

The GSM operators are moving toward the GPRS [2, 6] with data rates of 171 Kbps. TDMA (and some GSM) operators are planning for *Enhanced Data Rate for Global Evolution* (EDGE) [2, 3] (384 Kbps with full mobility). The IS-95 [3] CDMA operators are considering 1XRTT [8] (144 Kbps standard). The 1XRTT is the interim step towards cdma2000.

FIGURE 4.1
Evolution scenarios towards 3G networks.

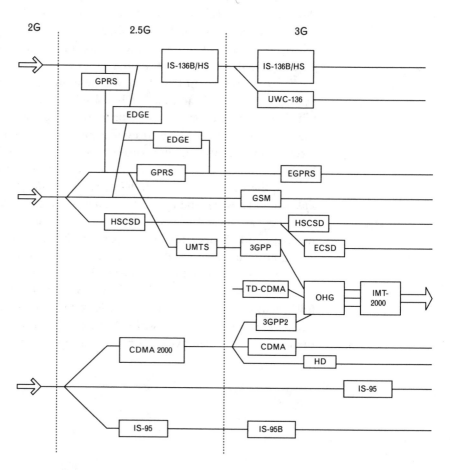

4.3   3G and Its Releases

The 3G will provide mobile multimedia, personal services, and the convergence of digitalization, mobility, the Internet, and new technologies based on global standards [3–5, 9]. The end user will be able to access the mobile Internet at the bandwidth (on demand) from 64 Kbps to about 2 Mbps. From a business perspective, 3G is the business opportunity of the twenty-first century.

The international standardization activities for 3G are mainly concentrated in the various regional organizations: the *European Telecommunications Standards Institute* (ETSI) *Special Mobile Group* (SMG) in Europe, *Research Institute of Telecommunications Transmission* (RITT) in China, *Association of Radio Industry and Business* (ARIB) and *Telecommunication Technology Committee* (TTC) in Japan, *Telecommunications Technologies Association* (TTA) in Korea, and *Telecommunications Industry Association* (TIA) and T1P1 in the United States.

Details of all proposals for IMT-2000 are available in [10]. The international consensus building and harmonization activities between different regions and bodies are currently ongoing. A harmonization would lead to a quasi-world standard, which would allow economic advantages for customers, network operators, and manufacturers. Therefore, two international bodies have been established.

First, 3GPP [11–13] is undergoing the harmonization and standardization of the similar ETSI, ARIB, TTC, TTA, and T1 WCDMA and related TDD proposals.

In the process of evolving to an IP core network, 3GPP decided to base this evolution on GPRS. The GPRS-based approach provides packet data access in 3GPP. The 3G.IP forum initiated the early work on an all-IP network in early 1999, but since then all the work has been moved to the 3GPP [11]. The UMTS network architecture is an evolution of the GSM/GPRS. The network consists of three subnetworks: *UMTS Terrestrial Radio Access Network* (UTRAN), *circuit switched* (CS) domain, and *packet switched* (PS) domain.

The UTRAN consists of a set of *radio network subsystems* (RNSs) connected to the *core network* (CN) through the Iu interface. If the CN is split into separate domains for circuit and packet switched core networks, then there is one Iu interface (Iu-CS) to the circuit switched CN and one Iu interface (Iu-PS) to the packet switched CN for that RNS, as shown in Figure 4.2.

An RNS consists of a *radio network controller* (RNC) and one or more node Bs. A node B is connected to the RNC through the Iub interface. Inside the UTRAN, the RNCs in the RNSs can be interconnected

FIGURE 4.2
UTRAN architecture.

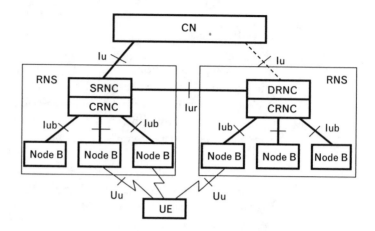

together through the Iur interface. The Iu and Iur are logical interfaces, which may be provided via any suitable transport network.

A node B can support one or more radio cells. A node B may support *user equipments* (UEs) based on FDD, TDD, or dual-mode operation. During macro diversity (soft handover), a UE may be connected to a number of radio cells of different node Bs and/or RNSs. Combining/splitting for soft handover may be supported within node B, drift RNC, and/or serving RNC. "Softer" handover provides better performance but is only possible within node B, between radio cells connected to that node B.

Each RNS is responsible for the resources of its set of radio cells and for handover decisions. The controlling part of each RNC (CRNC) is responsible for the control of resources allocated within node Bs connected to that RNC. For each connection between a UE and the UTRAN, one RNS is the Serving RNS. When required, Drift RNSs support the Serving RNS by providing radio resources, within radio cells connected to that Drift RNS.

Any RNC can take on the role as Serving RNC or Drift RNC, on a per connection basis for a UE. This supports macro diversity (soft handover) when the UE roams into another RNS. Eventually a relocation process (separate to handover) may be used to reroute the Iu connection to the new RNS, after which Drift RNS becomes the Serving RNS for the UE. *Radio access bearers* (RABs) are provided between the UE and core network (via the Uu radio interface, UTRAN internal interfaces, and Iu interface) for the transport of user data. Control plane protocols provide the control of these RABs and the connection between the UE and the network. Control plane protocols over Uu would be carried between *radio resource control* (RRC) entities in the UE and UTRAN.

During 2000, 3G was split by 3GPP (see Table 4.1) into two releases: R99 (now known as Release 3) and R00. Release 99 of the UMTS system supports WCDMA access and ATM-based transport. UMTS Release 2000 (UMTS R00) split into two releases (Release 4 and Release 5) [11], defines

TABLE 4.1   3G RELEASES

3G RELEASE	ABREVIATED NAME	SPEC. VERSION NUMBER	FREEZE DATE (INDICATIVE ONLY)
Release 6 (will be TR 21.104)	Rel-6	6.x.y	Scheduled June 2003
Release 5 (TR 21.103)	Rel-5	5.x.y	March 2002
Release 4 (TR 21.102)	Rel-4	4.x.y	March 2001
Release 2000	R00	4.x.y	See note below
		9.x.y	
Release 1999 (TR 21.101)	R99	3.x.y	March 2000
		8.x.y	

Note: The term "Release 2000" was used only temporarily and was eventually replaced by the term "Release 4."

two RAN technologies, a *GPRS/EDGE radio access network* (GERAN) and a CDMA RAN as in R3 UTRAN. Both types of RANs connect to the same packet switched core network (an evolution of the GPRS network) over an Iu interface. One main objective of UMTS R00 is to have the option of all-IP-based core network architecture, thus setting the tone for UMTS standardization in 2000 and beyond. Benefits expected from this approach include the ability to offer seamless services through the use of IP regardless of means of access, simultaneous multimedia services, and rapid service deployment, besides synergy with generic IP developments and reduced cost of service. However, the all-IP architecture in UMTS R00 (Release 4 and 5) must support services and capabilities of R99 and R00 and beyond. It must ensure an evolution path with sufficient backward compatibility.

The second international body is 3GPP2 [7], which was established for the cdma2000-based proposals from TIA and TTA. Technical specification work of cdma2000 standardization is being done within 3GPP2 in the following steps:

- The cdma2000 1x, which is an evolution of cdmaOne, supports packet data service up to 144 Kbps.

- The cdma2000 1xEV-DO introduces a new air interface and supports high-data-rate service on the downlink. It is also known as *high-rate packet data* (HRPD). The specifications were completed in 2001. It requires a separate 1.25-MHz carrier for data only. The 1xEV-DO provides up to 2.4 Mbps on the downlink, but only 153 Kbps on the uplink. Simultaneous voice over 1x and data over 1xEV-DO is difficult because of separate carriers.

• The cdma2000 1xEV-DV, which will introduce new radio techniques and an all-IP architecture for radio access and core network. The completion of specifications is expected in 2003. It promises data rates up to 3 Mbps.

SK Telecom and LG Telecom from Korea were the first operators to launch cdma2000 1x in October 2000. Since that time, only a few operators have announced cdma2000 1x service launches. Some operators recently announced setting up cdma2000 1xEV-DO trials [8].

The network architecture for a cdma2000 network is shown in Figure 4.3. The basic architecture is quite similar to the GSM/UMTS architecture. The main differences are in the packet domain where a *packet data switching node* (PDSN) is used. It has a similar role to the SGSN and GGSN in UMTS. Mobility management within 3GPP2, however, is based on Mobile IP (RFC 2002) instead of GPRS mobility management in GSM/UMTS PS networks. Furthermore, ANSI-41 MAP signaling is used instead of GSM MAP signaling. Activities have started in 3GPP2 for the evolution toward an all-IP network, similar to the IMS activities in 3GPP.

4.3.1   Release 3 (R3)

Release 3 or Release 99 (R3) is composed of the UTRAN attached to two separate UMTS core network domains as shown in Figure 4.4:

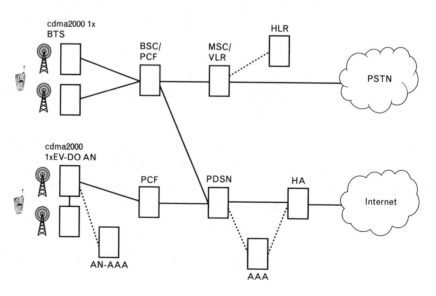

FIGURE 4.3
The cdma2000 1x and cdma2000 1xEV-DO network.

FIGURE 4.4
*Network architecture
of UMTS R3.*

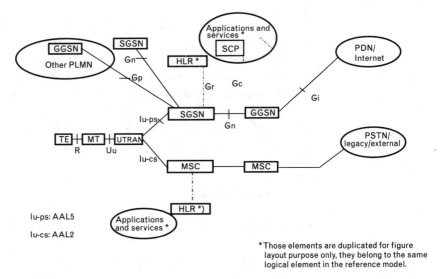

1. Circuit-switched domain based on *enhanced GSM mobile switching centers* (E-MSCs) consists of the following network elements:

   The *2G/3G mobile-services switching center* (2G/3G-MSC) including the *visited location register* (VLR) functionality;

   > *Home location register* (HLR) with *authentication center* (AC) functionality.

2. Packet-switched domain built on *enhanced GPRS support nodes* (E-GSNs) [10, 14] consists of (or involves) the following network elements:

   The *2G/3G serving GPRS support node* (2G/3G-SGSN) with *subscriber location register* (SLR) functionality;

   > *Gateway GPRS support node* (GGSN);

   > *Border gateway* (BG).

The HLR holds subscriber data and supports mobility in both domains. Two distinct instances of the so-called Iu interface are used between the access and the core network. The hybrid nature of UMTS R3 appears in several respects. It is most obvious in the transport and call control planes. From an end-to-end connectivity point of view, UMTS offers switched circuits toward the PSTN and *Integrated Services Digital Network* (ISDN), mainly to be used for voice communication. IP packet connectivity is provided as a pure network-layer service between a UMTS mobile station and an Internet host. The former is complemented by a sophisticated GSM/UMTS-specific service architecture based on IN principles, mainly for a wide range of supplementary and value-added voice services. In contrast, the latter is confined to cellular radio and mobility-enhanced bearers, although opening the stage for a variety of IP-based applications.

The circuit domain of UMTS R3 builds on the master/slave paradigm [11] of legacy GSM, inherited from the PSTN/ISDN, with the MSC acting as master and the mobile terminal acting as slave. The transport plane that physically transports voice is separated from *Signaling System #7* (SS7)-based call control or signaling plane that transports signaling messages and ensures advanced features to voice calls.

4.3.2   Release 4 (R4)

The main focus of R4 with respect to R3 is as follows:

- Hybrid architecture: ATM-based UTRAN (currently a workgroup is busy defining IP-based RAN) and IP/ATM-based CN;

- GERAN (support for GSM radio including EDGE);

- Enhanced services provided using toolkits (e.g., CAMEL, MExE, SAT, and VHE/OSA);

- Backwards compatibility with Release 99 services;

- Enhancements in QoS (real-time PS services), security, authentication, and privacy;

- Support for interdomain roaming and service continuity.

In R4 the circuit switched domain is split into a separate signaling plane (MSC server) and transport plane [*Media Gateway* (MWG)], which means the introduction of new entities. Standardized protocols will be used:

- Q.BICC or SIP-T (for inter-MSC signaling);

- H.248 / MeGaCo (for MSC server to MGW signaling).

This enables cost reduction by optimizing transport resources. The splitting of MSC into MSC server for signaling and MGW for transport makes R4 scalable, reliable, and cost-efficient with respect to R3 (see Figure 4.5). A MSC server is able to support several MWGs. If the serving MWG for a specific connection goes down, the MSC server is in state to reroute the traffic through a different MWG. As an implementation option it is possible to have a *many-to-many* (m:n) relationship between MSC Servers and MGWs, which allows for an efficient allocation of user plane resources as the MSC servers can load-balance between multiple MGWs.

FIGURE 4.5
*Network architecture
of UMTS R4.*

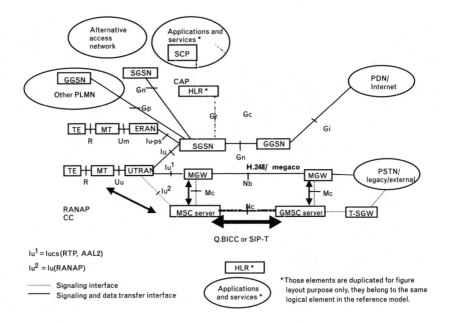

4.3.3   Release 5 (R5)

The R5 standardization mainly focuses on the new IP Multimedia Core Network subsystem, where IPv6 support is mandatory, in contrast to R4, where it was optional. The R5 solution will introduce the all-IP environment, including two major benefits:

1. *Transport:* utilization of the IP transport and connectivity with QoS throughout the network;

2. *End user services:* with SIP possibilities for offering a wide range of totally new types of services that are not possible to implement in R4 or earlier releases.

From the end user service perspective, the implementation of R5 into the network is an add-on. As new terminals are required for SIP services, it is clear that the old GSM and R3 services and subscribers still need to be supported. However, since the functionalities of classical SS7 call control and the IP call control are so different, it can be foreseen that it will not be possible to integrate these functionalities into the same network element. Thus, a *call state call function* (CSCF, which is basically a SIP server), needs to be introduced.

R5 architecture (see Figures 4.6 and 4.7) is still under discussion, but the general trend is to split the packet switched domain into control and transport planes. The benefits of splitting SGSN architecture are as follows:

FIGURE 4.6
Network architecture
of UMTS R5.

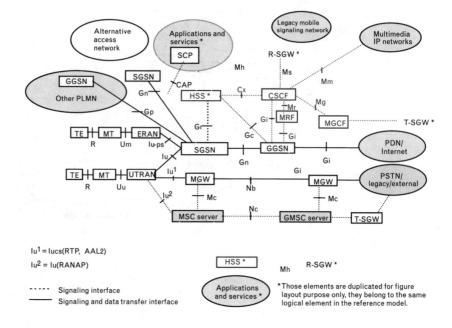

Iu¹ = Iucs(RTP, AAL2)

Iu² = Iu(RANAP)

- - - - Signaling interface
——— Signaling and data transfer interface

*Those elements are duplicated for figure
layout purpose only, they belong to the same
logical element in the reference model.

FIGURE 4.7
SGSN functionality
is split into SGSN
server for the control
plane and
SGSN-GW for the
transport plane.

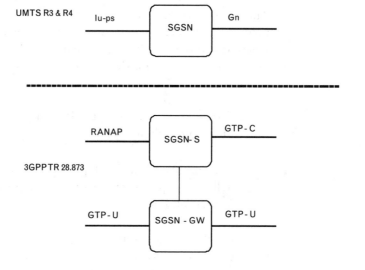

- Flexibility to allocate processing capacity for traffic and for control in different locations;

- Flexibility to independently scale the control plane and the user plane by increasing and/or decreasing the number of nodes required to handle the corresponding traffic;

• Allows an independent evolution and upgrade of nodes in the user plane and the control plane as the corresponding technology evolves.

As an implementation option it is possible to have an m:n relationship between SGSN servers and SGSN-GW, which allows for an efficient allocation of user plane resources as the SGSN servers can set up new PDP contexts to multiple SGSN-GWs for load-balancing.

4.4   3G Deployment Scenario

As mobile operators approach the current evolution toward 3G, many are examining the continued use of circuit switched technology within their core networks. Due to the current global economical recession, market uncertainties, and what is today regarded as significant business risks involved in 3G, most operators are trying to reduce and optimize the capital as well as operational investment in their next-generation networks. Even though a lot of money has been invested in the 3G licenses across Europe (i.e., sunk cost), there can be seen a hesitance in quickly investing too much money in what is a relative new technology where the standardization is barely stabilized. It is very interesting to realize that most 3G investments have been made before GPRS has become a commercial success. This is due to the relatively late deployment of commercial volume GPRS handsets/mobile stations and delayed introduction of mobile packet data into the market. This uncertainty alone puts a tremendous risk in all 3G business plans and should drive the 3G deployment scenario.

The 3G deployment scenarios are driven by two important assumptions:

1. There will be a significant shift from voice centric to data and multimedia centric services, which make up the traffic mix within these networks; this is illustrated by the current growth of SMS, and expected growth in WAP usage in conjunction with GPRS.

2. Operators can earn a profit selling 3G services.

Thus, it is clear that current radio access and circuit core network architectures and technologies do not provide an appropriate and efficient infrastructure for the delivery of bursty packet-based data services such as Internet, m-commerce, and corporate *virtual private networks* (VPNs).

To date, some mobile operators have created portfolios of traditional and next-generation services by building multiple networks. It is common to find a mobile operator using a *time division multiplex* (TDM) network to support voice, an ATM and/or frame relay network to support GPRS, and an IP network to support new features. Of course, using multiple network infrastructures to support multiple services is costly. In addition, building services that require combining diverse network technologies becomes exceptionally difficult and tedious because they must be manually provisioned.

The use of a single homogenous network based upon IP becomes the logical choice for the delivery of seamless services within mobile networks and allows these services to span the voice, data, and video domains—thus migrating the mobile network towards a true multimedia capability. The migration from circuits to packets is achieved by migrating all of the services and applications within the mobile network onto a packet-based (IP or IP+ATM) network; this is most logically achieved from the core network with migration outwards towards the RAN.

UMTS network architecture, AAL2-AAL5/ATM (Release 99) or IP (Release 2000) are used. Today it is not easy to determine at what point in time the all-IP model will be introduced. Furthermore, it is likely that R3/4 and R5 will co-exist in a large number of mobile operator's networks for quite some time. This is one of the reasons for deploying network architecture based on MPLS, where both IP and ATM are perfectly supported. Figure 4.8 shows a possible network deployment scenario for a 2G mobile operator. With MPLS it is possible to utilize a single physical infrastructure, switches, and transmission links for both types of traffic: native-ATM and IP. Network resource allocation (namely, switching capacities and transmission bandwidths) is controlled by a few network management commands. No forklift upgrades will be necessary when gradually moving from a R3 to

FIGURE 4.8
A network deployment scenario for a 2G operator.

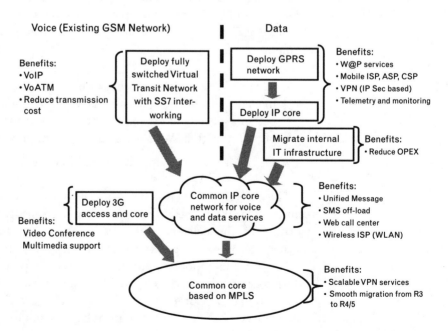

R5 architecture. This will protect mobile operators' initial investments and give total flexibility for the introduction of new services.

4.5   Conclusion: Impact on the Existing Network

In this section the impact of 3G on the existing legacy network architectures will be discussed. It is important to realize that even if most legacy networks support the same access technology, their architectures often are very different. Thus, it can be noted that existing mobile network architectures are very different in nature and can be characterized by human experience pool responsible for the architecture, network age and size, vendor or vendors, and whether they interface with fixed networks (either PSTN or IP). An understanding of these facets of the legacy networks (see Figure 4.9) will, together with the need for cost-efficient solutions, be a major driver for 3G deployment and the impact of 3G network rollout on the existing legacy networks. For example, if a mobile network operator already has an ATM backbone network and considerable experience with ATM engineering, it should use this synergy in order to reduce deployment cost and reduce investments in new transmission and switching equipment. A legacy network that does not have an existing ATM infrastructure could benefit from architecting an IP backbone, although ATM would still need to be supported for up to the Iu,cs and Iu,ps. It should also be noted that it is many times easier to find good IP engineering skills than people with ATM experience. For a single vendor legacy network, considerable synergy could be found in operational and maintenance experience as well as network monitoring systems by choosing the same vendor for its 3G network. The initial

FIGURE 4.9
A legacy 2G network.

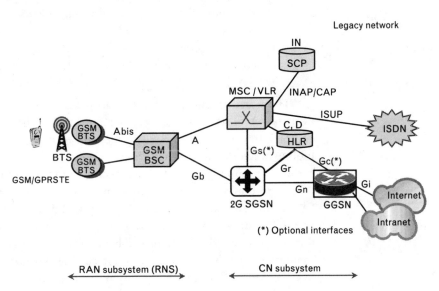

WIRELESS IP AND BUILDING THE MOBILE INTERNET

ease of integration and few interoperability issues would also be expected. The big question in this single vendor scenario is whether the equipment and service pricing will be attractive enough as compared to adding a new vendor for the 3G network. The legacy network will be a significant boundary condition for architecting the 3G network. Only Greenfield operations, where no legacy network is present or with very recent legacy operations, might deploy the theoretical architectures (i.e., such as all- or near-IP networks with the state-of-art IP QoS implementations) presented in standardization bodies.

With all the changes between GSM networks and 3G networks, the major impacts on the existing network are:

- *Handover*: It is assumed that the handover decision is always made inside GERAN. For inter-GERAN handover, functions to set up a path within the CN are required. Depending on the handover type, different changes in GSM are required. The backward handover, where the handover signaling is performed through the old base station, is similar to the current GSM handover. In the forward handover, the mobile station initiates the handover through the new base station. When trying to avoid the corner effect, in which the connection to the old station is lost, a very fast handover is required to prevent the blocking of the existing users in another cell. Forward handover requires a large number of changes in GSM.

- *Transmission infrastructure*: The transmission infrastructure has to meet new requirements imposed by wideband services. Since the data services are bursty and often asymmetric, the transmission solution has to be able to efficiently multiplex different types of information. ATM will provide efficient support for transmission of bursty wideband services. However, since ATM was originally designed for very high-speed transmission in the fixed network, some modifications may be needed to accommodate cellular-specific infrastructure requirements.

There are several possible scenarios for mobile operators to migrate from GSM/GPRS to UMTS. As mentioned earlier in this chapter, the complexities of the 3G migration depend to a high degree on the legacy mobile network and to what extent a single vendor environment is in place and will remain after the 3G migration. In a multivendor environment, the migration could be considerably more complicated due to mismatches in SW feature support and to which 3GPP technical specification release the various vendors adhere to. It is to be expected, therefore, that in a multivendor environment the operator will have to compromise the architecture and the services that initially will be launched. Furthermore, where the 2G and 3G vendors differ, interfaces might have to be reconsidered with the possible result of service touch-and-feel changes. A typical example is the interfaces between the HLR and the SGSN (Gr)/GGSN (Gc), and MSC/VLR and SGSN (Gs). Moreover, one vendor's software release package might differ substantially from another vendor's release (after all, with open interfaces features will be what differentiates the various vendors) and could allow for only basic services to be launched or result in significant development work to allow for feature match. A good example of this particular issue is in legacy networks with the open MAP interface between the MSC/VLR and the HLR (i.e., C & D interfaces).

Theoretically it is possible to interface vendor X HLR with a vendor Y MSC/VLR, but only the basic features could be explored due to feature mismatch. In practice an operator would always vendor match the MSC/VLR and HLR in order to get the maximum out of the architecture and network infrastructure:

1. To upgrade its existing GSM/GPRS CN for UMTS use. In this case, 2G and 3G networks share the same core infrastructure as shown in Figure 4.10. The possible impact on the existing network is as follows:

   Redimension of the existing CN to be able to support 3G broadband services;

   Optimize transmission network for a suitable traffic mix;

   Network management system.

2. To deploy an independent 3G CN from the 2G CN as shown in Figure 4.11. In this case, 2G and 3G networks will have minimum impact on each other. In a multivendor scenario, *interoperability tests* (IOTs) will be needed depending on the architecture, for example,

FIGURE 4.10
Common CN for both 2G and 3G.

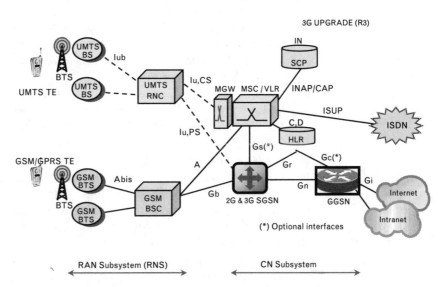

FIGURE 4.11 *Independent 2G and 3G networks.*

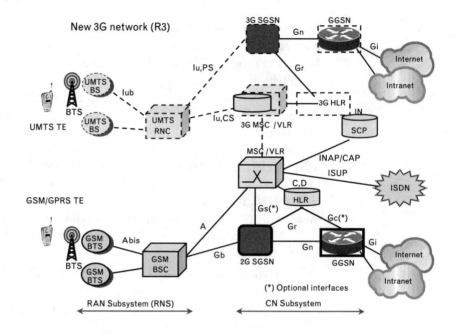

G_c (between HLR and GGSN), $I_{u,PS}$ (RAN and SGSN), and G_r (HLR and SGSN).

REFERENCES

[1]   Mehrotra, A. K., *GSM System Engineering*, Norwood, MA: Artech House, 1997.

[2]   Prasad, N. R., "GSM Evolution Towards Third Generation UMTS/IMT2000," *ICPWC'99*, February 17–19, 1999, Jaipur, India.

[3]   Ojanpera, T., and R. Prasad, *WCDMA: Towards IP Mobility and Mobile Internet*, Norwood, MA: Artech House, 2001.

[4]   Ojanpera, T., and R. Prasad, *Wideband CDMA for Third Generation Mobile Communications*, Norwood, MA: Artech House, 1998.

[5]   Prasad, R., W. Konhäuser, and W. Mohr, *Third Generation Mobile Radio Systems*, Norwood, MA: Artech House, 2000.

[6]   Prasad, N. R., "An Overview of General Packet Radio Services (GPRS)," *First International Symposium on Wireless Personal Multimedia Communications (WPMC'98)*, November 4–6, 1998, Yokosuka, Japan.

[7]   3GPP2, http://www.3gpp2.org.

[8]   http://www.cdg.org.

[9]   DaSilva, J. S., et al., "European Third-Generation Mobile Systems," *IEEE Communication Magazine*, Vol. 35, No. 10, October 1996, pp. 68–83.

[10] Huber, J. F., D. Weiler, and H. Brand, "UMTS, the Mobile Multimedia Vision for IMT-2000: A Focus on Standardization," *IEEE Communications Magazine*, 2000.

[11] 3GPP, http://www.3gpp.org.

[12] ETSI, "Requirements for the UMTS Terrestrial Radio·Access System," UMTS 04-01 UTRA, June 1997.

[13]   Chaudhury, P., W. Mohr, and S. Onoe, "The 3GPP Proposal for IMT-2000," *IEEE Communications Magazine*, December 1999.

[14]   ITU, http://www.itu.int/imt/2-radio-dev/rtt/index.html, July 1998.

CHAPTER 5

UTRAN Evolution to an All-IP Architecture

Dimitris Vasilaras, Georgios Sfikas, and Rahim Tafazolli

5.1 Introduction

A 3G mobile system for worldwide use is now being developed to enhance and supersede current systems. The UMTS will be an enhanced digital system and will provide universal personal communications to anyone, regardless of their whereabouts. At the same time, growth of the Internet is fuelling the demand for IP-based applications for the mobile market. Solutions based on 3G packet/IP-based networks will bring higher speed, consistent QoS, and coverage of voice, data, graphic, and voice-based information, allowing videoconferencing and Internet access to mobile users.

End users will be able to enjoy the ability to talk, send text messages, and share multimedia applications simultaneously on their wireless handsets with other end users. They will also be able to take advantage of new Internet-based supplementary services, make calls to today's PSTN phones, and set up sessions with tomorrow's fixed IP phones or IP servers. All of this is enabled by the higher data rates of the new 3G systems. Figure 5.1 summarizes the wireless all-IP subsystems.

The wireless all-IP network introduces new IP nodes as well as builds on existing radio access and backbone packet nodes. Software/firmware updates are required for the base stations to support integrated packet voice and data traffic on the 3G air interface. Software updates are required for the backbone packet network nodes (e.g., SGSN and GGSN) to support integrated packet voice and data traffic over the backbone packet network and out to the IP network. Software updates are also required for a *dynamic name server* (DNS)-like network element to support mapping of E.164 numbers to URLs or IP addresses. As it can be seen from Figure 5.1, the wireless all-IP

FIGURE 5.1
*Wireless all-IP
subsystems.*

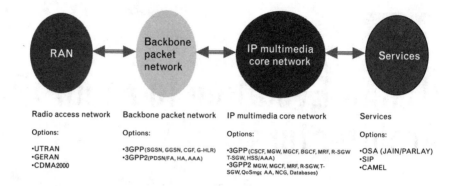

Radio access network	Backbone packet network	IP multimedia core network	Services
Options:	Options:	Options:	Options:
•UTRAN	•3GPP(SGSN, GGSN, CGF, G-HLR)	•3GPP(CSCF, MGW, MGCF, BGCF, MRF, R-SGW	•OSA (JAIN/PARLAY)
•GERAN	•3GPP2(PDSN/FA, HA, AAA)	T-SGW, HSS/AAA)	•SIP
•CDMA2000		•3GPP2 MGW, MGCF, MRF, R-SGW,T-	•CAMEL
		SGW, QoSmgr AA, NCG, Databases)	

architecture supports a layered network functionality. This allows for system flexibility and independent evolution of access, transport, application, and service creation domains.

Section 5.2 describes the 3GPP (Release 5) reference model, system architecture, and network configurations for the 3G wireless end-to-end IP multimedia system. Section 5.3 provides an overview of UTRAN. Section 5.4 focuses on the transport network by proposing solutions on how to introduce the IP as a transport bearer, in an all-IP-based UTRAN network. Section 5.5 outlines various differences in operating IP-over-SONET compared to running IP-over-ATM.

5.2   3GPP Reference Model

The all-IP architecture is an evolution from Release 99 specifications and is based on the cooperation of multiple subsystems, most of them based on packet technologies and IP telephony for simultaneous real-time and nonreal-time services. It aims to provide wireless mobility access based on GERAN and UTRAN with a common core network, based on evolution of GPRS, for both.

The characteristics of this network are as follows:

• Common network elements for multiple access types including UTRAN and GERAN;

• Packet transport using IP protocols;

• IP client-enabled terminals;

• Support for voice, data, real-time multimedia, and services with the same network elements.

Figure 5.2 shows the 3GPP reference model of the wireless all-IP network. Refer to [1] for more details on each interface and the elements discussed below. This architecture contains logical elements that have been defined based on network functions that have been grouped together to form the logical element. Actual implementation might merge logical elements in any combination.

Below is a description of the logical nodes (as presented in Figure 5.2) that constitute the 3GPP all-IP reference architecture.

- *UE:* The elements listed as UE are all the devices that can be attached to a wireless network [e.g., either a *mobile termination* (MT) or a combination of MT and *terminal equipment* (TE)].

- *TE:* This equipment may not be a wireless device but can be used to access a wireless network while attached to a wireless device. Examples of TE include laptops and PDAs.

- *MT:* This equipment is a wireless device that can be used to access a wireless network. A TE could be used in conjunction with this device to get access to wireless services. Examples of MT include mobile phones.

- *GERAN:* GERAN is the primary interface between the UE and the GSM/EDGE access network. GERAN consists of the *base transceiver system* (BTS) and *base station controller* (BSC). The Um interface is capable of supporting the GSM/EDGE air interfaces. GERAN is also connected to the 2G-SGSN via the Gb interface and to the 3G-SGSN via

FIGURE 5.2
The 3GPP reference model of the wireless all-IP network.

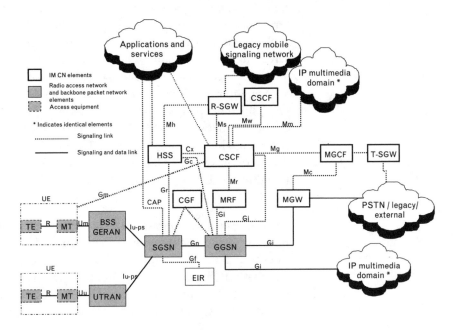

the Iu interface. The Gb interface is designed for the support of GPRS and EGPRS packet data services. Gb is not shown in Figure 5.2, since both signaling and traffic for *Voice over IP* (VoIP) and real-time data will be sent over the Iu interface.

- *UTRAN:* UTRAN is the primary interface between the UE and the UMTS access network. UTRAN is comprised of the node B and RNC. It supports the UMTS air interface. It is connected to the 3G-SGSN via the Iu interface. Signaling and bearer traffic for VoIP and real-time data will be sent over the Iu interface. The Uu interface is designed to support the UMTS WCDMA air interfaces. In particular for the deployment of VoIP and real-time data, UTRAN will provide various levels of QoS based on the application and the service level required.

- *SGSN:* The SGSN provides packet mobility management. Packet-capable mobiles attach (register) with the SGSN. Registration messages are transferred over the air interface and then through the Iu interface from the UE to the SGSN. The SGSN may receive messages destined for the MSC (CS domain). It does not, however, process these messages. Instead, the SGSN forwards them to the MSC/VLR to be processed (as happens in R3). Other functions of the SGSN include authentication, session management, accounting, billing, mapping of IP addresses to IMSI, maintenance of mobile state information, and interfacing with the GGSN. The SGSN is also capable of serving these mobiles that are operating outside their home systems.

- *GGSN:* The GGSN provides interworking between the UEs and external packet data networks using IP. The GGSN provides functions such as routing, packet session establishment and termination, message screening, service measurements, and billing. The SGSNs and GGSNs are connected via a backbone transport network through the Gn interface. User data is transferred between SGSNs and GGSNs using an encapsulation method known as *tunneling*. Tunnels are dynamically established between an SGSN and GGSN using the GTP and are deactivated when no longer needed. The GGSN is connected to packet data networks via the Gi interface, and to GPRS networks in different PLMNs via the Gp interface. Only IP will be supported at the Gi interface.

- *Charging gateway function (CGF):* This element collects the accounting records for the backbone packet network. This element communicates with the GGSN and the SGSN via the Ga interface. Moving to the all-IP architecture may require modification of the *charging data register* (CDR) contents (generated by the SGSN and GGSN, and collected by the CGF).

- *Equipment identity register (EIR):* Each terminal is identified by a unique *international mobile equipment identity* (IMEI) number. A list of IMEIs in the network is stored in the EIR. The status returned in response to an IMEI query to the EIR is one of the following:

  1. *White-listed:* The terminal is allowed to connect to the network;

  2. *Gray-listed:* Under observation from the network, possible problems;

  3. *Blacklisted:* The terminal has either been reported as stolen, or it is not type approved (the correct type of terminal for a UMTS network). The terminal is not allowed to connect to the network. Emergency calls from this mobile might be an exception.

- *Home subscriber server (HSS):* HSS is the main database that holds subscriber related information for a particular user.

- *CSCF:* CSCF is a crucial component of the all IP wireless network. CSCF mainly provides the following functions:

  1. Incoming call gateway function (routing of incoming calls, address handling);

  2. Call control function (accepts register requests and makes its information available through the location server, session control for the registered endpoint's sessions, interaction with services platforms for the support of services);

  3. Serving profile database function (receives profile information from the HSS in the home domain);

  4. Address handling function (analysis, translation, and mapping of alias addresses).

- *Media gateway control function (MGCF):* The MGCF terminates signaling between the 3GPP IM CN subsystem and the PSTN/PLMN. This provides the call control interface and translations between the PSTN and the IP network.

- *MGW:* The MGW is the transport termination point between a given IP network and the PSTN/PLMN.

- *Multimedia resource function (MRF):* MRF is used to provide support for tone and announcements, along with multiparty multimedia conference sessions.

- *Transport signaling gateway function (T-SGW):* The T-SGW provides transport level interworking between the PSTN/PLMN and 3GPP IM core network. T-SGW interfaces with the MGCF.

• *Roaming signaling gateway function (R-SGW):* R-SGW terminates signaling between circuit switched domain/GPRS domain and the IM CN subsystem.

The proposed architecture is designed to utilize emerging Internet standards and protocols. An example of this is the use of SIP for IM CN subsystem signaling for establishing a call. SIP [2] is a signaling protocol used for Internet conferencing and telephony. It is based on a client server architecture where the clients initiate the calls and the servers establish them. SIP is the preferred choice for the initiation of multimedia services instead of H.323 because of its simplicity and scalability.

5.3   UTRAN Overview

The current UTRAN model (formerly Release 99 and now known as Release 3) is shown in Figure 5.3. Its principal functions are to:

• Manage radio resources;

• Process radio signaling;

• Terminate radio access bearers;

• Perform call setup and tear-down;

• Process user voice and data traffic;

• Conduct power control;

FIGURE 5.3
UTRAN model.

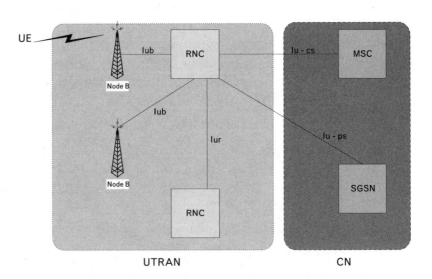

- Provide *operations, administration, management, and provisioning* (OAM&P) capabilities;

- Perform soft, intrasystem hard, and intersystem (UMTS and GSM voice or GSM GPRS) hard handovers;

- Support always-on packet data.

UTRAN provides for limited IP support and is only used as a transport bearer for the signaling traffic at the Iu interface (external to UTRAN; see Figure 5.3 for the definition of the interfaces). IP user traffic from the packet switched core network is tunneled from the SGSN to the RNC (Iu interface). The RNC transports the IP packets over AAL2 circuit connections towards the air interface. However, with growing demand for Internet-based services together with opportunity for VoIP, a significant percentage of traffic in the UTRAN will be IP.

For the support of IP in the UTRAN, at least two solutions can be provided. First, IP can be used as a transport bearer running on top of the existing R99 ATM transport network; ATM will provide the flexibility while the operators migrate to an all-IP solution. Second, IP is used as the principal transport protocol with any underlying layer 1 and layer 2 technologies.

Because the underlying ATM infrastructure is *Synchronous Optical Network* (SONET) or *Synchronous Digital Hierarchy* (SDH) deployed over wide area fiber links, interest has grown in running the IP directly over SONET/SDH, rather than using an ATM network, to increase bandwidth efficiency. These two options, however, have created a hot debate over which technology will provide the best solution. Furthermore, *multiprotocol label switching* (MPLS) [3] can be used in order to provide fast switching connection services to the IP layer by making use of any layer 2 transmission technologies. A brief discussion of SONET/SDH and ATM follows.

5.3.1   SONET/SDH

SONET/SDH is a high-speed TDM physical-layer transport technology, inherently optimized for voice. The basic transmission rate of SONET [4] (51.840 Mbps), referred to as Synchronous Transport Signal level 1 (STS-1), is obtained by sampling the 810-byte frames at 8,000 frames per second. It is normally represented as a two-dimensional byte-per-cell grid of 9 rows and 90 columns. The SONET/SDH frame is divided into transport overhead and payload bytes. The transport overhead bytes consist of section and line overhead bytes, while the payload bytes are made up of the payload capacity and some more overhead bytes, referred to as path overhead. The overhead bytes are responsible for the rich management capabilities of SONET/SDH.

Table 5.1 presents the corresponding transmission rates for SONET and SDH. SDH is the SONET-equivalent specification proposed by the ITU.

SONET/SDH is used as the bearer layer for higher-layer protocols such as ATM or IP/PPP, employed on devices that switch or route traffic to a particular endpoint. RFC 2615 specifies the use of PPP encapsulation over SONET/SDH links. PPP was designed for use on point-to-point links and is also suitable for SONET/SDH links.

5.3.2   ATM

ATM is a cell switching and multiplexing technology that combines the benefits of circuit switching (guaranteed capacity and constant transmission delay) with those of packet switching (flexibility and efficiency for bursty traffic). It uses a 53-byte fixed-length cell and *virtual circuits* (VCs) to transport data, voice, and video traffic between endpoints in the network. ATM defines both *permanent virtual circuits* (PVCs) and *switched virtual circuits* (SVCs). PVCs provide static (manually created) connections, whereas SVCs create dynamic connections established based on demand and current network state. The benefits of ATM are as follows:

- High performance via hardware switching;
- Dynamic bandwidth for bursty traffic;
- Class-of-service support for multimedia.

5.4   UTRAN Transport Network

The UTRAN transport network consists of nodes and links, for the purpose of transporting user, signaling, and management traffic, supporting at the same time different levels of QoS. Several implementations can fulfill the basic fundamental requirements of the UTRAN transport network. It relies on vendors, operators, and third-party service providers to determine the

TABLE 5.1   SONET DATA RATES

SONET	SDH	DATA RATES
STS-1	—	51.840 Mbps
STS-3	STM-1	155.520 Mbps
STS-12	STM-4	622.080 Mbps
STS-48	STM-16	2,488.320 Mbps
STS-192	STM-48	9,953.280 Mbps

best implementations for the transport network. The solution for the radio network control plane at the Iub and Iur interfaces is shown in Figure 5.4.

Currently there are two alternatives identified for the radio network signaling bearer in Iur for the Rel5 IP transport option. These alternatives are *signaling connection control part* (SCCP)/MTP3-User Adaptation Protocol (M3UA) [5] and SS7 *SCCP-User Adatation Layer* (SUA). M3UA is more likely to be the preferred choice for the radio netwc rk layer signaling bearer in Iur interface. A brief introduction to SCTP and M3UA is provided below.

SCTP [6] is an IETF protocol designed to transport PSTN signaling messages over IP networks. It offers the following services to its users:

1. Reliable delivery of user data (notification on lost packets);
2. In-sequence delivery of data;
3. Bundling of multiple user messages into a single SCTP packet;
4. Avoidance of the head-of-line blocking offered by TCP;
5. Message oriented data transfer.

M3UA is a protocol, currently being developed by IETF, for the transport of any SS7 MTP3-user signaling (e.g., ISUP, SCCP, and TUP) over IP using SCTP. The M3UA delivery mechanism provides the following functionality:

1. Support for transfer of SS7 MTP3-User Part messages;
2. Support for the management of SCTP transport protocol between a signaling gateway and one or more IP-based signaling nodes to ensure transport availability to MTP3 user signaling applications;
3. Support for the seamless operation of MTP3-user protocol peers;
4. Support for distributed IP-based signaling nodes;

FIGURE 5.4
Iub/Iur control plane protocol stacks.

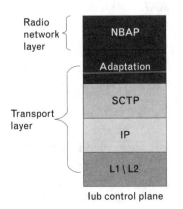
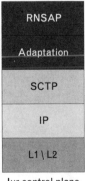

5. Support for the asynchronous reporting of status changes to management.

For the user plane several solutions are provided which are based on the transport of IP-over-SONET (LIPE, CIP, PPP/HDLC) or IP-over-ATM (PPPmux/AAL5, PPP/AAL2, CIP). These are fully described below:

- CIP and LIPE where segmentation, flow multiplexing, and QoS differentiation are performed above UDP/IP;

- PPPmux/AAL5/ATM, PPP/AAL2/ATM, PPP/HDLC, and MPLS solutions where they are performed below the IP layer.

User plane protocol stacks will be provided for each of the presented cases.

5.4.1 LIPE

Lightweight IP encapsulation (LIPE) [7] is a protocol designed by Lucent Technologies. LIPE is the IP adaptation layer, which multiplexes user frames into IP packets similar to AAL2 multiplexing frames into ATM cells. This new protocol provides the following services:

- *Bandwidth efficiency:* Efficient bandwidth usage in the transport network and especially on the last mile to the node B is directly linked to transport costs. Means shall be provided in the protocol stack, which reduce the packet header overhead.

- *Timing constraints:* To fulfill the timing requirements in the UTRAN, low transport delay is required, and possibly a distinction between different UTRAN service classes.

- *Channel addressing:* In the ATM solution a *channel identifier* (CID) is used in AAL2 to identify the transport bearer. Similar identification is also required for the IP solution.

- *Network element addressing:* There is no need to explicitly address network elements in the user plane, such as in a connection-oriented network like ATM, but this is required in connectionless networks like IP.

- *Segmentation:* On the last mile link between a node B and an edge router, one forwarded IP packet preempts the access to the medium for a duration proportional to the payload size. In order to guarantee some quality of service, a limit shall be set on the packet size, so that low-priority packets cannot block real-time packets.

• *Independence to the radio network layer:* Changes for the introduction of IP transport should only be made in the *transport network layer* (TNL). The radio network layer should remain unaffected by changes in the TNL.

The LIPE scheme uses either UDP/IP or IP as the transport layer. Each LIPE encapsulated payload consists of a variable number of *multimedia data packets* (MDPs). For each MDP, there is a *multiplexing header* (MH) that conveys protocol and media specific information. Figure 5.5 presents the format of an IP packet conveying multiple MDPs over UDP using a minimum size MH.

The details of the MH are shown in Figure 5.6. The MH comprises of two components: the extension bit (the E bit) and the MDP length field. Optional extension headers can be supported via the E bit. The E bit is the

FIGURE 5.5
LIPE encapsulation.

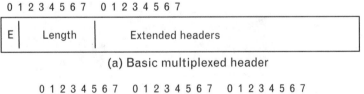

MH: Multiplexed header   MDP: Multiplexed data payload

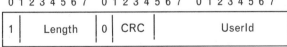

TID: Tunnel identifier

PPP/HDLC framing

FIGURE 5.6   *LIPE multiplexing header.*

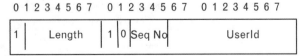

(a) Basic multiplexed header

(b) Extended multiplexed header with header CRC and UserID

(c) Extended Multiplexed header with sequence number and user id

least significant bit of the first byte of the MH header. It is set to one/zero to indicate the presence/absence of an extension header. If the E bit is set to one, the first header extension must be an extended header identifier field. The length field is 7 bits. This field indicates the size of the entire MDP packet in bytes, including the E bit, the length field, and optional extension headers (if they exist). Extension headers are used to convey user-specific information. It also facilitates the customization of LIPE to provide additional control information (e.g., sequence number and voice/video quality estimator).

5.4.2   CIP

The *Composite IP* (CIP) [8] protocol is similar to LIPE except for details of the packet header. CIP allows for efficient use of the bandwidth of the links by multiplexing CIP packets of variable size in one CIP container. The CIP scheme can be supported by two different link layers: CIP/UDP/IP/PPP/HDLC and CIP/UDP/IP/PPP/AAL5/ATM. The resulting protocol stack and packet structure are shown in Figure 5.7.

The CIP packet header is explained next:

- The CID section also contains CRC and flags, and it is used for multiplexing.

  1. The CRC protects the reserved flag, the segmentation flag, and the CID.

  2. The reserved flag is for further extensions.

FIGURE 5.7
CIP packet structure, packet header format, and protocol stack.

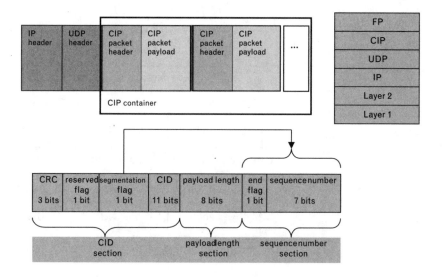

3. The segmentation flag indicates that the sequence number field and the end flag are present.

4. The CID is the channel ID. This is the identifier of the multiplex functionality.

- The payload length section is used for aggregation. This section is mandatory.

   1. The payload length is the length of the CIP packet payload, so, CIP packets, containing, for example, FP-PDUs with voice or FP-PDU segments with data, can be between 1 and 256 octets in size.

- The sequence number section, also containing the end flag, is used for segmentation and reassembly.

   1. The end flag marks the last segment of a packet in a sequence of segments.

   2. The sequence number is used to reassemble segmented packets.

PPP/HDLC

PPP is a point-to-point link layer protocol that provides the following functions:

- Encapsulates and transfers packets from multiple network layer protocols over the same link;

- Establishes, configures, and tests the link layer connection;

- Establishes and configures network layer protocols.

PPP uses an HDLC-like framing [9] mechanism for delineating its frames over underlying physical media The frame format for PPP in HDLC-like framing is shown in Table 5.2.

5.4.3 MPLS

MPLS is an IETF specification for attaching labels containing forwarding information to IP packets. Labels are analogous to the *data link connection indentifiers* DLCIs used in a frame relay network or the VPI/VCIs used in an ATM environment. Label switch/routers can read the labels much faster than they can look up destination information in routing tables. MPLS has been described as bringing ATM-like traffic management features to switched connections. By prefixing a connection label to the front of IP packets, more efficient packet switches can be made by avoiding the complex IP route table look-up operations. In MPLS, data transmission occurs

TABLE 5.2   PPP IN HDLC-LIKE FRAMING

Flag	8 bits
Address	8 bits
Control	8 bits
Protocol ID	1 or 2 bytes
Information	Variable
Padding	Variable
FCS	2 or 4 bytes
Flag	8 bits

on *label switched paths* (LSPs). LSPs are a sequence of labels at each and every node along the path from the source to the destination. Individual flows are grouped into streams. Streams of the same *forwarding equivalence class* (FEC), a subset of packets that are all treated the same way by a router will be assigned the same label and provided the same packet forwarding behavior in an MPLS domain. The assignment of a particular packet to a particular FEC is done just once, as the packet enters the network. Labels are assigned at the ingress node (LER) and swapped at each LSR with the selected label for the next LSR until the packet reaches the egress node where the label is stripped off. At subsequent hops, there is no further analysis of the network layer header of the packet. Rather, the label is used as an index into a table that specifies the next hop, and a new label. The old label is replaced with the new label (switched), and the packet is forwarded to its next hop. Figure 5.8 presents a UTRAN transport network based on MPLS.

It is possible to differentiate the service provided to particular flows by applying traffic engineering techniques based on user-defined policies. CR-LDP [10] and RSVP-TE [11] are two possible approaches to supply dynamic traffic engineering and QoS in MPLS. RSVP and LDP can be used to support MPLS header compression and path establishment during the session negotiation phase. The LDP case is described below.

MPLS header compression session negotiation can be accomplished with the LDP protocol, by adding a new FEC TLV (Type–Length–Value) that includes a source IP address, source UDP port, destination host address, and a destination UDP.

The compressor requests a label for a new 4-tuple combination (source IP address, source port, destination IP address, destination port), and the decompressor provides the MPLS label it wants to use for the IP address/UDP port back to the compressor.

The compressor LSR can then suppress the UDP/IP header and replace it with the appropriate MPLS label. When the decompressor LSR receives

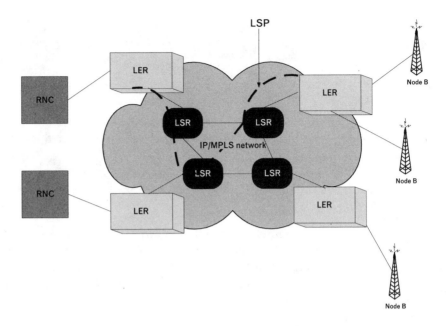

FIGURE 5.8
*MPLS-based transport
solution.*

the MPLS frame, it looks up the MPLS label in the mapping table and uses this information to restore the UDP/IP header.

5.4.4   PPPmux/AAL5/ATM

PPPmux provides a method to reduce the PPP framing overhead used to transport small packets. The idea is to concatenate multiple PPP encapsulated frames into a single PPP multiplexed frame by inserting a delimiter before the beginning of each frame. In this case PPP is carried over an ATM/AAL5 link layer [12]. The PPP layer treats the underlying ATM AAL5 layer service as a point-to-point link. In this context, the PPP link corresponds to an ATM AAL5 virtual connection. The format of the PPPmux frame is shown in Figure 5.9.

5.4.5   PPP/AAL2

PPP over AAL2 [13] addresses the bandwidth efficiency issues of PPP over AAL5 for small packet transport by making use of the AAL2 *common part sublayer* (CPS) to allow multiple PPP payloads to be multiplexed into a set of ATM cells. PPP over AAL2 defines an encapsulation that uses the *segmentation and reassembly service specific convergence sublayer* (SSSAR) for AAL type 2. The SSSAR sublayer is used to segment PPP packets into frames that can be transported using the AAL2 CPS.

An implementation of PPP over AAL2 may use a single AAL2 CID for transport of all PPP packets. A PPP over AAL2 implementation may also

FIGURE 5.9
PPPmux frame.

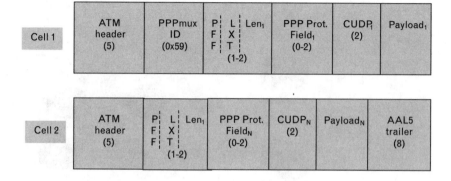

use multiple AAL2 CIDs to carry a single PPP session. Multiple CIDs could be used to implement a multiple class real-time transport service for PPP using the AAL2 layer for link fragmentation and interleaving.

The main differences of PPP/AAL2 over PPPmux/AAL55 are as follows:

- Better bandwidth utilization with PPP/AAL2;

- Since the multiplexing is done at the CPS layer, the delay for each individual packet is much shorter: as soon as an ATM cell is full, it is sent. With PPPmux, the multiplexing is done above ATM, so the packets that are multiplexed are delayed until the PPPmux packet is filled up. The same holds for delivery: all ATM cells used to transport the PPPmux packet must arrive before any of the multiplexed items can be delivered, where in PPP/AAL2 the packets are demultiplexed and delivered as cells arrive;

- Loss of one ATM cell causes the loss of the whole PPPmux packet, whereas in PPP/AAL2 only the packets included in that cell are lost.

5.5   Comparison of IP-over-SONET and IP-over-ATM

The following sections outline various differences in operating IP-over-SONET compared to running IP-over-ATM.

5.5.1   Protocol Overheads

IP achieves about 80% of the available line rate when operating over ATM, whereas it achieves 95% of the line rate when running over PPP/SONET/SDH. This is because, in the latter case, frames are transported directly into the SONET/SDH payload, thus eliminating the overhead required to support ATM (e.g., the ATM cell header, IP-over-ATM encapsulation, and the partial fill resulting from the fixed-length nature of ATM cells).

5.5.2 Bandwidth Management

ATM is providing bandwidth management by managing the bandwidth allocation to the various information streams (VCCs) flowing over a link. It assigns flexible bandwidth to these VCCs based on the required quality of service.

Within a connectionless *packet over SONET/SDH* (PoS) users network, all bandwidth is available by all customers and all applications at all times. There are no end-to-end traffic guarantees that allow the user to be sure that the data will arrive in a timely fashion. The IP layer has to schedule its packet transmissions to ensure that each information flow receives its fair share of link bandwidth. If a node or a link becomes congested, packets are often dropped in a random fashion (depending on the features employed on the router).

5.5.3 Network Management

In an IP-over-ATM network, network management can be provided by configuring policies for specific flows within the network management system and by preconfiguring long-hold switched VCs or by allowing the end user to negotiate admission to the network under a specific traffic contract as part of a user-initiated SVC.

The focus of network management within an IP/PPP-over-SONET/SDH environment is to manage a collection of point-to-point links and individual routing devices.

5.5.4 QoS

QoS relates to parameters such as end-to-end packet delay, jitter, loss, and throughput. User perceptions of voice quality depend on the timely delivery of VoIP packets and on packet-loss rates. The weaknesses of existing best-effort IP networks requires additional network QoS mechanisms to resolve these needs:

- To provide differentiated service to different classes of traffic;

- For basic mechanisms that measure use of network resources, and operator's service level agreements;

- For more predictable response in the face of real-time transient congestion;

• To dynamically request, or modify, end-to-end service quality.

ATM provides a rich set of QoS parameters that can be negotiated for each VCC. A list of these parameters is outlined below.

• *Cell loss ratio (CLR):* CLR is the percentage of cells not delivered at their destination because they were lost in the network due to congestion.

• *Cell transfer delay (CTD):* This is the delay experienced by a cell between network entry and exit points.

• *Cell delay variation (CDV):* CDV is a measure of the variance of the cell transfer delay.

• *Peak cell rate (PCR):* The maximum cell rate at which the user will transmit PCR is the inverse of the minimum cell interarrival time.

• *Sustained cell rate (SCR):* This is the average rate, as measured over a long interval, in the order of the connection lifetime.

• *Burst tolerance (BT)*: This parameter determines the maximum burst that can be sent at the peak rate.

PPP operates over a single point-to-point link and does not provide any QoS capabilities. As mentioned earlier, the IP layer has to manage its packet transmissions intelligently to ensure proper QoS for the information flows.

5.5.5   Flow Control

ATM uses functions, such as *call admission control* (CAC), traffic shaping, and *user parameter control* (UPC) or policing, to ensure that information flows stay within the boundaries of the negotiated traffic contract. Excess traffic is marked and is discarded under network overload conditions.

PPP provides no flow control mechanisms, so TCP's flow control operates directly over PPP links.

5.6   Summary

This chapter described the 3GPP reference model of the wireless all-IP network by focusing on the transport network in the UTRAN side. IP is rapidly becoming the network layer technology of choice for building packet networks and is also driving the growth of the worldwide Internet. Mass deployment, feasibility, scalability, and cost effectiveness of IP infrastructure is the primary reason for introducing IP as a transport option in the

UTRAN. Hence, any chosen architecture must maximize the benefit of IP technologies and infrastructure. Standardization of IP transport is intended to be layer 2 independent, and it is not clear yet what transmission technology will be used between IP and SONET. This chapter examined several alternatives for the transport of user data, all based on two solutions (ATM and PPP). Finally, some of the pros and cons of each respective solution in areas such as bandwidth management, quality of service, and flow control were examined.

Where raw speed is critical, IP-over-SONET is more attractive. Where flexibility in bandwidth management, quality of service, and network engineering is important, IP-over-ATM is a better solution. Issues relating to cost will override all of these concerns. Such costs include the cost of provisioning the service and the cost of maintaining it.

REFERENCES

[1] 3GPP TS 23.002, "Technical Specification Group Services and Systems Aspects."

[2] Handley, M., et al., "SIP: Session Initiation Protocol," RFC 2543, March 1999.

[3] Rosen, E., A. Viswanathan, and R. Callon, "Multiprotocol Label Switching Architecture," RFC 3031, January 2001.

[4] American National Standards Institute, "Synchronous Optical Network (SONET)—Basic Description Including Multiplex Structure, Rates and Formats," ANSI T1.105-1995.

[5] Sidebottom, G., et al., "SS7 MTP3-User Adaptation Layer (M3UA) Internet draft," February 2002, http://www.ietf.org/internet-drafts/draft-ietf-sigtran-m3ua-12.txt."

[6] Stewart, R., et al., "Stream Control Transmission Protocol," IETF RFC 2960, October 2000, http://www.ietf.org/rfc/rfc2960.txt?number=2960."

[7] Chuah, M. C., and E. J. Hernandez-Valencia, "A Lightweight IP Encapsulation (LIPE) Scheme," July 19, 2000.

[8] 3GPP TS 25933 v5.1.0 (2002-06), *IP Transport in UTRAN Work Task Technical Report*.

[9] Simpson, W., (Ed.), "PPP in HDLC-Like Framing," RFC 1662, *Daydreamer*, July 1994.

[10] Jamoussi, B. et al., "Constraint-Based LSP Setup Using LDP," RFC 3212, July 2000.

[11] Awduche, D., et al., "Extensions to RSVP for LSP Tunnels," RFC 3209, August 2000.

[12] Gross, G., et al., "PPP over AAL5," RFC 2364, July 1998.

[13] Thompson, B., T. Koren, and B. Buffam, "PPP over AAL2," draft-ietf-pppext-ppp-over-aal2-00.txt, February 2001.

CHAPTER 6

Beyond 3G: 4G IP-Based Mobile Networks

Donal O'Mahony and Linda Doyle

6.1 Introduction

The architecture of both the PSTN and also the first three generations of mobile telephone networks [1] were shaped by two main considerations. First, the networks were required to effectively provide a single service— circuit switched voice communications—across a network with low-capability edge nodes and a large amount of intelligence within the network. Any additional services that could be deployed on this infrastructure were looked upon as a bonus. Second, they required the operators to be able to charge for network usage based on the destination and duration of the call. Mobility between network operators was only worth doing if the home network operator stood to gain financially from this possibility.

The evolution of the Internet took a different path. In the early years, it was assumed that communities of users collaborated for the benefit of all. Since no organization wanted to act as a central network operator, the interior of the network was kept as simple as possible while allowing users to offer such services as they wished at the edges. The possibility of users moving around was catered for at the edges by Mobile IP and was implemented by agents external to the network itself, or at least located within edge routers.

In recent years, we have seen the beginning of convergence between the telecommunications and Internet communities. The PSTN has become the main Internet access network for residential users, and there is a huge industry focus on reengineering the channel structure of wireless telephony networks to accommodate data streams. In the other camp, the Internet telephony industry has proved that it is possible to build a production voice service on a packet-switched Internet-based network infrastructure, and the makers of large circuit switches have conceded that there are huge economic advantages to shifting over to packet switching.

The rapid adoption of mobile telephony, most notably in the Scandinavian countries, demonstrates another major trend that shows a demand on behalf of users to avail of the same communications on the move as were available through fixed lines.

In the midst of the great changes being brought about by the trends outlined above, the research community is considering what form the next (fourth) generation of fixed and mobile communications systems will take. A popular view in the telecommunications industry is that the 4G should evolve naturally from the 2G and 3G with incremental improvements being brought about without any fundamental architectural changes being necessary. We do not believe that this is the correct approach. In the following sections, we set out what we believe are the main drivers shaping the design of 4G systems. We briefly survey related work and then describe the progress we are making with our 4G test bed.

6.2   Drivers for the 4G Architecture

We adopt the view that the forthcoming 4G mobile systems offer us an opportunity to take a fresh approach in designing a network driven by services that people want now and in the future with less emphasis on the services that they wanted in the past. In the following sections, we examine five fundamental requirements underlying our concept of 4G mobile systems.

6.2.1   Support for IP-Based Traffic

It is clear from the rapid growth of the Internet in the late 1990s that IP forms an effective delivery mechanism for the kinds of network services that users are interested in availing of. The advent of VoIP has shown that person-to-person audio communication can easily be carried out across a packet-based IP network in spite of some difficulties in delivering consistently low end-to-end delays across the current infrastructure. This fact, when taken together with very low forecast growth rates for voice as compared with data, point to a future where voice makes up a very small proportion of total traffic. It is imperative, then, that the architecture of 4G networks is determined primarily by the need to deliver a quality IP service. The ability to handle voice and other streams with real-time constraints is a secondary (but essential) goal. The problem of delivering a predictable quality of service over wireless networks is a significant challenge in its own right and is fully addressed in Part II of this book.

6.2.2 Excellent Mobility Support

The adoption of mobile telephony services offered by 1G and 2G voice systems has surpassed many people's expectations. In many countries, most notably in the Scandinavian region, more than 60% of the overall population makes use of mobile telephones, and the trend elsewhere is to follow the same path. It has been forecast that there will be almost 1 billion subscribers to GSM, the most successful 2G system, by 2005.

The mobility facilitated by cellular telephones is often referred to as terminal mobility. An additional form of mobility envisaged for both the fixed and mobile phone networks is personal mobility, which allows a person to move from one terminal to another and have his or her calls and environment follow him or her as he or she registers on different mobile or fixed terminals. The Universal Personal Telephony service allows a user to be contactable via one personal number. Intelligence within the network, coupled with information on the users preferences and likely movement patterns, can forward the call to the appropriate terminal or to an automated response system.

In 4G systems, we must assume that all users of the network are potentially mobile with a sizeable proportion of them communicating via wireless terminals. Users must be contactable via a single (albeit multifaceted) identity. A means must be found to map from this identity to an address to which packets can be routed. Control of this mapping must lie firmly with the user who can modify the destination of the mapping and regulate the access that callers may have to it. In an environment where the path from source to destination may cross many different network domains, it would be unwise to associate this mapping with a single network operator. It is imperative that 4G networks provide a single consistent means of identifying users and allow this identity to be controlled by the user and efficiently mapped to a mutable destination.

Where 4G nodes are in motion, they may need to change their point of attachment to the network while a flow of packets is in progress. The degree to which this changes the path from source to destination depends on the topology of the network and the route chosen. The 2G and 3G cellular systems allow handoff where a mobile node shifts its point of attachment but stays within the same network domain. Typically, a change in domain is referred to as roaming, and active connections cannot be maintained under these circumstances. This implicit two-level tracking of a user's location (i.e., current domain and current position with that domain) can be generalized into an N-level mobility tracking scheme that can allow handoff in any circumstances where the network topology renders it possible.

6.2.3 Support for Many Different Wireless Technologies

The 1G, 2G, and 3G mobile systems relied on the use of spectrum reserved for public land mobile use and licensed by a very small number of network operators in each country. Differences in allocation timing and strategy of the spectrum have led to the need for multimode phones capable of adapting to use the spectrum available in a particular region.

In 4G systems, it is likely that a large number of different radio technologies may be used for network access. One trend that is very apparent is the use of radio spectrum in the less regulated *Industrial Scientific and Medical* (ISM) band. We have seen the emergence of Bluetooth [2] radio offering very short-range radio links of below 1 Mbps. Although this radio was originally developed as an easy way to replace cables interconnecting adjacent equipment, it has recently been standardized by the IEEE as standard number 802.15.1 as a more general purpose *wireless personal area network* (WPAN) technology. Also operating in the same band is the IEEE 802.11b WLAN radio system. This offers a throughput of 11 Mbps with a range of around 100 feet and lesser speeds over longer distances. This technology has been embraced by the market and is being used to provide high-bandwidth services to private users in buildings and campuses and also to public users in built-up areas and in airports. More advanced versions [3] of the standard are in preparation, including 802.11a operating in the 5-GHz band (which will deliver higher speeds of 54 Mbps) and 802.11g (which uses the same frequency band as 802.11b, but allows the data rate to be increased to 34 Mbps and higher).

At the other end of the spectrum, areas of very low population density or users at sea or in the air may be better served by satellite systems using a completely different set of radio technologies.

Existing 2G and 3G cellular may be useful in between these two extremes. A 4G node should be capable of adapting its radio capabilities to take maximum advantage of available spectrum.

6.2.4 Free from Unnecessary Operator Linkage

The GSM system was developed by European telecommunications companies in the mid- to late 1980s primarily as a mobile extension to the PSTN. The GSM model envisaged that users would subscribe to an operator who would build a cellular network infrastructure that would track the user as he moved from one location to another, making every effort to maximize the availability of service. In common with the dominant PSTN billing model in Europe at the time, all usage of GSM services was to be metered, and since the only way to pay for this was via the home operator, every action carried out by the mobile handset is with reference to the home network operator. Even when a user roams into a new domain, contact is made with the home network to establish a link with a billable entity before any calls can be made.

Two GSM handsets cannot communicate with each other directly; rather, they must each first authenticate themselves to the network, be linked with their billing details, and thereafter, operator mediated communication can take place. This mode of operation is consistent with the fact that the operator in effect owns the spectrum and is entitled to individually regulate and meter each access to it.

Where spectrum ownership is much more open, such as is the case in the ISM band, such restrictions are completely unnecessary and highly inefficient. Much of the dynamism that has taken place in wireless communications in recent years is the result of the easy access to such spectrum, and it is likely that this trend will continue in the future.

Where access to spectrum is not an issue, pairs or groups of nodes can form ad hoc networks to allow direct communication between nodes, and if appropriate, nodes can collaborate, relaying each other's traffic. Naturally, issues such as the need for user authentication occur in this kind of any-to-any direct communication, but arguably, these need to be solved in an operator-independent way in any case.

Once the special position of the operator is removed, there remains the problem that a wireless node wishing to communicate with a node outside its range cannot do so unless an intermediate node relays their packets to either another wireless node or on to the fixed network. If we could find a means of making real-time payments across a link, this would relieve us of the need to be associated with a billable entity and also allow us to motivate individual nodes to cooperate with us to relay traffic.

This motivation (financial or otherwise) could be used in a sparsely populated area to allow an individual wireless node to act as a packet relay between two out-of-range nodes. The payment method would allow the relay to be compensated for the battery drain and the usage of bandwidth that might otherwise be available to it.

In a heavily populated area, the same motivation could be used to encourage organizations to erect networks of wireless network access points in places like university campuses and shopping malls. Organizations that undertook this effort to any great scale would become the network operators of the 4G, but the fact that no special status is required to become an operator should ensure healthy competition.

6.2.5 Support for End-to-End Security

The security features inherent in 2G and 3G mobile systems are focused on two main services. First, the mobile users must be authenticated to the network operator. This authentication is generally limited to associating the user with a billing relationship that is operating satisfactorily. Where accounts are prepaid, this billing relationship often has no stored details on the identity of the user. There is no end-to-end exchange of credentials between a mobile user and their peer at the other end of the link.

The second service supported by 2G and 3G mobiles is content encryption. While this does deter attacks using simple scanning devices, it is no substitute for genuine end-to-end confidentiality.

In the 4G, mobile and fixed nodes will interact with each other without reference to their relationship to an operator. It is imperative that protocols and procedures are devised to allow the users of these nodes to authenticate one or more facets of each other's identity to a degree necessary to achieve the communication they desire.

6.3   4G Architecture and Research Issues

We envisage that the 4G mobile networks of the future will be based around an IP core network, which will, over time, completely displace both the fixed PSTN and also the 2G and 3G mobile voice networks. The architecture, as shown in Figure 6.1, will be based on delivering an IP transport service to a population where every user is potentially mobile and a large proportion of them make frequent use of wireless nodes to interact with people and services on the network.

Untethered users will gain access to the fixed network using a variety of different radio technologies. These will include short-range wireless systems such as Bluetooth and 802.11 as well as perhaps new access methods made available on the spectrum currently in use by 2G or 3G mobile voice systems. Individuals and organizations will install and operate wireless access points making use of a real-time payment system to provide such quasi-network operators with their financial incentive.

High-capability nodes will be equipped with software radios, which carry out as many radio functions as possible by executing signal processing

FIGURE 6.1
*Components of a
4G network.*

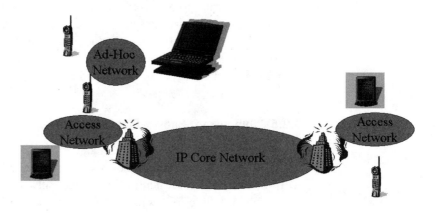

software on general purpose processors. Thus, the nodes will be able to communicate using whichever radio technology is expedient given its location and financial resources.

Ad hoc network protocols will be used to bind together groups of nodes for local communication and also to provide a link between each node and a range of geographically close fixed network access points. The motivation for this collaborative behavior may derive from the fact that the users have authenticated each other as belonging to the same group (e.g., workers in the same office or paramedics at the scene of an accident), or some kind of micro-payment method may be used to allow relaying nodes and network access points to profit from their activities.

Interactive links to other users will be established based on a user identity. In order to support both terminal and personal mobility, a directory service will be required to map between an identity or identity facet and an address to which packets can be routed. This service should be standalone and independent of any particular network infrastructure. The mapping may also be quite complex, allowing a user to build a profile of when and to whom he is available. In cases where confidentiality of location may be an issue, artificial relay points may be used to disguise the position of a node from the caller.

Before a call can be set up, each participant will need to authenticate themselves to the other parties within the call. Typically, users will have a multifaceted identity, and it may be inappropriate to make use of all of these facets for any given call. Facets may also depend on one another and progressive authentication may take place. For example, to gain entry to an office building, it may only be necessary to authenticate the fact that one is an employee. If an individual wants to open a safe containing financial documents, it may be necessary to also authenticate the fact that the individual's organizational role is financial director. Such identity facets would be inappropriate for use in a coffee shop where the same user wished to traverse a high-speed fixed network access point.

Once the parties have authenticated each other, end-to-end communication can take place with appropriate authentication and confidentiality measures being applied to the data traffic.

Where prebuilt access infrastructure does not exist, or where communication is taking place in a local region, nodes may resort to the formation of ad hoc networks. While this greatly enhances a node's ability to communicate, it also raises a number of additional research issues. Collaboration is the essence of ad hoc networking and will naturally take place where the members of the network belong to the same organization or have some other preestablished motivation to work together. In a public context—for instance, a number of wireless nodes encountering each other in a shopping mall—the motivation may be less obvious. There is a need for security protocols to allow nodes to selectively reveal information about different facets of their identity with an aim of maximizing the level of cooperation that is possible. If nodes are to offer relaying services to others, there will be a need for some kind of real-time payment method to allow a node to be compensated for the use of its resources (battery power and more).

Once an ad hoc network grows beyond a small number of nodes, there is a need to develop sophisticated routing protocols to allow nodes to relay for each other, thus maximizing their interconnectivity. It should also be possible for a wireless node that is within range of a fixed network access point to relay for other nodes within its ad hoc network.

Applications and services in the 4G will be built outside the network in keeping with the Internet tradition of keeping the core simple, fast, and efficient. Today, many of the Internet applications rely on entities contacting each other based on a network address. This is problematic for a number of reasons. First, it leads to a demand by end entities for the fixed assignment of addresses. If the address space is finite or inefficiently allocated (as is the case with IP version 4), this leads to address shortages and also causes problems when these end entities move with respect to the network. We envisage that in the future, applications will evolve to where addresses are deemphasized in favor of the use of more abstract names. Mobile users will be contacted by name, with appropriate servers to perform the mapping between address and the necessary routing information. Clearly such a mapping must be done in such a way that the user has control over what kind of routing information is held and who may gain access to it. Where a user is in motion, appropriate mechanisms must be devised to maintain a connection to a "name" as the route to the node is changing.

6.4   4G Research Efforts

Broadly speaking, research into the form of 4G mobile systems is being carried out by two somewhat overlapping communities. The first of these is the mainstream mobile telecommunications industry. The second community has come from a data networking background and is driven by technological progress in the area of wireless data networks leading to quite a different perspective on the path ahead. We will outline the direction being taken by both communities before elaborating further on our own thoughts on the fourth generation.

At the time of this writing, the telecommunications community's main source of revenue is based on a huge population of users of 2G mobile telephones. Now they are faced with the impending launch of 3G mobile systems. Most of the architectural decisions underlying the 3G were made with an eye to preserving the large investment in building the 2G network and retaining the customer relationship with the user population. It was thought that if the core network architecture was preserved, then new radio access methods (such as UTRAN) could be progressively introduced, and gradually, more support for data traffic and Internet access would be made available.

At the time of writing, prospects for the 3G look bleak. The mobile telephone market appears to be near saturation point, and the introduction of higher-speed data services has failed to produce the anticipated excitement and surge in user demand. Looking ahead to the data capabilities of 3G systems, it seems unlikely that these will match user expectations. In Europe, the auctioning off of spectrum at very high prices has placed a huge debt burden on aspiring 3G network operators that may be difficult to service.

Against this background, there is a reticence to contemplate revolutionary technologies that would divert attention away from the 3G, and consequently, new innovations are being discussed as add-ons to mature 3G networks rather than as technologies that would render it obsolete.

The two major proponents of 3G mobile systems—namely, the European UMTS Community and the U.S.-dominated cdma2000—have formed consortia (3GPP and 3GPP2, respectively) that are developing common standards for the provision of data services to users. In the case of 3GPP, the core network architecture is very much shaped by GSM with special purpose network entities added to support the transfer of IP datagrams between network endpoints or to the Internet. The 3GPP2, on the other hand, is much more Internet-like.

The Wireless World Research Forum, a consortium of many leading mobile telecommunications companies, has produced a "Book of Visions" [4], which summarizes the thinking of their member companies on what are important research topics for the future. This recognizes the need for greater support of the IP protocols and the importance of new radio access technologies. While there is recognition of the potential of ad hoc networking technology there is no mention of any architectural changes being made to accommodate this.

The rapid uptake of commodity WLAN products based on the IEEE 802.11b standard has given rise to many popular public wireless services that promise to deliver far superior data services to any currently envisaged in the 3G. In order to combat this threat, a number of organizations are attempting to integrate WLAN into 2.5G and 3G networks as simply a new radio access technique without any changes to architecture of the core network. This has led to a proposal to the 3GPP standardization effort, made in late 2001, that a study be commissioned to determine how WLAN technology such as IEEE 802.11 and HiperLAN2 could be integrated into UMTS. Attention to date has focused on a loose coupling where access to the WLAN is regulated with reference to the UMTS subscriber database. End-to-end traffic would pass from the WLAN to the Internet without going through the UMTS network entities.

The research community has been more daring [5] and has proposed an architecture where many heterogeneous networks, among them WLAN and UMTS, form subnetworks of a larger 4G system whose architecture is very Internet-like with mobility being handled at the IP level.

The second community of researchers that is interested in this area come from a wireless Internet perspective and have focused on the efficient transport of IP traffic over wireless channels with provision of varying qualities of service. These themes are elaborated on in other chapters of this book. In terms of mobility handling, there appears to be broad support for the idea that Mobile IP can be used to handle user movement between domains (roaming), while some alternative system such as Cellular IP, HAWAII, or TIMIP be used to deal with movements within a wireless domain. While the concept of handover is very much a part of these efforts, this community does not appear to be attempting to build a system that could aspire to displace the current global fixed and mobile telephony infrastructure. In a similar way, their efforts in the security arena do not appear to be targeted towards replacing telephone numbers by a single globally acceptable identity that could be cryptographically asserted.

6.5   The NTRG 4G Test Bed

At Trinity College in Dublin, Ireland, the *Networks and Telecommunications Research Group* (NTRG) has been investigating the form of 4G mobile systems since 1998. We have ongoing projects investigating many different aspects of 4G technology from applications though to physical layer issues. The individual projects are bound together by virtue of their individual contributions towards our 4G test bed.

Since we envisage that the 4G mobile nodes of the future will be constructed using commodity computing platforms, our target nodes are general purpose PC workstations, and where portability is important, laptop and palmtop variations of these. In order to keep a consistent operating environment across all platforms and to allow our work to integrate with popularly available applications, we have chosen to develop the Microsoft Win32 environment as supported by Windows 2000 on PC and laptop and make use of Windows CE in handheld and palmtop environments.

6.5.1   The Layered Architecture

Components of our 4G environment are implemented as standalone layers each realized by a single main thread. The interlayer interface is very simple, consisting of primitives to send information upwards or downwards through the stack and to attach a blackboard of parameters to each request that can be used by any layer through which the data passes. A sample layer stack is depicted in Figure 6.2.

6.5.2 Wireless Alternatives

Different radio hardware can be accommodated by writing a layer that interacts directly with the hardware and presents the simple interlayer transfer interface to whatever is above it. In this way, we have been able to perform our wireless experiments with links as diverse as Infra-Red (IRDA), Bluetooth, IEEE 802.11, and also a very simple half-duplex radio we constructed in-house, which uses amateur frequencies and allows us to experiment at a very low level.

6.5.3 Software Radio

Ultimately, we expect to be able to replace the real radios at the bottom of this protocol stack with a software radio operating in conjunction with a wideband front end, which would allow operation across a wide frequency band with the chosen form of modulation being performed in software. The individual building blocks of the software radio (such as channelization, coding, modulation, and demodulation) will be realized by individual layers with an appropriate stack being assembled to deliver the desired radio waveform. Thus far, we have receiver-only systems that implement a variety of forms of amplitude and frequency demodulation and can rapidly switch between the two if necessary.

FIGURE 6.2
The layered structure of the 4G test bed.

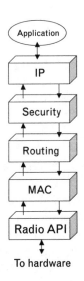

6.5.4 Routing Protocols

The subject of routing protocols enabling the formation of ad hoc networks of nodes has been under active study by the research community for some time now [6]. Although many different protocols have been simulated, very few implementations exist that can test these protocols across a real radio channel under different mobility scenarios.

Using our layered architecture, we initially implemented the *Dynamic Source Routing* (DSR) Protocol. This simple reactive protocol only begins to try to establish a route to a destination when there is data to be sent. This is in contrast to a proactive protocol, which attempts to maintain knowledge of the state of the network so that it is already in possession of sufficient routing routine information when data needs to be sent. We have also implemented a hybrid *Zone Routing Protocol* (ZRP), which is proactive for nodes that are in a nearby zone and reactive for those that are farther away. We anticipate that the use of these protocols across real wireless channels will give us a unique insight into their properties.

6.5.5 Emulation Facilities

When routing protocols or mobility aware applications are being developed and tested, one of the major problems is to expose the evolving software to particular mobility scenarios without conducting all debugging on the move and out of doors. Our initial approach to this challenge was to develop a layer (which we call the datagram layer) that emulates the radio broadcast environment across a collection of Internet links on the local area network.

Each of the emulated nodes is assigned an IP address, and the emulation layer in each node is told what other nodes are visible to it in radio terms. When a packet arrives to be sent on the emulated radio interface, it is encapsulated in an IP datagram and sent to each of the visible nodes.

This effectively allows us to construct ad hoc networks made up of processes running on any Internet-connected host. By constructing a relay layer that moves packets from one protocol stack to another, we can freely intermix nodes that are running on real wireless nodes and others that are sitting on the emulated radio layers. An example of this is shown in Figure 6.3.

While the above is an effective debug and test tool, emulated wireless nodes either see each other or they do not, and transmissions always get through without transmission errors. The ability of nodes to see each other or not is also statically configured.

In an attempt to improve our radio emulation environment, we have devised a system that we refer to as a reality emulator. In place of the radio layer in each node's protocol stack, we place a reality emulator client layer. Each of these clients connects via sockets to a server, as shown in Figure 6.4,

FIGURE 6.3
*An ad hoc network
emulated wireless
nodes of genuine and
emulated wireless
nodes.*

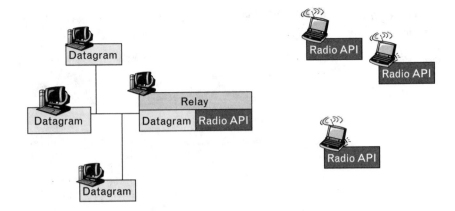

FIGURE 6.4
The reality emulator.

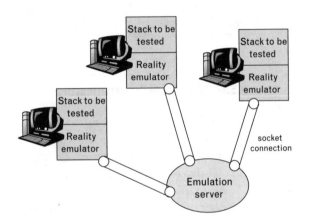

which emulates the way in which a radio channel will behave as nodes move with respect to one another and also encounter collisions in transmission.

When a node initializes itself, it connects to the emulation server where it is given an initial geographic position. By interacting with a graphical user interface on the server, a designer can control the transmission range of each radio and their movement with respect to one another. Figure 6.5 shows the server's view of a collection of emulated nodes moving around the Trinity College campus.

From the point of view of the client protocol stack, the behavior of the link is identical to that experienced by the node when running across a real radio link, and provided that the number of emulated nodes is kept reasonably small, the emulator can support substantial amounts of traffic including real-time voice streams. The system is particularly good at exposing the routing software to particular node movement scenarios that might otherwise be hard to achieve.

FIGURE 6.5   *The emulation server showing eight nodes and their radio ranges.*

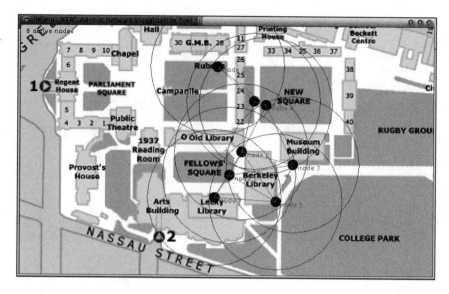

FIGURE 6.5   *The emulation server showing eight nodes and their radio ranges.*

6.5.6   The Security Architecture

We envisage that all users in the fourth generation are potentially mobile, and a large proportion of them will avail of wireless devices. This means that nodes will typically be interacting with peer nodes with little information on the identity represented by that peer.

Authentication in the physical world is typically achieved using a multiplicity of cards that people carry around in their purse or wallet. These may contain basic information about a person, such as might be present on a driving license. Other cards may contain information on an individual affiliation with a bank or perhaps more personal information such as a card detailing his or her blood type.

Individuals produce cards detailing different facets of their identity only as the need arises. For example, to make a cash withdrawal at a bank, they may need to produce both a driver's license and a bank card. They might hand over details relating to their medical records only to someone who had already proved he or she was a medical professional.

The above exchanges are characterized by individuals entering into a negotiation dialogue where credentials are exchanged in order of increasing importance as trust is progressively built between the individuals. This process concludes either when the shared trust has reached its highest level, or it has exceeded that required by the communication exchange.

At present, we are designing a system to do this kind of credential exchange between nodes in an ad hoc network. When a node wishes to avail of a service on another system, this will cause the credential exchange agents to enter into a dialog where details on different facets of a user's identity are exchanged according to a *credential release policy* (CRP).

Nodes may authenticate neighboring nodes up to a level at which they are happy to relay traffic through them. A considerably higher level of authentication may be needed before a pair of nodes may be willing to enter into a specified application dialog. Figure 6.6 depicts the security agent that carries out this authentication.

Based on the authenticated identity facets, which are shared between the two nodes involved in the trust negotiation, groups can be formed to share a common purpose. This will allow shared keys to be negotiated, which can be given to all members of the group to allow shared access to resources. Similar group formation systems have been developed for Internet multicast environments that can serve as a model for this [7]. This can be used for such applications as a walkie-talkie type system where all users hear all traffic or, indeed, any other form of data-based group collaboration.

6.5.7 Real-Time Payment

Most of the flaws in 2G and 3G mobile systems can be traced to the fact that the major technical decisions defining the architecture have been made based on one overriding concern, namely that of generating revenue for the network operators.

If the promise of the 4G is to be fulfilled, the focus must be on enabling connectivity between network users without tying this connectivity to a user-operator subscription. Clearly some other way needs to be found to allow providers of network infrastructure (even if this is just a single relay node) to be rewarded for making this available to the general population of nodes.

FIGURE 6.6
The 4G security agent.

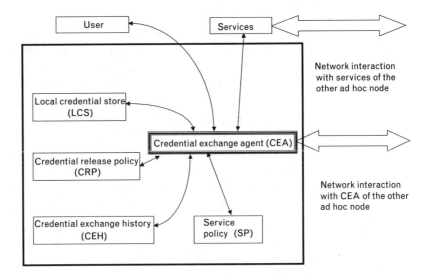

WIRELESS IP AND BUILDING THE MOBILE INTERNET

Figure 6.7 depicts a multiparty micro-payment system [8] developed within our research group that allows a stream of cryptographic payment tokens to be interspersed with the normal data packets that make up the end-to-end flow. Before this can commence, a pricing phase was undertaken where each party (or network operator) along the path offered their services for a particular price. Once the contract is agreed to, the payment tokens can be captured by nodes along the path and redeemed for an agreed proportion of the overall payment for the traffic. We plan to adapt this system to support ad hoc operation where the end-to-end route may change over the lifetime of the communication.

6.5.8   Applications

The principle application we have used on our 4G network testbed has been simple Web access, where pages containing HTML as well as complex multimedia data have been delivered to handhelds moving in a wireless network. We have also performed some experiments using the wireless link for video image data. More recently, we have developed a simple point-to-point telephony application using the Session Initiation Protocol for signaling. In the future we will modify this to integrate elements of our security architecture, in particular, to incorporate the notion of multifacetted user identities. There are a whole host of issues to be resolved before we can support mobility thorough the mapping of this identity information to addressing information that works well with both fixed and ad hoc networks.

6.6   Concluding Remarks

We have outlined above some of the problems inherent in the architectural design of 2G and 3G mobile systems. Arising from this, we have enumerated some of what we believe are imperatives for the design of a new 4G architecture. These ideas are being progressed by the mobile Internet research community and to a lesser extent, by researchers in mobile telecommunications. In order to progress our ideas for 4G systems, our research group has embarked on a number of distinct projects dealing with different aspects of the overall system. Each of these projects contributes to the construction of a test bed based on the use of commodity PC and PDA hardware running common operating systems. The test bed supports wireless mobility via a range of technologies from infrared to software radio and allows the experimentation with real ad hoc networks that are engineered to integrate with populations of fixed nodes. Security concerns are also catered for both in terms of node authentication and also as a means of payment for resources consumed. The utility of the 4G systems will be demonstrated

FIGURE 6.7
*Real-time, multiparty
micro-payment.*

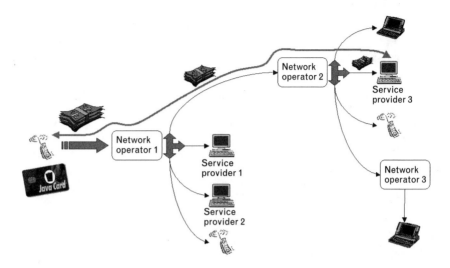

with noninteractive data-based applications as well as those, such as telephony, that involve continuous multimedia streams.

REFERENCES

[1]   O'Mahony, D., "Universal Mobile Telecommunications Systems: The Fusion of Fixed and Mobile Networks," *IEEE Internet Computing*, Vol. 2, No. 1, January/ February 1998, pp. 49–56.

[2]   Bisdikian, C., "An Overview of the Bluetooth Wireless Technology," *IEEE Communications*, Vol. 39, No. 12, December 2001, pp. 86–94.

[3]   Leeper, D. G., "A Long-Term View of Short-Range Wireless," *IEEE Computer*, Vol. 34, No. 6, June 2001, pp. 39–44.

[4]   Wireless World Research Forum, "Book of Visions," December 2001, http://www.wireless-world-research.org.

[5]   Bria, A., et al., "Fourth-Generation Wireless Infrastructures: Scenarios and Research Challenges," *IEEE Personal Communications*, Vol. 8, No. 6, December 2001, pp. 25–31.

[6]   Royer, E., and C. -K. Toh, "A Review of Current Routing Protocols for Ad Hoc Mobile Wireless Networks," *IEEE Personal Communications Magazine*, April 1999, pp. 46–55.

[7]   Waldvogel, M., et al., "The VersaKey Framework: Versatile Group Key Management," *IEEE Journal on Selected Areas in Communications*, Special Issue on Middleware, Vol. 17, No. 8, August 1999.

[8]   Peirce, M., and D. O'Mahony, "Flexible Real-Time Payment Methods for Mobile Communications," *IEEE Personal Communications*, Vol. 6, No. 6, December 1999, pp. 44–55.

CHAPTER 7

Ad Hoc Networks: A Mobile IPv6 Viewpoint*

Koen Cooreman, Liesbeth Peters, Bart Dhoedt, and Piet Demeester

7.1 Introduction

As a general trend, the importance of mobile networking is increasing every day. In the future, people will expect to be able to use their network terminals (laptops, PDAs, and mobile phones) anywhere and anytime. This chapter focuses on a special type of mobile network, namely, ad hoc networks [1], which do not need any fixed infrastructure to operate. More specifically, a solution is described as to how mobile ad hoc terminals can move while connectivity with a global IPv6 network [2] and other mobile ad hoc terminals can be supported and maintained. The fact that we are supposing a global IPv6 network and not an IPv4 network is important because specific features of IPv6 will be used to support global mobility of ad hoc terminals.

This chapter is organized as follows. This first section provides an introduction to ad hoc networks and their features and describes the relevant properties of IPv6. Section 7.2 will define the different levels of mobility that should be taken into account for mobile ad hoc networks. Section 7.3 gives an overview of the properties of existing ad hoc routing protocols. In Section 7.4, Mobile IP [3], the current standard to provide macro-mobility in an IP-based network, is briefly described. The principles of micro-mobility are described in Section 7.5, addressing the main features of Cellular IP [4]. In Section 7.6, we propose a mechanism that could be used to support communication and mobility of mobile ad hoc devices in an all-IPv6-based global network. Finally, Section 7.7 offers some concluding remarks.

* Part of this work has been funded by the Belgian Government (DWTC/SSTC) InterUniversity Poles of Attraction (IUAP) project: MOTION (Mobile Multimedia Communication Systems and Networks).

7.1.1   Ad Hoc Networks

Mobile ad hoc networks [1] have, unlike cellular networks (e.g., a UMTS network), no fixed infrastructure or backbone network to support the mobility of the terminals in the network. In cellular networks, the backbone network performs all networking operations, and mobile devices only communicate directly with one of the base stations of which they are in reach. This means that mobile devices in cellular networks can be rather simple devices: all network functionality is implemented in the backbone. Ad hoc networks, on the other hand, lack any infrastructure to support network functionality. In Figure 7.1, a small ad hoc network is illustrated. Having no supporting infrastructure means that all the network intelligence must be situated inside the mobile devices that make up the network. As a consequence, the routing of packets (we consider ad hoc networks to be packet switched and not circuit switched networks) from source to destination has to be accomplished by the network terminals, meaning that terminals in an ad hoc network do not only act as hosts but also as routers. Because the network topology can change quickly and unpredictably, an ad hoc network has to be self-configuring and should be very adaptable to changes (like hosts leaving the network, link breakage, and new hosts who attach to the network).

7.1.1.1   Applications

The main rationale for using ad hoc networks is not being reachable anytime and anywhere by anybody (as is the case of 2G and 3G cellular systems). Their major strength is actually the seamless connectivity with devices in the neighborhood and the fact that the complete network infrastructure is merely composed of the participating devices and that this network infrastructure is self-organizing. In the following paragraphs, several

FIGURE 7.1
An ad hoc network (left) versus a cellular network (right).

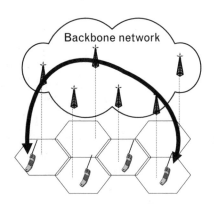

applications that can make use of this property of ad hoc networks are discussed.

Conferencing

One can think of conferences and meetings where almost everybody disposes of a laptop, notebook, or PDA. On such occasions, people would probably appreciate the fact that they can, seamlessly and without any configuration, exchange information, use a local printer, or connect to the Internet via a wireless Internet gateway. One could immediately download the current presentation being presented, browse through the slides on one's own device, print them on the local printer, or e-mail the presentation to a colleague who cannot be present. Using ad hoc networks to support this application implies that no infrastructure must be present and it even avoids the overhead (like, for example, the cost and routing within the infrastructure) of using an infrastructured network.

Rescue Operations

Since ad hoc networks can be deployed very quickly, they can be used to set up a network at a location where no infrastructure is present. They can be used, for example, during rescue operations at disaster sites, like earthquakes, fires, or floods. In such situations an infrastructured network is very likely to be damaged and made useless. As network applications become, even for rescue operations, more and more important, ad hoc networks can be used to support network connectivity without the need of any available fixed infrastructure.

Home Networks

Many families already have more than one computer at home. Since these computers are not necessarily located in the same room and connectivity between them is most likely desirable, the use of ad hoc networks is a very elegant solution. Most people are not network specialists, so the self-organizing and auto-configuring character of ad hoc networks is a very easy solution. Even if wires, instead of a wireless technology, are used to connect the different machines, ad hoc network technology can be used to support seamless connectivity between the different devices.

In the future, one can imagine that more and more devices at home will have a network connection and be remotely controllable. Some of these devices, like washing machines, will have a fixed location, but other devices, like a portable radio, can and will be easily displaced. For this last category, a wired connection is not very practical, and while moving those devices around, ad hoc network technology can be used to take into account the changing network topology.

Personal Area Networks

A *personal area network* (PAN) is a network of devices that are closely associated to one person. These kinds of devices may be attached to a person's clothing or carried around in a purse. In rather exotic visions of the future, these devices include virtual reality devices attached around the head and devices oriented toward the sense of touch. All the devices in a PAN may be attached to the Internet, but they will very likely have to communicate with each other. Because most devices in the same PAN will have an almost fixed position with respect to each other, mobility is not an important factor inside one PAN. However, when interactions between several PANs are needed, mobility can suddenly become much more important. To establish communication between moving PANs, ad hoc network technology can be used.

Sensor Dust

One could imagine a large collection of tiny and cheap sensors, forming an ad hoc network, to offer detailed information about terrain or dangerous environmental conditions. Via ad hoc network technology, the activities and the reports of these sensors can be coordinated remotely. Instead of sending emergency personnel into dangerous environments, sensors containing wireless transceivers can be distributed. Forming an ad hoc network, these sensors can cooperate in order to gather the desired environmental information. It is obvious that industrial and military applications are also of great interest.

Games

Naturally, ad hoc networks can be used not only for professional applications but for entertainment as well. Distributed network applications like games are supported as well, and people using ad hoc networks do not even have to know who they are playing with. They can play against the people that are accidentally within the neighborhood. One can think, for example, of trains, buses, airports, or even airplanes where people can spend quite of lot of time waiting. Using ad hoc network technology makes the possibilities for games inexhaustible, and costs are minimal because no network infrastructure of any network provider is used.

Military

Without any doubt, ad hoc networks can also be used for military purposes. As a matter of fact, it was the U.S. Department of Defense that sponsored the first research to enable packet switching technology to operate without the restrictions of fixed or wired infrastructure. In 1972, DARPA initiated research to develop and demonstrate a *Packet Radio Network* (PRN). Ad hoc networks can be utilized out of the military need for survivability and operation without preplaced infrastructure.

7.1.1.2 Issues in Ad Hoc Networks

The problem that will be addressed in more detail in this chapter is that of interconnecting ad hoc networks with the global Internet, and more precisely, the identification of mobile terminals and the correct routing of packets from and to each terminal while they are moving. For instance, when ad hoc networks get connected to the Internet via two or more gateways, the hierarchical addressing structure of the Internet causes problems: the boundaries of ad hoc networks do not have to correspond with the boundaries of the different domains in the Internet. Of course, next to interconnectivity with the Internet, other technical aspects of ad hoc networks are important.

Routing

Since the strength of ad hoc networks is the seamless connectivity with devices in the neighborhood without using any backbone infrastructure, packet routing is, of course, one of the hardest problems for these networks to face. Many different ad hoc routing protocols are proposed in the literature for performing this difficult task. The basic properties of the proposed and existing routing algorithms are described in Section 7.3.

Security

A wireless link is much more vulnerable than a wired link. Ad hoc networks have the disadvantage that the physical link used can easily be eavesdropped and manipulated, and in addition they are faced with the problem that a malevolent user can easily attach to the network. Such a user can easily load the available network resources, like wireless links and batteries of other users, and even disturb normal network operation.

Routing protocol packets carry important control information that governs the behavior of data transmission in the ad hoc network. Without proper protection, these packets can easily be subverted or modified. A malicious user can insert spurious information into routing packets and cause routing loops, long timeouts, and advertisements of false or old routing table updates. As such, research to secure ad hoc networks is definitely necessary. All proposed routing protocols place complete trust on the devices that make up the ad hoc network and are therefore vulnerable to attacks from malevolent users.

In practically every network, the fundamental security mechanisms are based on cryptographic keys. Since an ad hoc network is not always connected to the global Internet and a mobile device can easily be tampered with or leave the ad hoc network unpredictably, the distribution of public keys is difficult to accomplish. A (perhaps distributed) server for key management that is trusted by the mobile devices in the ad hoc network can fall under the control of a malicious party.

QoS

Just as in the case of conventional networks, users of ad hoc networks will expect some level of QoS. Since the topology of an ad hoc network can constantly change, reserving resources and sustaining a certain level of quality while network conditions constantly alter is very challenging and not straightforward. Several approaches exist to provide QoS in ad hoc networks. As an example, we can mention the INSIGNIA QoS framework [5], which is designed to support adaptive services. An adaptive service requires a minimum amount of bandwidth and is able to enhance the quality of the service when more resources become available. A clean separation is made between the routing protocol functionality and the QoS functionality, allowing several ad hoc routing protocols to be supported. The resource reservation signaling is performed in–band, which means that control information is carried along with the data packets allowing fast response to network changes.

Automatic Discovery of Available Services

In the example discussed above, where people can use a local device like a printer, the support for seamless networking is not sufficient. In that case, automatic discovery of available services is necessary as well. It would not be practical if each time a new service became available, an ad hoc networking device had to be configured to be able to use the new service. Therefore, a mechanism that allows discovery *and* use without any configuration should be used.

Billing System

In an ad hoc network, resources belonging to various users are necessary to support the ad hoc network functionality. Available bandwidth, battery power, and computing power of other users are consumed, and as a consequence, a billing system should be developed to pay or compensate for consumed resources.

7.1.2   IPv6

Throughout this chapter, knowledge of some specific features of IPv6 [2] is assumed. In order to understand this chapter, relevant features are briefly described in the following sections.

7.1.2.1   IPv6 Addressing Architecture

The most well-known feature of IPv6 is the use of 128-bit addresses instead of the 32-bit addresses of IPv4. The huge amount of possible addresses does not only allow many more devices to be connected to the network, but it also makes it possible for an interface to have more than one address, each within its own scope. The defined scopes are as follows:

- *Link-local address:* to be used for addressing on a single link;

- *Site-local address:* to be used for addressing inside a site;

- *Aggregatable global unicast address:* an address with global scope, which is comparable to an IPv4 unicast address.

Three different address types are defined:

1. *Unicast address:* identifies one single interface.
2. *Anycast address:* identifies a set of interfaces. When a packet is sent to an anycast address, it is delivered to one of the interfaces in the set.
3. *Multicast address:* identifies a set of interfaces. When a packet is sent to a multicast address, it is delivered to all the interfaces in the set.

Each interface is required to have at least one link-local unicast address and may be assigned multiple IPv6 addresses of any type or of any scope. The specific type and scope of an IPv6 address is indicated by the leading bits in the address, which is called the *format prefix* (FP). Several predefined multicast addresses exist, like, for example, the all-nodes multicast address or the all-routers multicast address with link-local scope. Figure 7.2 shows the IPv6 addressing architecture.

7.1.2.2   Router Advertisements and Solicitations

By sending router advertisements, routers advertise their presence on a link together with various link and configuration parameters, like address prefixes that can be used for stateless address autoconfiguration (see next section). These router advertisements are sent to the all-nodes multicast address, or they are sent to a specific node that has asked for a router advertisement by sending a router solicitation to the all-routers multicast address.

7.1.2.3   Stateless Address Autoconfiguration

Stateless address autoconfiguration is a mechanism by which a device is able to obtain addresses for its network interfaces automatically. To be able to perform stateless address autoconfiguration for an interface, an interface is required to have an interface identifier. Typically, a EUI-64–based interface

FIGURE 7.2
IPv6 addressing archi-
tecture. IID: interface
identifier; LLA:
link-local address;
SLA: site-local ad-
dress; GUA: aggre-
gatable global unicast
address. A host can
form its link-local ad-
dress by placing its in-
terface identifier into
the right-most bits of
the link-local unicast
address prefix
(FE80::). Site-local
and aggregatable
global unicast
addresses can be
formed by placing its
interface identifier into
the advertised site-local
and global unicast
addresses.

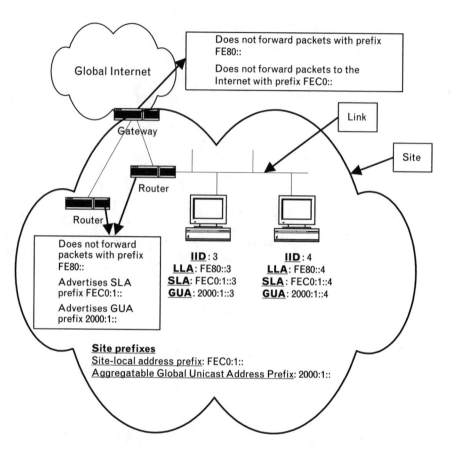

identifier[1], which is a unique 64–bit identifier, will be used. By placing its interface identifier into the right-most bits of the link–local unicast address prefix (which is FE80::[2]), a link-local address is formed. To ensure that the generated address is not already in use by another interface, a duplicate address detection mechanism exists. If this duplicate address detection mechanism indicates that the constructed link-local address is unique, site-local and global unicast addresses for an interface can be automatically constructed by appending the interface identifier to the prefixes that are advertised in router advertisements. If duplicate address detection fails, manual configuration of the specific interface is necessary.

1   An EUI-64 identifier is a 64-bit extended unique Identifier. A 48-bit MAC address $cccccceeeeee_{16}$, for example, can be transformed to an EUI-64 identifier by putting $FFFE_{16}$ in the middle of the MAC address, forming the EUI-64 identifier $cccccccFFFEeeeeee_{16}$. The letters "c" and "e" represent hexadecimal digits.

2   FE80:: means FE80:0:0:0:0:0:0:0.

7.1.2.4 Routing Header Option

This IPv6 header option is very similar to the IPv4 loose source and record route option and is used to indicate one or more intermediate nodes that must be "visited" by a packet along the way to its destination.

7.2 Mobility of Ad Hoc Devices

Mobility of ad hoc devices should be viewed at different levels of granularity. In this chapter we will distinguish three levels (see Figure 7.3):

1. *Macro-mobility*: With macro-mobility, we mean the mobility through a global network. While moving around in such a network, it should be possible that existing communications are not broken.

2. *Micro-mobility*: This is the mobility of a device in one single administrative domain of the global network. For most access networks, like cellular networks, this is the lowest level of mobility. In the case of ad hoc networks, however, a lower level of mobility exists: ad hoc mobility.

3. *Ad hoc mobility*: As mentioned in Section 7.1.1, mobility within an ad hoc network itself must be handled as well—the mobility of the devices constantly causes changes in the network topology. Because only direct peer-to-peer communication is possible, special routing protocols that can deal with these dynamics must be used. In the IETF Mobile Ad Hoc Networking Group (MANET) [6], several ad hoc routing protocols have been proposed.

Supporting macro-mobility for ad hoc devices is very straightforward: Mobile IP [3], which is described in Section 7.4, can be used to support this global mobility.

Supporting micro-mobility for mobile ad hoc devices, however, is less obvious. Existing solutions offering micro-mobility suppose that the mobile device is directly connected to an access router. Because of the multihop characteristics of ad hoc networks, this is generally not the case. Moreover, the boundaries of an ad hoc network do not necessarily coincide with the boundaries of administrative domains. This means that access routers of several different domains can be connected to one and the same ad hoc network.

If an ad hoc network is not connected to the global Internet, no special problems of addressing should occur. Using an existing ad hoc routing protocol should suffice to handle communications between the terminals in the network. Routing in general ad hoc networks, however, should not be a

FIGURE 7.3
Different levels of mobility for ad hoc terminals.

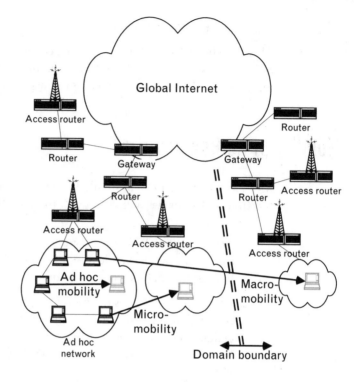

variant of aggregated routing because an ad hoc network cannot be sub-divided into subnets due to the mobility of the terminals.

In the next three sections, ad hoc mobility, macro-mobility, and micro-mobility solutions will be described.

7.3   Ad Hoc Mobility

In an ad hoc network it is very important to use a routing protocol that can rapidly adjust and adapt to topology changes without imposing high demands on the available resources of the network. Wireless links are, in comparison with wired links, limited in terms of bandwidth, and the mobile devices themselves very often have limited capacities (battery power, available memory, and computing power). These two factors, namely dynamic topology changes and limited resources, put heavy demands on the routing protocol. In the MANET group [6], several routing protocols have been developed or are under development, but *the* routing protocol to handle all situations has not yet been found. Many routing protocols perform very well under certain conditions but perform much worse in other situations. The developed ad hoc routing protocols can be roughly subdivided into five categories [1]:

1. Flooding;

2. Proactive routing protocols;

3. Reactive routing protocols;

4. Hybrid routing protocols;

5. Protocols using knowledge of the physical position of the devices that make up the ad hoc network.

The following sections briefly describe the properties of these five categories.

7.3.1 Flooding

Flooding is the simplest routing solution, being a brute-force routing strategy, not making use of any network topology knowledge. If a device A has a packet to send to device B, device A will simply broadcast the packet. Every device that is not device B will receive this packet and will broadcast the packet again so that its neighbors receive the packet too. If device B, possibly through several hops, is reachable from device A, it will eventually receive the packet, and if not, the whole network was loaded with a flood of useless packets. Flooding has the disadvantage that many devices receive and transmit a packet when it is not really necessary; thus, it consumes a large part of the limited available resources. The advantage of flooding is that it is very straightforward to implement and it hardly requires any computing power of the ad hoc devices.

7.3.2 Proactive Routing Protocols

In proactive routing protocols (also called *table-driven protocols*), all network nodes know the topology of the network. To realize this, adjusted versions of the classical distance vector and link state algorithms are used. Nodes will have to send control packets to their neighbors on a regular basis to inform their neighbors about possible topology changes. The rather large amount of bandwidth consumed by these control packets is not the only disadvantage. Also, every device in the network is obliged to have knowledge of the complete network topology. As the number of devices in the network increases, more memory is necessary and more control packets will have to be exchanged between the different devices. These protocols have the advantage that they are able to find a new route in the network rather quickly. *Destination-sequenced distance-vector* (DSDV) routing is an example of a proactive routing protocol for ad hoc networks. Also, Hiperlan1 uses a proactive protocol to support ad hoc networks.

7.3.3   Reactive Routing Protocols

In reactive routing protocols (also called *source-driven protocols*), the nodes in the network only store the routing information that is necessary at the moment. To discover a new route, these protocols use route detection. This route detection is mostly carried out by using the query/reply principle: if device A has packets to send to device B and A knows no route to device B, then A will try to discover a route to B by flooding a *route-request* (RREQ) into the network. How devices react on RREQs and how the route from A to B is finally determined depends on the routing protocol used. So, reactive routing protocols use a flooding mechanism to find a route, but the data packets are transmitted much more efficiently. Thus, with proactive routing protocols a route to the destination is immediately available, while reactive routing protocols induce a certain delay for route discovery. Due to the reactive nature of the routing protocol, an additional delay must be accounted for when a new path must be discovered. So, if the devices are very mobile, the control overhead of RREQs can be as high as in the case of a proactive routing protocol. They have the advantage, however, that only those routes that are currently necessary are kept in memory, implying that the mobility of nodes that are not actively participating in the ad hoc network have no influence on the network. With a proactive routing protocol, the complete network condition must be kept accurate among the network devices. Examples of reactive routing protocols are *ad hoc on-demand distance vector* (AODV) routing, *dynamic source routing* (DSR), and *associativity-based routing* (ABR).

7.3.4   Hybrid Routing Protocols

Proactive protocols are best in networks where the devices are close to each other and where only a few hops are necessary to connect two different devices. In larger networks with a larger hop-diameter, it may be better to use a reactive routing protocol. To make use of the benefits of both types of protocols, in the manner for which they are best suited, hybrid routing protocols were developed. These routing protocols have properties of both types of protocols. A hybrid routing protocol can, for example, proactively determine routes to devices that are close by and reactively compute routes to devices that are further away. In that case, each node only knows the details of the network topology in its neighborhood. Another example of a hybrid routing protocol is to form a hierarchical structure so that the flooding of data and control packets does not have to be accomplished by all the devices in the network but only by a set of devices selected for that purpose. This hierarchical structure, of course, must be formed and updated automatically and dynamically as the network topology changes.

7.3.5 Protocols That Make Use of the Known Physical Location

Protocols exist that make use of the known physical location of a device, meaning that these protocols must use a system that allows it to determine the exact locations [e.g., *Global Positioning System* (GPS)]. In that case, broadcasting of RREQs can be accomplished more efficiently and directed because the position of the device that must be reached is known. Flooding data packets towards the correct location of the destination of a packet is possible too.

Table 7.1 shows the most important ad hoc routing protocols. For most protocols, more information can be found in [6].

7.4 Macro-Mobility: Mobile IP

Mobile IP [3] is the current standard for supporting macro-mobility in IP networks. In IPv4, Mobile IP is specified apart from the IP specification, but in IPv6 it is an integral part of the IP-protocol specification. Its main characteristics will now be described.

Every mobile device that one wants to use for global communications is supposed to have a home domain, in which the device should have a unique global aggregatable unicast address (this is called the home address of the device). Via this home address, other devices (which can be mobile devices too) in the Internet can communicate with the mobile terminal. If this mobile device is in another domain than its home domain (this is called a foreign domain), this home address cannot be used because the classless interdomain routing of packets is based on aggregatable subnet prefixes.

TABLE 7.1   CLASSIFICATION OF AD HOC ROUTING PROTOCOLS

PROACTIVE PROTOCOLS	REACTIVE PROTOCOLS	HYBRID PROTOCOLS	USING PHYSICAL POSITION
Destination-sequenced distance-vector routing (DSDV) [1, 7]	Dynamic source routing (DSR) [1, 6]	Relative distance micro-discovery routing (RDMAR) [9]	Location-aided routing (LAR) [12]
Wireless routing protocol (WRP) [8]	Ad hoc on-demand distance vector routing (AODV) [1, 6]	Core-extraction distributed routing algorithm (CEDAR) [10]	Zone-based hierarchical link state routing (ZHLS) [13]
Optimized link state routing (OLSR) [6]	Temporally ordered routing algorithm (TORA) [1, 6]	Zone routing protocol (ZRP) [6, 11]	
Topology broadcast based on reverse-path forwarding (TBRPF) [6]		Landmark routing (LANMAR) [6]	

Therefore, the mobile device should obtain a *care-of address* (COA) with a subnet prefix that is used in this foreign domain. This address can easily be obtained by using IPv6 stateless address autoconfiguration (see Section 1.2). In Figure 7.4 mobile device A forms the COA 2000::10 using its interface identifier 10 and the prefix 2000:: (which is advertised by the router using router advertisements) of the foreign domain. Stateful address autoconfiguration (by using DHCP[3]) can be used as well. After obtaining this CAO, it should be "bound" to the home address of the mobile device. The mobile device accomplishes this by registering its care-of address with a router on its home link (which is the link it is attached to in its home domain), which is called the home agent of the mobile device. The COA is registered with the home agent by sending a packet to the home agent containing a *Binding Update* destination option. Although it is possible that the mobile device has more than one COA in the foreign domain (a router can advertise several prefixes that can be used on a link), only one of them can be bound to its home address. The home agent will encapsulate arriving packets that have as a destination the home address of the mobile device and send them to the mobile node's registered COA [see Figure 7.4(a)]. A *Binding Update* destination option, however, can also be used to register the COA with each device the mobile device communicates with and allows route optimization: By knowing the COA of the mobile device, other devices can send packets directly to the care-of address and not to the home address of the mobile device, avoiding triangular routing [see Figure 7.4(b)].

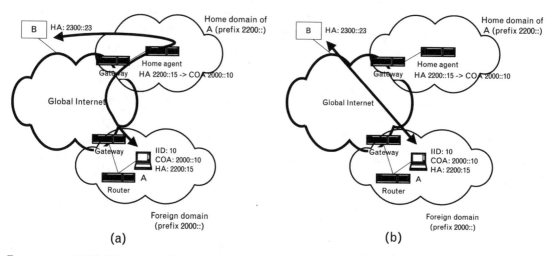

(a)   (b)

FIGURE 7.4   *Mobile IP operation: (a) without route optimization (triangular routing); and (b) with route optimization. IID: interface identifier; HA: home address.*

3   Dynamic Host Configuration Protocol (DHCP) is a protocol that provides dynamic configuration of Internet hosts. It allows for the dynamic assigning of IP-addresses to hosts in a network.

If a mobile device moves from one foreign domain to another, it will notify its home agent by sending a new *Binding Update* message. A mobile device can also send a *Binding Update* address to a home agent in the domain it has just left, which allows tunneling to the new COA of packets that have the old COA as a destination.

7.5 Micro-Mobility

Mobile wireless devices can connect with the global Internet through *access routers* (ARs). Using Mobile IP to support each change of AR would result in high overhead due to the frequent notifications to the home agent and high latency and disruption during handoff. Therefore, several mechanisms to support mobility in one administrative domain, like Cellular IP [4], HAWAII [14] and EMA [15], are proposed in which a visiting mobile device can use the same COA in the whole domain. All these mechanisms assume that a mobile device is connected directly with an AR. Of course, this assumption cannot hold if ad hoc networks are involved. A solution for ad hoc networks is presented in Section 7.6. In the following section, a brief description of Cellular IP is given to indicate how micro-mobility can be handled.

7.5.1 Cellular IP

As the name suggests, Cellular IP inherits cellular principles for mobility management such as passive connectivity, paging, and fast handoff control, but it implements them around the IP paradigm. A very important concept in the design of Cellular IP is its simplicity.

A Cellular IP network consists of ARs interconnected by wired links. Besides ARs, this wired network can contain nodes that have no radio interface but that merely serve as traffic concentrators or support mobility management functions (like node E in Figure 7.5). A Cellular IP network is connected to the Internet by a gateway router. All packets coming from mobile terminals will be routed from the AR they are connected to towards this gateway router, no matter the destination of the packet.

To make sure that all the nodes in the Cellular IP network have a correct routing entry towards the gateway router, this gateway router will periodically broadcast a beacon packet that is flooded through the Cellular IP network. The nodes in the network will record the neighbor they last received this beacon from and use it to route packets towards the gateway router.

FIGURE 7.5
A Cellular IP network. The entries in the routing caches are shown for each router in the access network.

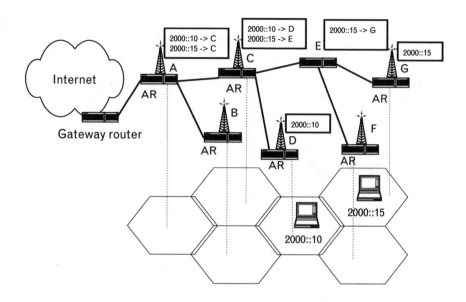

In order to be able to send packets towards a mobile device, the nodes in the network will use two different caches: a routing cache and a paging cache.

Routing caches are used to route packets towards a mobile terminal. Entries in the routing cache are timed out after a system-specific time value. When a data packet, sent by a mobile terminal, enters a node in the network, the node will put a mapping into its routing cache for the IP address of the mobile device and the neighbor node from which the packet entered. The data packet itself will be forwarded to the gateway router, using the routing entry towards the gateway router. In that way, every node on the path from the gateway router to the AR will have an entry in its routing cache for the specific mobile terminal. If a mobile terminal wants the routing caches to stay valid while it has no packets to send, it can periodically send special ICMP packets (route-update packets) to the gateway router.

Paging caches are very similar to routing caches and are used to find an idle mobile terminal if a packet has to be sent to it and no entry is available in the routing cache. The time-out interval of paging cache entries is larger than the routing cache time-out interval. To refresh paging entries, idle mobile terminals must send paging update packets at regular time intervals to the gateway router. If a node in the Cellular IP network has no entry in the routing cache for the mobile terminal it has a packet for and if it has no paging cache (it is not required that all nodes have a paging cache), then it will forward the packet towards all its downlink interfaces.

7.6   A Mechanism to Provide Global Connectivity for Ad Hoc Devices

In this section, a mechanism [16] is proposed that can support ubiquitous global mobility of ad hoc devices in an IPv6 network, using Mobile IP and one of the existing micro-mobility solutions, like Cellular IP. Hereby, the ad hoc routing protocol used will have to fulfill a few special conditions. In the following sections, all aspects of this mechanism are described.

7.6.1   Addressing

Although ad hoc routing is generally carried out at the IP layer, the ad hoc routing layer is viewed, in our mechanism, as a link-layer mechanism, making a complete separation between global Internet routing and ad hoc routing. Therefore, it is necessary to make a distinction between addressing at the ad hoc layer and addressing at the global IP-network layer. As with network interfaces in the global IP-network, each ad hoc interface of a mobile terminal (a terminal can have several interfaces: Bluetooth, IEEE 802.11, HiperLAN) should have a globally unique address at the ad hoc layer to uniquely distinguish the different interfaces in an ad hoc network. In order to obtain such unique identifiers, we suppose that the interface identifiers of ad hoc interfaces are unique so that the link-local addresses (see Section 7.1.2) that are formed from these interface identifiers are unique addresses as well. EUI-64-based interface identifiers, for instance, could be used for this purpose. We propose to use such unique link-local addresses as unique addresses for ad hoc interfaces so that no duplicate address detection must be performed at the ad hoc layer. Whenever ad hoc terminals come into reach of each other, the ad hoc routing protocol used can use the link-local addresses as unique identifiers for each ad hoc interface in the network. Note that the ad hoc interface (or interfaces) of an AR should have such a unique link-local address too and that it is an active part of the ad hoc network. In Figure 7.6, for instance, mobile terminal A forms the link-local address FE80::15 that can be used inside the ad hoc network.

Using link-local addresses at the ad hoc layer ensures that ad hoc terminals that are only intended to be used for ad hoc networking and not for global communication are not required to have a globally unicast home IP address. An ad hoc terminal that one also intends to use for global communication via the IPv6 network should have a globally unique IPv6 home address as well. Thus, these devices are addressable by two unique addresses.

FIGURE 7.6
Mobile terminal A
wants to connect with
a device B. IID: inter-
face identifier; LLA:
link-local address;
PR: prefix advertised
by AR; HA: home
address.

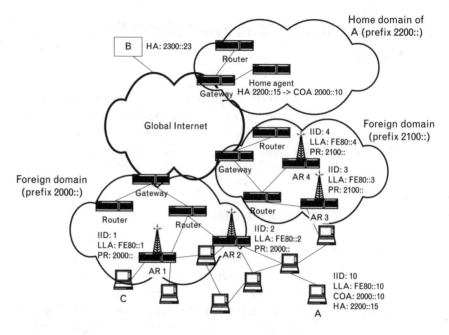

7.6.2   Finding an AR

In wireless access networks where mobile terminals connect directly to an
AR, mobile terminals can easily find out if an AR is in reach to connect to
by sensing a beacon sent out by the AR. If wireless mobile ad hoc access
networks are used, however, it is very possible that an ad hoc terminal cannot directly connect to an AR. Similar to this scheme, an AR should be able
to send a beacon towards the mobile ad hoc terminals to advertise its presence. In a multihop ad hoc network, flooding packets through the ad hoc
network can perform this. Because it is possible for ad hoc networks to be
spread out over large areas, such a beacon should just reach those ad hoc terminals that are within a certain hop limit from the AR. Because each additional hop in a path causes additional delay and higher transmission errors, it
is logical to limit the number of hops that an ad hoc terminal can be away
from an AR. Of course, we assume here that the ad hoc routing protocol
used supports packet flooding and that the hop limit field in the IP header
(similar to the TTL field in IPv4) is decremented every time a flooding
packet is broadcast by intermediate ad hoc terminals. When an ad hoc terminal receives beacon packets from different ARs, it should be able to
detect the closest AR. If the initial value of the hop limit is known in
advance, this should not be a problem. Otherwise, the initial hop limit
should be indicated in a special field in the beacon packet. Mobile ad hoc
terminal A in Figure 7.6, for instance, will choose to connect to AR 2 and
not to AR 3, because AR 2 is the closest AR. By determining the rate by
which these beacon packets are flooded into the network and the initial

value of the hop limit field, a trade-off should be made between perform-ance (flooding packets into an ad hoc network can be very bandwidth con-suming) and the rate by which ad hoc terminals can discover a better AR with which to connect.

7.6.3   Obtaining a COA

As mentioned in Section 7.1.2, IPv6 routers send router advertisements to the all-nodes multicast address on each of its interfaces it wants to be adver-tised as a router interface. The prefixes (site and global scope) that can be used by the terminals on the local link can be included as an option. These prefixes can be used by mobile terminals that want to connect with the global Internet to form an aggregatable global unicast COA. Because we assume that the link-local addresses are unique, constructing a unique COA is performed very easily by appending the interface identifier to one of the obtained prefixes. Note that these router advertisements should be flooded in the ad hoc network into the same region the beacon packets are flooded into the network. Therefore, the ARs will use the router advertisements as beacon packets. To improve the performance, the size of a router advertise-ment should be as small as possible. Therefore, each router advertisement should advertise exactly one address prefix, namely the prefix that can be used to form a unique global COA that is reachable from outside the visited domain. Mobile terminal A in Figure 7.6, for instance, forms the COA 2000::10 because AR 2 advertises the prefix 2000:: in its router advertisements.

To be able to use the same COA in a complete administrative domain, all the ARs of one domain should advertise the same prefix. As a conse-quence, this prefix can also be used by the ad hoc terminals to detect a domain change. When a domain change is detected, a new COA should be registered with their Mobile IP home agent.

Some remarks should be made:

- By using router advertisements as beacon packets, the rule that the hop limit field of a router advertisement should be 255, is violated. IPv6 prohibits the forwarding of packets sent to an address with link-local scope, but link-local scope is not well defined in ad hoc networks. As we consider ad hoc devices "on the same link," this rule is weakened here to allow an AR to flood router advertisements into an ad hoc network.

- Only router advertisements are used, no router solicitations. On a nor-mal IPv6 link, a new node on a link can send a router solicitation to obtain more quickly the necessary configurations. In our ad hoc layer, nodes depend on the rate by which routing advertisements are flooded into the network.

• Additional configuration data that is necessary for the specific micro-mobility solution used should be included in additional options in the router advertisements.

7.6.4   Communicating with an AR

Because global Internet routing and ad hoc routing are completely separated from each other, a mobile ad hoc device will have to use the link-local address of the AR to which it is connected in order to send packets to the Internet. This link-local address is obtained by the router advertisements, because the ad hoc interface of the AR must use its link-local address as source address to send packets into the ad hoc network. To send packets that are destined for the global Internet to its AR, an ad hoc terminal may use two different mechanisms:

1. Using a routing header destination option: This option is very similar to the IPv4 loose source routing and record route option. Via this option, a mobile device can specify the intermediate nodes that must be visited on the way to a packet's destination.

2. Encapsulating a global IP packet within an ad hoc routing IP packet, using the link-local address of the AR as destination.

Both approaches provide a way to tunnel global IP packets through the ad hoc network towards the AR. Note that an AR does not have to register a binding between the link-local addresses and the COA of the ad hoc devices that use the AR to connect to the Internet. The link-local address and the COA can be obtained from each other by switching between the domain prefix and the link-local address prefix (FE80::).

7.6.5   Switching Between Ad Hoc and Global Communication

The fact that ad hoc terminals can have two unique addresses can produce a situation where two ad hoc terminals that are in the same ad hoc network communicate via the global IP network and not via the local ad hoc network. By using route optimization (see Section 7.4), these terminals can find out each other's COA. From the COA, the link-local address can be easily obtained by replacing the subnet prefix with the link-local address prefix. A terminal that has this link-local address could be able, with the help of the ad hoc routing protocol, to find out whether the specific interface is within its reach directly through the ad hoc network. This depends, how-ever, on several parameters like the load of the ad hoc network and the number of hops that are between the two terminals and between the terminals and their ARs, respectively, whether communication should be performed better directly through the ad hoc network or via the global network. Because protocols like TCP cannot handle IP address changes (a communication is identified by the IP addresses and port numbers), mobile ad hoc terminals could anticipate a possible switch between link-local and global home addresses by using the global home addresses for a communication and use the tunneling mechanisms that are described in Section 7.6.4 if the communication is performed directly through an ad hoc network.

Having the same subnet prefix for their COA does not imply that two terminals are in the same ad hoc network. Likewise, having different subnet prefixes for their COA does not imply that two ad hoc terminals do not belong to the same ad hoc network. The boundaries of administrative domains and of ad hoc networks do not necessarily coincide.

7.6.6   Example

In this section, the described mechanism will be illustrated by an example (see Figure 7.6). In this example, a mobile terminal A (with interface identifier 10), which is not in its home domain, will try to make a connection with another device B (with home address 2300::23) somewhere in the global Internet. No specific micro-mobility solution or ad hoc routing protocol is assumed and the used addresses and prefixes serve only as an example.

Because AR 1 and AR 2 belong to the same administrative domain, they advertise the same prefix (2000::). AR 3 and AR 4 belong to another domain, thus they advertise another prefix (2100::). A will receive router advertisements from the ARs 1, 2, and 3 if we suppose that the hop limit of the beacon is at least 4 (the ARs are part of the ad hoc network). Because A will detect that AR 2 is the least hops away, it will decide to use AR 2 to connect to the Internet. By using the prefix that is advertised in the router advertisements sent by AR 2, A can form its COA to be used in this domain, namely 2000::10. A will also register the link-local address of AR 2 (FE80::2), which it can find in the source address field of the router advertisements sent by AR 2. Now A can send the necessary packets to AR 2 that are required by the used micro-mobility solution to connect with the domain. Note that a special field in the router advertisements could be used to indicate which micro-mobility solution should be used. When connected to AR 2, the micro-mobility solution used will make sure that within the domain, packets for the COA of A are routed correctly towards AR 2. In order to communicate globally, A will now register its COA to its home agent, so that the home agent can tunnel packets for the unique home address of A to its COA (using route optimization can avoid this tunneling). After these steps, A is ready to send packets to and receive packets from B.

If A has a packet to send to B, it will tunnel the packet through the ad hoc network to AR 2, using one of the mechanisms described in Section 5.4. To address AR 2 within the ad hoc layer, it uses the link-local address of AR 2 (FE80::2). When AR 2 receives a tunneled packet from A, it will route it to the correct destination. When AR 2 receives a packet that has the COA of A as its destination, it will tunnel it through the ad hoc network towards its destination. (The link-local address of A can be obtained from its COA by replacing the domain prefix 2000:: with the link-local prefix FE80::.)

While communicating with B, A can move and get closer to AR 1. If A detects, from the router advertisements it receives, that AR 1 has become the closest AR (in number of hops), it will start the necessary handover procedure to connect to AR 1, which is prescribed by the micro-mobility solution used. Because AR 1 is within the same domain as AR 2, the home agent of A must not be informed of the hand-over. Also, B will not notice the handover, because the COA of A does not change (if route optimization is used, B uses the COA of A).

When, while moving around, A detects that AR 3 is the closest AR, it will connect to AR 3. Depending on the micro-mobility solution that is used, A will detect the domain change by the advertised prefix in the router advertisements of AR 3. Because of the domain change, A will have to change its COA into 2100::10. Then, it will have to connect to AR 3, so that the new domain is able to route packets towards A. Finally, A will also have to register its new COA to its home agent and to B, such that it will use the new COA to send packets to A. It is possible that, during the AR change, A receives packets through the ad hoc network that have the old COA of A as destination.

7.6.7 Alternative Global Connectivity Proposal

In [17] a similar proposal has been formulated in which the focus is mainly on the discovery of Internet-gateways (which we called access routers) in an ad hoc network while using a reactive ad hoc routing protocol. Two methods to accomplish this are proposed, namely the extension of the route discovery messaging of a reactive routing protocol and the extension of router solicitations and advertisements (see Section 7.1.2). As in our mechanism described in Sections 7.6.1 through 7.6.6, the use of IPv6 and the fact that devices must have a globally unique address is assumed. A great difference with our mechanism is the fact that devices must search an Internet-gateway actively because an Internet-gateway will not announce its presence on its own initiative. In the following paragraphs, the two different methods are briefly described.

7.6.7.1 Changing Route Request of a Reactive Routing Protocol

To support this method of global connectivity, route request and reply messages of a reactive routing protocol should be modified to carry global prefix information and the gateway's IPv6 address. When a device wants to find an Internet-gateway, it must broadcast (flood) a route request to the INTERNET_GATEWAYS global multicast address or a global address of an Internet node. Once an Internet-gateway receives this request, it must reply with both the global prefix and its Internet-gateway address. As a requirement, the ad hoc routing protocol used must support some scheme for route reply to detect whether the route reply carries the Internet-gateway information and address. This can be done, for instance, via a flag that indicates whether the reply message includes Internet-gateway information.

7.6.7.2 Adjusted Router Solicitations and Advertisements

As was mentioned in Section 7.6.2, router advertisements and solicitations use link-local scope, which is not very well defined in ad hoc networks. Therefore, router advertisements and solicitations should be adjusted so that they can be used in ad hoc networks. In contrast to our mechanism, router solicitations are used here because an Internet-gateway will only send router advertisements in response to router solicitations that it receives. Thus, an Internet-gateway will not send periodically unsolicited router advertisements.

In our mechanism we have chosen to use unsolicited router advertisements in analogy with the beacons that are broadcasted by access points in single-hop centralized wireless technologies. In these technologies it is also the base station that notifies its existence by sending out a beacon, allowing each mobile device to determine the best access point to connect to in function of the signal quality of the beacons it received. Because we assume that the topology of an ad hoc network is under continuous change, we think that it is better to give the access router the responsibility of announcing its existence (and using the hop-count, announcing its distance to each node) to the mobile ad hoc devices in its neighborhood, than to allow each node separately to search periodically for the best access router by flooding router solicitations in the network.

Like in our mechanism, router solicitations and advertisements are forwarded by intermediate devices in the ad hoc network, violating the prohibition in IPv6 that they must not be forwarded. Of course, an Internet-gateway will not forward these router solicitations into the fixed network.

After one of the above operations, an ad hoc device knows a global prefix and an Internet-gateway address in the ad hoc network. This knowledge allows the device to form a global IPv6 COA using its 64-bit interface identifier, as is done in our mechanism. To send a packet to the global Internet, a device has to decide between two methods:

1. *With IPv6 routing header option:* the sender uses the Internet-gateway address in the destination address of the IPv6 header and the real destination address in the routing header.

2. *Without IPv6 routing header option:* the sender sends the packet to the global IPv6 address and relies upon next hop routing in the other nodes.

Using the first method allows a device to choose between gateways if more than one is present in the ad hoc network. With the second method, a device does not need to take care of the explicit route to the Internet-gateway, allowing each intermediate device to independently decide on how to best route the packet to the global Internet. In our proposal the sending device always decides which Internet-gateway is used (using IPv6 routing header or encapsulation; see Section 7.6.4) and likewise the first method (using IPv6 routing header) is recommended by this alternative proposal.

In conclusion, although the emphasis is on reactive routing protocols in the latter approach, both proposals are quite similar. The main difference is the fact that, here, an ad hoc device must actively search for an Internet-gateway if it needs one, which is a natural consequence of focusing on reactive routing strategy.

7.7   Summary

In this chapter, we presented mechanisms that can be used to provide global communication with ad hoc terminals that are connected, through an ad hoc network, to an access router that is connected to a global IPv6 communication network. In the proposed approach, existing solutions for handling macro- and micro-mobility are used, and a general ad hoc routing protocol that allows the flooding of packets in a limited region of the network is sufficient. The addressing and routing problem that occurs when an ad hoc network is attached to two or more gateways is managed by making a distinction between the ad hoc network layer and the global IP layer. By tunneling through the ad hoc layer, mobile terminals can get a virtual direct link to an access router, which is mostly required by existing micro-mobility solutions. While communication with ad hoc terminals is possible, the actual topology of the ad hoc networks is completely invisible for the global

IP layer and has no effect on addressing and routing within this global IP layer, allowing free mobility of ad hoc devices and ad hoc networks.

REFERENCES

[1]   Perkins, C. E., *Ad Hoc Networking*, Reading, MA: Addison-Wesley, 2000.

[2]   Deering, S., and R. Hinden, "Internet Protocol, Version 6 (IPv6) Specification," RFC 2460, December 1998.

[3]   Johnson, D. B., and C. Perkins, "Mobility Support in IPv6," Internet Draft, http://www.ietf.org, July 2001.

[4]   Campbell, A. T., et al., "Design, Implementation, and Evaluation of Cellular IP," *IEEE Personal Communications*, August 2000, pp. 42–49.

[5]   Lee, S. B., G. S. Ahn, and A. T. Campbell, "Improving UDP and TCP Performance in Mobile Ad Hoc Networks with INSIGNIA," *IEEE Communication Magazine*, June 2001.

[6]   IETF MANet Working Group, http://www.ietf.org/html.charters/manet-charter.html.

[7]   Perkins, C., and P. Bhagwat, "Highly Dynamic Destination-Sequenced Distance-Vector Routing (DSDV) for Mobile Computers," *ACM SIGCOMM '94*, 1994.

[8]   Murthy, S., and J. Garcia-Luna-Aceves, "A Routing Protocol for Packet Radio Networks," *MOBICOM '95*, 1995.

[9]   Aggelou, G., and R. Tafazolli, "RDMAR: A Bandwidth-Efficient Routing Protocol for Mobile Ad Hoc Networks," *Proceedings of the Second ACM International Workshop on Wireless Mobile Multimedia*, 1999.

[10] Sivakumar, R., P. Sinha, and V. Bharghavan, "CEDAR: A Core-Extraction Distributed Ad Hoc Routing Algorithm," *IEEE Journal on Selected Areas in Communications*, Vol. 17, No. 8, August 1999, pp. 1454–1465.

[11] Pearlman, M., and Z. Haas, "Determining the Optimal Configuration for the Zone Routing Protocol," *IEEE Journal on Selected Areas in Communications*, Vol. 17, No. 8, August 1999, pp. 1395–1414.

[12] Ko, Y., and N. Vaidy, "Location Aided Routing (LAR) in Mobile Ad Hoc Networks," *MOBICOM '98*, 1998.

[13] Joa-Ng, M., and I. Lu, "A Peer-to-Peer Zone-Base Two-Level Link State Routing for Mobile Ad Hoc Networks," *IEEE Journal on Selected Areas in Communications*, Vol. 17, No. 8, August 1999, pp. 1415–1425.

[14] Ramjee, R., et al., "IP Micro-Mobility Support Using HAWAII," Internet Draft, http://www.ietf.org, July 2000.

[15] O'Neill, A., G. Tsirtsis, and S. Corson, "Edge Mobility Architecture," Internet Draft, http://www.ietf.org, July 2000.

[16] Cooreman, K., B. Dhoedt, and P. Demeester, "Ad Hoc Networks: An IP Viewpoint," *Proceedings of the Fourth International Symposium on Wireless Personal Multimedia Communications (WPMC'01)*, Aalborg, Denmark, September 2001, pp. 1475–1480.

[17] Wakikawa, R., et al., "Global Connectivity for IPv6 Mobile Ad Hoc Networks," Internet Draft, http://www.ietf.org, November 2001.

Part II:
QoS and Resource Management

CHAPTER 8

Differentiated and Integrated Services for IP Applications over UMTS

Gabor Fodor, Birgitta Olin, Fredrik Persson, Christiaan Roobol, and Brian Wiliams

8.1 Introduction

As the standards for 3G mobile radio systems (within 3GPP/3GPP2) and for IP quality of service provisioning (within IETF) mature, there is a growing interest in the introduction of end-to-end QoS in the all-IP architecture. Indeed, in contrast to the 3GPP standard Release 99, the evolving Release R00/R01 (also referred to as Release 4/5, R4/R5) architecture includes the concept of the end-to-end QoS covering the full path between mobile Internet hosts and/or servers (Figure 8.1). QoS guarantees are generally considered to be one of the key components (together with wide area coverage and global roaming) in providing true multimedia mobile services to users [1]. (See Chapter 4 for a survey and summary of 3G standards and Chapter 5 for an overview of the evolution towards the all-IP architecture.)

We notice in Figure 8.1 that the so-called seamless all-IP architecture has three distinct, equally important facets:

1. New mobile Internet services, including the support for real-time, quality-assured conversational and streaming services that are made available through open interfaces to the globally roaming user.

2. The separation of the access part of the network from the core network via the open Iu interface allows independent evolution of the various access technologies from the development of the core net-

FIGURE 8.1
*The end-to-end all-IP
scenario.*

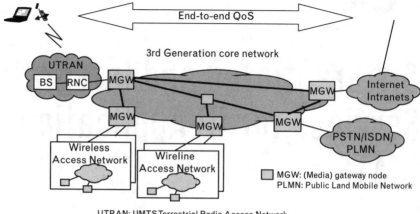

UTRAN: UMTS Terrestrial Radio Access Network
BS: Base station
RNC: Radio network controller

work. Generally, the architecture employs a number of open interfaces along the end-to-end path, including the Uu, Iu, Gn, and Gi
interfaces.

3. Increasing support for employing the IP technology at the transport
layer (L2 transport) of various parts of the network, including the *radio access network* (RAN) and the CN ensuring smooth interconnection with IP-based networks such as the Internet or different
intranets. [For instance, the use of IP as a transport technology for
signaling traffic has led to the development of a new IP transport protocol called *Stream Control Transmission Protocol* (SCTP). A short summary on the application of SCTP in the 3G architecture is provided
in Chapter 5.]

Providing the mobile Internet user with quality guarantees is an especially challenging task, because the requirements on end-to-end IP transparency and spectrum efficiency over the wireless link would often suggest
contradictory solutions.

For instance, IP transparency requires the transmission of IP packets
(including the packet header) all the way to the mobile station over the air
interface; while spectrum efficiency minimizes the number of transmitted
redundant bits and header information. Also, true IP transparency and access
technology independence imply that applications requiring a specific level
of service quality do not need to apply a wireless specific signaling and
resource reservation mechanism.

On the other hand, spectrum-efficient resource management of the
wireless link needs to be supported by fine granularity traffic and QoS
parameters that are specified by the application.

This trade-off is depicted in Figure 8.2, where the IP-over-radio trade-offs are visualized as a cube in a three-dimensional space that is spanned by the provided quality, the achieved spectrum efficiency, and the IP flexibility (i.e., how seamless the IP service to the end user actually is).

We also note that recent advances within the IETF *Robust Header Compression* (ROHC) and *Audio Video Transport* (AVT) working groups are paving the way to successfully address these trade-offs.

From an end-to-end perspective, it is clear that resource management in the different parts [wireless link, RAN, CN, and external (service) network] of the end-to-end path needs to be coordinated. The end-to-end reservation mechanisms need to allocate radio resources in the access part of the network and IP resources in the external network (core network) providing IP services. The appropriate resource management technique in the different parts of the network takes into account the following aspects:

- The *scarcity* of the resource to be managed. For instance, radio spectrum is generally considered to be a scarce resource requiring a high target utilization. In contrast, IP transmission resources in a backbone IP network are often regarded as a less scarce resource, allowing for some over-dimensioning.

- At what time scale the demand for the resources vary. Clearly, this aspect has an impact on whether per-call resource allocation signaling is beneficial or a quasi-static allocation of resources to semi-permanent connections is acceptable.

- Cost-complexity trade-off of managing the resources. This aspect determines what kind of scheduling, buffer management, and flow classification mechanisms are the most appropriate.

FIGURE 8.2
Trade-offs between IP flexibility, spectrum efficiency, and QoS.

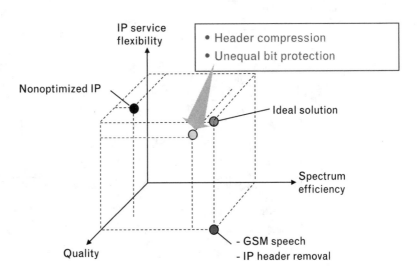

- Scalability and degree of traffic aggregation. These aspects influence the choice of the granularity of service differentiation (e.g., per-flow, per-code-point, aggregate, and so forth) and state maintenance in, for example, IP routers.

- Possibly other aspects, such as policy decisions.

To this end, the standards effort has proposed a layered service architecture (Figure 8.3), describing the following key elements [1]:

- Mapping of end-to-end service to services provided by the UE, UTRAN, core network, and external IP networks;

- Traffic classes and associated QoS parameters;

- Location of QoS functions;

- QoS negotiation;

- Multiplexing of flows onto network resources;

- An end-to-end data delivery model.

Thus, the end-to-end resource management for QoS provisioning uses different means in the different segments:

FIGURE 8.3
Mapping of end-to-end QoS requests to bearer services.

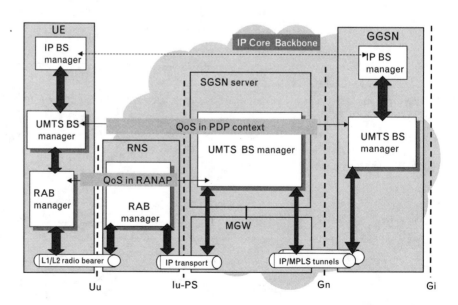

1. Within the UTRAN and CN, UMTS QoS control mechanisms support per-flow resource allocation and control employing the UMTS PDP context concept.

2. Beyond the UMTS gateway node in the external IP network, it is expected that the evolving IETF standards provide sufficient QoS control and traffic engineering for various services; for example, differentiated services, integrated services, MPLS, and other evolving mechanisms including *service level agreements* (SLAs) and SLS.

3. In the remote access network, access-specific optimized resource control ensures the required level of quality.

It is important to emphasize that in an all-IP scenario, the user-perceived end-to-end QoS must necessarily be defined at the IP layer, since it is this layer that is the common denominator, which provides end-to-end connectivity to user applications.

In this chapter we will consider some of the key challenges that need to be solved before the wireless/mobile Internet and the all-IP architecture can happen in an economically viable manner. It is important to realize that because of the expected rich variety of 3G terminals and QoS mechanisms, it is anticipated that there will be different end-to-end scenarios and associated QoS management mechanisms. The scenarios may differ in terms of the capabilities of the wireless terminal, the applied QoS signaling at the IP level and how efficiently resources are managed over the wireless access network. For instance, a 3G terminal that allows applications to specify their wireless QoS parameters may allocate wireless resources in a more spectrum-efficient way than a personal computer equipped with a wireless PCMCIA card providing full IP transparency to applications. We argue that in some cases further work within 3GPP and IETF is needed to address these issues, and we will specifically point out one such issue regarding the use of integrated services in the 3G environment.

We will start off by conducting a short a survey of end-to-end wireless IP scenarios that appear reasonable regarding the current IP and wireless standardization trends. We address some of the control and user plane issues that need to be understood in the context of QoS provisioning in each of these scenarios. Next, we take a closer look at providing the IETF standardized integrated services and discuss major problems and some ideas to provide solutions to these problems. We conclude that the convergence of the IP and wireless QoS mechanisms is a key enabler of the future mobile Internet. In particular, additional QoS parameters are necessary at the IP layer (and in particular in the IETF IntServ model) in order to operate efficiently and to provide QoS over various wireless accesses.

8.2   All-IP End-to-End Scenarios

8.2.1   Architectural Aspects

As shown in Figure 8.1 and discussed above, one of the key aspects of the 3GPP all-IP architecture is the separation of the radio access network from the core network containing a *circuit switched* (CS) and a *packet switched* (PS) domain. This separation and the standardization of the Iu interface supports the evolution of different access technologies (GPRS/EDGE, GERAN, UTRAN, and possibly others including WLAN and BRAN) that can use the same connectivity network to PSTNs and to the Internet.

Apart from the clear separation of the access part and the core network, there is a clear trend in separating (1) the content and user application layer, (2) the communication control layer, and (3) the connectivity layer. The content and application layer includes applications for information-centric services for which users pay. The communication control layer incorporates all the functionality needed to provide seamless and high-quality services across different public and private networks. The connectivity layer is primarily a transport mechanism that is capable of transporting voice, data, and multimedia information. This horizontally layered architecture that separates applications and services from the access and core networks facilitates the convergence of the mobile Internet and other communications services at the application level.

From the user's perspective, the horizontally layered architecture and the separation of the access network form the core network provides a personalized service environment, which is independent of the access type. Decoupling the applications from the underlying infrastructure is also expected to accelerate the further development of open standards at various interfaces including the Uu, Iu, and Gn interfaces, and not the least the application programming interfaces. To this end, it is anticipated that a subset of applications (typically those that are installed on a personal computer or on a laptop) running over IP will use the IETF standardized integrated services to specify their quality and resource requirements. In such a scenario, the necessary UMTS resource parameters must be derived from the traffic and QoS parameters specified at the IP layer. In what follows, we describe such end-to-end scenarios in details.

8.2.2   All-IP Scenarios

According to [2], the end-to-end QoS functional architecture for R5 consists of different domains, which need to coordinate in order to meet the end-to-end QoS requirements. An IP *bearer service manager* (BSM) is used to control the IP bearer service that is provided to the end user. It is important to point out that this IP BSM is unaware of the underlying transport technologies (IP transport in Figure 8.3) in the RAN and in the CN. It controls the Layer 3 (Figure 8.3) IP service only irrespective of whether the transport mechanism in the RAN and in the CN is ATM or IP. The IP BSM uses standard IP mechanisms (such as differentiated or integrated services) and thus facilitates an IP-based interface for the application programs (possibly via an operating system) for QoS control (Figure 8.4). The IP BSM exists both in the UE and in the GGSN. A mapping function provides the interworking between the mechanisms and parameters used within the UMTS and the external IP bearer service and interact with the IP BSM. The IP BSM in the UE and in the GGSN provide a set of capabilities for the IP bearer, as shown in Table 8.1.

3GPP analyzed six scenarios, depending on:

- Capability (RSVP/IntServ, DiffServ, PDP Context usage) in each device (UE, GGSN, external IP network);

- Location of the IP bearer service manager (UE, GGSN).

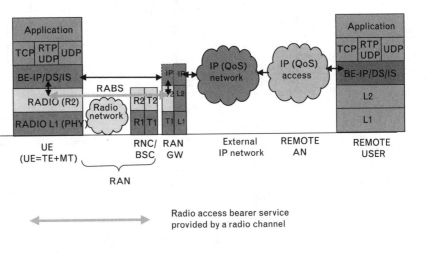

FIGURE 8.4
End-to-end protocol stack.

Radio access bearer service provided by a radio channel

TABLE 8.1   IP BEARER SERVICE MANAGER CAPABILITY IN THE UE AND THE GGSN

Capability	UE	GGSN
DiffServ edge function	Optional	Required
RSVP/IntServ	Optional	Optional
IP policy enforcement point	Optional	Required

The summary of these scenarios is provided in Figure 8.5.

8.2.3 All-IP Scenario Using Differentiated Services

The UE performs an IP *base station* (BS) function, which enables end-to-end QoS without IP layer signaling towards the IP BS function in the GGSN, or the remote terminal. The scenario assumes that the UE and GGSN support DiffServ edge functions, and that the backbone IP network is DiffServ enabled (Figure 8.6).

FIGURE 8.5
Summary of the 3GPP end-to-end all-IP conceptual models (scenarios).

Scenario 1	UE	Gateway (GGSN)	Extern. IP Net
PDP Ctxt.	X	X	
IP BS manager		X	
RSVP (Intserv)			
Diffserv		X	X

Scenario 2	UE	Gateway (GGSN)	Extern. IP Net
PDP Ctxt.	X	X	
IP BS Manager	X	X	
RSVP (Intserv)			
Diffserv	X	X	X

Scenario 3	UE	Gateway (GGSN)	Extern. IP Net
PDP Ctxt.	X	X	
IP BS Manager	X	X	
RSVP (Intserv)	X		
Diffserv	X	X	X

Scenario 4	UE	Gateway (GGSN)	Extern. IP Net
PDP Ctxt.	X	X	
IP BS Manager	X	X	
RSVP (Intserv)	X	X	
Diffserv		X	X

Scenario 5	UE	Gateway (GGSN)	Extern. IP Net
PDP Ctxt.	X	X	
IP BS Manager		X	
RSVP (Intserv)			
Diffserv		X	X

Scenario 6	UE	Gateway (GGSN)	Extern. IP Net
PDP Ctxt.	X	X	
IP BS Manager		X	
RSVP (Intserv)		X	
Diffserv		X	X

FIGURE 8.6
The end-to-end all-IP scenario using Diff-Serv in the UE.

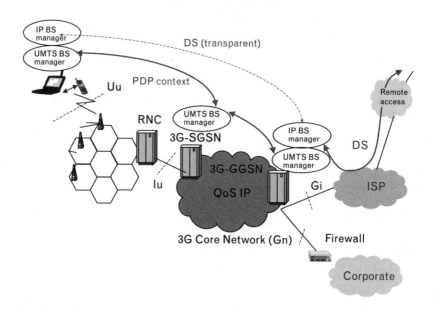

The application layer (e.g., SIP/SDP) between the end hosts identifies the QoS needs. The QoS requirements from the application layer are mapped down to the IP layer. In the downlink direction, the IP layer service requirements are further mapped down to the PDP Context parameters in the UE. In this scenario, the control of the QoS over the UMTS access network (from the UE to the GGSN) may be performed from the terminal using the PDP Context signaling. Alternatively, subscription data accessed by the SGSN may override the QoS requested via signaling from the UE. In this scenario, the terminal supports DiffServ to control the IP QoS through the backbone IP network. The IP QoS for the downlink direction is controlled by the remote terminal up to the GGSN. The PDP Context controls the QoS between the GGSN and the UE. The UE may apply DiffServ edge functions to provide the DiffServ receiver control. Otherwise, the DiffServ marking from the GGSN will determine the IP QoS applicable at the UE.

The end-to-end QoS is provided by a local mechanism in the UE, the PDP Context over the UMTS access network, DiffServ through the backbone IP network, and DiffServ in the remote access network in the scenario shown in Figure 8.7. The UE provides control of the DiffServ, and therefore determines the appropriate interworking between the PDP Context and DiffServ. The GGSN DiffServ edge function may overwrite the DSCP received from the UE, possibly using information regarding the PDP Context that is signaled between the UMTS BS managers and provided through the translation/mapping function to the IP BS manager. Note that DiffServ control at the remote host is shown in this example. Other mechanisms, however, may be used at the remote end, as demonstrated in the other 3GPP scenarios [2]. Figure 8.7 provides a summary of the simplified call flows in the DiffServ scenario. An application of the DiffServ scenario using selective packet prioritization for voice applications is presented in Chapter 10.

8.2.4 All-IP Scenario Using Integrated Services

Next, we consider the case where the IP BSM in the UE supports IP level signaling using RSVP and DS marking. (This is the case in Scenario 3 in Figure 8.5.) As indicated in Figure 8.8, the UE performs an IP BS management function, which enables end-to-end QoS using IP layer signaling towards the remote host. Observe however, that there is no IP layer signaling between the IP BSM in the UE and the GGSN: RSVP is transparent to the GGSN. The GGSN, however, may make use of information regarding the PDP Context, which is signaled between the UMTS BS managers and provided through the translation/mapping function.

This scenario assumes that the UE and GGSN also support DiffServ edge functions and that the backbone IP network is DiffServ enabled. In addition, the UE supports RSVP signaling that interworks within the UE to

FIGURE 8.7
Simplified call flows for the DiffServ scenario.

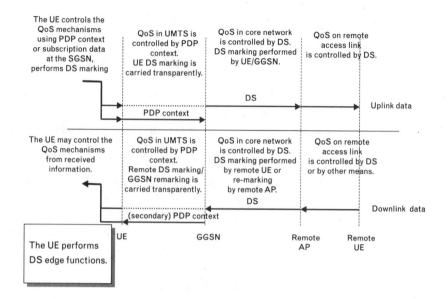

FIGURE 8.8
The end-to-end all-IP scenario using RSVP/IntServ in the UE.

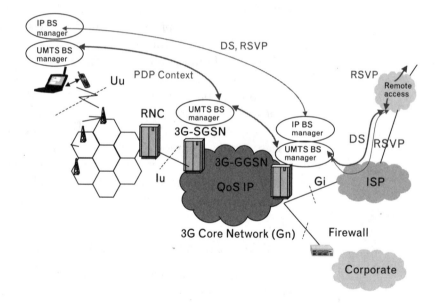

control the DiffServ. As an addition, the applications may use a session control protocol (e.g., SIP/SDP) between the end hosts to identify the QoS requirements. The QoS requirements from the user are subsequently mapped down to create an RSVP session. The UE establishes the PDP context suitable for support of the RSVP session.

In this scenario, the control of the QoS over the UMTS (from the UE to the GGSN) may be performed from the terminal using the PDP Context signaling. Alternatively, subscription data accessed by the SGSN may override the QoS requested via signaling from the UE.

In this scenario, the terminal supports signaling via the RSVP protocol to control the QoS at the local and remote accesses, and DiffServ to control the IP QoS through the backbone IP network. The RSVP signaling protocol may be used for different services. It is expected that only RSVP using the IntServ semantics would be supported, although in the future, new service definitions and semantics may be introduced. The entities that are supporting the RSVP signaling should act according to the IETF specifications for IntServ and IntServ/DiffServ interwork. The QoS for the wireless access is provided by the PDP Context. The UE may control the wireless QoS through signaling for the PDP context. The characteristics for the PDP context may be derived from the RSVP signaling information, or may use other information.

QoS for the IP layer is performed at two levels. The end-to-end QoS is controlled by the RSVP signaling. Although RSVP signaling can be used end-to-end in the QoS model, it is not necessarily supported by all intermediate nodes. Instead, DiffServ is used to provide the QoS throughout the backbone IP network.

At the UE, the data is also classified for DiffServ. Intermediate QoS domains may apply QoS mechanisms according to either the RSVP signaling information or DiffServ mechanisms. In this scenario, the UE provides interworking between the RSVP and DiffServ domains. The GGSN may override the DiffServ setting from the UE. This GGSN may use information regarding the PDP Context in order to select the appropriate DiffServ setting to apply, as shown in Figures 8.8 and 8.9.

The end-to-end QoS is provided by a local mechanism in the UE, the PDP Context over the UMTS access network, DiffServ through the backbone IP network, and DiffServ (or RSVP) in the remote access network in the scenario shown in Figure 8.8. The RSVP signaling may control the QoS at both the local and remote accesses. This function may be used to determine the characteristics for the PDP Context, so the UE may perform the interwork between the RSVP signaling and PDP Context. The UE provides control of the DiffServ (although this may be overwritten by the GGSN) and, in effect, determines the appropriate interworking between the PDP context and DiffServ. Figure 8.9 provides a summary of the simplified call flow in the IntServ scenario.

8.2.5 Using RSVP to Control the PDP Context

An alternative use case of using RSVP and IntServ to control the PDP Context is when the MT acts as an RSVP proxy that terminates the IP level signaling. In this scenario, the IntServ parameters provide input to the translation function in the UE, but RSVP is not used outside of the UE.

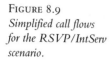

FIGURE 8.9
Simplified call flows
for the RSVP/IntServ
scenario.

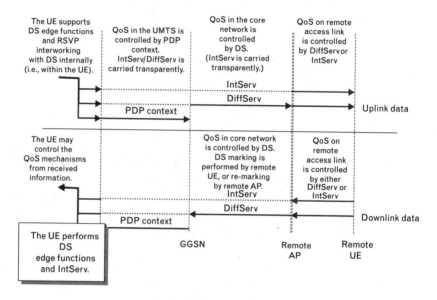

8.3   UMTS Service Classes and Parameters

8.3.1   Mapping the End-to-End Service to Local Bearer Services

Given the heterogeneity of access networks, and that the end-to-end connection typically traverses multiple access networks, it is clear that the end-to-end service typically maps to a concatenation of bearer services along the end-to-end path. A bearer service describes how a given network provides QoS, and is defined by the actual signaling protocol, user plane transport, and QoS management functions [1].

The current 3GPP architecture decomposes the end-to-end service into three main parts: the bearer service within the UE (TE-MT), the UMTS bearer service, and the external local bearer service. The TE-MT bearer service enables communication between the different components of the user equipment, namely the TE (which may be, for instance, a laptop computer), and the MT, which is responsible for the connection to the UTRAN and which could be, for instance, a PCMCIA card. The UMTS bearer service provides connectivity for the UE to the gateway (GGSN) node. The external local bearer service connects the UMTS core network and the destination (IP) node located in an external network, such as the Internet or an intranet.

To support different service qualities, four different UMTS bearer services have been defined. These bearer services are the conversational, streaming, interactive, and background classes, each one having a set of traffic and

QoS attributes that facilitate the mapping of the end-to-end service to the appropriate UMTS bearer service.

8.3.2 UMTS Traffic Descriptors

So far, the 3GPP defined traffic descriptors areas follows:

- The traffic class roughly defines the type of application that the radio bearer is optimized for. It also defines the set of attributes that are defined and applicable for that specific traffic class. This class parameter can, for example, be used for admission control (e.g., real-time traffic versus best-effort traffic). By including the traffic class itself as an attribute, UMTS can make assumptions about the traffic source and optimize the transport for that traffic type.

- Maximum (peak) bit rate is defined as the maximum number of bits delivered by UMTS between its endpoints (i.e., the RAN GW or the mobile terminal) within a period of time, divided by the duration of the period. Its purposes are (1) to limit the offered bit rate from applications or external networks and (2) to allow the maximum desired user bit rate to be explicitly defined.

- Guaranteed (mean) bit rate is defined as the guaranteed number of bits delivered by UMTS at an end point of the network (i.e., the RAN GW or the mobile terminal) within a period of time (provided that there is data to deliver), divided by the duration of the period. Guaranteed bit rate may be used to facilitate resource-based admission control within UMTS. It defines a "level of commitment" for the operator in the sense that the QoS guarantees are provided for this average user data rate.

- Maximum SDU size gives the maximum radio service data unit size. The maximum SDU size can be used for resource allocation and policing.

- SDU format information gives a list of the possible exact sizes of SDUs. In some cases, including the case when retransmission is not used, spectral efficiency and delay can be optimized, if the exact sizes of the radio SDUs are known. Also, mechanisms like unequal error protection/unequal error detection require that the internal payload format be known. The bearer can thus be less resource consuming, if the application can specify the payload formats and packet sizes.

- The source statistic descriptor can, by specifying the characteristics of the source of submitted radio SDUs (e.g., if the application provides speech traffic or not), improve the efficiency of the admission control algorithms.

8.3.3   UMTS QoS Attributes

There are also six attributes that describe the expected UMTS QoS:

1. The transfer delay indicates the ninety-fifth percentile of the distribution of the delay for all delivered radio SDUs during the lifetime of a radio bearer. The delay for a radio SDU is defined as the time from a request to transfer of a radio SDU at one endpoint to its delivery at the other end, including retransmission delay(s). It is used to specify the delay tolerated by the application, which allows UTRAN to set so-called transport formats and retransmission parameters.

2. Delivery order indicates whether the UMTS bearer shall provide in-sequence radio SDU delivery or not. Whether out-of-sequence radio SDUs are dropped or reordered depends on, for example, the specified SDU error ratio and residual bit error ratio (see below). By not having to provide in-sequence delivery, the required buffer sizes can be minimized.

3. Delivery of erroneous SDUs is used to determine whether error detection is needed or not, and it indicates whether radio SDUs detected as erroneous shall be delivered or discarded.

4. SDU error ratio indicates the fraction of radio SDUs lost or detected as erroneous. SDU error ratio is defined only for conformable traffic (i.e., traffic that keeps the agreed bit rate and maximum SDU size). It is only specified if error detection is used (see above).

5. Residual bit error ratio indicates the undetected bit error ratio in the delivered radio SDUs. If no error detection is requested, residual bit error ratio indicates the total bit error ratio in the delivered radio SDUs.

6. Traffic handling priority gives an internal priority handling for the interactive class. It specifies the relative importance for handling of all the radio SDUs belonging to one specific interactive bearer compared to the radio SDUs of other interactive bearers.

The attributes per traffic class are summarized in Table 8.2.

8.3.4   Conclusion

Since the UMTS resource allocation algorithms need information about both the offered traffic and the required QoS, a number of traffic and QoS parameters are associated with the UMTS bearer services. It is anticipated that many of these parameters may be useful for other wireless technologies that provide quality guarantees for various applications. Indeed, both the maximum and the mean average of the offered user traffic along with the

TABLE 8.2   UMTS TRAFFIC CLASSES AND THEIR ATTRIBUTES

TRAFFIC CLASS	CONVERSATIONAL	STREAMING	INTERACTIVE	BACKGROUND
Fundamental characteristics	Preserve time relation between information entities of the stream	Preserve time relation between information entities of the stream	Request–response pattern; preserve payload content	Destination does not expect the data within a certain time
Example of the application	*Voice*	*Streaming video*	*Web browsing*	*Background download of e-mail*
Maximum bit rate	X	X	X	X
Guaranteed bit rate	X	X		
Maximum SDU size	X	X	X	X
SDU format information	X	X		
Source statistics descriptor	X	X		
Transfer delay	X	X		
Delivery order	X	X	X	X
Delivery of erroneous SDUs	X	X	X	X
SDU error ratio	X	X	X	X
Residual bit error ratio	X	X	X	X
Traffic handling priority			X	

SDU format information and the description of the required quality (in terms of transfer delay, SDU error ratio, and residual bit error ratio) appear to be quite general parameters from the IP layer's point of view. We will refer to this observation in Section 8.5 where we discuss the proposed new parameters for the controlled load integrated service.

8.4   Suitability of Existing Integrated Services over RANs

Integrated services enhance the single best-effort service class model by introducing multiple service classes, including QoS classes [3]. IntServ allows the network nodes to perform explicit resource management at the IP level by allowing applications to characterize their resource requirements. IntServ is typically used together with RSVP signaling to allow the network to exercise admission control and traffic control and thereby ensure end-to-end QoS provisioning and efficient resource utilization. In the scenario of Figure 8.4, RSVP and IntServ are also used to establish and to characterize a radio bearer service. In this section we investigate the suitability of the IntServ QoS classes when operating over a (L2) RAN.

8.4.1 GQoS

The GQoS integrated service is defined to provide a service of guaranteed bandwidth with a bounded queuing delay [4]. The end-to-end behavior provided by a series of network elements provides an assured level of bandwidth that, when used by a policed flow, produces a delay-bounded service with no queuing loss for all conforming packets (assuming no failure of network components or changes in routing during the life of the flow). In wireless networks, the requirement to provision a fixed bandwidth is unrealistic. The performance of the radio channel may be influenced by many different factors including path loss, fading, and interference. These are most likely to change when the user is moving, but possibly also when the user is stationary. Any of these effects may reduce the capacity of the wireless network available to the user. Under these conditions, it is not possible to guarantee that a specific constant bandwidth can be maintained. GQoS permits the user to increase the bandwidth of the service to control the queuing delay, which is normally the only means to affect the end-to-end delay. For public wireless networks, it is important to be able to provide service to as many customers as possible. Therefore, operators will not allocate more resources than required for the traffic flow. Furthermore, a wireless network is one that is in practice too complex for attempting to generate the relevant characteristics. The minimum delay through the network varies depending on the exact parameters used. The queuing delay must then cover for the greatest delay that may be experienced above the minimum delay. (This is also dependent on the exact parameters of the radio bearer service.) In order to meet the service requirements, worst-case approximations would need to be used. For wireless access networks, the values that result would severely limit the usefulness of the service. Thus, the end-to-end service characteristics are heavily dependent on the exact radio parameters that are used. However, there is insufficient information in the GQoS service to determine appropriate settings for the radio parameters. Even if an indication that delay is important to the application is given, wireless networks require a refined definition of delay.

8.4.2 Controlled Load

The *controlled load* (CL) integrated service is defined to provide a service approximating that of a lightly loaded network [5]. The service should provide that a very high percentage of transmitted packets will be successfully delivered by the network, and the transit delay of most packets will not greatly exceed the minimum transmit delay of the routing vector. However, the concept of lightly loaded does not fit into the concepts of the radio network, and does not contain enough information to determine the performance. In the wireless environment, there are different characteristics that can be traded off, such as delay, bit error rate, and service cost. Furthermore, how these characteristics are varied depends both on the requested QoS parameters and the resource management in the RAN. Hence, although it is not feasible to appear lightly loaded as in CL, the wireless network can provide the means to control which characteristics of a wireless network are most important, and how close this can be approximated. CL, however, as it is defined today, does not provide the necessary means to manage the trade-off between these different characteristics.

8.4.3 Null Service

The *null service* (NS) allows applications to identify themselves to network QoS policy agents rather than requiring them to specify resource requirements [6]. It is important to note that the policy must be identified per user for each application. In a cellular network, the management of the policies for a large number of applications and users would be prohibitive. Furthermore, although many parameters for the radio bearer services could be identified based on the application, there are additional parameters that need more information about the actual specifics of the application session (for example, using a different codec may warrant a different bit error rate). Also, the user should be able to modify the parameters in order to control the network charges dependent on the importance of the specific session. That is not provided by NS. Therefore, even if some applications may make use of the NS over wireless accesses, most applications will need control over the QoS.

8.4.4 Conclusions

In this section we considered the problem of allocating network resources in a scenario where RSVP hosts access an IP network through a wireless network. This is a meaningful scenario that supports the separation of the applications from the wireless access technology. In such a scenario the wireless access network needs to allocate radio resources in order to provide QoS for the applications. While the specific radio related traffic descriptors and QoS parameters may vary from technology to technology, there is a number of traffic and QoS parameters that are needed in a wide range of RANs. We argued that in order to better support the spectrum-efficient RRM, the existing IntServ model needs to be extended such that it includes finer granularity of traffic and QoS parameters.

The suitability of the existing integrated services can be summarized as follows:

1. GQoS
   - Guaranteed bandwidth is not feasible due to the varying performance that occurs with wireless transmission.
   - Increasing the bandwidth to reduce the queuing delay will not be acceptable to network operators because of the effect on the number of served subscribers.
   - Providing correct information to cover for worst-case transmissions is complex, and the resulting service would be unsuitable for applications.
   - Sufficient information is not available to allow appropriate setting of the wireless parameters.

2. CL
   - If the wireless network is not lightly loaded, it cannot approximate a lightly loaded wireless network in all characteristics. More information is required to determine which characteristics of a lightly loaded wireless network are of greatest importance.
   - Sufficient information is not available to allow appropriate setting of the wireless parameters.

3. NS
   - Holding policy information for a large number of user and applications is unrealistic.
   - Additional information besides the application/subapplication would be required for the ideal setting of the wireless parameters.
   - The user requires control over the parameters to control network charges, but it is not supported.

Therefore, we conclude that existing integrated services do not lend themselves to support spectrum-efficient operation of wireless networks supporting explicit QoS provisioning.

8.5   Proposed Integrated Services Parameters and Mapping

As discussed above, the radio management functions often require detailed information about the media stream and its required QoS in order to optimize the QoS performance from the allocated radio resources.

(See Chapter 12 for a more detailed look at the radio resource management aspects in wireless IP networks.) Section 8.3 has examined the wireless parameters available in UMTS. It is also necessary to consider how information can be provided by the application to aid in setting these parameters. The controlled load service is intended to support a broad class of applications including adaptive real-time applications. The controlled load service is intentionally minimal, with no optional functions or capabilities. However, it is proposed here that the wireless network characteristics and requirements are sufficiently different from typical wireline interfaces that additional information is needed. Thus, it is proposed to extend the controlled load service with optional parameter information that will be useful for wireless networks to enable appropriate settings for the radio bearer characteristics. It is noted, though, that although this optional information is proposed for the controlled load service, it may also be applied to other services (e.g., GQoS).

The additional parameter information to be included must aid in determining the appropriate setting of the wireless parameters. Since the application may not know when a wireless link is involved in the connection, the additional information must not depend on the actual interface being used. Specifically, in the case of UMTS, the existing UMTS QoS handling at the PDP Context, RLC protocol, and the air interface level must not be affected by the introduction of these new parameters. Also, it must be straightforward for the application to determine appropriate values for these parameters. Determining the additional parameters ("wireless hints") in such a way that they can be useful for diverse access networks facilitate the seamless hand-over between different types of accesses. (Chapter 23 provides an overview of interworking and handover mechanisms between WLAN and UMTS networks.)

This section builds on the discussion on the wireless parameters of Section 8.3 and proposes a set of information aimed to allow the appropriate radio parameters to be determined, while being simple for the application to set correctly (that is to say, selection of the parameter value should be straightforward for the application).

8.5.1 Media Description Using MIME

Speech applications are typically an important application in cellular networks. For instance, the *adaptive multirate* (AMR) and the *wideband AMR* (AMR-WB) codecs have been designed with specific consideration given to spectrum usage, bit error rates, and performance for wireless. With the AMR codec, the media stream contains data of different classes with different levels of sensitivity to errors (from class A data, which is the most sensitive to bit errors, to class C, which is the least sensitive).

In order to provide optimal spectrum efficiency and performance, it may be necessary to provide different service to the different class data (*unequal error protection*, UEP). In such a case it is also important to specify the format of the media stream and the QoS associated with the different parts of the stream. For instance, the IETF RFC 3267 [7] proposes a standard *real-time protocol* (RTP) payload format to carry AMR and AMR-WB coded speech samples. We believe that the description of many media types by means of a few generic parameters is not realistic because of the large amount of parameters that would be required to characterize the media and the requested QoS. Also, the IETF RFC 2048 [8] indicates that the majority of the important real-time applications will have RTP encoding names as multipurpose Internet mail extensions (MIME) subtypes. Therefore, we propose the use of the MIME type registration to characterize media streams for which the CL integrated service is requested. Current ongoing work at IETF [9] defines the procedure to register RTP payload formats as audio, video, or other MIME subtype names. Furthermore, [8] registers all the RTP payload formats defined in the RTP profile for audio and video conferences as MIME subtypes. Some of these may also be used for transfer modes other than RTP. Using MIME as the media description format has the following benefits:

1. It provides a detailed description of the MIME registered media types. Apart from the Registration Template shown in Section 8.2.8 of RFC 2048, the MIME registration procedure allows the specification of additional parameters, including:

   - Published specification [RFC number];
   - Rate parameter: If the payload parameter does not have a fixed RTP timestamp clock rate, then a rate parameter is required. A particular payload format may have additional required parameters;
   - Optional parameters, like those mentioned in [7] (channel, ptime, or other payload format–specific optional parameters).

2. It is flexible in the sense that it facilitates the description of the relevant media types that are typically carried over the RTP protocol. We expect that most real-time applications will use RTP, so the MIME description is expected to be a useful hint on most of the relevant applications. For example, ongoing work within IETF proposes MIME type registration and associated RTP payload formats for AMR and AMR-WB speech encoded signals [7]. Standardizing the RTP payload formats and registering the media type as MIME types provides exact information about all possible SDI sizes that the wireless network will have to transport. In addition, standard RTP payloads make unequal error detection/protection mechanisms possible.

3. The MIME type media description is easy to map to fields in the SDP [10], which is commonly used to describe RTP sessions.

8.5.2 Proposed Additional Parameters for Real-Time Traffic

The following proposed parameters are qualitative hints rather than quantitative parameters.

SDU Format Information

As mentioned above, UEP plays an important role in spectrum-efficient resource management. UEP implies that different parts of an IP payload are associated with different error protection mechanisms when transferring over the air interface. In order to facilitate this mechanism, information about the payload format is necessary at the radio level. For some (but not all) MIME registered media types, parts of this piece of information may be available. In general, however, the specification of the SDU format as a service parameter allows any application to take advantage of the UEP mechanism. Furthermore, in some cases, the specification of the MIME type is not sufficient to perform UEP. The exact format of this parameter is for further study.

Bit Error Ratio

For real-time applications that use the UDP Lite transport protocol [11], it can be advantageous to provide rough information (hint) on the required (tolerated) *bit error ratio* (BER) over the wireless link. The requested BER reflects a trade-off between the quality degradation that real-time applications may suffer from bit errors caused by the wireless link and the wireless spectrum that the application actually consumes. A qualitative hint on the required BER may be low BER, medium BER, and high BER, which, together with the MIME media description, allows the wireless network to allocate the proper radio resources.

Maximum Transfer Delay (Expected Delay Bound)

The expected delay bound is an important parameter that allows the wireless network to configure the wireless bearer service. For instance, the appropriate interleaving depth is directly affected by the specified maximum delay requirement [12]. Also, this parameter is needed to determine the maximum number of retransmissions (if any) in the wireless link.

Packet Loss Ratio

The *packet loss ratio* (PLR) affects the subjective quality of many real-time applications including coded speech and video. On the other hand, the wireless network uses the target PLR value for admission control and to set some parameters of the wireless network (e.g., L2 buffer size).

8.5.3 Proposed Additional Parameters for Non-Real-Time Traffic

Packet Handling Priority

For interactive packet services, the *packet handling priority* (PHP) parameter helps the wireless network to provide QoS differentiation, especially in congestion situations without the need of complex reservation or scheduling mechanisms.

PLR

The PLR parameter can be useful even for nonreal-time applications. For instance, for TCP applications, limiting packet loss means predictable performance, while for the wireless segment this parameter is used as an input to admission control. (Chapter 14 provides modeling and performance aspects of the TCP protocol over the next generation of broadband wireless access networks.)

8.6   Numerical Examples

In this section we consider some of the available UMTS link level simulation results, which provide quantitative arguments for the proposed IP level (IntServ) parameters. The simulation program used in the link level performance evaluations was the *Communications Simulation and System Analysis Program* (COSSAP) by Synopsys, Inc. The COSSAP simulation program is a stream-driven simulation program, with support for asynchronous operation since no timing control between the simulation elements is required. The simulated channel model is a "Pedestrian A (3 km/h)." In this section we will consider how the radio *frame error rate* (FER) depends on the radio channel quality, which is characterized by its signal-to-interference ratio (C/I) value.

8.6.1 SDU Format Information to Facilitate UED/UEP

Ordinarily, the bits in a frame from a cellular voice codec are divided into three classes: 1a, 1b, and 2. Bit-error sensitivity varies between these classes: class 1a includes the most sensitive bits; class 2 includes the least sensitive bits.

In a typical third-generation system, the bits in class 1a are covered by a cyclic redundancy code (CRC) that checks for errors in the frame. Thus, we say that the voice frame uses an *unequal error detection* (UED) scheme. If information on the different classes of bit-error sensitivity cannot be transferred from the codec to the radio-access network, or if the bits in the voice frame are not organized into classes, then the UED scheme cannot be used.

Instead, an *equal error detection* (EED) scheme—a CRC that covers the entire voice frame—is introduced. To have the quality in these two cases, each must receive the same number of frames with a bad CRC.

Where circuit-switched traffic is concerned, only frames with a bad CRC contribute to the frame error rate. But in the all-IP architecture, frames with a bad CRC, frames that are lost due to jitter, and fatal errors in the IP header all contribute to the frame error rate. By fatal errors in the IP header we mean UDP checksum errors, errors in the link layer, and header decompression errors.

The calculation of the BER only includes those errors that occur in bits not protected by the CRC. The residual errors in bits protected by the CRC must be as close to zero as possible. If any residual bit errors exist in class 1 bits, the voice decoder may produce noticeable artifacts.

The SDU format information parameter describes the structure of the voice frame and thereby provides the prerequisite for performing bit-classification and UEP/UED. Recognizing this, ongoing work at IETF aims to standardize the RTP payload format for the AMR and AMR-WB codecs [7].

If UEP is not available (but UED is), the lowest FER requirement of class 1 and the BER requirement of class 1b constitute the radio channel requirements. The impact of using UED/UEP on the required C/I is illustrated in Figure 8.10. Minimizing the required C/I for a given performance is important, since the target C/I has direct impact on the spectrum efficiency and the capacity of the system. In this example we considered an AMR 12.2 codec with maximum and guaranteed bit rate of 13 Kbps, the maximum transfer delay of 80 ms, and the maximum SDU size of 32 octets. The number of class 1a, 1b, and 2 bits is [89, 103, 64], and the residual BER is 10^{-6}, 10^{-3}, 5.10^{-3}, respectively. The required SDU error ratio for class 1a bits was set to 0.7% and unspecified for class 1b and class 2 bits.

We can observe in Figure 8.11 that for a required FER of around 0.7%, approximately 20% lower C/I value is sufficient if SDU format information is available and UEP is exercised.

8.6.2 (Residual) BER

The proposed BER parameter, as a new IP level parameter, specifies the target residual BER at the radio link level. The target residual bit error ratio, in turn, determines the appropriate number of CRC bits that ensure the required FER. It is important to minimize the number of CRC bits, since longer CRCs than necessary require a higher C/I value.

Indeed, in Figure 8.11 we observe the difference between the required C/I for a prescribed FER in the cases when the number of CRC bits is 8 and 24.

FIGURE 8.10
Voice FER as the function of the C/I with/without SDU format information.

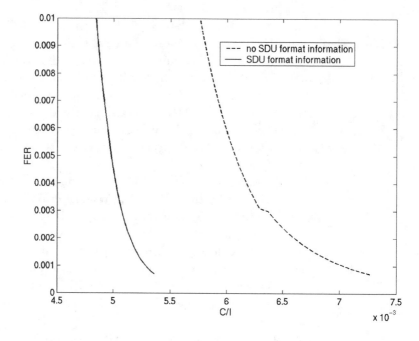

FIGURE 8.11
Voice FER as the function of C/I and the applied CRC length.

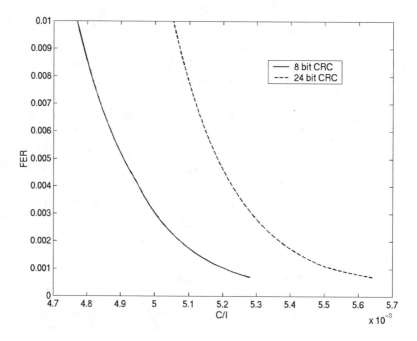

8.6.3 Maximum Transfer Delay

To understand the importance of the maximum transfer delay, as a proposed new IntServ parameter, we need to discuss the services and functions provided by the RLC protocol at the radio layer. The RLC protocol provides three different types of services to upper layer (in this case IP) PDUs:

1. *Transparent data transfer:* This service transmits upper layer PDUs without adding any protocol information, possibly including segmentation/reassembly functionality.

2. *Unacknowledged data transfer:* This service transmits upper layer PDUs without guaranteeing delivery to the peer entity. It provides detection of erroneous data; that is, it delivers only those SDUs to the receiving upper layer that are free of transmission errors by using the sequence-number check function. The receiving RLC entity delivers an SDU to the upper layer as soon as it arrives at the receiver.

3. *Acknowledged data transfer:* This service transmits upper layer PDUs and guarantees delivery to the peer entity. If the RLC is unable to deliver the data correctly, the user of the RLC at the transmitting side is notified. For this service, both in-sequence and out-of-sequence delivery is supported. In this service, the error-free delivery is ensured by means of retransmissions. The receiving RLC entity delivers only error-free SDUs to the upper layer.

If the application can indicate (even if only qualitatively) its delay requirement, the radio resource management function can select the appropriate transfer mode for the service. This is important because the usage of the acknowledged mode (and retransmissions) allows the system to use a lower C/I value for the radio channel.

In Figure 8.12 the RLC PDU error ratio is plotted as the function of the C/I. If the delay requirement at the IP layer is such that the RLC layer may use retransmissions, then a relatively high error ratio is acceptable at the RLC PDU, since the RLC layer will use retransmissions of the PDUs before delivering them to the IP layer.

To further highlight the importance of the delay parameter, consider Figure 8.13. If the required maximum delay at the IP layer is low, then a high C/I value is necessary at the radio layer, whereas less stringent delay requirements allow lower C/I values and thereby help optimize radio resource usage.

8.6.4 PLR

The PLR parameter at the IP layer helps the radio layer to determine the appropriate target RLC PDU error ratio. For instance, we note in Figure 8.12 that a target PDU error ratio of 1% would require approximately two times as large a C/I value than a target PDU error ratio of around 8%.

FIGURE 8.12
RLC PDU error ratio
as the function of the
C/I value.

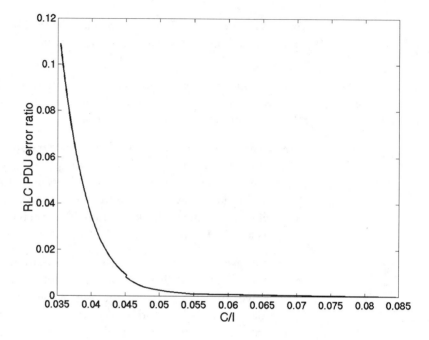

FIGURE 8.13
Required maximum
SDU delay as the
function of the C/I
value.

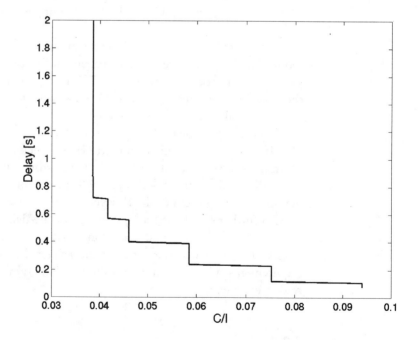

8.7 Conclusions

In this chapter we focused on the QoS aspects of the evolving all-IP architecture of third-generation mobile systems. As the Internet evolves from a single class (best-effort) network towards a multiservice network (where some classes provide QoS guarantees), third-generation systems are expected to provide access to various Internet-based services with a relatively high QoS granularity.

Also, in the all-IP architecture, mobile users could ideally get access to the Internet via various access networks through open interfaces. On the other hand, since wireless resources are generally much more scarce than IP resources, there is a need to maintain spectrum efficiency in the RANs. Therefore, in this chapter we have argued and showed through simple numerical examples that wireless QoS support in the form of a few "wireless hint" parameters at the IP layer can significantly decrease the required signal power that is necessary to maintain IP level (and end-to-end) QoS and thereby help to increase spectrum efficiency.

The design of the core network is a major challenge, since in all-IP networks various service classes—some with strict QoS guarantees—are present. Therefore, new IP design methodologies will play an important role in the core design. We believe that the strong traffic engineering support of the MPLS technology is the most likely candidate for this part of the all-IP architecture. The design of MPLS core networks consists of UMTS-specific and general MPLS steps.

As the rollout of these new-generation networks starts, the research community has begun to explore the successor technologies that will help develop the fourth-generation networks. It is likely that 4G networks will offer at least an order of magnitude higher bit rates than that of 3G networks, even in a high-speed mobility environment. Also, 4G systems will likely connect humans, machines, and various appliances. Therefore, the ratio of data traffic to voice traffic will significantly increase, which further emphasizes the need for the research issues discussed in this chapter.

REFERENCES

[1] Dixit, S., Y. Guo, and Z. Antoniou, "Resource Management and QoS in Third-Generation Wireless Networks," *IEEE Communications Magazine*, Vol. 39, No. 2, February 2001, pp. 125–133.

[2] 3GPP TS, "End-to-End QoS Concept and Architecture," 3G 23.207.

[3] Wroclawski, J., "The Use of RSVP with IETF Integrated Services," IETF RFC 2210, September 1997, http://www.ietf.org/rfc/rfc2210.txt.

[4] Shenker, S., C. Partridge, and R. Guerin, "Specification of Guaranteed Quality of Service," IETF RFC 2212, September 1997, http://www.ietf.org/rfc/rfc2212.txt.

[5] Wroclawski, J., "Specification of the Controlled Load Quality of Service," IETF RFC 2211, September 1997, http://www.ietf.org/rfc/rfc2211.txt.

[6] Bernet, Y., A. Smith, and B. Davie, "Specification of the Null Service Type," IETF RFC 2997, November 2000, http://www.ietf.org/rfc/rfc2997.txt.

[7] Sjoberg, J., et al., "RTP Payload Format and File Storage Format for AMR and AMR-WB Audio," IETF RFC 3267, June 2002, http://www.ietf.org/rfc/rfc3267.txt.

[8] Fred, N., J. Klensin, and J. Postel, "Multipurpose Internet Mail Extensions (MIME): Registration Procedures," RFC 2048, November 1996.

[9] Casner, S., and P. Hoschka, "MIME Type Registration of RTP Payload Formats," IETF draft, work in progress, http://www.ietf.org/internet-drafts/draft-ietf-avt-rtp-mime-06.txt.

[10] Handley, M., and V. Jacobson, "SDP: Session Description Protocol," RFC 2327, April 1998.

[11] Larzon, L.-A., M. Degermark, and S. Pink, "The UDP Lite Protocol," IETF draft, work in progress, http://www.ietf.org/internet-drafts/draft-ietf-tsvwg-udp-lite-00.txt.

[12] Ojanpera, T., and R. Prasad, (Eds.), *Wideband CDMA for Third Generation Mobile Communications*, Norwood, MA: Artech House, 1998.

CHAPTER 9

Provisioning QoS in 3G Networks with RSVP Proxy

Balázs Benkovics

There is a growing number of real-time, bandwidth demanding applications requiring QoS guarantees from the operating environment. In addition, more and more terminals will access the Internet using wireless networks. Emerging 3G (UMTS) systems [1] are capable of providing IP-based real-time multimedia services for mobile users. QoS can only be achieved by means of QoS management on an end-to-end basis (i.e., end user to end user), as well as potentially across many domains since the Internet is a concatenation of many autonomous systems, usually managed by different operators. The requirement for end-to-end QoS management does not necessarily mean that the same resource reservation signaling protocol must be applied all along the route between communicating partners. In fact, it is most likely that the end-to-end QoS management architecture will consist of many interoperable and concatenated QoS management architectures rather than one global end-to-end QoS infrastructure.

This segmented approach enables the operators to provision QoS in their system by using a solution that best suits their infrastructure. Moreover, it has great significance for mobile operators since the deployment of all-IP 3G wireless systems [2] presents new challenges for QoS signaling protocols current implementations cannot address.

The most well known IP-based resource management protocol designed for fixed networks is the RSVP [3]. RSVP is capable of dynamic resource management in one-to-one or one-to-many configurations. The protocol messages are object based, which means every message consists of one or more basic object and other optional objects. This approach results in a flexible behavior, but rather large message sizes; moreover, the protocol uses periodic refresh messages to maintain reservation states along the data path. In mobile environments the bandwidth is rather low compared to fixed networks, so these relatively large messages take up valuable resources from the applications themselves. Another serious problem derives from the high frame error rate value of the radio links: frames may get damaged more often due to error bursts over the air link. This causes a higher packet loss ratio and can result in connection breakdowns since, in the absence of RSVP refresh messages, the reservation state automatically times out and gets deleted. To sum up, RSVP requires several modifications to use it efficiently in a wireless environment. On the other hand, endpoints in particular segments of the Internet will most likely use RSVP for resource reservation, and these RSVP-capable nodes are willing to exchange QoS assured data with IP terminals in wireless networks. Moreover, some wireless-enabled IP terminals are quite likely willing to run the same IP multimedia applications (probably using RSVP signaling) regardless of whether they connect to the Internet via a fixed or wireless access.

There are two basic approaches to improve the efficiency of the protocol in mobile networks. The first one recommends modifications in the protocol functionality such as changing the refresh time or bundling more messages into one [4]. The common problem with these proposals is that major changes are required in the protocol functionality and thus in the end stations and routers.

The other way is to use a separate resource reservation protocol in the mobile core and access network that is designed for mobile networks, and apply a protocol converter device (an RSVP proxy) in different places in the system. Such a proxy converts the mobile-specific signaling messages to RSVP messages and vice versa. Such a resource management signaling solution for mobile packet switched networks is the PDP context signaling introduced by 3GPP [5]. This latter approach could solve the problems presented above, considering bandwidth utilization and reliability, while providing transparent operation to communicating partners.

Another important issue in cellular systems is the maintenance of QoS for ongoing connections during handoffs. There exist several methods that aim to provide seamless handoffs for such systems. The simplest one is based on packet buffering in the old RNC and packet forwarding to the target RNC [6]. This method, however, requires that the target RNC have enough free capacity to handle the call. Therefore, admission control is performed upon the request, and if that fails the call is blocked. This situation is usually more annoying than blocking of a new call, therefore it is a common method that a certain amount of resources in each cell is reserved for handoff calls exclusively. Based on the amount of resources reserved, the experienced blocking probabilities for handoff calls can be an order of magnitude lower than that of new calls. The number of reserved channels can be fixed or can vary over time based on the velocity, heading, and other parameters of users in neighboring cells [7]. Despite this, it can occur that there are insufficient resources in the target cell and the required QoS can not be provided by RSVP. It is not a problem in our proposal—as in case of insufficient resources, the call will be blocked by the network infrastructure and thus RSVP signaling will time out and reservation states tear down along the data path.

In this chapter the options for using RSVP-capable terminals for accessing the Internet over a UMTS system are investigated. The impact on reliability and performance of using an RSVP proxy as mentioned above is also studied. Several aspects of assuring QoS are considered regarding scenarios where the mobile terminal communicates with a remote host through the UMTS core network, an external IP network (such as the Internet), and the remote access network.

The remainder of this chapter is organized as follows. Section 9.1 briefly presents RSVP. Section 9.2 summarizes the idea of deploying a proxy in an internetwork. Section 9.3 deals with the basics of end-to-end QoS provisioning in a 3G network and identifies the possible role of an RSVP proxy in the UMTS architecture. Simulation results are presented in Section 9.4. Finally, before the concluding remarks, a short summary of the status of relevant standards is discussed in Section 9.5.

9.1 RSVP

RSVP is a receiver-oriented, soft-state protocol that is capable of communicating resource allocation requests between endpoints. The protocol also supports reservations in one-to-many configurations; thus, multicast applications can also employ it for their QoS session management. Multicast operation, however, is out of the scope of this chapter.

Figure 9.1 shows the logical components playing an important role in provisioning QoS for a specific data flow. The application requests a guaranteed quality session by using the appropriate system call of the operating system. This request is passed to the local RSVP process, which determines whether the node has sufficient available resources to supply the requested service by forwarding the QoS parameters to the admission control module. The policy control module determines whether the user (i.e., the application) has administrative permission to make the reservation. If both checks succeeded, parameters are set in the packet classifier and in the packet scheduler, otherwise an error message is returned to the application. The packet classifier determines the packets in the data flow that will receive the requested QoS, while the packet scheduler is responsible for sending data packets according to the requested service. It is also the task of the RSVP process to reserve the appropriate resources along the path to the destination.

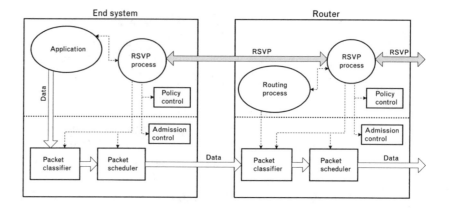

The two fundamental protocol messages are *path* and *resv*, which can be applied to set up and maintain or modify resource reservations between communicating partners.

As depicted in Figure 9.2, the *path* message generated by the source describes the traffic intended to be sent over the newly established connection. Each networking element on the way towards the destination registers the new flow and forwards the *path* message. The destination answers with a *resv* message containing not just the traffic descriptor but also the QoS requirement of the new flow, which is determined at the receiver by processing the information found in the *path* message.

The *resv* message follows the route of the *path* message in the opposite direction back to the source. The actual resources in the network elements are committed upon receipt of the *resv* message. If there are no resources available to satisfy the QoS requirement specified in the *resv* message, the network element sends a *resverr* message back. Resource reservations can be confirmed by sending a *resvconf* message back to the receiver.

If the source or destination wishes to release the connection, it can be done using *pathtear* or *resvtear*, respectively. Resource reservations need to be refreshed periodically, which is done by sending *path* and *resv* messages

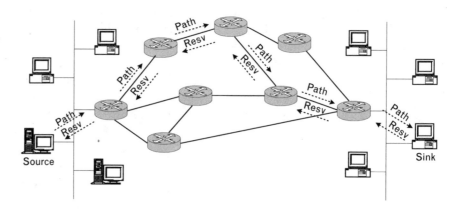

regularly. If no refresh message arrives before the expiry of a timer supervising the process, the resources associated with that particular connection are released. This soft-state operation provides robustness in fixed networks. If a link fails, the next pair of refresh messages will automatically be routed around the fault, thus creating a new reservation on an alternative path. Moreover, this soft-state approach provides an easy way of changing resource requirements of an already established connection. The protocol supports merging reservations on a given network interface to reduce processing overhead in network devices. It is accomplished by aggregating the resource requirements of flows, which originate at the same source, and forwarding refresh messages containing the modified resource requirement to the next hop.

It is important to note that RSVP is designed for unidirectional resource reservation. Establishing bidirectional reservations requires two, independent unidirectional reservations.

RSVP messages have an object-based structure, which means that every message consists of several protocol objects (some are mandatory, while others are optional) that contain specific information on the following:

- Session;

- Protocol time values;

- Traffic specifications;

- QoS parameters;

- Errors in resource reservations.

An RSVP session (i.e., data flow) is defined by the destination IP address (and the port, optionally) and the transport layer protocol. Every valid protocol message must contain a session object. The refresh messages between adjacent nodes are sent periodically based on the settings specified in the time values object. As these values may change during the lifetime of the connection, every refresh message is required to carry such information. Traffic specifications include the characteristics of the data flow (*TSpec*), as well as a filter (*FilterSpec*) that determines packets for which the QoS has to be provisioned. Packets addressed to a particular session but not matching any filter criteria are treated as best-effort traffic. Although RSVP provides means of transmitting QoS parameters in its messages, it is important to clarify that this information is opaque to the protocol, and therefore, the actual rules for comparing these parameters must be defined outside RSVP.

This object-based structure provides easy future expandability of the protocol, and moreover, it helps to reduce processing overhead in network elements by introducing optional objects. On the other hand, it results in rather high message sizes that, in conjunction with the periodic sending of refresh messages, make it difficult to use the protocol in low-bandwidth systems.

Berger et al. [4] proposes modifications to RSVP to address problems with refresh volume and reliability issues. A *bundle* message is defined as one that consists of a variable number of standard protocol messages. It helps to reduce overall message processing load, as well as bandwidth consumption by aggregating multiple RSVP messages within a single PDU. When bundling messages, network elements must take into consideration that so-called trigger information (i.e., messages indicating change of route or resource requirement) should be delayed a minimal amount of time. The proposal also introduces a *message_id* object to help reduce refresh message processing by allowing the receiver to more readily identify an unchanged message. The *message_ack* entity can be used to detect message loss and support reliable RSVP message delivery on a per-hop basis. A summary refresh (*srefresh*) message is defined to enable refreshing *path* and *resv* state without the transmission of entire refresh messages, while maintaining the ability of RSVP to indicate when state is lost and to adjust to changes in routing. It is performed by marking the states by sending a preceding *message_id* object in the appropriate *path* or *resv* message, and including the identifier of that message in the *srefresh* message. Similar to the *bundle* message, a summary refresh message can also carry refresh information for more data flows.

Figure 9.3 shows the benefits of using refresh overhead reduction extension of the protocol on link load. There are two ongoing RSVP connections between hosts *A.1* and *B.3* and hosts *A.4* and *B.4*, respectively. Routers *R1, R2, R4,* and *R6* are capable of using the refresh overhead reduction extension, while *R5* is not. (*R3* is not important for us, since it is not in the data path). The default behavior of the network elements is to use the overhead reduction extensions if possible. As router *R5* is not capable of processing such messages, the default *path* and *resv* messages are transmitted along the path between *R2, R5,* and *R4*. Note that the standard RSVP

FIGURE 9.3
*Refresh overhead
reduction scenario.*

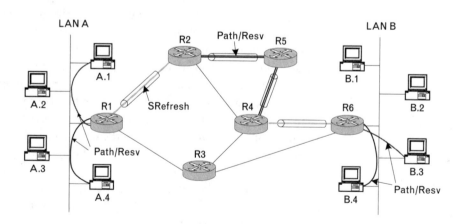

merging function cannot be used in this scenario since the source nodes of the flows are different.

These modifications are applicable in RSVP routers, where many parallel RSVP reservations need to be established and maintained, and they are not really beneficial for individual terminals, where typically only a couple of parallel sessions exist.

9.2 Performance-Enhancing Proxies

Performance-enhancing proxies are often used in internetworks to compensate for the inefficiencies stemming from differences in link layer characteristics [8]. Applying a proxy typically means that connections between endpoints are realized in two legs: the first leg is terminated by an intermediate device (proxy), and there is a separate new connection established between the proxy and the other endpoint. The solution enables a link-optimized protocol stack to be used over each leg. One important aspect is the transparency of the proxy operation, which means that no modification of the endpoints is required in order to benefit from the proxy.

Applying RSVP in wireless (and, more specifically, in 3G) environments raises the following concerns:

- The high frame error rate over the radio channel might cause too many reservation and refresh messages to be lost. This would result in increased connection setup time or erroneous release of connections.

- Regular refresh messages consume a lot of bandwidth relative to the capacity of the wireless channel.

- RSVP is quite complex to implement mainly because it supports multicast reservations with different reservation styles. Supporting multicast sessions, however, is not typically required in wireless systems.

- RSVP supports unidirectional reservations. This is not suitable for conversational services (e.g., telephony), which represent a significant portion of traffic in a 3G system.

Even if the RSVP modifications mentioned in the previous section alleviate these problems, there still remains the question whether the traffic and QoS parameters defined in existing RSVP are appropriate for specifying resource requests in a wireless network. A possible solution might be to place a proxy at the border of the fix and wireless network as shown in Figure 9.4. This would act as a sender proxy that issues RSVP messages

FIGURE 9.4
RSVP proxy between a fix and a wireless network.

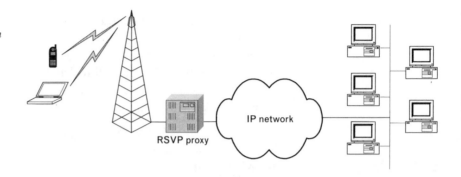

upon receipt of radio network–specific resource reservation signaling and vice versa.

9.3   Enabling End-to-End QoS for Packet Switched Services in UMTS

In Figure 9.5 the mobile terminal communicates with a remote host through the UMTS network, an external IP network (such as the public Internet), and the remote access network. In order to constitute an end-to-end, QoS-enabled service, multiple QoS mechanisms interact with each other along the end-to-end path. The end-to-end path is composed of a local bearer service interconnecting the terminal equipment and mobile terminal, a UMTS bearer service traversing the UMTS network between the TE and the GGSN, and an external bearer service that carries the data between the GGSN and the remote TE. UMTS QoS concept and architecture specifications [9, 10] describe the provisioning of QoS-enabled UMTS bearer services and the mapping and translation functionality needed between UMTS bearer service and the local and external bearer service.

FIGURE 9.5
A 3G end-to-end scenario with RSVP QoS signaling.

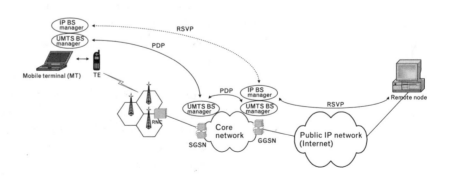

Internet applications do not directly use the services of UMTS, but they use Internet QoS definitions and attributes. The application requests an end-to-end bearer service using the IP BSM. This entity utilizes a translation function and it maps the IP bearer service request to a UMTS bearer service request. Finally it asks the UMTS BSM to establish the required UMTS bearer service capable of satisfying the QoS needs of the application.

The UMTS BSM is responsible for resource allocation in the UMTS network coordinating the establishment of underlying bearers in the radio access network and in the core network. In order for a TE to be able to send data across a UMTS network, a PDP context shall be activated. To establish such a context, to modify it according to the needs of the applications, and to request special QoS treatment for certain traffic flows, there is a UMTS-specific protocol [5], which controls the resource reservation across the UMTS system. This is a hard-state protocol optimized for the wireless environment both in terms of its operation and in terms of the QoS parameters it supports.

Given that UMTS has its own resource control signaling method, some considerations are needed to find out how to support native IP application with RSVP signaling capability. The remainder of this section describes end-to-end bearer establishment scenarios where at least one of the communicating endpoints is a UMTS TE and RSVP is also involved in the control signaling. These scenarios are collected in [10]. The remainder of this section concentrates on UMTS bearer establishment procedures; therefore, the local and external bearer control procedures are not shown, and the TE and the MT are represented as a single entity called the user equipment.

9.3.1   Resource Reservation with End-to-End RSVP

Figure 9.6 depicts one possible signaling sequence for the case when the UE initiates RSVP messages and this signaling is carried to the remote endpoint. Alternative signaling sequences are also possible: the Create PDP Context Request message can be sent after the *path* message, after the *resv* message, and it can also be sent after the RSVP controlled reservation is completed. In case the GGSN is RSVP aware, it also processes the RSVP messages.

The UE implements an IP bearer service manager function, which enables end-to-end QoS using IP layer signaling towards the remote end. However, the UE relies on this end-to-end communication being utilized by the UMTS bearer service manager in order to provide the QoS over the UMTS segment. The backbone IP network is RSVP and/or DiffServ [11] enabled. If the backbone relies on DiffServ mechanisms, the interworking between RSVP and DiffServ can be provided either by the UE or by the GGSN.

The application layer (e.g., SIP/SDP [12, 13]) between the end hosts identifies the QoS requirements. This information (e.g., TS23.228 [14] describes interworking from SIP/SDP to QoS requirements) is mapped down to create an RSVP session. The UE shall also establish the PDP context suitable for support of the RSVP session.

In this scenario, the terminal supports signaling via the RSVP protocol to control the QoS across the end-to-end path. The GGSN may also support RSVP signaling, and it may also use this information, not just the PDP context, to control the QoS towards the remote terminal. Although RSVP signaling occurs end-to-end in the QoS model, it is not necessarily supported by all intermediate nodes.

The end-to-end QoS is provided by a local mechanism in the UE, the PDP context over the UMTS access network, DiffServ through the backbone IP network, and RSVP in the remote access. The RSVP signaling may control the QoS at the local access. This function may be used to determine the characteristics for the PDP context, so the UE may perform the interwork between RSVP and the PDP context.

Alternatively, the UE may support both differentiated services and RSVP functionality. At the UE, the data is also classified for DiffServ. Intermediate QoS domains may apply QoS according to either the RSVP signaling information or DiffServ mechanisms. In this scenario, the UE is providing interworking between the RSVP and DiffServ domains. The GGSN may override the DiffServ setting from the UE.

The main drawback of this option is that all RSVP signaling is carried over the UMTS network, which represents a significant overhead.

9.3.2   Service-Based Local Policy and RSVP Sender/Receiver Proxy

Figure 9.7 depicts a scenario where the use of RSVP may be intended to enable the external network provider to support traffic engineering, efficient resource management, and call blocking if needed to handle temporary overloaded conditions. If the RSVP connection setup is unsuccessful, it is the operator's choice to have, for example, the GGSN initiate a PDP context deactivation. The scenario assumes that the GGSN supports DiffServ edge functions, and the backbone IP network is DiffServ enabled.

This section provides the flows for bearer establishment, resource reservation, and policy control with PDP context setup and RSVP interworking. The UE performs an IP BSM (BS) function, which enables end-to-end QoS without IP layer signaling and negotiation towards the IP BS function in the GGSN, or the remote host. The P-CSCF provides the authorization token to the UE during the SIP session setup process, and the UE provides the authorization token to the GGSN in the PDP context activation/modification message, in order to enhance the interworking options to an RSVP function in the GGSN. The IP QoS bearer service towards the remote host is controlled from the GGSN.

FIGURE 9.6
*Mobile originated re-
source reservation with
end-to-end RSVP.*

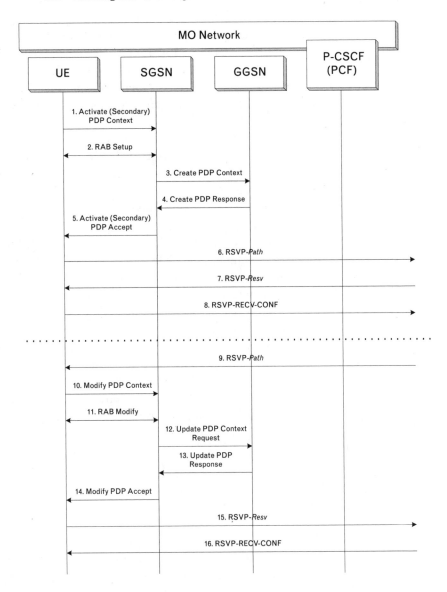

FIGURE 9.6
*Mobile originated re-
source reservation with
end-to-end RSVP.*

The GGSN uses the authorization token to obtain a policy decision from the P-CSCF(PCF) which will be used to derive IP level information (e.g., destination IP address, TSpec, FilterSpec, PolicyData). This is done via the standardized interface between the PCF and GGSN. In addition, IP level information may also be derived from PDP context (e.g., QoS parameters).

The application layer (e.g., SIP/SDP) between the end hosts identifies the QoS needs. The QoS requirements from application layer are mapped down to the IP layer and further down to the PDP context parameters in the UE. The GGSN may use the IP level information to invoke RSVP messages to set up the uplink as well as the downlink flows in the backbone IP network up to the remote host.

In the uplink direction, the GGSN acts as an RSVP sender proxy and originates the *path* message on behalf of the UE. It must also periodically refresh the *path* message and correctly terminate the *resv*, *resvtear*, and *patherr* messages for the session.

In the downlink direction, the GGSN acts as an RSVP receiver proxy and generates the *resv* message on behalf of the UE. The GGSN should install a *resv* proxy state and act as if it has received a *resv* from the true endpoint (UE). This involves reserving resource (if required), sending periodic refreshes of the *resv* message, and tearing down the reservation if the path is torn down.

The QoS for the downlink direction is controlled by the PDP context between the UE and the GGSN. The GGSN terminates the RSVP signaling received from the remote host and may use the IP level information to provide the interworking with RSVP towards the remote host.

The end-to-end QoS is provided by a local mechanism in the UE, the PDP context over the UMTS access network, DiffServ through the backbone IP network, and RSVP in the remote access network.

An important advantage of this scenario is that no RSVP signaling is carried over the UMTS network. To satisfy the needs of RSVP-capable applications running in the UE, a second proxy can be introduced in the UE that is responsible for regenerating the RSVP signaling towards the application based on the requested changes in the PDP context parameters.

9.4   Simulation

The control plane functionality of the RSVP and the RSVP proxy functionality is implemented in an event-driven simulator. This section presents simulation results for assessing the benefits of the proxy architecture in terms of robustness of the connections and the amount of signaling load carried over the air interface.

The simulation setup is the one depicted in Figure 9.4: a mobile terminal generates resource reservation request towards an endpoint in the fixed network. This request travels through a radio link, where it might be lost. The error model is a simple, frame level model. A two-state Markov chain is embedded in the link, which has a *drop* and a *no-drop* state. The new state is calculated upon the arrival of a new message. In *drop* state all packets are lost, while in *no-drop* state, the packet remains intact when transferred through the link. On the other side of the radio link, there is a network element, which either forwards the resource reservation messages transparently toward the destination, or it acts as an RSVP proxy. The network element is

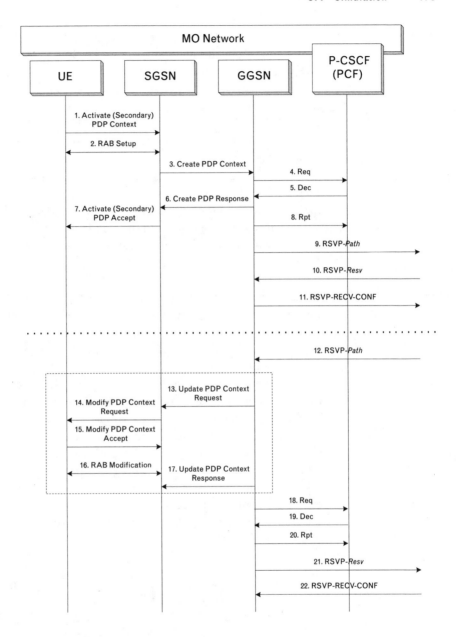

FIGURE 9.7
Mobile originated service-based local policy and RSVP sender/receiver proxy.

connected to the endpoint in the fixed network via a link, which has enough capacity that reservation message loss does not occur due to buffer overflow.

In the fixed network, RSVP timeout values as recommended by IETF are set: refresh messages are generated in every 30 seconds, and the protocol has to tolerate three successive losses, so the cleanup timer, which supervises the release of outdated reservations is set to 95 seconds. On the wireless side, different settings were tried in order to evaluate the trade-off between signaling overhead and robustness. With each of the timer settings several simulations were run changing the packet loss ratio from 10% to 60%, in 10% steps.

The RSVP sessions were generated according to a Poisson process and their holding time was exponentially distributed with mean value of 90 seconds. RSVP message sizes were set as stated in [3] when no optional objects are present in them—that is, the *path* and *resv* messages are 40 and 64 bytes long, respectively.

The first scenario demonstrates that a stand-alone RSVP should not be used as a signaling protocol over the wireless access network. The benefits of spacing out the refresh messages on the wireless network load are analyzed with varied packet loss ratio. All the simulations were run with and without the proxy to observe the differences. In this setting the role of the proxy is only to allow configuring different cleanup timer values for the fixed and wireless side of the connection. Figures 9.8 and 9.9 depict the number of erroneously released connections as a function of the frame error rate over the air interface. "Erroneous" release here means that the network element tears down an active connection earlier than its intended release by the application due to the loss of too many refresh messages and the corresponding expiry of the cleanup timer.

As one can see in Figure 9.8, increasing the frequency of reservation messages actually leads to worse performance with the given error model. The reason is that refresh messages sent in a quick succession are more likely to fall into a common burst of error over the wireless link. Figure 9.9 demonstrates that configuring a less stringent cleanup timer helps a lot, especially when the frame error rate is very high. Configuring RSVP reservations to tolerate the loss of five consecutive refresh messages halves the amount of erroneous connection releases when the frame error rate is over 30%. Based on these results, the conclusion is that it is more beneficial to increase the

FIGURE 9.8
Number of erroneously released connections without proxy.

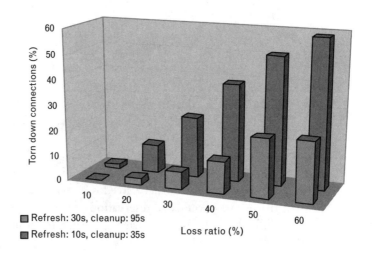

FIGURE 9.9
*Number of erroneously
released connections
with proxy.*

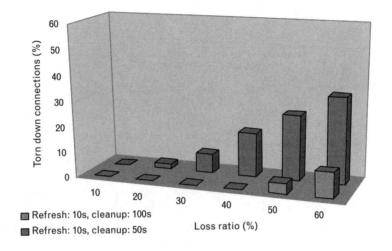

number of tolerated refresh packet losses than to configure more frequent refresh messages when the goal is to increase the robustness of RSVP reservations.

In the second scenario, it is assumed that a hard-state, wireless-specific protocol such as PDP context signaling is run between the mobile terminal and the edge of the UMTS network (GGSN). The role of the proxy is to convert this protocol to soft-state RSVP toward the endpoint in the fixed network. The scenario is depicted in Figure 9.10.

Two simulations are run to determine the signaling load on the wireless link. RSVP with default refresh (30 seconds) and cleanup (95 seconds) timers is used on the wired side of the proxy in both cases, while RSVP with the same default timer settings or a hard-state protocol is used on the wireless side of the connection. Table 9.1 highlights the huge difference between the signaling load caused by a soft-state and a hard-state protocol over the radio link.

9.5   A Short Summary of Related Standards

Standardization of the latest UMTS releases is a result of a coordinated effort of the 3GPP and the IETF. The 3GPP is responsible for specifying the system architecture, the corresponding interfaces, and all aspects of the radio

FIGURE 9.10
*Protocol conversion
with RSVP proxy.*

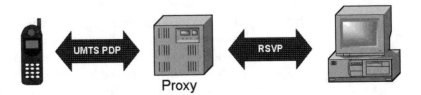

TABLE 9.1 COMPARING THE SIGNALING LOAD USING SOFT- AND HARD-STATE PROTOCOLS

PROTOCOL	REFRESH	CLEANUP	SIGNALING LOAD
Soft state	30 seconds	95 seconds	15
Hard state	∞	∞	4

network and mobility management. If a technology or a solution standardized by IETF is appropriate to be used in UMTS, 3GPP refers to the relevant IETF specification. Further on, if modification or enhancement of any IETF solution is needed or there is a new requirement that the IP networking technology should satisfy in order to be used in UMTS, 3GPP communicates the requirement to the IETF standardization process.

When it comes to running RSVP-enabled applications over a UMTS network, the following 3GPP and IETF standards are relevant. The UMTS QoS architecture and the QoS parameters that are used in the system are specified in 3GPP TS 23.107. The mechanisms required to enable using the UMTS system as a segment of an end-to-end QoS path are described in 3GPP TS 23.207. The signaling sequences for establishing and maintaining multiple PDP context with different QoS capabilities are described in 3GPP TS 23.060. Finally, the relationship between session layer signaling, application layer QoS requirements, UMTS and IP QoS requirements are detailed in 3GPP TS 23.228. The above documents are all stable versions approved by the 3GPP Technical Specification Group.

The RSVP protocol is a proposed standard and it is specified in IETF RFC 2205. RSVP refresh overhead reduction extensions is also a proposed standard (IETF RFC 2961). The RSVP proxy itself has not yet been specified in a standards track RFC.

9.6 Summary

In this chapter RSVP was introduced and concerns were raised about its applicability in UMTS systems as a resource control signaling solution. The basics of the UMTS QoS provisioning mechanism were presented and the obvious requirement of transparently supporting RSVP-capable native IP multimedia applications accessing the Internet through a 3G network has been identified. To reconcile these two contradicting aspects, we considered the deployment of an RSVP proxy, which is capable of translating between a wireless-optimized signaling protocol and RSVP at the border of the wireless network. Finally, the results of some basic simulations were presented to demonstrate the benefits of the proxy approach in terms of decreasing the signaling load on the wireless link and improving the robustness of the resource reservation.

REFERENCES

[1] Korhonen, J., *Introduction to 3G Mobile Communications*, Norwood, MA: Artech House, 2001.

[2] 3GPP Technical Specification 23.922, "Architecture for an All-IP Network," Version 1.0.0, 1999.

[3] Braden, R., et al., "Resource Reservation Protocol (RSVP)—Version 1 Functional Specification," IETF RFC 2205, September 1997.

[4] Berger, L., et al., "RSVP Refresh Overhead Reduction Extensions," IETF RFC 2961, April 2001.

[5] 3GPP Technical Specification 23.060, "General Packet Radio Service (GPRS); Service Description; Stage 2," Version 4.1.0, 2001.

[6] 3GPP Technical Report 25.936, *Handovers for Real-Time Services from PS Domain*, Version 4.0.1, 2001.

[7] Benkovics, B., "Traffic Management in 3rd Generation Mobile Networks," MSc. thesis, Budapest University of Technology and Economics, Department of Telecommunications, 2001.

[8] Border, J., et al., "Performance Enhancing Proxies," IETF RFC 3135, June 2001.

[9] 3GPP Technical Specification 23.107, "QoS Concept and Architecture," Version 5.2.0, 2001.

[10] 3GPP Technical Specification 23.207, "End-to-End QoS Concept and Architecture," Version 5.1.0, 2001.

[11] Blake, S., et al., "An Architecture for Differentiated Services," IETF RFC 2475, December 1998.

[12] Handley, M., et al., "SIP: Session Initiation Protocol," IETF RFC 2543, March 1999.

[13] Handley, M., and V. Jacobson, "SDP: Session Description Protocol," IETF RFC 2327, April 1998.

[14] 3GPP Technical Specification 23.228, "IP Multimedia Subsystem; Stage 2," Version 5.3.0, 2002.

SELECTED BIBLIOGRAPHY

Armitage, G., *Quality of Service in IP Networks*, London: Pearson Higher Education, 2000.

Ojanpera, T., and R. Prasad, (Eds.), *Wideband CDMA for Third Generation Mobile Communication Systems*, Norwood, MA: Artech House, 1998.

Springer, A., and R. Weigel, *UMTS — The Universal Mobile Telecommunication Systems*, Heidelberg, Germany: Springer-Verlag, 2002.

CHAPTER 10

QoS Support for VoIP over Wireless

Henning Sanneck, Werner Mohr, Nguyen Tuong Long Le, Christian Hoene, and Adam Wolisz

10.1 Introduction

10.1.1 Motivation

Cellular wireless systems like GSM, as well as wireless telephony systems for local/indoor use like DECT, have found widespread use in recent years. These systems offer a level of speech quality that is acceptable for most users. In the future, however, it will be important to enable voice services based on IP (VoIP) in wireless systems for the following reasons:

- *Bandwidth savings by employing packet switching:* Wireless telephony systems like DECT and GSM are based on circuit switched technology and need to reserve a fixed amount of bandwidth for each call. However, a speech transmission consists of talk spurts and periods of silence. Transmitting no information during silent periods reduces the mean bandwidth per voice flow and reduces interference with other transmitters (discontinuous transmission). When multiple voice flows are transmitted over the same (shared) wireless medium, the "active" and "silent" periods are statistically distributed among the different flows. Only packet switched systems (e.g., those based on IP) can thus achieve a statistical multiplexing gain by allocating resources dynamically only to active flows. The achieved savings in bandwidth are particularly important in wireless networks due to the scarcity and thus cost of spectrum.

- *Provisioning of traditional and emerging Internet services in wireless systems:* The growing demand for Internet services requires an efficient integration of these services with voice services using IP. Future wireless systems (beyond the third generation) will even be designed to support only packet switched operation, because all services can be delivered in an integrated manner using IP as the unifying platform. Then, it will no longer be necessary to maintain a circuit switched mode concurrently with packet switching, which reduces terminal complexity and simplifies the access network architecture, thus reducing costs.

10.1.2   Challenges

To really exploit the described benefits of statistical multiplexing and service integration, it is necessary that VoIP services can be seamlessly provided with good QoS over several different network technologies. Voice, however, being a real-time service, has very strict delay and packet loss requirements when being transmitted over a packet switched network. These requirements are difficult to fulfill even in wireline network environments because of network congestion, which may cause excessive delay, jitter, and loss. Additionally, in a wireless environment the transmission channels have a fluctuating quality and are highly error-prone because of physical layer impairments. The exhibited unpredictable behavior makes guarantees on the transmission quality very difficult. Furthermore, wireless links typically offer lower bandwidth compared to wireline links. Thus, wireless links (typically as the first and/or last hop) tend to be the limiting bottleneck on the end-to-end path.

To overcome these difficulties, an *integrated approach* covering the application (voice codec) level, the network layer QoS scheme, and the wireless networking technology considered here needs to be explored. On the one hand (seen from a user point of view), this is due to the large user community that is already very much accustomed to using ubiquitous wireless voice communications in second-generation cellular networks like GSM (Figure 10.1, evolution path 1). This means that VoIP over wireless must provide at least similar if not better perceived quality at the user level while still being spectrum efficient. On the other hand (seen from a perspective of network evolution), the third generation of wireless networks and even more the generation beyond ([1], see. Chapter 6) are moving towards an IP-oriented architecture (Figure 10.1, evolution path 2). Other network technologies (e.g., the IEEE 802.11 WLAN standard) have been designed for data traffic and are therefore already well suited for IP transport, but do not specifically support real-time services like voice.

Figure 10.2 (adapted from [2]) summarizes the challenges posed by the described evolution paths as a cube with three dimensions. While acquiring the advantage of the flexibility offered by IP-based services, additional technologies must be included to assure the service quality and the spectrum efficiency by employing IP QoS technology and speech/header

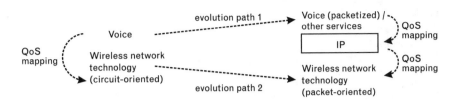

FIGURE 10.1
*Dual evolution path
for wireless VoIP.*

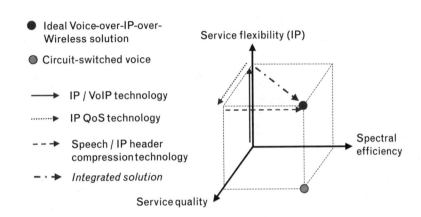

FIGURE 10.2
*VoIP over wireless
problem dimensions
[2].*

compression technology, respectively. In Figure 10.2 this is depicted by the arrows parallel to the respective dimensions of the cube. While combining separate IP QoS and speech/header compression solutions is feasible, an integrated approach symbolized by the diagonal arrow promises to be more effective.

Figure 10.3 shows the simplified architecture of the transport part of a wireless VoIP system to be used as a reference model for this chapter. Emphasis is placed on basic issues of QoS support; therefore, components that provide mobility management and call setup signaling are not considered here. Clearly, however, there are interactions between those components and the QoS support scheme, which need further consideration—for example, emergency calls that need fast call setup signaling, particular good QoS support, and preference in the mobility management (to minimize interruptions/distortions due to handover).

Basically, the architecture introduced above can be partitioned into three levels:

1. Layers 4 and above: this level contains all end-to-end-related (payload-specific) QoS functionality plus the mapping to the generic layer 3 QoS support scheme;

2. Layer 3 QoS support scheme (e.g., differentiated services);

FIGURE 10.3
Transport part of a VoIP over wireless system.

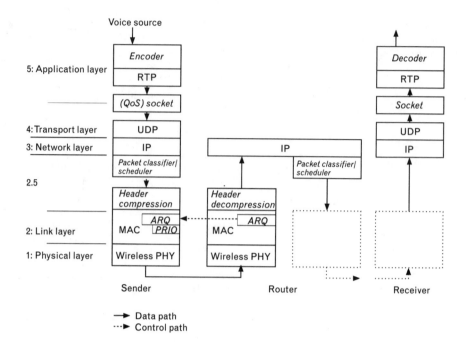

3. Layer 2 and below: this level contains all link related (network technology-specific) QoS functionality plus the mapping between the layer 3 QoS support scheme and the respective network technology, sometimes referred to as layer 2.5.

In the following section, the most important factors for speech quality in a wireless VoIP system are identified (speech coding, packet loss caused by bit errors, delay caused by the medium access). Then, related work from the areas of WLANs and cellular networks is discussed. Finally, an example for an integrated QoS support scheme to alleviate the problem of packet loss is presented, where information about the expected loss concealment performance (first level) is mapped to packet prioritization (second level), and finally to selective link layer retransmissions (third level). The evaluation shows that even just a single possible retransmission try for selected packets enhances significantly the perceptual quality in the presence of bit errors. The proposed scheme thus avoids a significant decrease in perceptual quality as well as the explosion in the number of retransmissions (which occurs if every packet is eligible for retransmission) at the same time.

10.2 Factors Influencing the Speech Quality in a Wireless VoIP System

In a wireless VoIP system the speech quality is impaired due to several causes, as illustrated in Figure 10.4. Additional impairments not depicted are due to codec tandeming/transcoding (i.e., performing several en-/decoding processes on a single voice flow) and echo effects.

For the evaluation of speech quality, the *mean opinion score* (MOS) is a widely used method. This method is based on the assignment of a subjective rating level by the user, based on an absolute scale divided into intuitively clear categories (Table 10.1).

Toll quality, which is the quality the user experiences during a telephone call, has an MOS value of 4. On cellular systems the user is accustomed to a quality of about 3.5. Lower values are often rejected as annoying and should be avoided.

Because of the time and money consumption for subjective tests, objective quality methodologies are required while developing and evaluating telecommunications systems. With this kind of method, speech quality is

FIGURE 10.4
*VoIP quality
impairments .*

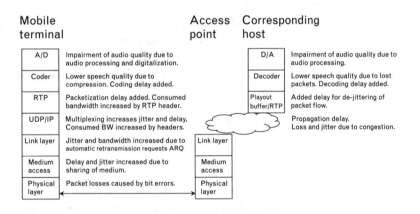

TABLE 10.1   MOS RATINGS

RATING	SPEECH QUALITY	LEVEL OF DISTORTION
5	Excellent	Imperceptible
4	Good	Just perceptible, but not annoying
3	Fair	Perceptible and slightly annoying
2	Poor	Annoying but not objectionable
1	Unsatisfactory	Very annoying and objectionable

evaluated by comparing the decoded speech signal to a reference speech signal without human intervention by applying some psycho-acoustical transformation to both signals (see Chapter 27). Recently, several methods have been developed [3] and standardized in the ITU-T.

10.2.1 Speech Codecs for Wireless VoIP

In order to reduce bandwidth consumption in the transmission of speech signals, speech coding is employed to compress the speech signals, (i.e., to use as few bits as possible to represent them while maintaining a certain desired level of speech quality). Efficient speech coding is, on the one hand, particularly important for wireless VoIP to decrease the amount of payload data and thus increase the spectral efficiency. On the other hand, speech compression needs to go hand-in-hand with header compression, because otherwise—assuming that a certain packetization delay (i.e., a certain number of speech frames per packet) is maintained—with increasing compression ratio the actual compression gain will decrease. This is the case because the size of the header becomes the dominant component of the overall packet size.

In general, speech coding techniques are divided into three categories: waveform codecs, voice codecs (vocoders), and hybrid codecs. Advanced speech coding requires a relatively high computational complexity, although today it can usually be performed with inexpensive hardware in the end systems.

Table 10.2 gives a brief overview on bit rates, applications and perceptual quality on an MOS scale.

In the following, two of the currently most popular codecs are explained in more detail.

TABLE 10.2   MOS RATINGS FOR DIFFERENT SPEECH CODECS

VOCODER	BIT RATE (IN KBPS)	SPEECH QUALITY (MOS)	APPLICATION
G.723	5.3 or 6.8	3.8	Videotelephony, VoIP
G.729	8.0	4.0	Mobile telephony, VoIP
G.711	64.0	4.5	Fixed telephone systems
GSM Half-Rate	5.6	3.5	GSM/2.5G networks
GSM EFR	12.2	4.0	GSM/2.5G networks
GSM	13.0	3.5	GSM networks
AMR	4.75–12.2	3.5–4.0	3G mobile networks

G.729

The G.729 is a hybrid codec, specified by the ITU, and employs the *Conjugate Structure Algebraic Code Excited Linear Prediction* (CS-ACELP) algorithm. The speech is partitioned into units called frames. For the G.729 these frames are 10-ms long, corresponding to 80 bits. The encoder accepts 16-bit linear PCM data sampled at 8 kHz as input data and produces 8-Kbps coded data. The codec includes loss concealment at the receiver side to cope with occasional frame losses. Concealment algorithms try to appropriately fill gaps caused by losses. This can be done, for example, by repeating the previous sounds or by interpolation. In the mean, speech quality of G.729 is about 3.7, 3.2, and 2.8 for frame loss rates of 0%, 3%, and 5%, respectively.

AMR

The AMR speech codec was originally developed for GSM but has become the mandatory codec of 3G WCDMA systems. Similar to the G.729, it provides toll quality and employs CELP-based coding. In addition, it allows the bit rate of the encoding to be dynamically changed. Thus, it can adapt to the capacity of the transmission channel. The coding rate ranges from 12.2 Kbps down to 4.75 Kbps, and frames sizes can vary between 244 and 95 bits correspondingly. The codewords of the AMR within a frame are ordered into three classes with regard to their significance. Therefore, even if some part of the frame is corrupted, other parts may be successfully decoded. In UMTS the bit rate of AMR and the wireless channel coding are adapted jointly. A lower AMR bit rate can be transmitted with a higher amount of forward error correction data, making the flow more robust against errors on the wireless link. A higher AMR bit rate can be transmitted with less redundancy, thus improving the speech quality. AMR can be employed not only for UMTS but for VoIP as well. It has a receiver side loss concealment to cope with packet losses and a voice activity detection to limit the bandwidth during silence. However, Internet protocols do not currently provide mechanisms to support the partial decoding of corrupted frames.

10.2.2   Transmission Impairments

In the following paragraphs, the major impairments for voice with regard to the transmission over wireless media are explained.

10.2.2.1   Packet Loss/Loss Correlation

Packet loss in IP-based networks often occurs when a router becomes congested—that is, it receives more packets to forward than it can process. Another reason for losses are transmission errors (bit errors) of the underlying medium. Typically, the BER is extremely low for wireline networks, but it can be significant for wireless networks. Applications can use the message sequence number of transport protocols such as RTP to detect a packet loss. In order to provide an acceptable quality, loss recovery/control must be performed. While coding schemes can exploit the redundancy within the speech efficiently for compression, together with packet loss, compression can lead to even more significant degradations of the output speech quality as for *pulse code modulation* (PCM) speech. When using, for example, a backward adaptive waveform coding scheme like *adaptive differential PCM* (ADPCM), the decoding of the next arriving packets after the loss can lead to significant distortion due to the potentially large changes in the signal amplitude. Vocoders and hybrid coders use even more adaptivity for compression; however, since the decoder state is not directly coupled to the amplitude as in ADPCM, the distortions are less dramatic. But, they might persist longer until the decoder has resynchronized with the encoder.

To summarize, it can be said that redundancies within a speech signal can be exploited both for compression and loss resilience. The higher the compression of the signal, the lower the intrinsic loss resilience. Obviously the time interval in which the decoder does not receive data from the network is a crucial parameter with regard to user perception (if a loss is perceived not at all, as a glitch, or as a dropout). The time interval at the user level translates to the burstiness of loss (loss correlation) at the packet level.

10.2.2.2   Delay

Studies have been undertaken, which verify that users expect a particular response time from their telephone partners. If the response time is not within a certain time window, the talker perceives the delay as annoying. The delay from the speaker's mouth to the ear of the listener is called one-way delay. A delay between 50 ms and 150 ms is within an acceptable range. Lower response times are not perceptible to the user. A delay between 150 ms and 400 ms shows marginal acceptance, and delay above 400 ms is prohibitive (for 3G cellular networks, 3GPP defined 400 ms as upper limit for one-way delay).

The components of the delay that a packet in a wireless VoIP system experiences can be described as follows:

- *Propagation delay (physical layer):* the time that the physical signal needs to travel across the wireless link. Propagation delay represents a physical limit given by the physical bandwidth and noise conditions in the used frequency band that cannot be reduced.

- *Serialization delay (physical layer):* the time needed to send a packet completely out on the link. This delay component is thus dependent on the packet size and on the actual link speed (which is in turn determined by the channel capacity).

- *Channel coding delay (physical layer):* the processing time needed for additional coding/decoding procedures to protect the signal from distortion on the wireless link. Additional time is consumed by (bit level) interleaving procedures that spread adjacent information bits within a frame as well as over frame boundaries to trade off the burst error probability against delay.

- *Medium access delay (MAC/RLC layer):* the time until the shared medium is declared free for transmission of a frame (which can be caused by contention or by signaling procedures). The medium access delay points to the fundamental trade-off in MAC design: for a centralized MAC approach even at low load levels, capacity is wasted for signaling. At high load levels, however, this wasted capacity is typically less than the capacity wasted by collisions during the contention phases of a distributed MAC. With a centralized MAC, the delay can be independent of the number of participating stations; however, in a distributed MAC the delay is clearly dependent on that number. This leads also to an increased value for the jitter (introduced below) for the latter case. An additional component of the medium access delay is the time needed for retransmissions that are necessary when MAC frames arrive corrupted at the receiver or collisions on the medium occur.

- *Forwarding delay (network/link layer):* the time a router or bridge takes to forward a packet.

- *Queuing and (de-)multiplexing delay (network layer and other layers):* the time a packet has to spend in queues between the protocol layers (particularly between layer 2 and 3) before it can be processed.

- *Packetization/depacketization delays (application layers):* the time needed to build data packets at the sender (await the arrival of a sufficient amount of data from the application or the upper protocol layer, compute and add headers at the respective layer), as well as to strip off packet headers at the receiver.

- *Algorithmic delay and look-ahead delay (application layer):* the time it takes to digitize speech signals and perform voice encoding at the sender. Typically, encoding works on a sequence of PCM samples (frames) so that first enough samples have to arrive. Some codecs also need to buffer data in excess of the frame size (look-ahead).

- *Decoding delay (application layer):* the time needed to perform decoding and conversion of digital data into analog signals at the receivers.

Jitter is the variability of delay (not necessarily being the delay variance). A major reason for jitter is the variation of the queuing delay component when several packets in a router compete for the same outgoing link. In wireless networks, the medium access delay as introduced above is another major contributor to the delay jitter. It should be noted that all delay components introduced above may exhibit some variations. The reasons for this may range from variable radio conditions to variable execution times for network protocol software within the operating system.

10.3   Related Work

In the previous section it was shown that speech quality impairments may arise basically at all layers of the protocol stack, though some of these impairments (like losses caused by bit errors on the physical layer and delay caused by the medium access) are dominant in wireless networks. Therefore, in this section some related work is discussed which focuses on these key problems. While there is a rich literature on QoS support for packet voice [4, 5], only recently have extended investigations in the area of wireless VoIP begun.

10.3.1   VoIP over WLANs

During recent years, IEEE 802.11 WLAN has gained increased popularity and is increasingly deployed because the price of equipment has dropped and it is easy to install and maintain. WLAN might be a cost-saving and local alternative to cellular wireless networks. The main use of wireless LAN, however, has been limited to pure data services like Web access, e-mail, and file transfers. Voice transmission over WLAN has not been considered to the same extent.

The IEEE 802.11 specification defines different modes of operation to support data and voice transmission at the same time, but until recently no commercial product supported both *distributed coordination function* (DCF) for best-effort traffic and the *point coordination function* (PCF) for prioritized data like voice.

In [6] the authors present an analytical evaluation of the performance of packet voice transmission using the PCF of an 802.11 WLAN. They use the PCF polling mode because it offers a packet switched but connection-oriented service, which seems to be a more suitable approach for telephony than a connectionless, contention-based mode like the DCF. Veeraraghavan, et al. show that [while the PCF is a *constant bit rate* (CBR) transmission mode which does not exploit voice activity detection schemes], a reasonable number of concurrent voice calls can be accommodated at an access point. They also point out, however, the need for error recovery by FEC, *automatic repeat requests* (ARQs), or loss concealment due to the high error rates experienced.

Köpsel and Wolisz [7] find that the DCF function provides a sufficiently low channel access delay for voice if the link rate is at least 11 Mbps (but not necessarily for lower link rates). The incurred jitter, however, is significant and thus (dependent on the playout buffer algorithms) might lead to a high resulting end-to-end delay. The authors showed, however, that a local priority scheduling in connection with prioritized medium access could significantly improve these delay values. In terms of absolute delay and jitter, better results are achieved using the PCF rather than DCF. The overall goodput for link rates below 11 Mbps is lower, however, than that for the DCF, whereas the opposite is true for higher link rates. In [8] the same authors determine when to switch from DCF to PCF mode of operation depending on the offered load. The reason for a potentially low goodput using PCF is unsuccessful polling events (i.e., a station is polled but has no data to send). As a remedy, Köpsel and Wolisz propose to add and remove stations from the polling list depending on the state (talk spurt, silence) of the VoIP flow and demonstrate some enhancement because of this scheme.

While it has been shown that the PCF leads to lower delay values than the DCF, DCF-based approaches still seem attractive due to their distributed nature, which can be essential, especially in ad hoc network environments. Furthermore, using differentiation at the MAC level between traffic classes in connection with DCF can even lead to a better delay performance when compared to using PCF for all traffic. This has been shown in [9] where the minimum size of the *contention window* (CW) is changed depending on the priority. This approach corresponds to the enhanced DCF (EDCF) proposed within the 802.11 working group. Aad and Castellucia [10] and Barry et al. [11] also present MAC-level prioritization schemes that are realized by modifying the scaling of the contention window, the *DCF interframe spacing* (DIFS) time interval and the maximum supported frame length without fragmentation dependent on the priority. In [10] it is shown that while all three approaches are comparable in their network utilization, the DIFS-based prioritization is superior in terms of throughput stability and operation when the bit error rate is significant.

Hoene, Carreras, and Wolisz [12] present an approach to map the importance of voice packets onto the behavior of the 802.11 MAC. In addition to the retransmission of higher priority packets at the MAC layer, they also employ redundant transmissions at the application level as well as a combined solution. In contrast to the two QoS mappings proposed in Figure 10.1, a direct mapping of the application-level preference to the link level is proposed. This avoids any modification (priority marking) to the IP packets (it is a transparent "protocol booster" approach). Hoene et al. also present measurement results for the DCF mode in a configuration where losses are introduced by actual channel errors as well as a simulated play-out buffer (not by contention at the MAC level), also demonstrating a performance gain when employing selective packet protection.

10.3.2   VoIP over Wireless WANs (Cellular Networks)

Schieder and Ley [13] present an analysis of VoIP in a GPRS environment focusing on the achievable minimum end-to-end delay. To send data in GPRS, it is necessary to establish a *temporary block flow* (TBF) by an exchange of signaling messages. However, the timers for the release of the TBF have been adjusted for non-real-time applications typically generating large data packets (i.e., the TBF may even be established/released on a per-packet basis). Therefore, for a VoIP flow using a low-bit rate codec with silence detection showing on-off-behavior, it frequently happens that the TBF is released between and even within talk spurts due to the lack of data and then needs to be reestablished, thus generating a lot of delay as well as signaling overhead. This is particularly the case when the available data rate (the rate with which the sender queue is cleared) is significantly higher as the required bit rate for the VoIP flow, because the TBF release timer starts earlier. Because establishing a TBF takes around 100 ms, the bound on one-way delay introduced in Section 10.2.2.2 is frequently exceeded. To overcome the described problem, the authors propose a delayed release of the TBF or a "semipermanent" TBF, where the established TBF is kept for duration of a call and resources may be dynamically assigned to such an established TBF. Another important measure is admission control to limit the number of voice sessions in a cell limiting the end-to-end delay to a reasonable value. But at the same time, additional best-effort traffic should be accepted to increase the overall network utilization. Then, it must be assured that the best-effort traffic does not interfere too much with the VoIP traffic by an appropriate multiplexing/prioritization scheme.

Svanbro et al. [2] highlight the trade-off between the IP service flexibility on the one hand and the thus imposed new challenges of assuring a certain speech quality and spectral efficiency on the other hand. They simulate different header compression schemes and compare their performance relative to a UMTS circuit switched voice service. The result is that without any header compression the relative capacity drops to about 50%, whereas with header compression it can be maintained at 80% to 90%. In addition, the authors emphasize the importance of a traffic description, which can then be appropriately mapped onto the attributes of a UMTS radio bearer (this problem is addressed in detail in [14]). Furthermore, they propose to multiplex different flows onto the same radio bearers, thus reducing the overall number of bearers. Hereby, all flows within the aggregate should belong to a certain class of traffic (real-time, non-delay-critical signaling, delay-critical signaling). The class properties can then be reflected by certain properties of the radio bearer (e.g., no or very few retransmission for the real-time traffic class).

Poppe et al. [15] compute the mouth-to-ear delay for a VoIP call that traverses a UMTS network, an IP backbone, and is then routed to the

PSTN. Their evaluation is based on the "E-model" which assesses jointly the impact of loss, delay, noise, and echo on user perception. The authors express the E-model quality as a function of the mouth-to-ear delay, where the latter is dependent on the UMTS interleaving span (transmission time interval), the number of voice frames per packet, and the (non-)presence of header compression. Thus, it is possible to find an operating point as a compromise between quality and bandwidth efficiency. Poppe et al. find the fact that header compression is enabled or not influences the bandwidth efficiency more than the number of voice frames per packet.

10.4 QoS for Wireless VoIP Using Selective Packet Prioritization

Because of the complexity of the QoS support problem for wireless VoIP, which has been described in the previous sections, it can be concluded that a QoS support scheme needs to take the whole protocol stack (Figure 10.3) into account. First, the resiliency of applications to loss/delay should be exploited. Then, QoS requirements of the application (the codec) should be known at lower layers, although this should be done as generically as possible. This allows a mapping to several wireless network technologies. To demonstrate a particular example of an approach with the described properties (focusing on the packet loss problem), QoS support for wireless VoIP using selective packet prioritization is introduced in this section. Selective packet prioritization can be regarded as one form of unequal error protection (see Chapter 8).

10.4.1 Analysis of the G.729 Frame Loss Concealment

In addition to packet losses caused by the properties of the wireless link (contention for media access, bit errors), speech packets can be discarded when routers or gateways are congested, as well as when they arrive late at the receiver (i.e., their playout time has already passed). Considering highly adaptive coding schemes like those introduced in Section 10.2.1, packet loss results in loss of synchronization between the encoder and the decoder. Thus, degradations of the output speech signal occur not only during the time period represented by the lost packet, but also propagate into following segments of the speech signal until the decoder is resynchronized with the encoder. To alleviate this problem, the introduced category of codecs typically contains an internal (codec-specific) loss concealment algorithm at the decoder.

The studies here are limited to one codec, G.729. The reason for this restraint was that the G.729 has already been investigated to a certain extent.

Rosenberg [16], Sun [17], and Sanneck [18] used it for similar measurements. Considering only one codec allowed them to increase the accuracy by conducting more measurements with the same parameter sets. If the measurements are repeated using other codecs that belong to the same class of codecs [based on *linear prediction* (LP), particularly CELP-based ones], they will likely lead to comparable or similar results [17].

To cope with occasional frame losses, the G.729 codec includes at the receiver side loss concealment. Concealment algorithms try to appropriately fill gaps caused by losses. This can be done, for example, by reusing the previous internal decoder state (extrapolation). The applicability of the G.729 concealment has been shown to be dependent on the location of the loss within the speech signal. In [18] the resynchronization time of the decoder after a number of consecutive frames are lost has been measured. The position of the frame loss is varied to cause a number of consecutive voiced/unvoiced frames to be lost and then count the number of the following frames until the *signal-to-noise ratio* (SNR) exceeds a certain threshold. An appropriate value for this threshold is considered to be 20 dB. That is, the G.729 decoder is said to have resynchronized with the G.729 encoder after the loss of a number of frames when the energy of the error signal falls below 1% of the energy of the decoded signal without frame loss. Figure 10.5 shows the resynchronization time plotted against the loss position. The speech sample is produced by a male speaker where an *unvoiced-voiced* (uv) transition occurs in the eighth frame.

Figure 10.5 shows that a loss of a consecutive number of frames at different positions has significantly different levels of impact on the error introduced into the speech signal and thus on speech quality. The loss of unvoiced frames seems to have a rather small impact on the speech quality, and the decoder recovers the state information fast thereafter. However, the loss of voiced frames causes a larger degradation of the speech quality, and

FIGURE 10.5
Resynchronization time (in frames) of the G.729 decoder after the loss of k consecutive frames (k ∈ [1,4]) as a function of frame position.

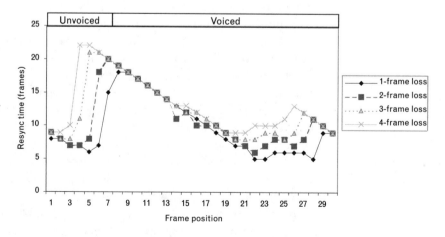

the decoder needs more time to resynchronize with the sender. Moreover, the loss of voiced frames at an unvoiced/voiced transition leads to a very significant degradation of speech quality. The two above experiments have been repeated for different male and female speakers with similar results. The above phenomenon could be explained as follows:

- Because voiced sounds have a higher energy and are also more important to the speech quality than unvoiced sounds, the loss of voiced frames causes a larger degradation of speech quality than the loss of unvoiced frames.

- Because of the periodic property of voiced sounds, the decoder can successfully conceal the loss of voiced frames once it has obtained sufficient information on them.

- The decoder fails to conceal the loss of voiced frames at an unvoiced/voiced transition because it attempts to conceal the loss of voiced frames using the filter coefficients and the excitation for an unvoiced sound. Moreover, because the G.729 encoder uses a moving average filter to predict the values of the line spectral pairs and only transmits the difference between the real and predicted values, it takes a lot of time for the decoder to resynchronize with the encoder once it has failed to build the appropriate linear prediction filter.

10.4.2 Selective Packet Marking/Prioritization

The result described on the ability of the G.729 decoder to conceal packet loss is exploited to develop a new packet marking/prioritization scheme called *Speech Property-Based Selective Packet Marking* (SPB-MARK), which employs two priorities. The SPB-MARK scheme concentrates the high-priority packets on the frames essential to the speech signal and relies on the decoder's concealment for other frames.

Figure 10.6 shows the simple algorithm written in a pseudo-code that is used to detect a uv transition and protect the voiced frames at the beginning of a voiced signal. In the algorithm, the procedure *analysis()* is used to classify a block of k frames as voiced, unvoiced, or uv transition. The procedure *send()* is used to send a block of k frames as a single packet with the appropriate priority (either +1 or 0). N is a predefined value that defines how many frames at the beginning of a voiced signal are to be protected. Simulations have shown that the range from 10 to 20 are appropriate values for N (depending on the network loss condition). In the simulation presented in Section 10.4.4, $k = 2$ has been chosen, a typical value for interactive speech transmissions over the Internet (20 ms of audio data per packet). A larger number k would help to reduce the relative overhead of the protocol header but would increase the packetization delay and make sender classification and receiver concealment in case of packet loss (due to a large loss gap) more difficult.

The network priorities are then enforced by the appropriate traffic shaping/policing mechanisms at the network nodes and/or are mapped to available lower layer traffic control mechanisms. Here, a mapping on error control mechanisms of a wireless link layer is considered.

10.4.3 System Architecture

Section 10.4.1 has shown that some segments of the signal are essential to the speech quality, while others, in the event of a packet loss, can be extrapolated at the receiver from data received earlier. This knowledge has been exploited by the design of the SPB-MARK algorithm, as described in the previous section. Thus, the requirements of the voice application from the network in terms of the reliability of packet delivery can be reduced. This appears to be particularly useful for wireless networks, where losses may be caused by channel fading in addition to congestion (due to contention for the media access).

Therefore, it is important that a voice application can make its QoS requirement known to the network on a per-packet (rather than per-flow) basis. However, this should be done in a generic way at the network layer. This allows the conveyance of QoS requirements known only at the source to other QoS enforcement entities in the network (particularly a wireless last hop). Additionally, it is possible to map the per-packet QoS requirements to different networking technologies. The differentiated service architecture developed within the IETF provides such a QoS assurance on a per-packet basis. In addition to fulfilling the requirement stated above, scheduling classes of packets instead of individual flows inside the network has better scaling properties by only maintaining state and enforcing QoS for traffic aggregates.

The approach introduced in Section 10.4.2 can be seen as an approximation to a joint source/channel coding system. In such systems, if the source coding (e.g., speech) and channel coding on the wireless link are developed hand-in-hand, the resulting transmitting quality is superior to any general-purpose wireless system, which splits source and channel coding. The trade-off here is in the flexibility of the (layered) architecture (supporting several applications with generic QoS mechanisms) versus the achievable performance (a system where the channel coding is optimized for the only existing application running directly on top of the wireless link is better performing than an IP-based system).

FIGURE 10.6
SPB-MARK
pseudo-code.

```
protect = 0
foreach (k frames)
        classify = analysis(k frames)
        if (protect > 0)
                if (classify == unvoiced)
                        protect = 0
                        send(k frames, "0")
                else
                        send(k frames, "+1")
                        protect = protect-k
                endif
        else
                if (classify == uv_transition)
                        send(k frames, "+1")
                        protect = N-k
                else
                        send(k frames, "0")
                endif
        endif
endfor
```

Figure 10.7 shows the considered system architecture[1] with the parts modified from Figure 10.3 shaded, and it also depicts the considered interlayer interfaces. Note that the figure only shows the application of the proposed scheme (SPB-MARK, ARQ) at the first hop, while the ARQ scheme will typically also operate at a wireless last hop.

In the architecture, DiffServ packet priorities are mapped to QoS control mechanisms of the layers below. This is particularly important for wireless link layers, because prioritized packets not only need to be protected against other flows using the shared medium (PRIO interface in Figure 10.7) but also need to be protected against channel errors using ARQ and/or FEC. (Only channel errors are considered here: see, for example, [10, 11], for mapping DiffServ to link-level priority in an 802.11 WLAN to prioritize the media access.) This mapping then in turn has the implication that the real-time constraint of the application has to be taken into account (particularly when retransmitting packets). Thus, a comprehensive approach covering an application-, internetworking-, and link-level view is enabled.

10.4.4 Simulations

Figure 10.8 shows the simulation setup consisting of a sender and receiver connected over a single wireless link. It comprises the following major components (note that only a small subset of the functionality of the

---

1   Header compression is not considered here. Clearly, some knowledge of the varying importance of packets could also be exploited to adjust the robustness of a header compression algorithm.

FIGURE 10.7
System architecture.

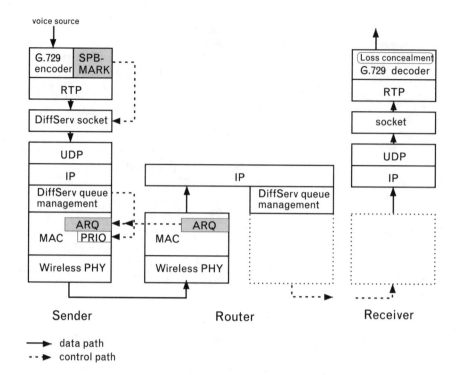

components and protocols shown in brackets needs to be implemented for the simulation):

- The ITU-T reference implementation of the G.729 encoder and decoder[2];

- The SPB-MARK algorithm;

- The ARQ algorithm: at the wireless link layer, only the packets selected by the prioritization scheme (SPB-MARK) are retransmitted. Therefore, this scheme is called SPB ARQ. The results are compared to the FULL ARQ method (i.e., every packet is eligible for retransmission). For both schemes a simple send-and-wait ARQ mechanism with a limit on the maximum possible number of retransmissions of a single packet due to the tight real-time constraint of voice is considered;

- An error model for a single wireless link: a simple Bernoulli model for bit errors, which lead to frame corruptions and finally packet losses (the effects of packet loss due to buffer overflow in routers are studied

2   http://www.itu.int/itudoc/itu-t/rec/g/g700-g799/software/g729.

in [18]) is considered. The packet loss probability can be computed as follows:

$$p_L = 1 - (1 - \text{BER})^{s/\text{bit}}$$

BER is the bit error rate, and s is the voice packet size at the PHY level. For the simulations an s value of 944 bits has been used: the header sizes are 24, 34, 20, 8, 12 bytes at the PHY, MAC, IP, UDP, and RTP-layer, respectively (no header compression is used). The packet payload comprises 20 bytes corresponding to $k = 2$ voice frames (see Figure 10.6).

· Objective speech quality measurement for a comparison between the decoded voice data distorted by the voice encoding/decoding procedure with and without the simulated channel errors. Novel objective quality measures are employed that attempt to estimate the subjective quality as closely as possible by modeling the human auditory system. Thus, the necessity for extensive subjective testing can be avoided. In the *evaluation the enhanced modified bark spectral distortion* (EMBSD) [3] and the *measuring normalizing blocks* (MNBs) described in Appendix II of the ITU-T Recommendation P.861 are used. These two objective quality measures are reported to have a very high correlation with sub-

FIGURE 10.8
Simulation setup.

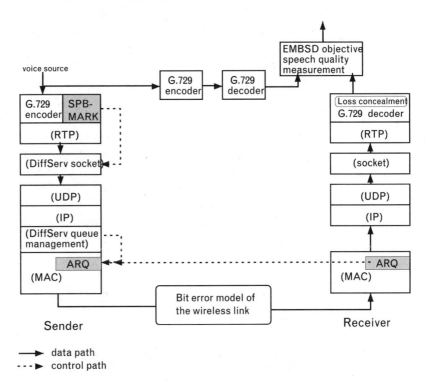

WIRELESS IP AND BUILDING THE MOBILE INTERNET

jective tests. Their relation to the range of subjective test result values (MOS) is close to being linear, and they are recommended as being suitable for the evaluation of speech degraded by transmission errors in real network environments such as bit errors and frame erasures. Only EMBSD results are shown; however, the results using the MNB measure are virtually identical.

The simulation setup allows us to map a specific PCM signal input[3] together with network model parameters to a speech quality measure. While using a simple loss characterization for the wireless link, a large number of loss patterns are generated by using different seeds for the pseudo-random number generator (for the presented results, 300 patterns are used for each simulated condition). This procedure takes into account that the input signal is not homogenous (i.e., a packet loss burst within one segment of the speech signal can have a largely different perceptual impact than a loss burst of the same size within another segment).

Figure 10.9 shows results for the perceptual distortion (i.e., the user-level quality), where the BER and the maximum possible number of retransmissions (r_{max}) of a single packet are varied. Also, a scale is shown for the corresponding MOS value for subjective tests ranging from 1 (unsatisfactory quality) to 5 (excellent quality).

Judging from the vertical distance between the curve for $r_{max} = 0$ (no retransmissions) and the curves for the ARQ schemes, it can be observed that retransmissions generally can improve the perceptual quality significantly. For $r_{max} = 1$, the performance of SPB ARQ is very similar to that of the FULL ARQ scheme. By increasing r_{max}, the performance gap between SPB ARQ and FULL ARQ for higher bit error rates increases. This gap, however, must be seen in relation to the necessary number of retransmissions to achieve the perceptual quality figures shown. Figure 10.10, therefore, gives the number of actual retransmissions normalized to the total number of packets of a flow (100% would mean that the total number of packets transmitted including retransmissions has doubled).

Here it can be seen that the number of retransmissions for SPB ARQ for low bit error rates is similar to the FULL ARQ method. For an increasing BER, however, the increase in the number of retransmissions is much lower than with FULL ARQ. Also, when increasing r_{max}, the necessary overall number of retransmissions for SPB ARQ is much lower than for the FULL ARQ scheme.

Summarizing, it is suggested that for both ARQ schemes at least one retransmission should be allowed for the higher priority packets. For SPB

---

3   The used speech material contains different male and female voices. The length is 11.25 seconds. The sample can be obtained at http://www.dvsinc.com/speech/orig_dam.zip.

FIGURE 10.9
Perceptual distortion (EMBSD) of the ARQ schemes.

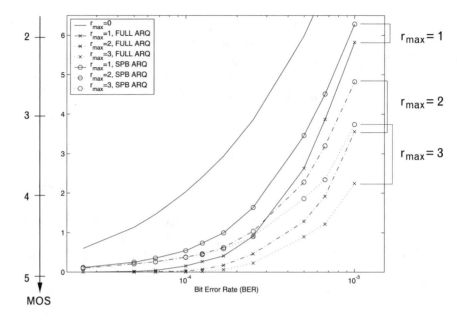

FIGURE 10.10
Relative number of retransmissions for the ARQ schemes (percentage).

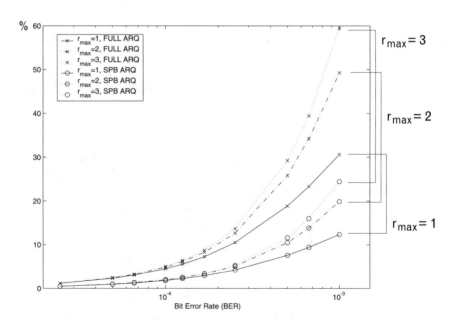

ARQ, r_{max} can be safely set to a value of 3 also for higher BER values (close to 10^{-3}). For such high bit error rates, this avoids the significant decrease in perceptual quality (which occurs for low values of r_{max} with SPB ARQ) as well as the explosion in the number of retransmissions (which occurs for high values of r_{max} with FULL ARQ). Thus, SPB ARQ offers a reasonable cost (additional processing to derive the packet priority, necessary retransmissions) versus quality trade-off, particularly for high bit error rates. It should also be noted that a lower number of retransmissions reduces the probability of congestion in the wireless network and improves the delay performance by avoiding unnecessary retransmissions, which would block the transmission of the next high-priority packet.

10.4.5   Conclusions

The impact of packet loss at different positions within a speech signal on the perceived quality for the G.729 codec has been investigated. It has been shown that the loss of voiced frames after an unvoiced/voiced transition leads to a significant degradation of the speech quality, while the loss of other frames is concealed rather well by the decoder's concealment algorithm.

Based on this result, a *selective packet prioritization scheme* (SPB-MARK) has been developed that protects the packets that are essential to the speech quality by marking them with a higher DiffServ priority, while relying on the decoders concealment in case other low-priority packets are lost. The priorities are applied at the wireless link layer to identify and retransmit only these essential packets. This architectural separation allows the conveyance of QoS requirements known only at the source to other QoS enforcement entities in the network (in particular, a wireless last hop can thus apply similar error control mechanisms as the wireless first hop).

Simulations using a simple network model and subsequent evaluation using objective speech quality measures show that even just a single possible retransmission enhances significantly the perceptual quality in the presence of bit errors. Particularly for higher BERs, the proposed selective ARQ scheme (SPB ARQ) avoids a significant decrease in perceptual quality as well as the explosion in the number of retransmissions (which occurs if every packet is eligible for retransmission). Thus SPB ARQ offers a reasonable cost versus quality trade-off.

Generally, the properties of the proposed system architecture can be summarized as follows: by employing DiffServ as the layer 3 QoS interface, it is possible to map per-flow and/or per-packet QoS requirements to different networking technologies with different QoS/MAC concepts. It would, for example, be possible to use the layer 3 priority as an input to a channel-state dependent packet scheduling algorithm to match application with link characteristics. Thus, it can be said that an abstraction from specific applications as well as specific link layers is possible. If VPN- or Internet-wide DiffServ is available, the mapping can also be exploited in the fixed network. Furthermore, it is possible to locally differentiate (at layer 3) between flows of one mobile/access point using simple scheduling/queue management mechanisms.

REFERENCES

[1] Wireless World Research Forum (WWRF), Alcatel, Ericsson, Motorola, Nokia, Siemens, http://www.wireless-world-research.org/, 2002.

[2] Svanbro, K., J. Wiorek, and B. Olin, "Voice-over-IP-over-Wireless," *11th IEEE International Symposium on Personal, Indoor and Mobile Radio Communications*, London, England, September 2000.

[3] Yang, W., M. Benbouchta, and R. Yantorno, "Performance of the Modified Spectral Distortion as an Objective Speech Quality Measure," *Proc. ICASSP*, Vol. 1, Seattle, WA, May 1998.

[4] Perkins, C., O. Hodson, and V. Hardman, "A Survey of Packet-Loss Recovery Techniques for Streaming Audio," *IEEE Network Magazine*, September/October 1998.

[5] Kostas, T., et al., "Real-Time Voice over Packet-Switched Networks," *IEEE Network Magazine*, Vol. 12, No. 1, January/February 1998.

[6] Veeraraghavan, M., N. Cocker, and T. Moors, "Support of Voice Services in IEEE 802.11 Wireless LANs," *Proc. Infocom*, Anchorage, AK, April 2001.

[7] Köpsel, A., and A. Wolisz, "Voice Transmission in an IEEE 802.11 WLAN-Based Access Network," *Proc. WOWMOM 2001*, Rome, Italy, July 2001, pp. 24–33.

[8] Köpsel, A., J.-P. Ebert, and A. Wolisz, "A Performance Comparison of Point and Distributed Coordination Function of an IEEE 802.11 WLAN in the Presence of Real-Time Requirements," *Proc. 7th Intl. Workshop on Mobile Multimedia Communications (MoMuC 2000)*, Tokyo, Japan, October 2000.

[9] Lindgren, A., A. Almquist, and O. Schelén, "Evaluation of Quality of Service Schemes for IEEE 802.11 Wireless LANs," *Proceedings of the 26th Annual IEEE Conference on Local Computer Networks (LCN 2001)*, Tampa, FL, November 2001.

[10] Aad, I., and C. Castellucia, "Differentiation Mechanisms for IEEE 802.11," *Proceedings IEEE Infocom*, Anchorage, AK, April 2001.

[11] Barry, M., A. Campbell, and A. Veres, "Distributed Control Algorithms for Service Differentiation in Wireless Packet Networks," *Proceedings IEEE Infocom*, Anchorage, AK, April 2001.

[12] Hoene, C., I. Carreras, and A. Wolisz, "Voice over IP: Improving the Quality over Wireless LAN by Adopting a Booster Mechanism—An Experimental Approach," *Proc. ITCOM 2001*, Denver, CO, August 2001.

[13] Schieder, A., and T. Ley, "Enhanced Voice over IP Support in GPRS and EGPRS," *Proceedings IEEE Wireless Communications and Networking Conference (WCNC)*, Chicago, IL, September 2000, pp. 803–808.

[14] Crow, B. P., et al., "Investigation of the IEEE 802.11 Medium Access Control (MAC) Sublayer Functions," *Proceedings IEEE Infocom*, April 1997.

[15] Poppe, F., D. De Vleeschauwer, and G. Petit, "Choosing the UMTS Air Interface Parameters, the Voice Packet Size and the Dejittering Delay for a Voice-over-IP Call Between a UMTS and a PSTN Party," *Proceedings IEEE Infocom*, Anchorage, AK, April 2001.

[16] Rosenberg, J., *G.729 Error Recovery for Internet Telephony*, Technical Report CUCS-016-01, Columbia University, 2001.

[17] Sun, L., et al., "Impact of Packet Loss Location on Perceived Speech Quality," *2nd IP-Telephony Workshop*, New York, April 2001, pp. 114–122.

[18] Sanneck, H., et al., "Intra-Flow Loss Recovery and Control for VoIP," *Proceedings ACM Multimedia 2001*, Ottawa, Ontario, Canada, September 2001.

CHAPTER 11

Delivering QoS in Mobile Ad Hoc IP Networks

Gerben Kuijpers, Thomas Toftegaard Nielsen, and Ramjee Prasad

11.1 Introduction

An ad hoc network is a dynamic multihop wireless network that can be established by a group of mobile nodes without the aid of any preexisting network infrastructure or centralized administration. Although an ad hoc network can operate independently of any fixed network infrastructure, it can have access to the fixed network infrastructure via one or more wireless access points.

The largest part of the current research on ad hoc networks deals with mechanisms to provide connectivity between the different mobile nodes, similar to how routing protocols in the fixed Internet provide connectivity.

IP forwarding in both ad hoc networks and the fixed Internet offers an unreliable, connectionless network-layer service that is subject to packet loss, reordering, and packet duplication. Because of the lack of any guarantees, the traditional IP delivery model is often referred to as providing best-effort service. While this service model works well for legacy Internet applications such as e-mail, remote access, and file sharing, for more demanding applications, such as voice over IP, audio and video streaming, it will not suffice. Each application has its own set of requirements on available bandwidth (i.e., bit rate), packet loss, latency, and jitter. Thus, contrary to the best-effort approach, the network will have to provide the appropriate service to fit the application's requirements. For those more demanding applications, the network needs to be able to provide some form of resource assurance. Moreover, the network needs to treat packets of different applications in a different way. This is referred to as service differentiation. The capability of providing resource assurance and service differentiation in a network is often referred to as QoS.

Since it is expected that more demanding applications will find their way to ad hoc networks as well as to the fixed Internet, QoS mechanisms need to be available for both. During the last decade much research has been going on to provide QoS in the fixed Internet, while the same has only recently started for ad hoc networks. Any solutions for providing QoS in ad hoc networks that are connected to the fixed Internet must be able to cooperate with the QoS mechanisms in the Internet. This is illustrated in Figure 11.1. The partial QoS mechanisms on the path between source and destination need to cooperate to be able to provide end-to-end QoS for the connection. Mapping of the ad hoc QoS mechanisms to the specific link layers is an important QoS issue. However, link layer QoS and its interaction with IP layer QoS is not within the scope of this chapter.

This chapter is meant as a tutorial on QoS support in IP-based ad hoc networks. The remainder of the chapter is divided into four parts:

1. Section 11.2 introduces ad hoc networks, including ad hoc network characteristics, routing protocols, and addressing issues.

2. Sections 11.3, and 11.4 deal with QoS issues in general and applied to ad hoc networks, where:

   • Section 11.3 introduces the general frameworks for QoS support, also referred to as QoS models and their applicability to ad hoc networks.

   • Section 11.4 deals with QoS signaling and the issues with signaling in ad hoc networks.

3. Section 11.5 describes routing mechanisms that search for routes in ad hoc networks according to certain QoS requirements, so-called QoS routing mechanisms.

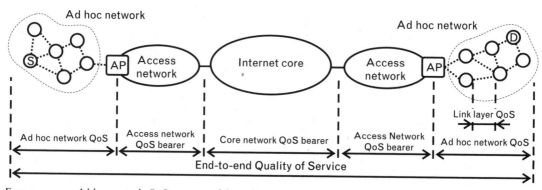

FIGURE 11.1   *Ad hoc network QoS as a part of the end-to-end QoS for ad hoc networks that are connected to the fixed Internet infrastructure by means of an access point (AP). In order to provide end-to-end QoS on the path from node S to node D, QoS at different levels of the network architecture is involved. The different partial QoS mechanisms need to cooperate to provide end-to-end QoS.*

4. Section 11.6 and 11.7 conclude the chapter with a discussion on future directions and a summary.

11.2 Mobile Ad Hoc Networks

A mobile ad hoc network consists of a number of mobile nodes that have one or more wireless interfaces, possibly based on different technology. The ad hoc network is an autonomous system of mobile nodes that can operate independently of any fixed network infrastructure. The network may, however, be interconnected with the fixed Internet. Mobile nodes with multiple wireless interfaces can interconnect the different wireless networks. Similar to how IP interconnects different networks in the fixed Internet, it can also provide connectivity in heterogeneous wireless ad hoc networks. An example is shown in Figure 11.2, where WLAN and Bluetooth are used as ad hoc wireless networks and UMTS is used as a wireless access network.

11.2.1 Mobile Ad Hoc Network Characteristics

The following characteristics are typical for ad hoc networks:

- *Multiple wireless links:* A path between a mobile node in the ad hoc network and any other node typically consists of multiple wireless links. Each link is highly bandwidth constrained and highly prone to errors. Lower layer techniques such as forward error coding, interleaving, and link-layer retransmissions can reduce the error rate of a wireless link, but only at the cost of additional delay.

FIGURE 11.2
Illustration of how IP can provide end-to-end connectivity in heterogeneous wireless networks using Bluetooth, WLAN, and UMTS. The nodes with multiple wireless interfaces interconnect the different link layer technologies to one IP network. The gateways provide access to the fixed Internet.

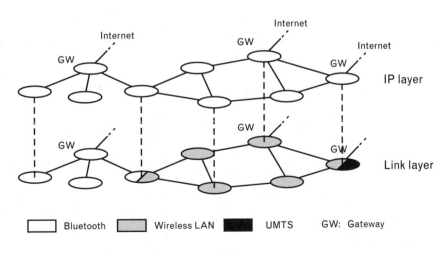

- *Dynamic topology:* Since each mobile node is free to move, the network topology may change randomly. As a result, optimum paths between mobile nodes can vary constantly, putting high demands on routing protocols.

- *Energy constrained operation:* The mobile nodes typically rely on batteries for their operation. Thus, conservation of energy is a very important design issue in mobile ad hoc networks.

- *Limited resources:* In addition to limited battery power, smaller mobile nodes, such as telephones, also have critical limitations concerning processing power and memory capacity. As a consequence, the network functionality implemented in those nodes needs to be as simple as possible.

Due to the characteristics mentioned, it is not trivial to provide connectivity between nodes in ad hoc networks, let alone provide QoS support. The nodes can have different radio ranges, link capacity, and access techniques. Moreover, the battery power, processing power, and memory capacity of the mobile nodes can vary greatly.

11.2.2   Ad Hoc Routing Protocols

Routing protocols provide connectivity in IP networks. Numerous routing protocols have been designed for ad hoc IP networks. They can be categorized into proactive and reactive protocols. In the proactive approach, each mobile node maintains a route to every other node in the ad hoc network at any moment. The *Optimized Link State Routing* (OLSR) Protocol [1] is an example of a proactive routing protocol. The reactive protocols set up routes on–demand, at the moment packets have to be sent. An example of a reactive routing protocol is *Ad Hoc On-Demand Distance Vector* (AODV) routing protocol [2].

The main advantage of proactive routing protocols is that routes to every node are instantly available, while reactive protocols introduce extra delay since routes are discovered on demand. Reactive protocols, however, are more efficient in their signaling than proactive protocols; this is especially true for lightly loaded networks. Which of the two approaches performs best depends on the specific scenario with respect to traffic distribution and node mobility.

Current ad hoc routing protocols focus on providing connectivity between nodes in the network. Typically, shortest paths are found in terms of the number of wireless hops. No QoS parameters, such as bandwidth, delay, and delay jitter, are taken into account by those protocols.

11.2.3 Addressing Issues and Mobility Management

In an IP-based ad hoc network that has no connections to external IP networks, each node needs a unique IP address, which serves as a node identifier. Such addresses can be fixed for each node in, for example, private ad hoc networks. All ad hoc routing protocols described in Section 11.2.2 provide per-node routes, so for those routing protocols it is sufficient that the uniqueness of the IP address of the nodes is assured.

When the ad hoc network is connected through one or more gateways to the fixed Internet, uniqueness of the IP addresses of the ad hoc network nodes is not sufficient. The addressing architecture of the fixed Internet is hierarchical, and an IP address functions both as an endpoint identifier and as a routing directive. This is no problem in fixed networks, since the identity of a node maps to a fixed position in the network. Mobile nodes, however, will need a new, topologically correct IP address every time they connect to a different network. This change of IP address will break any TCP connections running between the mobile node and some other Internet node. Moreover, the mobile node will not be reachable anymore for any other node that sends packets to the previous IP address of the mobile node.

Mobile IP [3, 4] solves this problem for a single mobile node, connected via a single-hop wireless link, by using a two-tier addressing scheme (see Figure 11.3). The mobile node has a permanent IP address, its home address, and a temporary address, the topologically correct COA. An entity in the home network of the mobile node, the home agent, takes care of the mapping between the two addresses and the forwarding of packets to the current location of the mobile node. Each time the mobile nodes moves to a new network and gets a new COA, it registers this address with its home agent. The message flow of Mobile IP is shown in Figure 11.3.

11.3 QoS Models

A QoS model specifies the architecture of a network with respect to what services can be provided. It can also be referred to as the QoS framework, describing how resource assurance and service differentiation are dealt with in the network. This encompasses the different QoS aspects, such as QoS signaling, QoS routing, and interaction with link layer QoS. QoS signaling and QoS routing are the subject of Sections 11.4 and 11.5, respectively, while interaction with link layer QoS is not within the scope of this chapter. The separation between QoS signaling, QoS routing, and the QoS model is based on [5].

The datagram model of the Internet is based on simplicity and the ability to adapt to changes in network topology, but it does not make any

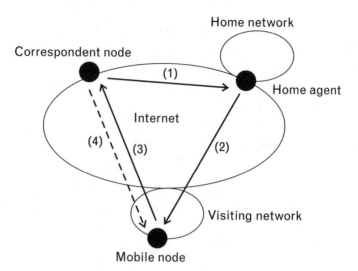

FIGURE 11.3   *Mobile IP packet flows. The mobile node is assumed already to have registered its current position with its home agent. Traffic destined for the mobile node is intercepted by the home agent (1) and tunneled to the mobile node's current location (2). The mobile node can directly send packets to the correspondent node (3). Optionally, the mobile node can update the correspondent node about its COA, so consequent routing through the home agent can be avoided (4).*

guarantees on the delivery of packets or on the service that those packets receive in terms of delay, delay jitter, or cost. This is commonly referred to as the best-effort service model. The best-effort model works well for most of the early applications such as file transfer or e-mail. However, newer and upcoming applications such as IP telephony, audio and video streaming, and interactive multimedia applications require different, more stringent services from the network. Therefore, two new architectures were developed during the last decade in the Internet community: integrated services and differentiated services. In the remainder of this section these two service architectures will be described and their applicability to ad hoc networks will be discussed. Finally, a service model that combines features from both integrated services and differentiated services and is designed for ad hoc networks will be introduced.

11.3.1   Integrated Services

In the early 1990s the IETF started the integrated services (IntServ) working group to standardize a new resource allocation architecture and new service models for the Internet. The IntServ model that resulted from this work [6] includes best-effort service, real-time service, and controlled link sharing.

The reference implementation framework for the IntServ model includes four components: the packet scheduler, the classifier, the admission control routine, and the reservation setup protocol. This is shown in Figure 11.4. In the following clarification a flow is defined as a distinguishable stream of related packets that results from a single user activity and requires the same QoS. The router function that creates different qualities of service is called traffic control. Traffic control consists of three components:

- *Packet scheduler:* The packet scheduler manages the forwarding of different flows using a set of mechanisms like queues and timers. A part of the scheduler is called the estimator, which measures outgoing flows to develop statistics that control packet scheduling and admission control.

- *Classifier:* The classifier maps each incoming packet into some class. All packets in the same class get the same treatment from the packet scheduler. A class may correspond to a broad category of flows or may hold only a single flow.

- *Admission control:* Admission control implements the decision algorithm to determine whether a new flow can be granted to the requested QoS without impacting earlier guarantees. Admission control is invoked at each node to make a local accept/reject decision at the time a host requests a service along some path. Admission control should not be confused with policing, which is a packet-by-packet function at the edge of the network to ensure that a host does not violate its promised traffic characteristics. Policing is considered one of the functions of the packet scheduler.

The final component of the IntServ implementation framework is the reservation setup protocol. This protocol is necessary to create and maintain the flow-specific state in the end-hosts and routers along the path of a flow, according to the QoS demands of the specific flow. The most widely used

FIGURE 11.4
Integrated services architecture including a packet scheduler, classifier, admission control routine, and a reservation setup protocol.

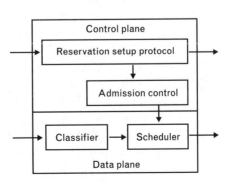

resource reservation signaling protocol today is the RSVP, which will be discussed in more detail in Section 11.4.1.

11.3.2   Integrated Services in Ad Hoc Networks

This section discusses the applicability of the IntServ architecture in mobile ad hoc networks, taking into account the characteristics described in Section 11.2.1.

11.3.2.1   Functionality

To be able to support integrated services, each router has to implement all elements of the architecture: packet scheduler, classifier, admission control, and RSVP. Nodes in an ad hoc network function as both host and router, so each node will have to implement all integrated services functionality. This results in large processing overhead for the mobile nodes, which is undesirable because of their limited resources concerning memory capacity, processing power, and battery lifetime.

11.3.2.2   Bandwidth Constraints

Bandwidth is always a scarce commodity in wireless communication systems, and hence, signaling overhead should be minimized wherever possible. This is in conflict with the per-flow reservation signaling of integrated services in a network with dynamic topology. Frequent updates need to be sent, made even worse by the probability of signaling packet loss because of the relatively large error rates of wireless systems. Integrated services can also reserve resources for aggregates of flows. The number of flows carried by each link, however, will be limited due to bandwidth constraints.

11.3.2.3   Resource Reservation

In order for a mobile node to be able to determine whether or not to accept a resource reservation for a particular flow, the node needs to know what resources are still available. The link layers of some ad hoc networks with multiaccess link layers do not provide any mechanisms to reserve resources. When all nodes are within each other's reach, this can be solved by a so-called *subnet bandwidth manager* (SBM) [7]. The SBM manages the multiaccess link, and each node that wants to reserve bandwidth has to request this from the SBM. In ad hoc networks this is more complicated. This is explained in Figure 11.5. The SBM functionality will have to be implemented in a distributed way, and nodes will have to reserve resources with all SBMs that are in range. In the example of Figure 11.5, if nodes B and C

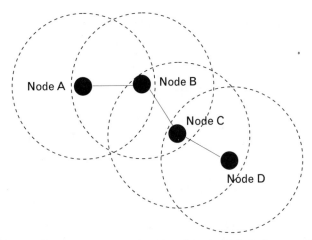

FIGURE 11.5 *Shared resources in a wireless ad hoc network. Node B shares a multiaccess link with node A and also shares a multiaccess link with node C, but node D is out of reach. However, the available bandwidth for the link from node B to node C depends also on the traffic sent from D to C, since node C can receive traffic from both. One centralized SBM requires a node that can reach all other nodes. In this example there is no such node. However, if node B and C implement distributed SBM functionality, the network can be covered. When node B wants to reserve resources on the link from B to C, it has to reserve bandwidth with its own SBM module and with the SBM of node C.*

implement distributed SBM functionality, the ad hoc network can be covered.

11.3.2.4 IP Mobility Issues

In a situation where a mobile node moves between different access points and uses Mobile IP to remain reachable and continue its communication sessions, all traffic destined to the mobile node's home address will be tunneled by the home agent to the current location of the mobile node. RSVP uses an option in the IP header, the router alert option, to signal to RSVP-enabled routers that a packet contains resource reservation information. When packets from the home agent to the mobile node are tunneled, the RSVP routers do not get this information and the path between the home agent and the mobile node can only provide best-effort service.

11.3.2.5 Scalability

The per-flow reservation mechanism, together with the overhead of reservation messages in a dynamic topology limit the scalability of integrated services in ad hoc networks. In cases where applications have strict delay constraints, however, QoS in ad hoc networks is probably only feasible when the network is relatively small. In such a scenario, the scalability problems of per-flow resource reservation of integrated services could be of minor severity.

11.3.3 DiffServ

DiffServ [8] was developed as an alternative resource allocation scheme, because by 1997 it was felt that IntServ was not ready for large-scale deployment, while the need for an enhanced service model had become more urgent. The Internet community started the search for a simpler and more scalable approach that could offer a better than best-effort service. A new working group was established in the IETF with the goal to develop a framework for allocating different levels of service in the Internet.

DiffServ provides service differentiation by dividing traffic into different classes at the edge of a network by so-called boundary nodes. A boundary node classifies each incoming packet into a certain traffic class that is identified by a single *DS code point* (DSCP). The routers within the core of the network—the interior nodes—forward packets according to a certain *per-hop behavior* (PHB) that is associated with the DSCP (i.e., packets associated with one DSCP are prioritized over packets associated with another DSCP). No state needs to be kept at each router, which allows for a large number of flows without the scalability problems of IntServ. The DiffServ network architecture is illustrated in Figure 11.6(a).

DiffServ does not make any per-flow reservations, but rather provides QoS for aggregates of flows. Resources are assured through provisioning and prioritization for traffic classes. Provisioning of resources for the different traffic classes is based on SLAs between users and the network providers. An important part of an SLA is the *traffic conditioning agreement* (TCA). The TCA includes the parameters that describe the allowed traffic per class in traffic profiles and policing actions in case the traffic profile is violated. Traffic policing is used at the boundary nodes to ensure the traffic conforms to the SLA. SLAs can be either static or dynamic and are typically determined from long time scale traffic aggregate statistics.

FIGURE 11.6
(a) Boundary nodes and interior nodes in DiffServ. (b) Boundary node functionality, including traffic classification and traffic policing. Traffic policing uses metering, marking, shaping, and dropping.

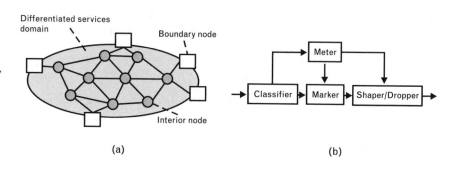

Traffic policing is a crucial part of DiffServ. Otherwise, whenever too much traffic is carried in a certain class, congestion will occur in that class, seriously degrading the service quality. In Figure 11.6(b) the functionality of a DiffServboundary node is shown, consisting of the following:

- *Classifier:* The classifier maps incoming packets into a certain traffic class that is identified with a single DSCP.

- *Meter:* The meter measures the traffic flow and compares this to the traffic profile in the SLA.

- *Marker:* The marker sets the appropriate DSCP for each packet. In case the meter detects a violation of the traffic profile, re-marking of packets is an option.

- *Shaper/Dropper:* The shaper/dropper shapes or even drops the traffic if the traffic profile is violated. A traffic shaper can, for example, be implemented as a rate limiter.

11.3.4   DiffServ in Ad Hoc Networks

In this section the applicability of DiffServ in mobile ad hoc networks is discussed.

11.3.4.1   Functionality

In a fixed DiffServ network, only nodes at the edges of that network need to implement sophisticated classification, marking, policing, and shaping functionality. Interior nodes in a fixed DiffServ network only need to implement PHBs. In an ad hoc network, however, the concept of interior nodes and network edge has only limited meaning. Due to node mobility, the network edge and interior nodes are dynamic. Moreover, each node in an ad hoc network typically acts as a host and a router, originating as well as forwarding traffic. Such nodes will have changing or even multiple roles in the DiffServ framework. This is illustrated in Figure 11.7, where several traffic flows are identified in an ad hoc network. As a result, each node will need to implement complete DiffServ functionality.

11.3.4.2   Class-Based Resource Allocation

In ad hoc networks, with its limited bandwidth, congestion can easily occur. It is thus very important to allocate resources in an efficient way. In fixed networks the allocation of resources to different service classes is based on SLAs that, for example, can be derived from statistics on traffic aggregates over a relatively long time scale. With a dynamically changing network

FIGURE 11.7
Dynamic roles of Diff-
Serv nodes in an ad
hoc network. Nodes 2
and 3 both originate
and forward packets
and thus act as interior
and boundary nodes at
the same time.

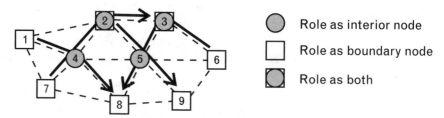

topology, the SLA and, more specifically, the traffic profiles need to be dynamically negotiated. This is not an easy task and may be rather inefficient, especially when node mobility is high.

Resource allocation is even more complicated in ad hoc networks based on multiple access technology, where each individual node can have difficulties in determining what resources are available, since the resources are shared. The problem of determining what resources are available when multiple nodes share one link was discussed in more detail in Section 11.3.2.3.

11.3.4.3 Bandwidth-Constrained Operation

QoS signaling in DiffServ is limited to the DS code point set in each packet. The DS code point introduces no extra overhead, since existing IP header fields are used for this purpose. This is an advantage for a bandwidth-constrained environment such as a mobile ad hoc network. However, for DiffServ to function in ad hoc networks, dynamic traffic profile negotiation is necessary, which will most probably result in large signaling overhead.

11.3.4.4 IP Mobility Issues

The QoS mechanism provided by DiffServ does not keep per-flow state. The DS code point alone determines packet-forwarding prioritization, not any flow descriptor such as an IP address. When nodes change IP address, because they connect from the ad hoc network to the Internet through a different gateway, this will have no influence on the DiffServ prioritized forwarding. Changing IP addresses might, however, complicate the dynamic traffic profile negotiation.

11.3.4.5 Scalability

The two main advantages that differentiated services has over integrated services in fixed networks are better scalability and less signaling overhead.

Both advantages are not valid in ad hoc networks. First, all nodes in a Diff-Serv network will have to implement complete DiffServ functionality. Furthermore, it is unclear how much lower the signaling overhead is for DiffServ in ad hoc networks, since it is necessary to do dynamic traffic profile negotiation.

11.3.5 A Flexible QoS Model for Mobile Ad Hoc Networks

A *flexible QoS model for mobile ad hoc networks* (FQMM) is proposed in [9]. FQMM considers the characteristics of mobile ad hoc networks and combines the per-flow service granularity of IntServ with the service differentiation of DiffServ.

Distinction is made between three types of nodes: ingress, interior, and egress nodes. FQMM defines an ingress node as the node that originates packets, interior nodes as the nodes that forward packets for other nodes, and egress nodes as the node for which the packets are destined. Note that each node can have multiple roles at the same time, and these roles change dynamically based on the nodes movement with respect to the network and the traffic characteristics (see Figure 11.7).

The allocation of resources in FQMM is done using a hybrid scheme of per-flow allocation as in IntServ and per-class allocation as in DiffServ. Per-flow classification is only done for traffic of highest priority, while the rest of the traffic gets per-class treatment. By applying per-flow granularity to only a small portion of the traffic, the scalability issues of IntServ can be avoided. FQMM uses relative and adaptive traffic profiles. Since the effective bandwidth of a link between two nodes in an ad hoc network is time varying, the traffic profile for each class is defined in terms of a relative part of the available bandwidth. The adaptivity of the traffic profile means that the parameters describing the aggregate traffic are adjusted for the current link capacity. Traffic conditioners are placed at the ingress nodes that are responsible for re-marking, shaping, or discarding packets according to the traffic profile.

FQMM combines some of the advantages of both DiffServ and IntServ and at the same time partially gets rid of some of the disadvantages. Several issues remain, however. It is not clear what part of the traffic can receive per-flow treatment and how to enforce this. Furthermore, the issue of dynamic traffic profile negotiation is not addressed. Nodes also need to implement complete IntServ and DiffServ functionality, using up a lot of valuable resources. Although it is not mentioned in [9], FQMM can be expected to cooperate rather easily with QoS mechanisms in the fixed Internet, since it is based on both IntServ and DiffServ. This is a clear advantage for ad hoc networks that are connected to the Internet through access points.

11.4   QoS Signaling

QoS signaling deals with communicating control data for QoS mechanisms through the network. This includes requests for resource reservation and release of those resources as well as messages for setting up and tearing down flow specific paths. QoS signaling systems can be divided into in-band signaling and out-of-band signaling. In in-band signaling the control information is carried together with the data packets, while out-of-band signaling uses separate control packets for this purpose. In the remainder of this section two signaling protocols will be described: RSVP, which is the de facto standard resource reservation protocol used for IntServ, and INSIGNIA, which is a signaling protocol that is specially designed for use in ad hoc networks.

11.4.1 RSVP

The most widely used resource reservation signaling protocol today is the RSVP [10]. RSVP is used by a host to request specific qualities of service from the network for particular flows and by routers to deliver these QoS requests to all nodes along the path. Resources are requested for unidirectional flows (i.e., only reservations are made in one direction. RSVP identifies a communication session by the combination of destination IP address, transport layer protocol type, and destination port number). The main messages used by RSVP are the PATH and RESV messages. In Figure 11.8(a), the use of the PATH and RESV messages to reserve resources is explained.

The reservation states in RSVP are kept as soft state [10], which have to be updated regularly by periodical retransmissions of the PATH and RESV messages. RSVP provides additional messages for explicit deletion of QoS state.

11.4.1.1 RSVP in a Dynamic Topology

First consider a scenario where the mobile nodes in the ad hoc network remain connected to the same access point for connections with the Internet, while each mobile node is free to move within the ad hoc network. In this situation the IP addresses of the mobile nodes do not need to change. However, paths between sender and receiver can constantly change, and thus, resources on different paths need to be reserved. To cope with the changing topology, the PATH and RESV messages of RSVP have to be refreshed at least as often as the rate of change of the topology. Although the reservation state kept in each mobile node times out after a certain time, the time-out value is typically set to several times the refresh rate of PATH and RESV messages, to cope with possible loss of one of those messages. For a highly dynamic ad hoc network this will result in a waste of resources because of stale resource reservations. This is illustrated in Figure 11.8(b). However, RSVP includes messages for explicit deletion of QoS state that can be initiated both towards the sender and receiver. In a highly dynamic ad hoc network this additional signaling is necessary to prevent large numbers of outdated reservations. Unfortunately, the mobile node might already be out of range, making it impossible to send such a QoS deletion message.

When a mobile node moves between different access points and uses Mobile IP to remain reachable, no reservations can be made on the path between the home agent and the mobile node, as was explained in Section 11.3.2.4. Thus, on that path only best-effort service can be provided.

In short, the applicability of RSVP in ad hoc networks is limited. Since the refresh period of the resource reservations needs to be of the same order as the rate of change of the network topology, the overhead of RSVP in highly dynamic networks is considerable. Moreover, when RSVP is used together with Mobile IP, no reservations can be made on the path between home agent and mobile node.

11.4.2 INSIGNIA Signaling

The INSIGNIA signaling system [11] is designed with the characteristics of mobile ad hoc networks in mind. The signaling is kept lightweight, minimizing the bandwidth used and is capable of reacting quickly to the network dynamics such as node mobility and wireless link quality. Like RSVP, INSIGNIA provides per-flow resource reservation management; however, contrary to RSVP, INSIGNIA uses in-band signaling. Signaling is included

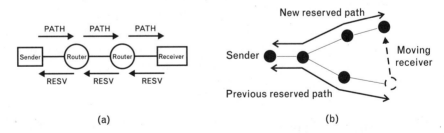

(a) (b)

FIGURE 11.8 *(a) RSVP message flow. The sender transmits a PATH message containing a description of the data flow. Each router that receives the PATH message installs reverse routing state. The receiver responds with a RESV message. The role of the RESV message is to carry a reservation request to the routers along the reverse path, which apply admission control to the requests, and, if accepted, the parameters of the scheduler are set according to the reservation specification. (b) Stale reservations after topology change. A route with reserved resources is available between the sender and receiver. After the receiver moves, it is only reachable via another path. New resources are reserved along this new path, but the old reservation is not timed out yet, thus network resources are wasted, as the old reservation is unused.*

in an IP option (INSIGNIA option) in the IP packet header that is used for flow reservation, restoration, and adaptation mechanisms. INSIGNIA distinguishes between routing, forwarding, and resource reservation. Any ad hoc routing protocol can be combined with INSIGNIA signaling.

A source node that wants to make a flow reservation sets a bit in the INSIGNIA option (reservation mode bit) and sends the packet to the destination. The intermediate nodes check whether or not the reservation can be accepted. If the request can be accepted, resources are committed and subsequent packets receive the appropriate treatment. Contrary to RSVP, if the reservation is denied, no rejection is sent back to the source, but packets receive best-effort service from this node and subsequent nodes. When the destination node receives the reservation request, it replies to the source with a so-called QoS report, specifying the status of the flow reservation.

Reservations made at the intermediate nodes are kept as soft state, so reservations are automatically removed when a flow travels along a different path. The soft state timers are refreshed each time a packet associated with the appropriate flow is forwarded.

The destination node actively monitors the quality of the flows and can take action to adapt flows under certain observed conditions. This is based on a certain adaptation policy that can be different for each application that is running between source and destination. Through QoS reports, the destination node can ask the source node to adapt the reservation for the flow (e.g., in case the observed quality of the flow is low, due to a highly loaded network).

The INSIGNIA signaling system is promising for ad hoc networks. It reduces signaling overhead significantly and provides mechanisms to adapt quickly to changes in network topology. A possible concern is the flow state that needs to be kept in the mobile nodes. For large ad hoc networks this might be a scalability problem. Furthermore, no interaction with IntServ and DiffServ is specified, which is of major importance when a QoS connection needs to be established between a node in the ad hoc network and a node somewhere else on the Internet.

11.5   QoS Routing

Routing protocols for the fixed Internet and the routing protocols described in Section 11.2 of this chapter determine routes based on the destination and typically a simple link metric such as the number of hops. QoS routing protocols search for paths from source to destination that meet certain end-to-end QoS requirements, such as delay, delay jitter, bandwidth, or a certain combination of metrics. QoS routing does not reserve resources; QoS signaling and resource management do this.

In [12] QoS metrics are divided into concave and additive metrics by the following definition. Let $m(i, j)$ be a QoS metric for a certain link (i, j). For a path $P = (s, i, j, \ldots, k, t)$, the metric $m(P)$ of the total path P is concave if

$$m(P) = \min\{m(s, i), m(i, j), \ldots, m(k, t)\} \tag{11.1}$$

The metric $m(P)$ of the total path is additive if

$$m(P) = \min\{m(s, i) + m(i, j) + \ldots + m(k, t)\} \tag{11.2}$$

An example of a concave metric is bandwidth—that is, a certain bandwidth is required on each link along a path. Delay and delay jitter are examples of additive metrics.

The dynamic nature of the nodes in an ad hoc network makes it difficult to maintain the precise link state information that is needed for QoS routing. Furthermore, many wireless link layers are based on a multiple access medium, which complicates the link state determination of each individual node. Another issue is the short end-to-end path lifetime. When QoS routing has found a suitable path, fulfilling the QoS constraints, node mobility and wireless link degradation may quickly make this path insufficient and an alternative path has to be found. Finally, QoS routing requires all nodes to disseminate rather detailed link state information throughout the network.

Many QoS routing mechanisms have been proposed for the fixed Internet [13]. Despite the difficulties of providing QoS routing in mobile ad hoc networks, several solutions have been proposed that will be introduced in this remainder of this section.

11.5.1 Core-Extraction Distributed Ad Hoc Routing

The *core-extraction distributed ad hoc routing algorithm* (CEDAR) is described in [14] and provides QoS routing that can adapt to the dynamics of ad hoc networks. The three main components of CEDAR are (1) core extraction, (2) link state propagation, and (3) route computation. CEDAR provides QoS route computation based on a minimum bandwidth requirement.

11.5.1.1 Core Extraction

QoS routing requires the nodes of a network to flood local link state throughout the network. In a dynamically changing network, frequent updates are necessary, which introduces a lot of signaling overhead. Moreover, [14] states that flooding in ad hoc networks is highly lossy, making it undesirable that all nodes participate in route computation. Therefore, CEDAR selects a subset of the nodes of the ad hoc network as core nodes, which form the core of the network. The core set is an approximation of a *minimum dominating set* (MDS) that is the minimum set of a *dominating set* (DS) of the network. A dominating set of a network is a set of nodes, DS, such that every node in the network is either in DS or is a neighbor of a node in DS. CEDAR uses a distributed algorithm to select core nodes, which can quickly react to node mobility. A node *D* that is selected to be part of the DS by a certain node *S* is called the dominator, *dom(S)* of node *S*. In Figure 11.9 an example of a selected core in CEDAR is shown.

Only the nodes that are part of the core set compute routes, both for themselves and on behalf of neighboring nodes that are not part of the core set.

11.5.1.2   Link State Propagation

The core nodes use whatever link state information they have about the network for QoS route calculation. One extreme could be to only store local topology information in each core node. This will result in a rather poor routing algorithm [14], but is has very low overhead for dynamic networks. At the other extreme, the entire link state of the ad hoc network could be stored at each core node. Using all the link state information optimal routes could be computed, at the cost of considerable overhead and the risk of finding stale routes based on outdated link state information when the network is highly dynamic. CEDAR requires the core node to have up-to-date information about its local topology and also maintain link state of stable high-bandwidth links further away in the network. This ensures adaptability in highly dynamic networks and approaches optimal route computation in highly stable networks.

The goal of stability and bandwidth based link-state propagation is achieved by CEDAR using so-called increase and decrease waves. An increase wave propagates an increase of bandwidth, and a decrease wave propagates a decrease of bandwidth on a certain link. Waves are generated whenever the available bandwidth changes by a threshold value with respect to the last time a wave was generated. The maximum distance (in terms of number of hops) a wave can travel is a function of the available bandwidth of

FIGURE 11.9
*Example of a
CEDAR core set;
black nodes are part of
the core.*

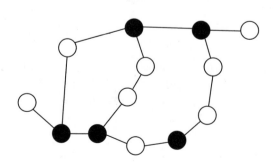

the link. Moreover, increase waves are slowly forwarded through the network while decrease waves are forwarded fast. As a consequence, for unstable links, the increase waves are quickly killed by decrease waves, preventing unstable link state to spread through the network.

11.5.1.3 Route Computation

The QoS route computation used in CEDAR is an on-demand routing algorithm. When a source node S wants to establish a connection to a destination node D, with a required bandwidth B, it sends the triplet S,D,B to its dominator node $dom(S)$. If $dom(S)$ can compute a route that complies with the bandwidth requirement using its local state, it replies directly to S. Otherwise, if $dom(S)$ has a core path to $dom(D)$ it continues with the QoS route establishment phase. If $dom(S)$ does not know the location of node D, it starts a core broadcast to discover D and $dom(D)$, and simultaneously establishes a core path to $dom(D)$.

After the core path establishment, $dom(S)$ knows $dom(D)$ and the core path from $dom(S)$ to $dom(D)$. This core path provides a good directional guideline for possible QoS paths. $Dom(S)$ has partial knowledge of the ad hoc network topology, which consists of up-to-date local topology and some possibly out-of-date information about remote links (obtained from increase/decrease waves). $Dom(S)$ uses the local topology to find a path from the source node S to the furthest possible core node, say $dom(T)$, that is on the core path from $dom(S)$ to $dom(D)$ that can provide at least the requested bandwidth B. Among all possible paths, the shortest-widest path (least number of hops, largest available bandwidth) is picked. $Dom(S)$ now forwards the connection request of node S to $dom(T)$ and $dom(T)$ starts QoS route computation using its own local state. Eventually, either a suitable path to destination node D is found, or the local path calculation will fail to provide a path with the requested QoS. In case of success, the concatenation of the partial paths calculated by the core nodes provides the end-to-end QoS path.

11.5.2 Ticket-Based Probing

In [12] a multipath distributed QoS routing scheme is proposed, called ticket-based probing. The QoS routing is done based on a per-connection basis; therefore, routing overhead is a major concern. Nodes are assumed to keep up-to-date state information of all local links. This state includes delay, unused bandwidth, and cost, which can be simply one (as a hop count) or a function of the link utilization. Furthermore, nodes are assumed to maintain end-to-end state information for every possible destination in terms of end-to-end delay, maximum bandwidth, and cost. This information is updated periodically by a distance vector protocol for ad hoc networks (see, for example, [2]) and is inherently imprecise in an ad hoc network, since the network state and topology are dynamic.

In *ticket-based probing* (TBP) a ticket represents the permission of searching one path. When a QoS route needs to be determined, a source node S issues a number of tickets based on the available link state information. More tickets are issued for connections with tighter QoS requirements. Probes (routing messages) are sent from the source towards the destination to search for a path that satisfies the QoS requirements. Each probe is required to carry at least one ticket. An intermediate node receiving a probe with more than one ticket is allowed to split the probe into multiple ones, each forwarded on a different subpath. In the splitting process, the more residual bandwidth a link has, the more tickets are assigned to the probe that is forwarded on that link. When the destination node receives a probe message, a possible path is found. The routing process is terminated when all probes have either reached the destination node or have been invalidated by the intermediate nodes (because the QoS requirements could not be fulfilled). To detect the invalidated probes, TBP requires those tickets to be forwarded to the destination, so eventually all tickets will arrive at the destination, excluding tickets that are lost due to the network dynamics. If only invalidated tickets arrive at the destination, a message is sent back to the source with a rejection of the QoS request. If multiple probes with a valid ticket are received, the probe with the least cost is selected as the primary path, while the others are the secondary paths. A probe stores the path in the probe itself while it travels through the network. A confirmation message is sent back to the source using the route stored in the probe that is resulted in the main path, reserving resources on its way. In Figure 11.10 an example (from [12]) of TBP is shown.

The TBP mechanism is a general QoS routing scheme, which can handle different QoS constraints. In [12] its application to both delay-constrained routing and bandwidth-constrained routing are explained in detail. TBP specifies several techniques to maintain the established path in the dynamic network environment: rerouting, path redundancy, and path repairing. TBP can provide multiple QoS paths, without flooding the network. For ad hoc networks that connect to the Internet, however, interaction with, for example, IntServ and DiffServ will have to be specified, in order to provide end-to-end QoS.

11.6   Future Directions

It is expected that a wide variety of wireless devices will be able to participate in mobile ad hoc networking in the near future. This, combined with the increasing demand for Internet connectivity for more and more devices,

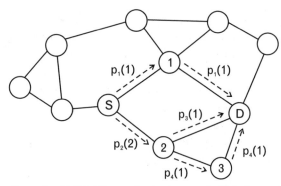

FIGURE 11.10   *Example of TBP. The number in the parentheses is the number of tickets a probe carries. Two probes are sent from source node S, p_1 and p_2. At node 2, the probe is split into two, p_3 and p_4, each carrying one ticket. Since the source issued three tickets, at most three probes are active at a time and at most three paths to the destination node D are searched. Here these paths are: {S, 1, D}, {S, 2, D}, and {S, 2, 3, D}.*

means that future QoS mechanisms in ad hoc networks will have to include interaction with QoS mechanisms of the fixed Internet, such as IntServ and DiffServ. Only then, end-to-end QoS can be provided for ad hoc network nodes that connect to nodes outside the ad hoc network.

Results from research on routing protocols for ad hoc networks point out that the performance of a routing protocol heavily depends on the scenario that is considered. The degree of node mobility, traffic distributions, and link-layer capabilities influence the performance. In cases with high mobility and a lot of traffic flows, proactive protocols perform better, while for lightly loaded networks and a small number of traffic flows reactive protocols are preferable. A routing protocol that would perform well in all situations should adapt itself to the mobility and traffic conditions. This can also be applied to QoS provisioning. The CEDAR approach applies this to a certain extent, adapting the link state exchange to the node mobility. Future QoS mechanisms should be highly adaptable to a range of parameters such as node mobility, traffic conditions, wireless radio range, battery lifetime, and cost.

Depending on the number of nodes in the ad hoc network, a per-flow QoS treatment could gradually change to a per-class treatment. Moreover, the amount of QoS signaling can be adjusted to the current traffic and the node mobility, making a trade-off between signaling overhead and reservation time-outs. When a wide range of wireless devices are capable of participating in mobile ad hoc networking, it is very important that those devices can negotiate QoS parameters and set up QoS connections to each other and through each other to other nodes and networks. Since the devices can belong to different administrative domains and have a wide variability in capacities, general QoS mechanisms for ad hoc networks based on such nodes must be highly adaptive.

Recently, the IETF has started work on how to signal QoS to be able to provide end-to-end QoS, with a focus on interworking between domains in which different QoS solutions are deployed. This work is done in the newly formed *Next Steps in Signaling* (NSIS) working group. Both wired and wireless networks will be considered, as well as the coexistence of QoS signaling and mobile IP. At this stage it is not possible to say whether future results of this working group can be directly used in ad hoc networks, but the results will have impact on QoS mechanisms in ad hoc networks that are connected to the Internet.

11.7 Summary

Providing QoS in mobile ad hoc networks is far from a trivial task. In this chapter it was made clear what problems occur when directly trying to apply QoS mechanisms from the fixed Internet, IntServ and DiffServ, to ad hoc networks. Different proposals from the literature for QoS support in ad hoc networks were described. FQMM, a QoS model that was designed for ad hoc networks, combines the per-flow granularity of IntServ with the service differentiation of DiffServ. Moreover, INSIGNIA, a QoS signaling mechanism for ad hoc networks was described, which solves several of the problems with RSVP signaling in such environments. Finally, two proposals for QoS routing in ad hoc networks, CEDAR and TBP, were described.

REFERENCES

[1]   Jacquet, P., et al., "Optimized Link State Routing Protocol," draft-ietf-manet-olsr-07, July 2002, work in progress.

[2]   Perkins, C., E. M. Belding-Royer, and S. R. Das, "Ad Hoc On-Demand Distance Vector (AODV) Routing," draft-ietf-manet-aodv-11,7 June 2002, work in progress.

[3]   Perkins, C., "Mobility Support in IPv4," RFC 3344, August 2002.

[4]   Johnson, D. B., C. Perkins, and J. Arkko, "Mobility Support in IPv6," draft-ietf-mobileip-ipv6-18, June 2002, work in progress.

[5]   Wu, K., and J. Harm, "QoS Support in Mobile Ad Hoc Networks," *Crossing Boundaries—The GSA Journal of the University of Alberta*, Vol. 1, No. 1, November 2001, pp. 92–106.

[6]   Clark, D., et al., "Integrated Services in the Internet Architecture: An Overview," RFC 1633, June 1994.

[7]   Yavatkar, R., et al., "SBM (Subnet Bandwidth Manager)," RFC 2814, May 2000.

[8]   Blake, S., et al., "An Architecture for Differentiated Service," RFC 2475, December 1998.

[9]   Xiao, H., et al., "A Flexible Quality of Service Model for Mobile Ad hoc Networks," *Vehicular Technology Conference 2000 Spring*, Tokyo, Japan, May 2000.

[10]   Braden, R., et al., "Resource Reservation Protocol (RSVP)—Version 1 Functional Specification," RFC 2205, September 1997.

[11]   Lee, S. B., et al., "INSIGNIA: An IP-Based Quality of Service Framework for Mobile Ad Hoc Networks," *Journal of Parallel and Distributed Computing*, Vol. 60, 2000, pp. 374–406.

[12]   Cheng, S., and K. Nahrstedt, "Distributed Quality-of-Service Routing in Ad Hoc Networks," *IEEE Journal on Selected Areas in Communications*, Vol. 17, No. 8, August 1999.

[13]   Clark, D. D., and W. Fang, "An Overview of Quality of Service Routing for the Next Generation High-Speed Networks: Problems and Solutions," *IEEE Network*, Special Issue on Transmission and Distribution of Digital Video, November/December 1998.

[14]   Sinha, P., et al., "CEDAR: A Core-Extraction Distributed Ad Hoc Routing Algorithm," *IEEE Infocom '99*, New York, 1999.

Radio Access Control in Wireless IP Networks

Vinh Phan-Van and Savo Glisic

12.1 Introduction

Recently, a variety of emerging technologies and protocol enhancements has been investigated and designed for providing IP-based multimedia services to mobile users. These research and standardization efforts are summarized in Chapter 4 and in [1, 2]. Evolutions of today's cellular mobile communications and currently developed 3G systems, such as UMTS, are believed to be the most significant segments of future wireless IP networks for reliable mobile access and operations. The radios of those cellular systems are likely to remain unchanged or to have only minor modifications because of the economic reasons. We have witnessed the radio spectrum auctions for UMTS in Europe reaching hundreds of billion of Euro. This sum, added to the costs of building new networks, certainly has significant impacts on future network operations, service extensions, and service charges. The "anytime, anywhere" theme and "fairness" among users might be compromised with service level agreements based on subscriber and network profile characteristics. Future network operators need to have an in-hand tool to manage their network operation and performance more or less similarly to an airline's customer-service system where passengers are categorized into classes receiving different quality of service on any required routes, but all sharing expensive limited space of planes. Thus, for practical deployment of future multimedia wireless networks, there is a need for efficient radio resource management that allows optimizing spectrum efficiency and customizing various user and network profiles for network operation, while satisfying QoS requirements for a variety of user applications and demands. The key in providing QoS for the users while efficiently sharing scarce radio resources to optimize *grade of service* (GoS) for the networks is the access control at the air–interface acting as ticket–selling and check–in processes of the airline's service system. The access control herein includes *call admission control* (CAC) for the real–time (RT) connection–oriented service domain and *packet access control* (PAC) for the non–real–time connectionless service domain. These mechanisms must be simple and flexible in order to guarantee fast response to multiple and diverse QoS requests from mobile users in multimedia wireless IP networks. GoS definition is often associated with call–blocking and call–dropping or handoff–failure probabilities of the real–time traffic, which is given in detail later.

In this chapter, we address some challenges and some practical solutions of CAC and PAC for the need of future network operators, and provide an accurate and traceable analytical method for evaluating, dimensioning, and optimizing teletraffic performance of multimedia wireless systems. Section 12.2 presents a short overview of the radio access control problem under the RRM framework [3] for wireless IP networks. Section 12.3 presents a comparative study of CAC policies and QoS differentiation paradigms for the connection–oriented service domain. The focus is on the *uplink* (UL) of cellular networks, although the analogy can be applied for the *downlink* (DL) and other access networks as well. The highlight of this section is the introduction of *soft-decision CAC* (SCAC) for interference-limited environments, such as spread-spectrum CDMA radio access. The idea behind SCAC is to exploit UL interference distributions to compensate for fluctuations of the local-average *signal-per-interference-plus-noise ratio* (SINR) dominating QoS while expanding the call-admissible region to maximize the system capacity. This method requires neither complex measurements nor mutual exchange of information between adjacent cells about states of the network. Therefore, SCAC is especially applicable in wireless IP networks, consisting of number of segments and domains where simplicity and scalability requirements favor such localized or distributed RRM. Section 12.4 deals with PAC for connectionless service domain. The challenge is how to control non-real-time packet radio access in UL so that it can adapt to the remaining capacity left by the real-time traffic with optimal throughput-delay trade-offs. First, a precise dimensioning of available UL resources for packet radio access is presented based on asymptotic and quasi-stationary analysis of real-time traffic load. Then, average upper-limit UL data throughput for different bit rates and packet-lengths are estimated that can be used for the design of optimal PAC schemes. Dynamic feedback information–based multiple access (DFIMA) schemes are introduced, which have potential in providing differentiated QoS for non-real-time data users. Such PAC schemes are increasingly important for the efficiency of *radio resource utilization* (RRU) in serving mobile messaging services that have been taking the consumer market by storm recently. Throughout Sections 12.3 and 12.4, numerous modeling and implementation issues are addressed. The performance characteristics, including call-blocking, handoff-failure, and QoS-loss probabilities, as well as GoS and UL data throughput, are derived so that they can be used extensively for network dimensioning and planning frameworks. Section 12.5 provides some concluding remarks and future directions.

12.2   Overview of the Radio Access Control Problem

This section gives an overview of the radio access control problem, focusing on cellular segments of the wireless IP networks, where limited and extremely expensive radio resources are the bottleneck of the system capacity.

12.2.1   Service Models

The increasing bit rates supported in 3G radios together with the increasing processing power of mobile terminals gives opportunities for introducing various real-time and interactive multimedia applications and services to mobile users. Towards a seamless convergence of fixed and mobile services and a coexisting of voice/video and elastics applications like file transfer or Web browsing over common IP-based platform, it is important to have a robust QoS classification that is capable of providing a fine QoS resolution. In IP-based networks, different applications have different requirements concerning delay-tolerant and error-tolerant characteristics. For example, voice applications can tolerate low packet delay and moderate BER, while the opposite is the case for messaging services. However, it can be expected that QoS parameters of wireless users are more complicated than that of wired users for the same application due to restrictions and limitations of the air-interface. The delay that a packet experiences during its delivery in the network is one of the most important QoS parameters for both wired and wireless packet access. The BER parameter is very important for wireless environments, which is often used as the performance measure of a wireless link. Other QoS parameters can be, for example, minimum-guaranteed bit rate, maximum bit rate, maximum packet size, delay variation, and so forth. There are several standardized QoS classifications, for example, from 3GPP and ETSI for UMTS [4, 5], or from IETF for integrated and differentiated IP service models [6, 7]. These can be unified considering two main service classes: real-time and non-real-time. Each can further be divided into subclasses depending on QoS parameters (i.e., packet delay and BER requirements). For example, [4, 5] define four QoS classes: conversational or class A, streaming or class B, interactive or class C, and background or class D. The first two classes correspond to real-time connection-oriented services with strict delay or delay-variation constraints, for example, 20- to 50-ms packet delay for class A and 1-ms delay variation for class B. The last two classes are for non-real-time connectionless services, where bounded delay constraint is applied for class C (e.g., utmost 1s), and low BER is required for class D (e.g., at least 10^{-8}).

Providing multimedia services to mobile users in IP-based packet radio access environments faces many challenges due to restrictions (e.g., power constraints), limitations (e.g., bandwidth), and high uncertainty (e.g., fading impacts) of radio channels. The access control mechanisms at the air-interface including CAC and PAC are responsible for making use of available radio resources under those conditions in such a way that users can be satisfied with provided services and optimal system capacity and performance can be achieved. By translating QoS parameters and user profile characteristics into required radio resource consumption and serving priority, together with keeping track on available radio resources, access control mechanisms can make proper decision upon reception of QoS requests. The access control problem is formulated in the following section.

12.2.2 Problem Formulation

In general, an optimum use of frequency-reuse patterns and propagation path-loss characteristics for minimizing the carrier-to-interference ratio (hence increasing the utility of radio spectrum) is the most restraining factor on overall capacity and performance of cellular systems. Let us consider a cellular system as a segment of wireless IP networks that consists of N_b *cellular access points* (CAPs) or base stations. There are N_c waveforms (or channels) available; N_m possible packet transmission modes, each corresponding to a different bit rate; and N_u active users in the system. Let $\mathbf{B} = \{1, 2, ..., N_b\}$; $\mathbf{C} = \{1, 2, ..., N_c\}$; $\mathbf{M} = \{1, 2, ..., N_m\}$; and $\mathbf{U} = \{1, 2, ..., N_u\}$. In order to establish a radio link for any communication services in UL direction, for instance, each active terminal $u_i \in \mathbf{U}$ needs to be assigned to a serving CAP $b \in \mathbf{B}$, suitable waveform or channel(s) $c_i \in \mathbf{C}$, a mode $m_i \in \mathbf{M}$, and transmitter power $P_{tx,i}$. This resource assignment is restricted by limitations of available waveforms or channels in $\mathbf{C}$, transmitter power $P_{tx,i}$ together with tolerable interference level of CAP, and allowable modes in $\mathbf{M}$. The main tasks of RRM are to allocate and maintain adequate radio resources for all active radio connections in the system so that the required QoS of mobile users moving around service area can be met once they are admitted into the system. This means that SINR, dominating the quality of radio link, has to be kept above a target level depending on transmission mode. Thus, the following inequality must hold for the radio connection between user u_i and serving CAP b using waveform c_i [3]:

$$\gamma_i = \frac{P_{ix,i} G_{b,i}}{\sum_{n \neq i} P_{ix,i} \theta_{i,n} G_{b,i} + N_{0b}} \geq \gamma_{i\_target} \tag{12.1}$$

where γ_i and γ_{i_target} denote the actual and target SINR of u_i connection, respectively; $G_{b,i}$ is the path-loss between u_i transmitter and b receiver for all $u_i \in U$; $\theta_{i,n}$ denotes the cross-correlation between waveform c_i and c_n for all active users; and N_{0b} is the thermal noise power received at b.

As the number of users and demands for services increase, the number of available waveforms or channels will become insufficient to support QoS requests of all users complying with (12.1). The access control provides the means to access the resources of each cell efficiently and assign the available resources to achieve the highest spectrum efficiency. The access-control strategies can be classified into static, dynamic, and flexible classes. In the static class, the resource allocation is often based on a priori knowledge of available network resources and resource consumption of users for requested services. This is mainly done during the planning stage (e.g., with assigning a set of channels to each cell permanently according to frequency-reuse pattern, then letting the user borrow one or several channels from that set for its requested services). The design for static access control is often simple and distributed, but it has to take into account the worst-case scenarios of network and channel conditions, as well as resource consumption in order to guarantee the required QoS. In the dynamic class, the resource allocation is supposed to make use of available resources fully. There is no need for planning frequency-reuse pattern. The design in most cases relies on complicated on-line measurements and excessive signaling to keep track of changes of network conditions and user resource consumption for accommodating as many service requests as possible. In practice, adopted access-control strategies often combine aspects of both static and dynamic ones, forming the flexible class. For example, radio resources can be allocated in the static manner up to a certain level, the rest is utilized in the dynamic manner; or a flexible access-control strategy can be based on both static modeling and dynamic measurements of resource and traffic statistics.

12.3 CAC

CAC denotes a decision process for accepting or rejecting a new connection depending on available radio resources for required QoS, traffic and user profile characteristics, and effects on existing calls imposed by new call. CAC is commonly used in wireless personal communication networks as an effective method to manage the radio resources adapting to traffic and channel variations.

12.3.1 Overview

Real-time connection-oriented multimedia services including voice and video calls are expected to represent a major contribution to traffic load of wireless IP networks. Success of these services depends on the network capabilities in providing differentiated QoS to meet a wide range of customer demands. CAC policies are vital for the two-fold objective: (1) to deliver the required QoS and (2) to provide operators with freedom in controlling payload and group behavior of traffic classes regarding required services as well as subscriber potential classes for optimizing network utility and revenues. CAC policies are often divided into two categories: static or dynamic. From the implementation point of view, these can be further divided into modeling based or measurement based, centralized or distributed, and interactive or noninteractive policies. As argued in the access-control classification in Section 12.2, however, there should be also a flexible class for practical implementations of CAC.

In static CAC systems, a call request is accepted only when sufficient resources are available to meet its required QoS and to maintain the QoS of ongoing calls. RRM entity must decide which resource combinations can be accepted into the network. In orthogonal FDMA/TDMA cellular systems, fixed channel assignment–based CAC policies [3] are often adopted because of the simplicity and the capacity maximization in line with a fixed threshold of the number of channels (cell-capacity). In interference-limited CDMA cellular systems, such a fixed threshold is hard to determine efficiently because qualitative and quantitative features of CDMA systems appear to be heavily intertwined and mutually supportive.

In dynamic CAC systems, a call request can be rejected even if a set of new calls meets their QoS requirements. On the other hand, a new call can be accepted into the network even if the instantaneous QoS may be violated, but when averaged over states, the service quality is met. Therefore, the network utility can be increased, and various cost-effective and traffic-regulating policies can be adopted. The reinforcement-learning or measurement-based adaptive CAC is an effective solution for the dynamic class that can be described with a semi-Markov decision process [8, 9]. The price, however, of better performance offered by such adaptive schemes can be the disgracefully increasing complexity of implementations in both hardware and software. One should keep in mind that in wireless IP networks CAC mechanisms must be simple and scalable because only this condition can ensure system efficiency and guarantee fast response to diverse demands of users. In general, CAC policies can be evaluated with respects to the following metrics:

- *Efficiency:* based on trade-offs of simplicity in implementation and ability in autoconfiguration with various traffic, user, and network profiles for maximizing system capacity and revenues. The measures are the probabilities of call blocking and handoff failure or some cost functions (e.g., GoS defined in detail later);

- *Reliability:* measured by the probability of losing communication quality during services for given offered multimedia traffic intensities;

- *Robustness:* seen against the need for redesign due to uncertainty and changes of the system parameters.

In the sequel, an elaboration of three different CAC policies—namely, static modeling-based complete-sharing CAC (MdCAC), dynamic measurement-based CAC (MsCAC) and flexible statistics-based soft-decision CAC (SCAC)—for the UL of CDMA cellular access segments of wireless IP networks is presented. QoS differentiation paradigms are introduced and integrated into CAC mechanisms, in which load-based thresholds or fracturing factors are defined for preshaping the traffic (i.e., hard or soft blocking of less important calls). These provide a flexible means for operators to prioritize different traffic classes regarding QoS requests and subscriber profiles for desired network operation patterns or serving customs. The elaboration is carried out to demonstrate the design of an effective SCAC scheme and to provide a traceable and accurate analytical method for evaluating the performance of CAC schemes for multimedia cellular access networks.

12.3.2 Performance Evaluation

12.3.2.1 System Modeling

Principles of MdCAC, MsCAC, and SCAC Systems

Let us consider the UL of a cell in a DS/CDMA cellular system consisting of multiple and identical cells and supporting M connection-oriented RT service classes. Let set $\mathbf{K} = \{1, 2, ..., M\}$. For all $k \in \mathbf{K}$, let r_k be the bit rate, γ_{target_k} the target $E_b N_0$ SINR of class-k, and g_k the processing gain of class-k radio transmissions that depends on spreading factor, data-source activity factor, diversity-combining gain, and so forth. Define:

$\mathbf{n} = (n_k, \ k \in \mathbf{K})$ is the occupancy vector, where n_k is the number of class-k calls in progress;

$\mathbf{w} = (w_k, \ k \in \mathbf{K})$ is the load vector, where w_k is the average load factor of a class-k call, representing the average resource consumption of radio connections onto system spectrum or radio resources and given by $w_k = (1 + g_k / \gamma_{target\_k})^{-1}$ [9, 10].

C_a – the average normalized UL capacity. In the ideal situation, 100% of available radio spectrum can be utilized, resulting in $C_a = 1$. In real situations, C_a is strongly dependent on thermal noise and interference from nearby cells. The quantity of actual capacity is a *random variable* (RV), and C_a itself can be considered as a bounded stationary RV as well. The "traditional" average system capacity corresponding to each service class —defined as the largest number of users that the system can handled properly— is given by the ratio C_a/w_k. The average system capacity C_a is often predicted by using the load equation in the boundary condition of tolerable interference level as in [8–15]:

$$C_a = \frac{1-\eta}{1+f} \tag{12.2}$$

where η is the ratio of thermal noise density and UL maximum tolerable interference level; f is the coefficient representing the ratio between other- and own-cell interference (for both mean and variance values), which varies with changes of propagation parameters, *transmitter power control* (TPC) inaccuracy, and traffic distributions. For instance, simulation results in [10] show that f is between 25% and 55% in micro-cellular environments using omni-directional antennas. The fluctuations of this parameter have essential impacts on system capacity resulting in the "soft-capacity" feature of CDMA systems.

The most straightforward CAC policy is to share "fixed" C_a units of radio resources in static complete-sharing fashion among active users based on their QoS requests, resulting in MdCAC policy. In an MdCAC system, a call request for class-k service is accepted immediately if utmost $(C_a - w_k)$ units of resources are occupied, otherwise it is rejected. This system may face the following problems. First, the QoS guaranteed scenario results in the worst case of system capacity and RRU. Second, there is a need for redesign of the system-capacity constraint C_a for changes of CDMA radio channel parameters and traffic statistics. Third, advantages of CDMA techniques, such as the soft capacity feature, cannot be exploited.

Results of recent research have emphasized that robust CAC policies may require on-line measurements [11–14], resulting in far more complex and high-cost software/hardware implementations [9, 10]. The principle of MsCAC is that a new call is accepted immediately into the system if at least the required resource consumption of its service class is available, otherwise it is rejected. The decision is made based on on-line measurements of related statistics. The later analysis and simulation results show that when UL traffic is less bursty, the MsCAC has no capacity gain over MdCAC described above. It may even suffer degradation due to measurement errors, which agrees with the results presented in [12, 13, 15].

In order to harmonize the advantages and overcome the problems of static MdCAC and dynamic MsCAC with flexible SCAC, let us first reconsider the resource consumption by using the well-known effective-bandwidth concept. The parameters γ_{target_k}, w_k, and f are now treated as stationary RVs (we use the same fonts to denote their mean values.) It is assumed that QoS parameters of the requested services and adequate TPC mechanism can provide and ensure a priori knowledge and stationary behavior of the resource consumption. Thus, based on the knowledge of local-average SINR (i.e., log-normal RV with standard deviation σ in range of 2 dB [9, 10, 16]) and cell interference distributions (i.e., Gaussian [9, 10, 15]), the resource consumption can be modeled as follows. For all $k \in \mathbf{K}$, resource consumption of a class-k connection is a stationary, independent, and bounded normal RV having mean w_k, variance σ_k^2, and falling in between a max- and min-guaranteed resource consumption $w_{u,k}$ and $w_{l,k}$. The system capacity is accordingly modeled with a bounded Gaussian RV having mean C_a, variance Θ^2, and falling in between C_l and C_u. In theory, C_u can be set up to the isolated-cell capacity of $(1 - \eta)$. In general, the mean, variance, and boundary values of link resource consumption and cell capacity can be estimated by using the definition of load vector $\mathbf{w}$ above and (12.2), together with mean, variance, and boundaries of SINR and f. Due to the interference-limited nature of CDMA systems, the cell shall meet their QoS constraints if

$$c_{other} + c_{own} \leq (1 - \eta) \tag{12.3}$$

where c_{other} represents the equivalent UL load factor produced by other-cell interference, and c_{own} is the load factor generated by active users in own-cell. Define:

$c = \mathbf{nw}$ the system load state that is identical to c_{own}, where $\mathbf{n}$ is the occupancy vector and $\mathbf{w}$ is the load vector defined above.

From (12.3) and with Gaussian interference and log-normal SINR, the QoS-loss probability (local outage) conditioned to the system in load state c can be determined by

$$\Pr\{QoS\,loss|c\} = Q\left(\frac{1 - \eta - E[Z|c]}{\sqrt{Var[Z|c]}}\right) \tag{12.4}$$

where $E[Z|c] = c(1 + f)\exp[(\varepsilon\sigma)^2 / 2]$; $Var[Z|c] = c(1 + f)\exp[2(\varepsilon\sigma)^2]$; $Q(x)$ is the standard Gaussian integral function; σ is the standard deviation of log-normal SINR in decibels; and constant $\varepsilon = \ln(10)/10$. In the long run, with respects to Gaussian distributed intercell interference, we have

$$\Pr\{c_{other} \le 1 - \eta - c\} = 1 - Q\left(\frac{1 - \eta - c - fE[c]}{\sqrt{fVar[c]}}\right) \tag{12.5}$$

This relation implies that intercell interference has a significant impact on the soft capacity and over all communication quality in the long run. Therefore, an optimal way to benefit from this soft-capacity feature is to use flexible SCAC mechanisms having decision functions adapted to intercell interference distribution to compensate fluctuations of local-average SINR and hence maintain the link quality while expanding the admissible region. In such SCAC systems, instead of using only average system capacity C_a as in MdCAC, we utilize both upper and lower limits C_u and C_l in making decisions for improving the system capacity and the robustness as follows. A new call of class-k shall be:

- Admitted immediately if the current system load state c is utmost $(C_l - w_k)$;

- Admitted with a probability $\pi_k(c)$ defined below based on (12.5) if c is utmost $(C_u - w_k)$ but larger than $(C_l - w_k)$;

- Rejected immediately otherwise.

The handoff calls are admitted in nonpreemptive priority discipline [i.e., a handoff failure of class-k call occurs only when c is larger than $(C_u - w_k)$]. The state-dependent admission probability $\pi_k(c)$ is given by

$$\pi_k(c) = \begin{cases} 1, if\ c + w_k \le C_1 \\ 1 - Q\left(\frac{1 - \eta - c - w_k - fE[c]}{\sqrt{fVar[c]}}\right), if\ C_l < c + w_k \le C_u \\ 0, otherwise \end{cases} \tag{12.6}$$

Additional options can be used for the decision function that increases the flexibility of SCAC in tuning the system performance. For instance, to avoid an overestimate of C_u, which may cause an increase of local-outage probability, some adjustments of the mean and variance in $Q(x)$ function, or other functions such as incomplete-Gamma $\Gamma(v, x)$, can be used for holding back the offered traffic further in high-load states. For the latter option, $\pi_k(c)$ for $C_l < c + w_k \le C_u$ becomes

$$\pi_k(c) = 1 - \Gamma(v, \alpha(1 - \eta - c - w_k)) \tag{12.7}$$

where α and v are implicitly defined as $fE[c] = v/\alpha$ and $fVar[c] = v/\alpha^2$. In general, $E[c]$ and $Var[c]$ can be calculated by using steady-state solutions and inversion techniques. To reduce the computational complexity, the corresponding mean and variance values of the MdCAC system can be reused in the analytical method. This is because MdCAC systems well represent the stationary behavior.

SCAC policy is proposed to gain advantages over both MdCAC and MsCAC policies in simplicity, robustness, and better RRU. SCAC can also be combined with measurement-based techniques to reduce the complexity of estimators, to compensate bias of measurements, and to enhance performance. Implementation issues will be discussed later on in more detailed. Furthermore, SCAC provides better traffic shaping (regulation) gain than MdCAC and MsCAC. It gives more chances for calls of high resource-consuming classes to access the system when traffic intensity of lower classes is heavy. This cannot be improved with the other two CAC policies.

QoS Differentiation Paradigms

To regulate the operation of multiservice cellular systems as described in the introductory section so that operators can take full control of system resources and serving customs, the access control has to consider not only characteristics of requested services but also subscription profiles of users. This will provide fair decisions in resource allocation for agreed QoS upon accommodating a new call request. QoS differentiation paradigms specify serving traffic patterns depending on the offered traffic intensity of each traffic class during different busy-hour periods of the day for maximizing GoS and revenues. These have to be simple and easy to reconfigure and extend for a variety of user and network profile characteristics. The QoS differentiation for real-time calls can be viewed as traffic prioritization in CAC. This can be realized by defining multiple load-based thresholds or fracturing factors for hard or softer blocking of call requests, similar to the well-known threshold dropping or weighted fair queuing control mechanisms without actual queues. Thus, the system resources are now shared noncompletely among users and real-time traffic classes in blocked-call-cleared fashion.

Suppose there are J different user classes sorted in the decreasing order of priority—for example, 1 is gold, 2 is silver, and so forth. Let $\mathbf{J} = \{1, 2, \ldots, J\}$. Together with M different service classes sorted in the increasing order of resource consumption, it can be considered as there are $J \times M$ prioritized traffic classes, which form different group behaviors according to user or service classes. Note that hereafter the term "traffic class" or class-(j, k) is used when QoS differentiation is applied to distinguish from service class-k alone in complete-sharing scenarios. Let us introduce a $J \times M$ prioritized admission probability table given in matrix-form as follows:

$$A_0 = [a_{jk}(c)] \, with \, j \in \mathbf{J}, k \in \mathbf{K}, 0 \leq a_{jk}(c) \leq 1 \tag{12.8}$$

where $a_{jk}(c)$ is the prioritized admission probability of a new call request from class-j user for class-k service at a given system load state c.

In the hard threshold blocking case, it is straightforward that $J \times M$ thresholds may be needed, each corresponding to a traffic class. Define a threshold table:

$$L = [l_{jk}] \, for \, all \, j \in \mathbf{J}, k \in \mathbf{K}, l_{jk} \geq l_{uv} \, if \, j \geq u \, and \, k \geq v$$

Thus, $a_{jk}(c)$ in (12.8) can be given by:

$$a_{jk} = \begin{cases} 1 \, if \, c + w_k \leq l_{jk} \\ 0 \, otherwise \end{cases} \qquad (12.9)$$

The number of necessary thresholds could be reduced significantly if only group behaviors were of interests (e.g., J or M thresholds would be needed instead of $J \times M$. For example, upper bound C_u of the system capacity can be used for the gold class access in SCAC system, whereas the bronze class has a blocking threshold of lower bound C_l regardless of requested services.

In the softer blocking case, $a_{jk}(c)$ can take any value in the [0,1] interval depending on the system load-state and the priority of requested traffic class. This results in load-based fracturing factors for each traffic class or group behavior. The fractional paradigm on the one hand gives operators better flexibility to tune GoS, and on the other hand allows to unify analysis of a class of guard resource (or guard channel) schemes for QoS differentiation and CAC. The hard threshold blocking case described previously is, in fact, a special variation of this fractional rule.

Using QoS differentiation, the admission probability of a class-(j, k) call for all $j \in \mathbf{J}$ and $k \in \mathbf{K}$ in SCAC system, conditioned to the system load state c, is given by

$$A_s = [\pi_k(c) a_{jk}(c)] \qquad (12.10)$$

where $\pi_k(c)$ is defined in (12.6) or (12.7) alternatively; and $a_{jk}(c)$ in (12.8).

Up to this point, handoff calls have been assumed to have the highest priority regardless of their associated traffic classes. If not so, prioritization of handoff calls can be handled similar to that of new calls. QoS differentiation of the background non-real-time packet switched services can be done not only in blocking fashion, but also in granting different throughput-delay connections.

12.3.2.2 Derivation of Performance Characteristics

Traffic Model

To study the performance of CAC policies with respect to teletraffic performance, the following simple traffic model is used. Assume that class-k calls arrive at the corresponding cell according to Poisson process with rates $\lambda_{l,k}$ and $\lambda_{hl,k}$ for new and handoff calls, respectively. In scenarios with QoS differentiation, let $\lambda_{l,jk}$ and $\lambda_{hl,jk}$ be arrival rates of traffic class-(j, k) for new and handoff calls, respectively. The average call duration given that there are no handoff failures is $1/\mu_1$, and the outgoing handoff rate per cell per call is μ_2, which are commonly valid for each and every call. The call duration and the time interval between two successive handoffs of a call are exponential-distributed RVs. These compose the cell-resident time, which is also an exponential-distributed RV with a mean of $1/\mu = 1/(\mu_1 + \mu_2)$.

In general, μ_2 is a function of cell-sizes and user-mobility parameters. The system is assumed to be in the condition of user mobility equilibrium (i.e., the mean number of incoming mobile terminals equal to the mean number of outgoing mobile terminals per time-unit per cell). There are two simple and accurate models to approximate μ_2 reported by Thomas in [17]. Let ν be the speed of mobile terminal, L the cell perimeter, and A the cell area.

In the macro-cellular environment, terminals usually move with high speed along the cell in one direction. The linear model can be used to approximate μ_2 as follows:

$$\mu_2 = E[\nu] / L \tag{12.11}$$

In the micro-cellular environment represented by a two-dimensional model with user random movement, μ_2 is given by

$$\mu_2 = E[\nu]L / \pi A \tag{12.12}$$

The arrival rate of class-k handoff calls can be approximated by

$$\lambda_{hl,k} = \lambda_{lk}(1 - B_k)(\mu_1 / \mu_2 + F_k)^{-1} \tag{12.13}$$

where B_k is the new-call blocking probability and F_k is the handoff failure probability of class-k.

In scenarios with QoS differentiation, we have

$$\lambda_{hl,jk} = \lambda_{ljk}(1 - B_{jk})(\mu_1 / \mu_2 + F_{jk})^{-1} \tag{12.14}$$

where B_{jk} is the new-call blocking probability and F_{jk} is the handoff failure probability of class-(j,k).

Performance Measures

The performance measures of interests are the probabilities of new-call blocking, handoff failure, and QoS loss. To represent the overall teletraffic performance of the system, GoS as a cost function of network operation is defined using a combination of those probabilities that need to be optimized. The weighting factors for each component in the cost function are set based on the relative importance of new or handoff calls, user classes, resource consumption, and cost of service. For single-service systems without QoS differentiation, GoS can be defined by $GoS \equiv (10F+B)$ as in [3], where F is the handoff-failure probability and B is the call-blocking probability. For multiservice systems with/without the QoS differentiation, GoS can be defined as follows. For all $j \in \mathbf{J}$ and $k \in \mathbf{K}$, denote:

B_{jk}—the new-call blocking probability of class-(j, k);

F_{jk}—the handoff failure probability of class-(j, k);

B_k—the new-call blocking probability of class-k calls;

F_k—the handoff failure probability of class-k calls;

P_l—the QoS-loss probability.

Taking into account the multiple traffic-classes and QoS-loss probability, GoS is quantified by

$$GoS \equiv (1 - v_0 P_l) \sum_{j \in \mathbf{J}} \sum_{k \in \mathbf{K}} \kappa_j \xi_k (v_1 B_{jk} + v_2 F_{jk}) \tag{12.15}$$

where the weighting factors are set according to applied QoS differentiation paradigms. For instance, $v_0 = 0.95$ for a 5% allowable QoS-loss rate during the calls. Parameter $v_1 = 0.33$ for new call blocking and $v_2 = 0.67$ for handoff failure since handoff has higher priority than new call. Parameter κ_j for the gold user class is set to 0.45, the silver to 0.33, and the bronze to 0.22, meaning that the gold users need pay more for being served prior to the others. Finally, parameter $\xi_k = w_k / \sum_{k \in \mathbf{K}} w_k$ represents the relative resource consumption of each service class, which is proportional to the cost of service given that all calls have the same duration. Thus, GoS is dependent on offered *multimedia traffic intensity profiles* (MTIP) taking values in [0,1] and needs to be minimized. It is obvious that there is a trade-off between QoS-loss rate and GoS. The system Erlang capacity can be determined for given percentages of those probabilities or a certain GoS threshold.

Analytical Method

Results of extensive studies on stochastic loss-networks and knapsack-packing problems over the years can be applied for developing a traceable and accurate analytical method to derive the performance characteristics with our system and traffic models described above. Let us reconsider a single-link, nonbuffering and complete-sharing loss-network system with system capacity denoted by C. Let the offered traffic intensity of class-k be $\alpha_k = \lambda_k / \mu_k$. Define the set of all possible system states:

$$\Omega = \left\{ \mathbf{n} : \sum_{k \in \mathbf{K}} n_k w_k \leq C \right\} \quad (12.16)$$

The steady-state probability of system being in state $\mathbf{n}$ has a product-form

$$p(\mathbf{n}) = \frac{1}{G} \prod_{k \in \mathbf{K}} \frac{\alpha_k^{n_k}}{n_k!} \mathbf{n} \in \Omega \ \text{with} \ G = \sum_{\mathbf{n} \in \Omega} \prod_{k \in \mathbf{K}} \frac{\alpha_k^{n_k}}{n_k!} \quad (12.17)$$

The blocking probability of class-k calls can be determined theoretically by

$$B_k = \sum_{\mathbf{n} \in \Omega : \mathbf{nw} + w_k \geq C} p(\mathbf{n}) \quad (12.18)$$

For a large set of system states (i.e., large size of set $\mathbf{K}$ and C/w_k), the cost of computation with the above formulas is prohibitively high. This problem has been considered by many authors and has resulted in elegant and efficient recursion techniques for calculating the blocking probabilities. The most suitable technique for practical evaluation of various load-based and cost-effective CAC policies is the stochastic knapsack approximation as described in [18, 19]. Define the set of all feasible system load states:

$$\psi = \{ c : c = \mathbf{nw}, \mathbf{n} \in \Omega \} \quad (12.19)$$

The steady-state probability of the system load state c is given by

$$s(c) = \frac{q(c)}{\sum_{c \in \psi} q(c)} \quad (12.20)$$

with $q(c)$ in a recursive form: $q(c) = \dfrac{1}{c} \sum_{k \in \mathbf{K}} w_k \alpha_k q(c - w_k) \ \text{for} \ c \in \psi^+$; $q(0) = 1$ and $q(-) = 0$.

The blocking probability of class-k calls is given by

$$B_k = \sum_{c \in \psi : c > C - w_k} s(c) \tag{12.21}$$

Static MdCAC System

The above results can be applied directly for this system with the following parameter setting: C replaced with C_a; and $\alpha_k = (\lambda_{l,k} + \lambda_{hl,k}) / \mu$. B_k and F_k are identical and obtained by using (12.21). The equilibrium QoS-loss probability can be calculated by

$$P_l = \left(\sum_{c \in \psi} cs(c) \right)^{-1} \sum_{c \in \psi} cs(c) \Pr\{QoS\,loss | c\} \tag{12.22}$$

where $\Pr\{QoS\,loss | c\}$ is given by (12.6).

The mean and variance of the own-cell load are given by

$$E[c] = \sum_{c \in \psi} cs(c) \tag{12.23}$$

$$Var[c] = \sum_{c \in \psi} c^2 s(c) - E^2[c] \tag{12.24}$$

In this MdCAC system, for a given offered traffic intensity of each service class, the system performance strictly depends on the values of C_a and **w**. Thus, overestimates or underestimates of these parameters may result in wasting or insufficiently allocating resources, hence increasing GoS or lowering the perceived QoS, respectively. Moreover, in the presence of additional uncertainty causing the changes of traffic descriptor parameters, the system cannot provide appropriate QoS to the users, and therefore, C_a and **w** need to be redesigned. With a reliable modeling, however, good overall statistical multiplexing gain and performance can be expected. Note that C_a and **w** can be improved (maximizing C_a while keeping **w** as low as possible for the specified QoS requirements) by using techniques like sectorization, multiuser detection, and smart antennas.

Dynamic MsCAC System

In the MsCAC system, the network side attempts to learn statistics of traffic based on on-line measurements. This overcomes the nonrobust problem of the MdCAC system above. The goal of this subsection is to illustrate the MsCAC system behavior with its estimation-error sensitivity, and to lay groundwork for studying the scaled aggregate UL load fluctuations. Herein, we use the multivariate Gaussian approximation to dimension the admissible region. Let ζ be the acceptable equilibrium outage probability of the

MsCAC system. The stationary constraints of the system capacity can be defined by

$$\sum_{k \in \mathbf{K}} n_k w_k + Q^{-1}(s) \left(\sum_{k \in \mathbf{K}} n_k \sigma_k^2 \right)^{1/2} \leq C_a + Q^{-1}(s)\Theta \qquad (12.25)$$

where Θ^2 and σ_k^2 are the variance of bounded Gaussian system capacity and connection-basis resource consumption, respectively. The parameters of (12.25) defined above now rely on on-line estimations. The additive uncertainty due to measurement errors may have significant impacts on the system performance. In addition, the need for reliable estimations of various statistics may result in far more complex hardware/software implementations compared to the MdCAC systems. The over-bounding hyper-plane of the admissible region can be determined by the intersection of the axes in M-dimensional Euclidean space. The intersection point $(0, 0, \ldots, c_k, 0, \ldots, 0)$ corresponds to the single-class capacity region of class-k that is given in estimation by

$$c_k = \hat{C}_a / \hat{w}_k + Q^{-1}(s)\left(\hat{\Theta} - \hat{\sigma}_k / \hat{w}_k \sqrt{\hat{C}_a / \hat{w}_k} \right) \qquad (12.26)$$

The performance characteristics of either a memoryless or auto-regressive MsCAC system will be obtained by simulations. It has been shown in [13], however, that MsCAC policies are best suited for serving quite bursty traffic. Their performance is almost the same, and none of them are capable of meeting the loss targets accurately. One can reach the same conclusion from the above analysis.

Flexible SCAC System

The capacity of the SCAC system is extended to C_u replacing C in (12.16) and (12.19).

In scenarios without QoS differentiation, we invoke (12.20) with modification as follows:

$$q(c) = \frac{1}{c} \sum_{k \in \mathbf{K}} w_k \alpha_k (c - w_k) q(c - w_k) \; for \, c \in \psi^+ ; q(0) = 1 \, and \, q(-) = 0 \quad (12.27)$$

with state-dependent offered traffic intensity

$$\alpha_k(c) = [\lambda_{hl,k} + \lambda_{l,k\pi k}(c)](\mu_1 + \mu_2)^{-1} \qquad (12.28)$$

where $\pi_k(c)$ is the admission probability of a new call given by (12.6) or (12.7). The performance measures are obtained as follows:

$$B_k = \sum_{c \in \psi : c > C_1 - w_k} s(c)[1 - \pi_k(c)] \tag{12.29}$$

$$F_k = \sum_{c \in \psi : c > C_u - w_k} s(c) \tag{12.30}$$

The equilibrium QoS-loss probability P_l can be determined by using (12.22).

In scenarios with QoS differentiation, (12.27) can be used as such with the following modification of (12.28):

$$\alpha_k(c) = \left[\sum_{j \in J} \lambda_{hl,k} + \lambda_{l,jk} \pi_k(c) a_{jk}(c) \right] (\mu_1 + \mu_2)^{-1} \tag{12.31}$$

The admission probability of a new class-(j, k) call is equal to $_k(c) a_{jk}(c)$ as showed in (12.10). The performance measures of interests for this system are given by

$$B_k = \sum_{c \in \psi} s(c) \left[1 - \pi_k(c) a_{jk}(c) \right] \tag{12.32}$$

$$F_{jk} = \sum_{c \in \psi : c > C_u - w_k} s(c) \tag{12.33}$$

Because of the handoff priority, $F_{jk} \equiv F_k$ is independent of user class in **J**. By summing up B_{jk} over **J** or **K** set, we can obtain the corresponding loss probability for group behaviors. P_l can be obtained by using (12.22). GoS in the SCAC system is significantly improved over that in MdCAC and MsCAC systems. The operators can easily control and tune GoS of their networks by adjusting or updating the admission probability table, $A_0 = [a_{jk}(c)]$ in (12.8).

12.3.2.3 Performance Comparison and Discussions

Next, numerical and simulation results are presented in order to (1) quantify the performance of different CAC policies as well as the benefits of using flexible SCAC and QoS differentiation; and (2) demonstrate the flexibility and the accuracy of the analytical method used to evaluate the teletraffic performance. The results are obtained for a cellular system supporting three RT service classes: class-1 is 12.2-Kbps *enhanced full rate* (EFR) voice, class-2 is 64-Kbps video, and class-3 is 144-Kbps multimedia calls. The parameters are given in Table 12.1. The offered traffic intensities of the three classes are considered in the following proportions: 7:2:1 and 4:3:3 for class-1, class-2 and class-3, respectively. These are called *multimedia traffic intensity profiles* (MTIP). In other words, let λ be a so-called common divisor of the three-class offered traffic; then 7:2:1 MTIP for instance means that the offered traffic of class-1 is 7λ [Erlang], class-2 is 2λ [Erlang], and class-3 is λ [Erlang]. Thus, the total offered traffic is 10 [Erlang] if $\lambda = 1$.

TABLE 12.1 SUMMARY OF THE SYSTEM PARAMETERS

NAME	DEFINITION	VALUES
W	CDMA chip rate	3.84 Mcps
f	Coefficient of the average other-to-own cell interference	40±15% (micro-cellular)
η	Constant coefficient of the thermal noise density and maximum tolerable cell interference level	−10 dB
C_a	Average of system capacity	0.6429
C_l	Lower-limit of system capacity	0.5806
C_u	Upper-limit of system capacity	0.7200
Θ^2	Variance of system capacity	9e-03
σ	Standard deviation of local-average SINR	1.5 dB
L	Cell perimeter (micro-cellular)	200m
A	Cell area with hexagon shape	2.5981e4 m^2
$E[v]$	Mean value of user's velocity	1 m/s
r_1	Bit rate for class-1 service	12.2 Kbps (EFR voice)
$\gamma_{target-1}$	SINR target for class-1 QoS	6±1 dB for BER = 10e-03
ρ_1	Voice-source activity factor	0.4
w_1	Mean of class-1 *connection effective load factor* (CELF)	0.0050
$w_{l,1}$	Lower limit of class-1 CELF	0.0040
$w_{u,1}$	Upper limit of class-1 CELF	0.0063
σ_1^2	Variance of class-1 CELF	2.69e-06
r_2	Bit rate for class-2	64 Kbps (video)
$\gamma_{target-2}$	SINR target for class-2 QoS	2.75±0.75 dB, BER = 10e-05
ρ_2	Video-source activity factor	1

TABLE 12.1 (CONTINUED)

NAME	DEFINITION	VALUES
w_2	Mean of class-2 CELF	0.0304
$w_{l,2}$	Lower limit of class-2 CELF	0.0257
$w_{u,2}$	Upper limit of class-2 CELF	0.0360
σ_2^2	Variance of class-2 CELF	3e-05
r_3	Bit rate for class-3	144 Kbps (multimedia)
$\gamma_{target-3}$	SINR target for class-3 QoS	2±0.5 dB, BER = 10e-05
ρ_3	Data-source activity factor	1
w_3	Mean of class-3 CELF	0.0561
$w_{l,3}$	Lower limit of class-3 CELF	0.0503
$w_{u,3}$	Upper limit of class-3 CELF	0.0625
σ_3^2	Variance of class-3 CELF	7e-05
$1/\mu_1$	Mean call-holding time	120 seconds
T_p	TTI of packets	10 ms, 40 ms, 200 ms
R_p	Bit rate	32 Kbps, 64 Kbps, 384 Kbps
γ_p	SINR for the above bit rates	3 dB, 2 dB, 1 dB

In scenarios with QoS differentiation, the customers are assumed to be divided into two user classes: business ($j = 1$) and economy ($j = 2$) having equal service demands: $\lambda_{l,1k} = \lambda_{l,2k}$ for $k = 1, 2, 3$. The offered traffic patterns given by MTIP above (e.g., 4:3:3) is split in half for each user class resulting in 2:2:1.5:1.5:1.5:1.5 MTIP of six traffic classes. The business class is served as long as enough resources are available. In a hard threshold blocking scenario, the economy users are served only if less than 70% of effective resources are occupied—that is, $a_{2k}(c)$ is given in (12.9) with l_{2k} equal to 0.5 for all k. In a soft fractional blocking scenario, the economy class can share the resources equally with the business class if less than 65% of effective resources are occupied—that is, c is less than 0.47. Else if c is less than C_l, invoke (12.8) with $a_{21}(c) = 0.8$, $a_{22}(c) = a_{23}(c) = 0.6$. Else if c is less than C_u, $a_{21}(c) = 0.4$, $a_{22}(c) = a_{23}(c) = 0.3$. Otherwise, $a_{2k}(c) = 0$ for all k. GoS is defined in (12.15) with weighting factors given as follows: $v_0 = 0.95$, $v_1 = 0.33$, $v_2 =$

0.67; $\kappa_1 = 0.7$ for business and $\kappa_2 = 0.3$ for economy class; $\xi_1 = 0.06$, $\xi_2 = 0.33$, and $\xi_3 = 0.61$.

Figure 12.1 presents GoS for various CAC systems versus common divisor of the real-time offered traffic in 4:3:3 [Figure 12.1(a)] and 7:2:1 MTIPs [Figure 12.1(b)], respectively. This shows the following. First, there is no capacity gain by using MsCAC over MdCAC in serving constant bit rate real-time services. Second, the performance of MsCAC systems is sensitive to measurement errors. Third, the SCAC systems offer a significant improvement of GoS and so the Erlang capacity. Their performance can easily be tuned and stabilized by using either threshold hard blocking or fractional soft blocking QoS differentiation paradigms. Foureal-timeh, MTIPs have significant impact on the scale of GoS behavior and improvement in general. GoS of static complete-sharing systems can worsen dramatically in heavily loaded situations where the intensity of the less resource-consuming class is much higher than that of the others (e.g., in 7:2:1 MTIP). This problem has been addressed before as the traffic shaping capability.

The MdCAC and SCAC policies are simple to implement without need for special software/hardware. However, the modeling parameters are required a priori, without which the CAC mechanisms cannot operate. Due to the diverse nature of different traffic sources and their often-complex statistics, some of the parameters might be hard to determine efficiently. The soft-decision solutions are believed to offer more flexibility in determining the modeling parameters, thus achieving good multiplexing gain and robustness. On the other hand, implementations of the MsCAC policy require advanced hardware and software to ensure the reliability of measurements. For this reason, it is not a cost-effective solution. Moreover, estimation errors in some circumstances may cause significant degradations

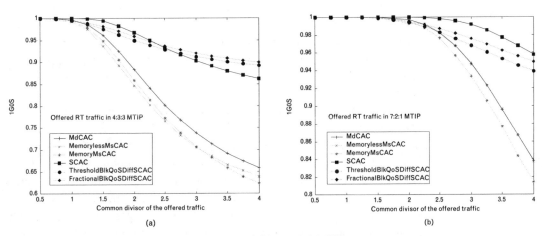

FIGURE 12.1 *GoS comparison (a) in 4:3:3 MTIP and (b) in 7:2:1 MTIP.*

of the system performance. But the advantage of MsCAC is that it seems insensitive to the traffic nature and the operation is robust. The network can learn and adapt to the statistics of traffic even when the burstiness of traffic is considered as out of control for the modeling-based systems. To gain the best from all design criteria, some parameters needed for soft-decision functions (i.e., means and variances of UL load factors) can be estimated from measurement results of UL received interference that are needed anyway for operation of TPC mechanisms.

12.4 PAC

In this section, PAC is discussed mainly for the challenge of controlling non-real-time packet radio access in UL so that it can adapt to the free capacity left by the real-time traffic with optimal throughput-delay trade-offs. Based on the understanding of the dynamics of real-time traffic and free (available) radio resources in UL, effective PAC schemes, such as DFIMA, can be designed, which are vital for the efficiency of RRU.

12.4.1 Limits of UL Radio Resources for Non-Real-Time Packet Radio Access

12.4.1.1 Assumptions

Non-real-time packet access in UL should be controlled so that optimal throughput-delay trade-off can be achieved while QoS and GoS of the real-time traffic is not affected. For the design of such dynamic control mechanisms, there is a need for a precise estimate of available UL radio resources to schedule non-real-time packet transmissions properly. Next, a reliable method, based on asymptotic and quasi-stationary analysis of the real-time traffic, is provided to predict free capacity and average upper limits of UL data throughput. First, let us make some assumptions for modeling issues.

Non-real-time services are considered to have lower priority than real-time services. The resource consumption of the non-real-time traffic is never to exceed the free resources left by the real-time traffic. Non-real-time packet transmissions share common radio channel in UL using different DS/CDMA code sequences. The amount of UL resources consumed by a packet transmission can be determined similarly to the resource consumption of a real-time connection (with much shorter service time). The parameters needed for packet transmissions (e.g., bit rate, packet-length in bits, target SINR) can be predefined or set by the control entities. The time axis is divided into T_s-long time-slots for packet transmissions, which can be equal to one or multiple radio-frame duration of 10 ms as defined in 3GPP standards. In scenarios without QoS differentiation, packet transmissions are synchronized starting at the beginning and finishing at the end of the same time-slot. For a given time-slot, all transmissions have the same bit rate and SINR target, which may be changed on slot-by-slot basis. In scenarios with QoS differentiation, different maximum delays and minimum data rates are guaranteed for different user classes. Thus, packet transmissions may have different parameters and resource consumption, which are handled by suitable PAC schemes. Let R_p be the bit rate, L_p the packet-length in bits, γ_p the target SINR to meet quality requirements, and T_p the *transmission time interval* (TTI) of a packet in UL, $T_p = L_p/R_p$. In scenarios without QoS differentiation, TTI is assumed to be a time-slot, $T_p \equiv T_s$.

12.4.1.2   Estimation of UL Resource Availability and Upper-Limit on Throughput

Let $c(t)$ be the real-time system load state at time t; $C(t)$ the system soft capacity at time t; and $z(t)$ the free resources left by the real-time traffic in the system at time t. The quasi-stationary behavior of $c(t)$ or $z(t)$ over a sustained period of time are of interest. In the equilibrium condition, we have $z = (C_a - c)$. The process of z and c itself is not Markovian in general, but for larger C_a/w_k it behaves as an approximate Markovian process. For instance, to obtain the equivalent one-dimensional birth-death process for c in complete-sharing systems, we can use the Pascal approximation as follows. First, we need to scale the system capacity and resource consumption of each real-time service class into integers. To do so we assume $\min\{w_k, k \in \mathbf{K}\} \equiv w_1$; then define the scaled load vector, the scaled system state, and the scaled average system capacity as follows:

$$\mathbf{w}^* = \{w^*_k, k \in \mathbf{K}\} \equiv \{\lfloor w_k / w_1 \rfloor, k \in \mathbf{K}\}; \qquad c^* \equiv \lfloor c / w_1 \rfloor; \qquad \text{and}$$

$C^*_a \equiv \lfloor C_a / w_1 \rfloor$, where $\lfloor x \rfloor$ is the maximum integer not exceeding the argument. The set of equivalent system states is therefore

$$\psi^* \equiv \{c^* : c^* = 0, 1, 2 \ldots, C^*_a\}$$

Let the normalized mean service time be $1/\mu \equiv 1$. The equivalent birth-death process at steady state c^* has a death-rate of c^* and a birth-rate of:

$$\lambda(c^*) = v^2 / \omega^2 + c^*(1 - v/\omega^2) \text{ with } v = \sum_{k \in \mathbf{K}} w^*_k \alpha_k \text{ and}$$

$$\omega^2 = \sum_{k \in \mathbf{K}} (w^*_k)^2 \alpha_k \tag{12.34}$$

where normalized $\alpha_k = (\lambda_{l,k} + \lambda_{hl,k})$.

The steady-state probability $s(c^*)$ satisfies:

$$c^* s(c^*) = \lambda(c^* - 1)s(c^* - 1), c^* \geq 1 \, and \sum_{0}^{C_a^*} s(c^*) = 1 \qquad (12.35)$$

The quasi-stationary probability of being in load state c^* over a period of time τ can be defined by

$$p(c^*, \tau) = \lim_{t \to \infty} \Pr\{c^*(t + \tau) = c^* | c^*(t) = c^*\} \qquad (12.36)$$

This can be approximated by an exponential function as follows:

$$p(c^*, \tau) = e^{-\tau[\lambda(c^*) + c^*]} \qquad (12.37)$$

Equation (12.37) for $\tau = T_p$ provides the equilibrium probability that there are maximum $z^* = C_a^* - c^*$ units of resources available for packet transmissions. Let $S(c^*)$ be the state-dependent upper-limit number of successful packet transmissions in a given time-slot; thus:

$$S(c^*) = p(c^*, T_p) \frac{\left(C_a^* - c^*\right)w_1}{R_p \gamma_p / W} \qquad (12.38)$$

Equation (12.38) can be used to study trade-offs of the design parameters for optimal PAC schemes. For example, assume that there are N data users in the system, each generating data packets according to a Poisson process with rate Λ per time slot, $\Lambda \leq 1$. Thus, there are $N_p = \lfloor N\Lambda \rfloor$ average number of active data sources per time slot. If each of them attempts to transmit their packets at the beginning of a given time slot with the probability of min(1, $S(c^*)/N_p$), the successful packet transmission probability [(i.e., Pr{having less than $S(c^*)$ transmissions actually initiated} according to the binomial distribution)], is at least 1/2. This is very reasonable compared to ALOHA random access. The average upper-limit of UL packet-data throughput denoted by S is given by

$$S = (1 - P_l) \sum_{c^* \in \psi^*} s(c^*) S(c^*) \qquad (12.39)$$

where P_l is the QoS-loss probability defined in the previous section. By using Little's formula, the average lower limit of packet delay is given by S/N_p time-slots.

Figure 12.2 presents the average upper-limit UL data throughput of packet transmissions, as well as the impacts of real-time traffic, bit rates, and TTI on the throughput. One can see that throughput characteristics are more sensitive to bit rate than to TTI. Such results can give valuable quantitative input data for the design of effective PAC schemes. An overview of DFIMA schemes is presented next, where feedback channel state information from CAP is used to provide 1:1 mapping of optimal transport formats for packet transmissions in next time-slot.

12.4.2 Dynamic Feedback Information Multiple Access (DFIMA)

Non-real-time packet access in UL needs to adapt to a quasi-stationary stochastic process of free capacity left by the real-time traffic. DFIMA schemes are believed to be promising candidates to optimize UL packet transmission characteristics and RRU. In DFIMA schemes, data terminals attempt to transmit their packets with a so-called *transmit permission probability* (TPP), also with bit rate and TTI changing dynamically on a slot-by-slot basis according to feedback-control information from the network. Similar schemes have appeared in earlier literature with different contexts [20]. The implementations can vary from one to another, but they all agree that feedback-control schemes are simple and effective. Based on the results from Section 12.4.1.2, optimal TPP and bit rate for packet transmissions in the next time-slot can be predicted. In case of QoS differentiation, TTI or packet-length can be changed as well. These parameters can be transmitted in the content of feedback information.

FIGURE 12.2
Throughput estimation of non-real-time packet access.

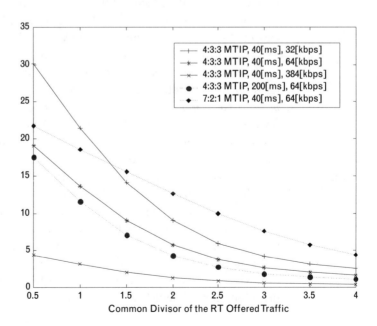

Let us use an example for illustration. It is assumed to serve as many users simultaneously as possible with optimal throughput-delay trade-off. QoS differentiation is not applied, and TTI is assumed to be a time slot, $T_p \equiv T_s$. Thus, only TPP and bit rate parameters need to be determined for optimal UL packet transmissions. The number of active data users in next time slot $N_p = \lfloor N\Lambda \rfloor$ and load state c^* of the real-time traffic at the end of current time-slot are supposed to be known. First, based on (12.37) and (12.38), bit rate can be chosen so that $S(c^*)/N_p$ ratio is as close to 1 as possible. Then, TPP for users attempting to transmit their packet in next time-slot is set equal to $\min(1, S(c^*)/N_p)$. Thus, a maximum number of users can be served with optimum UL throughput and reasonable average packet-delay of S/N_p, where S is given by (12.39). The performance of DFIMA can be optimized with respect to TPP, bit rate, TTI, and QoS differentiation paradigms that are very flexible and effective.

One notices, however, that if a terminal has a packet to send, it needs to wait until the next time-slot to attempt its transmission. Therefore, access delay to the first attempt per packet can be up to one time slot. To reduce the access delay, packet transmission can be synchronized to the beginning of a mini-slot (e.g., 10-ms radio frame can be further divided into 15 mini-slots as in 3GPP standards). This results in a spread-slotted asynchronous multiple access system with feedback control, which is similar to the schemes reported in [20] for data-only CDMA packet radio systems. DFIMA can be extended to provide flexible means of QoS differentiation. This can be done through setting different TPP/TTI/bit rate for different user classes in feedback information depending on the load state of the real-time traffic and offered load of the non-real-time traffic. Detailed analysis of these access control schemes is still to be studied. However, DFIMA can be expected to outperform the well-known ALOHA and *carrier sensing multiple access* (CSMA) in throughput-delay trade-off to the same extent as shown in [20]. DFIMA also overcomes the hidden-terminal problem of CSMA systems by using feedback control, and it is simpler and more flexible than *packet reservation multiple access* (PRMA). Moreover, the estimations of real-time system load states used in CAC mechanisms can be reused for implementations of the DFIMA schemes as well. In addition to that, non-real-time traffic demands need be measured or estimated to predict optimal parameters (e.g., TPP, bit rate, and TTI) to be used as the feedback information. To minimize the size of feedback information, hence reducing processing delays and signaling overhead, fast look-up tables can be preconfigured in both user and network sides for providing 1:1 mapping between feedback information and optimal transport format combination. It is believed that 1-octet feedback-information size is enough to ensure sufficient exchange of control information. Overall, DFIMA schemes are very promising for efficient PAC and RRU in future wireless IP networks.

12.5 Concluding Remarks

This chapter has demonstrated that CAC and PAC are the effective means of RRM and traffic engineering for providing QoS and optimizing GoS in future wireless IP networks serving various user and network profiles. Using flexible access control schemes to accommodate QoS requests adapting to interference-fluctuation statistics can result in not only efficient RRU but also better teletraffic regulation and policing capabilities. Examples are SCAC and DFIMA schemes, which are simple, robust, and effective for RRM in future wireless IP networks.

The modeling, design considerations, and analytical methods provided in this chapter can be used extensively in network planning and dimensioning frameworks to study trade-offs of system parameters, multimedia capacity, and optimal network operations. This is of significance especially for practical deployment of wireless IP networks.

Future wireless IP networks are based on a variety of emerging mobile services and radio access networks with different coverage, spectrum, radio resource allocation, and handoff strategies in order to provide high capacity and ubiquitous wireless communications. Fully exploit available radio resources in a region with many different independent radio access networks may require common RRM entity and integrated access control strategy. This can be realized with a virtual control architecture that enables seamless integration of different radio access systems. Flexible CAC and PAC with soft and quick decision-making capability are very important in such environments.

REFERENCES

[1] Perkins, C., "Mobile Networking Through Mobile IP," *IEEE Internet Computing*, Tutorial, January/February 1998.

[2] Macker, J. P., V. D. Park, and M. S. Corson, "Mobile and Wireless Internet Services: Putting the Pieces Together," *IEEE Communications Magazine*, June 2001.

[3] Zander, J., and S. L. Kim, *Radio Resource Management for Wireless Networks*, Norwood, MA: Artech House, 2001.

[4] 3GPP TS 23.107, "UMTS: QoS Concept and Architecture," V4.3.0 Release 4, January 2002.

[5] 3GPP TS 22.105, "UMTS: Service Aspects, Services and Service Capabilities," V4.2.0 Release 4, June 2001.

[6] Braden, R., D. CLark, and S. Shenker, IETF RFC 1633, "Integrated Services in the Internet Architecture: An Overview," June 1994.

[7] Blake, S., et. al, IETF RFC 2475, "An Architecture for Differentiated Services," December 1998.

[8] Dziong, Z., J. Ming, and P. Mermelstein, "Adaptive Traffic Admission for Integrated Services in CDMA Wireless-Access Networks," *IEEE J. Select. Areas Comm.*, Vol. 14, December 1996, pp. 1737-1747.

[9]   Viterbi, A. J., *Principle of Spread Spectrum Communication*, Reading, MA: Addison-Wesley, 1995.

[10] Holma, H., and A. Toskala, *WCDMA for UMTS Radio Access for Third Generation Mobile Communications*, New York: John Wiley & Sons Inc., 2000.

[11] Shin, S. M., C. H. Cho, and D. K. Sung, "Interference-Based Channel Assignment for DS-CDMA Cellular Systems," *IEEE Trans. Veh. Techn.*, Vol. 48, January 1999, pp. 233–239.

[12] Grossglauser, M., and D. Tse, "A Framework for Robust Measurement-Based Admission Control," *IEEE ACM Trans. Netw.*, Vol. 7, June 1999 pp. 293–309.

[13] Breslau, L., S. Jamin, and S. Shenker, "Comments on the Performance of Measurement-Based Admission Control Algorithms," *IEEE InfoCom Conf. Proc.*, 2000, pp. 1233–1242.

[14] Dimitriou, N., R. Tafazolli, and G. Sfikas, "Quality of Service for Multimedia CDMA," *IEEE Comm. Mag.*, July 2000, pp. 88–94.

[15] Ishikawa, Y., and N. Umeda, "Capacity Design and Performance of Call Admission Control in Cellular CDMA Systems," *IEEE J. Select. Areas Comm.*, Vol. 15, October 1997, pp. 1627-1635.

[16] Liu, Z., and M. E. Zarki, "SIR-Based Call Admission Control for DS-CDMA Cellular Systems," *IEEE J. Select. Areas Comm.*, Vol. 12, May 1994, pp. 638–644.

[17] Thomas, R., H. Gilbert, and G. Mazziotto, "Influence of the Movement of the Mobile Station on the Performance of a Radio Cellular Network," *Third Nordic Seminar*, Copenhagen, September 1988.

[18] Kaufman, J. S., "Blocking in a Shared Resource Environment," *IEEE Trans. Comm.*, Vol. 29, No. 10, October 1981, pp. 1474–1481.

[19] Ross, K. W., and D. H. K. Tsang, "The Stochastic Knapsack Problem," *IEEE Trans. Comm.*, Vol. 37, No. 7, July 1989, pp. 740–747.

[20] Phan, V. V., and S. Glisic, "Estimation of Implementation Losses in MAC Protocols in Wireless CDMA Networks," *Inter. Jour. Wireless Info. Net.*, Vol. 8, No. 3, July 2001, pp. 115–132.

CHAPTER 13

RRM in Multicarrier Allocation–Based Systems

Zhu Lei

13.1 Introduction

The trend towards convergence has propagated to the wireless communication and computer industries. As a result, wireless IP is becoming a cutting-edge technology to achieve a truly wireless Internet [1–3]. Conceptually, the terminology "wireless IP" consists of two fundamental aspects: backbone network architecture(s) and radio access networks. The first item, called the IP backbone network infrastructure, has been rapidly expanding over both private and public networks, which offer interconnection among PCs over heterogeneous networks (e.g., Ethernet, ATM, LAN, ISDN, and PSTN) [4, 5]. The latter one transmits IP packets over radio, hence enabling mobile extension of wired Internet access. Efficient radio transmission technologies are crucial to cope with wireless barriers and accomplish high transmission speeds over desired coverage areas. From the radio access perspective, this is a main focus of current wireless data services. Focusing on the performance enhancement in terms of high user peak rate, high spectral efficiency, and wide coverage, each technical approach provides its own roadmap for approaching future wireless networks. Let us take a look at standardization statues and research focuses of current wireless access technologies.

The standardization process towards wireless IP involves two cooperating organizations that working towards linking the Internet backbone network and wireless multiple access networks. The 3GPP is a collaboration agreement that was established in December 1998. The collaboration agreement brings together a number of telecommunications standards bodies that are known as organizational partners [2]. The Internet Engineering Task Force is a large open international community of network designers, operators, vendors, and researchers concerned with the evolution of the Internet architecture and the smooth operation of the Internet [3]. The Internet is supported by a backbone network connecting many computer networks and based on a common addressing system and communications protocol called TCP/IP. Since its creation in 1983, it has grown rapidly beyond its largely academic origin into an increasingly commercial and popular medium. By the mid-1990s, the Internet connected millions of computers throughout the world. Many commercial computer networks and data services also provided at least indirect connection to the Internet. The original uses of the Internet were e-mail, file transfer (using the File Transfer Protocol), bulletin boards and newsgroups, and remote computer access (telnet). The World Wide Web (WWW), which enables simple and intuitive navigation of Internet sites through a graphical interface, expanded dramatically during the 1990s to become the most important component of the Internet [4]. Following the convergence of wireless and IP, 3GPP and IETF become closely cooperating partners to realize wireless IP.

The key problem for wireless IP networks is spectral efficiency, due to spectrum scarcity. Therefore, solving the paradox of high spectral efficiency and radio link quality interrupted by heavy cochannel interference is a problematic issue. In CDMA type systems (e.g., WCDMA and cdma2000), unit frequency reuse has been employed. For users in a bad radio link situation, the higher spreading factor (lower data rate) is chosen, a so-called trading user data rate for radio link quality within a given frequency band. In TDMA/ FDMA type systems (e.g., GPRS/EDGE), a more aggressive reuse pattern (reuse 1/3 for traffic channels) is considered. To cope with radio link quality variation, variable modulation and coding schemes are used; in addition, incremental redundancy (diversity technique on time domain) is employed in EDGE [6]. In general, these approaches are attributed to a single carrier solution in which variable data rate is achieved by link adaptation and incremental redundancy, and high user peak rate is achieved by timeslot aggregation and/or code aggregation to a user. In this case, fast rate adaptation and fast power control are required to achieve good capacity on fast fading channels. The realization of better link adaptation performance is relay on accurate link quality estimation, although high data rate transmission results in fast changing link conditions make it more difficult to accomplish perfect link adaptation. In contrast to the single carrier approach, a multicarrier approach offers possible solutions on frequency domain. Instead of fast link adaptation, slow link adaptation in combination with frequency diversity can combat with fast fading channels. It is commonly believed that multicarrier is a good candidate of achieving high data rate transmission (e.g., in cdma2000, WLAN, DAB and DVB) due to its natural advantage of overcoming *intersymbol interference* (ISI). Currently, such multicarrier systems provide transmission data rate up to 10 Mbps or more [7–9]. One possible direction is the harmonization of current personal communication systems that enable connection at high data rate to efficiently enable IP over wireless, as well as a good technology that has properties of future wireless networks.

Since the main theme of this chapter is RRM of *multicarrier allocation* (MCA) system for wireless IP networks, a catalog of MCA- and non-MCA-based systems is defined and described under a multidimensional concept, where a non-MCA-based system indicates a multislot and/or a multicode system. Examples of MCA and non-MCA systems will be given. Looking from the property of a single radio frequency (also called a carrier), a higher data rate, which implies less spreading delay, can be tolerated by the radio channel; as a consequence, a wider frequency bandwidth is required to insure quality of the transmitted signals, otherwise ISI occurs. Definitions and extensive discussions about MCA and non-MCA-based systems are given in the next section, followed by a section about the impact of MCA-based systems on wireless IP networks.

After reviewing MCA- and non-MCA-based systems under a multidimensional concept of radio channels in Section 13.2, the impact of MCA-based system on future wireless IP networks is discussed in Section 13.3. Using system model and performance measures in Section 13.4, investigations are done for diversity, power control, and carrier grouping techniques in Sections 13.5, 13.6, and 13.7, respectively. Concluding remarks are offered in Section 13.8.

13.2 A Multidimensional Concept of Radio Channels in Future Wireless IP Networks

This section gives a fundamental definition for a channel in future wireless IP networks, which contains frequency and time domains (and could be extended to the code domain as well). Using such a basic definition, a framework called multidimensional channel allocation is proposed in which systems are catalogued into MCA based and non-MCA based. We then illustrate and discuss existing systems as specific examples, with emphasis on MCA-based systems.

A specific example of non-MCA-based systems is GPRS/EDGE type systems, in which data transmission is achieved by means of multiple time-slots over a single carrier, hence fast link adaptation (and/or fast power control) is required to enable radio link transmission of different data rates based on different link quality. Although due to bursty traffic properties, interference changes very fast when compared to circuit switched systems, and therefore, long-term prediction that could be used, for example, for accurate closed loop power control is difficult. Ways to solve such a problem so that network spectral efficiency can be enhanced is either to use fast link adaptation/power control on a single frequency or slow link adaptation/power control on multifrequency. To take full utilization of channel fluctuation and to transmit as much information as possible over a channel, fast link adaptation tries to trace and predict the fast changing channel conditions in a shorter time scale. Hence, it needs information on interference level, path gain, transmit power level, and channel availability on each time slot. Previous research work shows that link quality measurement (e.g., SNIR or interference level estimations) should be fast and accurate. In modern wireless high-speed data services, shortening the time from link quality measurement to the time of transmission decision made is a critical issue for implementing such an approach.

The strategy of an MCA system is to give a user more channels on the frequency domain, in combination with slow link adaptation and slow power control schemes. The cdma2000 and other orthogonal frequency division multiplexing (OFDM)-type systems (e.g., WLAN and Bluetooth) belong in this category. The next section will give extensive discussions on the impact of an MCA system, while a comprehensive survey on multi-carrier CDMA is in given in [8]. The application MCA system (e.g., cdma2000) has been specified in 3G standards (see Chapter 4).

Looking from the aspect of RRM, definitions of a channel are very interesting to trace. In 1G analog systems (e.g., NMT and TACS), a channel usually meant a narrow frequency band [i.e., frequency division multiplexing (FDM) type systems]. In 2G digital systems, a channel meant one time slot in a *time division multiplexing* (TDM) type system, or one code in a *code division multiplexing* (CDM) type system. In contrast, future wireless technology is aiming at high-speed data transmission and high spectral efficiency for multimedia services. The wireless Internet has been a driving force in the development of research work on high-speed downlink data transmission. Consequently, it brings up the context of multidimensional radio resource management. Consequently, a channel should be defined in future wireless IP networks (i.e., frequency, time, and code domains). Also, in contrast to conventional single channel assignment for voice service, multichannel assignment is applied to fit variable data transmission for future wireless multimedia traffic properties. The multichannel assignment framework consists of frequency, time, and code domains, and so-called multicarrier, multislot, and multicode approaches. The main idea behind this is to offer user variable data rates (so-called multirate) in a more flexible way and to give more freedom on system design so that high spectral efficiency and high user peak rate can be accomplished. Not only in modern wireless systems but also in conventional analog and digital systems is cochannel interference a main obstacle to realize high spectral efficiency. Especially in wireless multimedia services is it becoming more challenging since the interference appears bursty, with fast changing and hardly predictable properties.

To offer a high-speed data transmission on downlink, a framework called multichannel assignment in packet-based wireless systems is proposed, under certain QoS constraints (e.g., traffic class based time delay, packet loss rate). Moreover, the channel is defined on both the frequency and time domain. Hence, the term "multichannel" is interpreted as multiple carriers (multicarrier) and/or multiple time-slots (multislot) in this chapter. It should be noted that a multiple code (multicode) approach is essential as well. The core problem here is using interference handling techniques dealing with cochannel interference in an interference limited cellular data system (e.g., GPRS/EDGE type system) to maximize the possible network throughput with the given bandwidth. Under this general scope, the research interest is narrowed down to discover cochannel interference handling techniques in multichannel assignment areas. Then the possible directions for combining scheduling schemes is discussed. Scheduling is a good method to reduce interference, especially on media traffic loads, by avoiding links with heavy interference from interfering each other so that limited radio resources can be allocated in an efficient way.

The fundamental definition for channel allocation studies is a channel. As shown in Figure 13.1, the definition of a channel is a burst at a certain carrier assigned to a BS. Bursts are queued on the time domain, although transmission could be done over channels on multicarrier and/or multislot fashion. This definition forms the basic frame of MCA and non-MCA systems.

Consider a finite cellular network consisting of N cells, corresponding to N BSs located at the center of each cell. The definition of a channel is system level orientated. Consider a finite system consisting of N cellular cells. A BS, connected to the backbone network, supports each cell. The available frequency band is denoted by a set of carriers. The number of users supported by such a network is a set of $K = \{(k)\} = \{1,2,3...N_k\}$. Assume that each BS is perfectly synchronized with its closer BSs. Such an assumption has been commonly used in similar type system modeling. The cochannel interference occurs among the same channels using the same frequency and the same time slot in different BSs.

The available radio resource to be assigned here is channel. Each channel is defined on both the frequency and time domain, as shown in Figure 13.1. A channel is $CH_{t,f,k}$ where f denotes the carrier selected and t denotes the time slot selected by user k. For each user admitted by the network, a group of available K channels can be used by a user. Look into the problem within one time frame over all the available carriers in a BS. A single channel quality is computed by

$$\gamma i, f, t = \frac{\hat{S}_i}{\hat{I}_i + n_i} = \frac{P_{i,f,t} \cdot G_{ii}}{\sum_{j \neq i} P_{j,f,t} G_{ji} Q_{j,f,t} + n_i} \tag{13.1}$$

FIGURE 13.1
The definition of a channel in a three-dimensional model (four-dimensional with code domain).

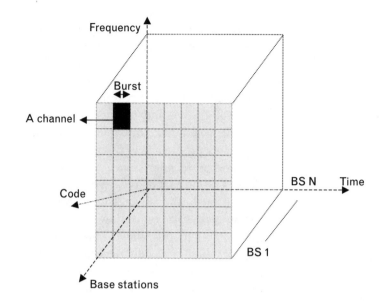

where $\hat{S}_i$, $\hat{I}_i$ and n_i and denotes estimated signal, cochannel interference, and noise strength received by user i. Assume the constant transmission power $P_{j,f,t} = P_o$ to keep neat the parameter settings at the current stage and to simplify the research problem. The channel quality information (e.g., SNIR) is connected to transmission rate. In case of multichannel allocation, focusing on a specific user, the channel quality status defined by a vector to express the quality of all K channels a user has

$$\overline{\gamma}_{(k)} = \{\gamma_{i,f,t} | CH_{j,f,t} \in CH_{(k)}\} \tag{13.2}$$

The basic disciplines for assigning channels are SNIR based (where SNIR information is quite difficult to obtain), interference based [like *dynamic packet assignment* (DPA) and *dynamic channel assignment* (DCA)], and pass gain based. The current method for channel assignment is to use historical knowledge about channels, with the assumption that at the time when a channel(s) was assigned to a user, the historical channel information obtained at the moment of measuring still works (or at least helps). The investigations done in Sections 13.5, 13.6, and 13.7 have been done based on this conventional assumption, commonly used in circuit switched networks.

13.3   The Impact of MCA Systems on Wireless IP Networks

In brief, a multicarrier system divides the complex problem of wideband transmission into a set of simply conquered narrowband transmission problems. Shannon's classic paper from 1948 identified multicarrier transmission as the optimum method to solve complex transmission problems. He pointed to multicarrier transmission is a good method to handle ISI over linear channels. The idea of subdividing a signal frequency band into a set of contiguous bands was recognized very early in the fields of signal processing and data communications to be a powerful technique for achieving efficient system realizations. Now such a method has been extended to wireless systems like multicarrier CDMA [8]. In contrast to WCDMA, multicarrier CDMA divides a single frequency band in to M subcarriers for the downlink. The driving force for such a waveform design is the ease with which one can convert from a narrowband CDMA signal to a wideband CDMA signal, by means of taking M narrowband CDMA carriers and assigning them all to one user [10]. Hence, the usable frequency band of a user can be increased by a factor of M. An overview of multicarrier CDMA systems has been given in [11].

In the wireless world, high spectral efficiency and flexible data rate access are the focus of future wireless IP networks. The current research work is aimed at maximizing spectral efficiency by means of channel allocation schemes, partially involved in the frequency plan approach. The conventional channel allocation concept where only one channel is assigned to the uplink or downlink between an MS and its connected BS, has been extended to three-dimensional (or more) channel allocation to support high data rate services. This means that more than one channel is assigned to an MS from its serving BS. These channels could be time-slots and/or frequency bands. With multiple frequency allocation, there is a possibility of a frequency diversity gain when the signal is transmitted over multipath radio channels. The conventional channel allocation can be catalogued into *fixed channel allocation* (FCA), *dynamic channel allocation* (DCA), *hybrid channel allocation* (HCA) of FCA and DCA, and *random channel allocation* (RCA) [12]. These channel allocation schemes try to maximize the total capacity of wireless systems assuming constant data rate per user. When variable data rates are needed, as is the case in multimedia communications, a new dimension should be added into the allocation domain. A channel allocation scheme should not only maximize the capacity in terms of number of users per cell but its throughput as well. Obviously, the design of the system becomes more complicated as more variables exist. A variable rate system can be achieved through the allocation of multiple slots, multiple carriers, and/or through link adaptation by using variable modulation and coding. Previous studies showed that multiple channel allocation combined with link adaptation could improve the system throughput as compared to single channel allocation without affecting the system coverage [13].

Aiming at transmitting IP packets over a wireless bearer efficiently, the following sections give a glance at current wireless access technologies for wireless IP, and then from the aspect of RRM they introduce an MCA approach. The rest of this chapter provides deep insight into basic properties of an MCA system for future wireless IP networks, in terms of spectral efficiency (average user rate) and system fairness. As noted, enhancing spectral efficiency implies dense frequency reuse that causes heavy cochannel interference in such a wireless network. So, dealing with cochannel interference is an efficient way to improve downlink throughput. In the following sections, more about this topic, with an emphasis on channel allocation, frequency diversity with maximal ratio combining, power control, and carrier-grouping methods will be discussed. The results show a promising improvement on downlink. As a reference from the network perspective, these results are usable to determine buffer size, reasonable network load, or further intensive work both on the network level and the link level. In general, it offers properties of an MCA system so that IP packets can pass through the wireless barrier more efficiently.

13.4   System Model and Performance Measures

This section describes a system model of an MCA system in which a group of carriers is assigned to an MS from its connected BS. To improve system capacity and promote coverage within a system, the MCA system is used in combination with frequency diversity, which will be described in Section 13.5.

Consider a wireless network with a total of C orthogonal frequencies. Each frequency provides a constant information data rate of R bps. These frequencies could be the subcarriers of an OFDM scheme or simply the carriers in an FDM scheme. These frequencies are used by the network to cover an area with N BSs in set B. The network infrastructure of an MCA system is that each user is assigned a fixed group of K carriers for its information transmission. As a basic tool, Figure 13.1 clearly shows that these carriers on frequency domain are assigned to a user, while those on time domain and BS domain are kept the same. Thus, depending on the interference situation, every active user can use up to K carriers simultaneously. The average throughput of a user is then obtained as the sum of the effective data rate in K frequencies. Notice that the number of users per cell is not affected by this modification. In other words, by using this structure, the system will keep the same number of users, but each user can improve its peak throughput up to K/R bps.

One way to adapt this system to these new requirements is to transform it into a system with universal reuse and assign a fixed group of K frequencies to any given user admitted into the system with $K = \kappa$. The system is characterized by its reuse distance and average channel outage probability P_{out}, defined as $P_{out} = \Pr\{\Gamma_{i,j} < \gamma_o\}$. The channel outage probability represents the fraction of time where the radio channel is not usable. The parameter γ_o represents the required threshold for good reception quality (related to the type of modulation and coding scheme used in the system), and $\Gamma_{i,j}$ represents the total SNIR.

For circuit switched networks, the outage probability should be kept small to ensure a good connection (2% outage can be considered a good value for channel outage probability). Even though it provides low outage probability, such a design procedure is not attractive for multimedia type systems where the traffic is bursty and the user data rate is changing. In such situations it will take a long time to transfer files from one point to another and will further require a large buffer at the transmitter. Therefore, some modifications are needed to increase the throughput and reduce user outage probability (represents system fairness).

Let us consider a system with M users and K frequencies, where each user has a constant data rate R per frequency. The effective data rate of user n is denoted as

$$R_{e(n)} = \left(\sum_{i=1}^{K} X_{n,i} \right) R \tag{13.3}$$

where the index is

$$X_{n,i} = \begin{cases} 1 & \Gamma_i > \gamma_0 \\ 0 & elsewhere \end{cases} \tag{13.4}$$

The performance measure for average user transmission ability is the average throughput per user, defined as

$$\lambda = \frac{\sum_{n=1}^{M} E\{R_e(n)\}}{M} \tag{13.5}$$

and $E\{\}$ represents the time average. Another performance measure for spectral utilization is spectral efficiency, defined as

$$\rho = \frac{\lambda}{K} = \frac{\sum_{n=1}^{M} E\{R_e(n)\}}{M.K} \tag{13.6}$$

For measuring the system fairness, user outage probability is introduced, which states the probability that an arbitrary user arriving in such a network cannot get any data transmission under a certain traffic load q.

$$\Omega_{out} = \Pr\{R_e(n) = 0|q\} \qquad (13.7)$$

Here, traffic load q, represents the ratio of the average number of arrival MSs and the number of maximum supportable users in one cell. The user outage will be used in measuring performance impacts of power control and carrier grouping.

The performance enhancement by means of frequency diversity, power control, and carrier grouping are discussed further in Sections 13.5, 13.6, and 13.7. The impacts of these techniques are illustrated by conducting simulations.

13.5   Frequency Diversity for an MCA System

To improve system capacity and promote fairness, this MCA scheme is used in combination with frequency diversity using *maximal ratio combining* (MRC). Thus, a minimum required throughput can be guaranteed within the system even to users in bad fading positions. The previous work on link level showed that frequency diversity and MRC gave a significant improvement in terms of average throughput per user and user outage probability, compared to the one without frequency diversity and MRC [6, 14].

By applying frequency diversity and MRC in MCA, all users close to their BSs will be able to use all the frequencies assigned to them, and users far away from their cells may be able to use only part of the K frequencies. A good degree of fairness can also be provided within the system by means of frequency diversity. Users who are unable to extract the full rate from each of their carriers can use transmitter (frequency) diversity and achieve their connections with a lower data rate. The investigations are done using parameters listed in Table 13.1, how much is gained by using frequency diversity from network aspect. Using the model proposed in Sections 13.5.1 and 13.5.2, an investigation of this comparison is performed, which constant transmitted power is assumed.

13.5.1   MCA Scheme Without Frequency Diversity

In this scheme, a group of frequencies is first assigned to a user followed by computation of the SNIR experienced by each frequency within the system. The central node starts by shutting off frequencies that have the lowest SNIR, one by one until the point where all the remaining links have good connections. Since all frequency groups are used in every cell, a better system throughput is expected for the same capacity of C/K users/cell. This scheme, however, requires a lot of signaling where all link information needs to be reported to one node. Its computational complexity also

TABLE 13.1   PARAMETERS USED FOR STUDIES OF AN MCA SYSTEM WITH AND WITHOUT FREQUENCY DIVERSITY

Cell number	$N = 49$
Carriers for one user	$1 \leq K \leq 6$
Micro-cell radius	$R = 800\text{m}$
SIR threshold(s)	10, 7, 6 dB
Shadow fading correlation	0.5
Propagation constant	4
Channel active factor	100%
Simulation confidence	2,000 times

increases with the number of users within the system. One way to reduce this complexity is to use the scheme described in the next section, where the selection of the appropriate frequencies for all MSs is done locally.

13.5.2   MCA Scheme with Frequency Diversity

In this scheme, each base station acts independently of all other BSs. Considering the downlink case (from the BS to the mobile), each MS measures the link gain and the total interference plus noise power appearing during demodulation of each frequency within its group. With this information, the mobile station computes the SNIR of each frequency. For instance, a user is using K carriers simultaneously. Diversity is only applied adaptively when a single link cannot reach the quality threshold. SINR of each frequency is then compared with the channel quality threshold. When a frequency cannot fulfill the channel quality threshold, the frequency diversity is applied. This frequency will be randomly combined with other frequencies lower than the quality threshold within the same group. The diversity procedure starts by combining two frequencies and ends up with three. The SNIR quality threshold varies due to the number of combined frequencies.

For example, if two frequencies are used to carry the same information, and assuming independent flat fading on each frequency, the combined signal will provide a diversity gain of order two. With this diversity gain, the average bit error probability of the link is improved and the required threshold for good quality of reception is lowered as compared to the single frequency quality threshold. Combining more than two frequencies will require lower thresholds. The required thresholds are used as frequencies for the same information signal with MRC; the investigation is carried out for two and three frequencies combination, respectively. Based on this diversity combining, the multicarrier allocation scheme can take advantage of both the multipath fading channel and the interference situation within the system. Here, transmission decisions made by an MS. Note that the diversity combining can also be used in the first scheme, the centralized fashion. It has been found, however, that such a combination increases the complexity of the first scheme considerably, making it too difficult to use.

Figure 13.2 illustrates the average throughput of the first and the second multicarrier allocation schemes as a function of the number of frequencies per group. Note that the average throughput per user increases with the number of frequencies per group, K. It is also observed that the second scheme performs better than the first scheme. The reason for this is that the second scheme takes advantage of the diversity gain obtained through the combination of multiple frequencies instead of switching them off. Of course, adding this feature to the first scheme will provide good gains in its throughput as well. However, as mentioned earlier, such a feature will increase the complexity of the first scheme considerably.

Figure 13.3 illustrates the normalized spectral efficiency for the two schemes as a function of the number of frequencies per group, K. It is observed that the second scheme has a worse normalized spectral efficiency for $K = 1$ but outperforms the first scheme for higher values of K. This is natural because transmitter (diversity) can be applied when K is larger than one. Note that the normalized spectral efficiency of the first scheme is almost independent of K. Thus, for a cellular system with the parameters given in Table 13.1, the average throughput per user is a linear function of K with a factor of 0.4 approximately. The same tendency can also be observed for the second scheme but with a much better throughput of approximately $0.7 \ K$.

FIGURE 13.2
Average throughput per user in an MCA system.

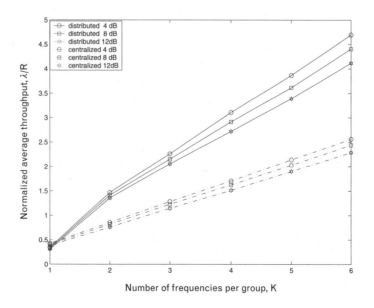

FIGURE 13.3
*Spectral efficiency in
an MCA system.*

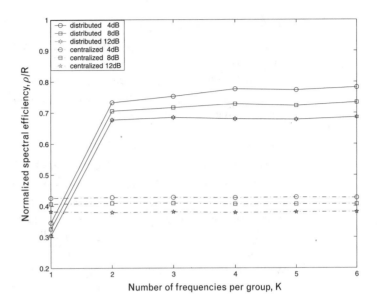

In fact, the second scheme provides a gain of about 50% in spectral efficiency. With its simplicity, this second scheme seems attractive and suitable for future multimedia applications. One might think of using power control especially with this second scheme. With a proper adjustment of the transmitted power in the different links, better interference management can be achieved, and as a result, better throughput could be obtained. This point is left for further investigations in Section 13.6.

13.6   Power Control for an MCA System

This section shows that the MCA system using frequency diversity gives better performance. In spite of the carrier allocation approach, one may naturally think about using power control to enhance system performance. The consequence of doing so is to obtain frequency diversity gain and multipath gain, and reduce the SNIR quality threshold. Moreover, coupling power control into the previous one yields more downlink enhancements [15].

Numerous research works in power control can be catalogued into centralized and decentralized ways. The experience from previous work reveals that the core problem of a distributed power control scheme is a trade-off between minimizing outage probability and increasing convergence speed [16]. System capacity was further enhanced by using both channel allocation and power control schemes [17, 18].

In this section, the performance of power control applied to an SNIR-based close loop strategy is described. The advantage of doing so is to use the downlink SNIR report for both channel assignment and power setting decision, hence saving information transmission for channel situation. Logically, allocating more carriers to a user and using micro-diversity, a better fairness (lower user outage) can be accomplished. Consequently, the relation between the spectral efficiency and the system fairness is indeed interesting in that a trade-off might appear. This issue will be described at the end of this section.

Consider a downlink power control performed incorporating with channel allocation in centralized and distributed fashion. Scheme 1 performs power control and channel allocation on the network level, while the distributed scheme does so in each MS locally. It has been shown in previous work that the distributed fashion gives better system throughput and spectral efficiency because of its gain from micro-diversity. In addition to its easier deployment in terms of system design, it was emphasized in the research work on the distributed MCA-based power control schemes to further improve system performance. However, scheme 1 is also proposed as a reference of comparison.

The power control is closed loop SNIR based, assuming accurate SNIR estimation and the unchanged status from the time of estimation to the time of performing power control. Note that channel and power settings share the same information. This gives the advantage of saving information needed on control channels. The closed loop SNIR-based power control is assumed to have no dynamic range and no updated step limitation (in a real system, 30-dB dynamic range and minimum 1-dB power update step are commonly used); the power control adjustment is based on the estimated downlink SNIR. Assume that all the MSs change their powers once in a time interval. Moreover, these changes will be done one by one, which implies that the actions of the previous MSs have impacts on the decision of the next one. Its corresponding transmission power vector that is going to be changed at a certain time interval $t \rightarrow t + \Delta t$ is denoted as

$$\widetilde{P}_n^{(t)} = \left[P_{1,in}^{(t)}, P_{2,in}^{(t)}, \dots P_{K,in}^{(t)} \right] \tag{13.8}$$

Assume that $\widetilde{P}_n^{(0)} = P_{\max}$, which represents the pilot power level. The initiate power levels on traffic channels are based on

$$P_{j,in}^{(1)} = \frac{P_{j,in}^{(0)}}{\Gamma_{j,in}^{(0)}} \cdot \gamma_0 \quad for \ j \in [1, K] \tag{13.9}$$

where γ_0 is the SNIR quality threshold for single carrier transmission. The transmitter power on traffic channel of nth user at tth time instant is given by

$$P_{j,in}^{(t)} = \min\left\{ P_{\max}, \gamma_d \frac{P_{j,in}^{(t-1)}}{\Gamma_{j,in}^{(t-1)}} \right\} \text{ for } j \in [1, K] \tag{13.10}$$

where $t > 1$ and $n \geq 1$, both t and n are integer values, and γ_d is the SNIR quality threshold for a certain error probability requirement, which depends on the number of branches used to transmit the same information signal with maximum ratio combining, expressed as

$$\sum_{j=1}^{d+1} \Gamma_{X_j,in} \geq \gamma_d \tag{13.11}$$

where $X_j \in [1, K]$ represents any channel involved in one maximum ratio combining among K carriers, assigned to user n. Contrary to that described in Section 13.5, where the number of carriers involved in one MRC is two or three, we now look at the possible combination of 2, 3, 4, 6, and 8 carriers, so that user outage probability can be minimized. In parallel with the previous section, two schemes in both centralized and distributed fashion are considered. The parameters used are listed in Table 13.2.

13.6.1 Scheme 1: MCA with Power Control

The channel and power assignment is in the network level. This scheme performs based on the predicted downlink SNIR. A transceiver lowers the

TABLE 13.2   PARAMETERS

NUMBER OF CELLS IN THE NETWORK	$N = 49$
Number of carriers used by a user	$K \in [4, 6, 8]$
Cell radius (microcell)	$R = 800m$
Shadow fading deviation	$\sigma_s = 8$ dB
SNIR thresholds γ_d	$\gamma_0 = 10$
	$\gamma_1 = 7$
	$\gamma_2 = 6$
	$\gamma_3 = 6$
Shadow fading correlation	$\rho_{ss} = 0.5$
Propagation constant	$\alpha = 4$
Channel activity factor	100%
System traffic load	0.3, 0.5, 0.7, 1, 2
Results confidence corresponding to traffic load	10,000 simulations

transmission power when SNIR is higher than the threshold, and shuts off the worst channel (the one with the lowest SNIR) in the network. Such a process will be performed continuously until all the remaining channels are qualified in terms of SNIR. In this scheme, the power control scheme mentioned above will always be used with $\gamma_d = \gamma_o$.

13.6.2   Scheme 2: MCA with Frequency Diversity and Power Control

Contrary to scheme 1, the channel and power assignment decisions on each MS are used together. In addition, micro-diversity is applied here to increase system coverage in terms of user outage probability. Each mobile is trying to assign as high as possible data rate by starting from data rate R for each channel, then $R/2$, $R/3$, or $R/4$ (depending on number of carriers per user, K). At the same time, all the assigned channels are trying to be used, instead of using some and switching off the others.

The investigation of the above schemes will be performed in a serving area covered by N omni-regular hexagon cells. The parameters used in the simulation environment are listed in Table 13.2. Comparing scheme 1 and scheme 2, the average throughput is evaluated, as is spectral efficiency and user outage probability, defined in Section 13.3. Among these, average throughput and spectral efficiency illustrate the total system throughput, and user outage probability demonstrates the degree of fairness of such a system.

Different carrier combinations based on different number of carriers is investigated. For instance, when four carriers assigned to a user, look at good channels that are able to transmit signals on one branch (i.e., channels with SNIR above the first target). If this fails, look at the second target that combines two branches. If this still fails, finally look at the third target that combines four branches. For six carriers per user, instead of looking at four carriers combination, look at three carriers, then six carriers combination. For eight carriers per user, steps are the same as for four carriers per user, and add the final step that looks at the eight carriers combination. For carriers that do not succeed, the power of these branches will be shut off, and this user will be counted into outage.

In Figure 13.4, scheme 2 has a higher average throughput than scheme 1, due to frequency diversity. The results show that power control improves the average throughput, since it reduces the cochannel interference and results in more qualified and usable channels. It is observed that the average throughput increases while assigning more carriers to each user, which is easily explained by the fact that the more carriers given to a user, the more freedom to use these carriers. The average throughput increases as traffic load gets higher.

Figure 13.5 shows the spectral efficiency. First, the spectral efficiency of scheme 1 is almost the same as scheme 2 without power control. Second, scheme 2 is better than scheme 1 in the sense of spectral efficiency. As traffic

FIGURE 13.4
*The average through-
put in an MCA sys-
tem.*

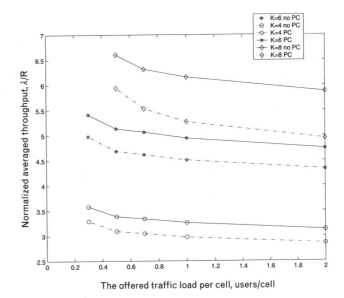

FIGURE 13.5
*The spectral efficiency
in an MCA system.*

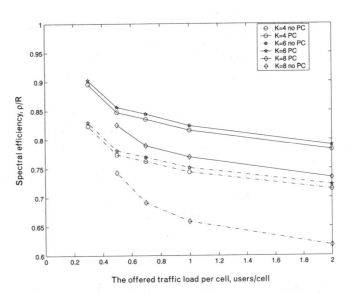

load becomes higher, the spectral efficiency becomes lower. The interpreta-
tion of this is that the more users, the more interference for each user to
cause lower average throughput per user.

Figure 13.6 not only depicts the user outage probability of proposed
schemes, but also shows the trade-off for having average throughput and
spectral efficiency at one side. By summarizing Figures 13.4, 13.5, and 13.6,
it is noticed that due to the low interference at low traffic load, the average
throughput is higher and outage is lower compared with that at the high
traffic case. One interesting phenomenon is that increasing carriers per user

FIGURE 13.6
*The user outage in an
MCA system.*

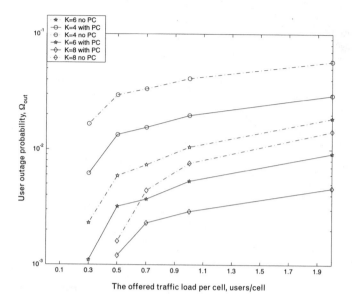

from four to six gives improvement on both system throughput and fairness. However, increasing carriers per user from six to eight caused improved fairness but the price to pay is spectral efficiency. Generally, scheme 2 gives about 8% spectral efficiency improvement and half user outage of the no power control cases, and this gain is independent of traffic load and the number of carriers per user. It is also noticed that six carriers per user gives the highest spectral efficiency. The user outage at four carriers per user is about five times that of six carrier per user, and about 10 times that of eight carrier per user.

It has been shown that power control is a robust way to combat with cochannel interference. Moreover, another technique to deal with co-channel interference is employing interference averaging technique, called carrier-grouping technique in coming section.

13.7   Carrier-Grouping for an MCA System

As we enhance spectral efficiency, the outage is another prime. User outage is the probablility that a user in the network receives no data transmitted in a certain time interval. Channel outage is the probability that a channel is lower than the SNIR threshold. User outage indicates the network coverage, while channel outage indicates the number of retransmissions of a channel. In this section a method of channel allocation is proposed to achieve interference diversity in a multicarrier-based system shown in Figure 13.7, and the different ways of transmitting data on a group of sub-carriers are studied and discussed. Such an approach provides a good

FIGURE 13.7
*Carrier-grouping
method: (a) fixed-
grouping method, and
(b) random-grouping
method.*

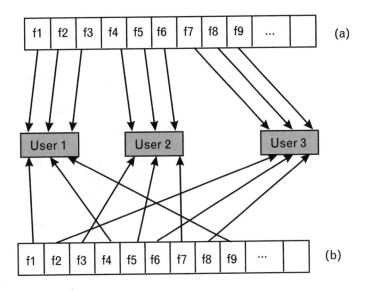

research direction for multicarrier receiver design. Since orthogonality can hardly be preserved when OFDM signals are transmitted over radio channels, adjacent channel interference will cause serious degradation on SNIR value and reduce system throughput. By using the channel-grouping method, each user gets a group of carriers that are not neighboring ones, so enhancing performance by reducing adjacent channel interference. Of course, from a network point of view, as traffic gets lower, there is more gain from both interference diversity and illuminating adjacent channel interference.

In Sections 13.5 and 13.6, a group of frequencies was assigned to an arrival user, in which the frequencies in a certain subcarrier group were the same in every cell at each time, defined as fixed carrier-grouping [see Figure 13.7(a)]. Each time a user arrives, the network chooses one subcarrier group from these three groups and assigns it to the user. The chosen group is fixed, and all the frequencies in the chosen group will experience the same cochannel interference. The disadvantage of the fixed carrier grouping is that there is no link quality variation among frequencies, hence no interference diversity within a subcarrier group. Alternatively, an interference averaging technique can be used into carrier-grouping methods, so that link quality variation occurs within a subcarrier group. Figure 13.7(b) shows random carrier-grouping. Taking an example of nine carriers, for each user who requests K channels, there are unoccupied ones among nine carriers; then randomly pick three from the unoccupied ones, for instance, {f1, f4, f9} {f3, f5, f7} {f2, f6, f8}. Due to changing frequencies in one group, interference diversity is achieved. This decreases the user outage probability, especially in a low traffic load case. As traffic load gets higher in the network, the link quality variation gets smaller compared to the low traffic case, hence interference diversity gain gets smaller. From the view of the whole radio network, frequencies are used in a random manner by means of random carrier-grouping; such average cochannel interference over all frequencies is called the interference-averaging approach. This is a commonly used technique for frequency hopping or CDMA type systems. The benefit of using the interference-averaging technique is that no channel status information needs to be transmitted, hence shortening the time and simplifying the process of channel assignment. The price to pay is that the performance will not be as good as if the interference-avoidance technique were used, where the channel assignment is based on the information of channel interference levels, signal strength, and so forth.

Comparisons between the spectral efficiency and user outage probability of fixed and random carrier-grouping are presented in Figure 13.8. In general, random carrier-grouping outperforms the fixed one in terms of user outage probability, due to interference diversity gain. It is also observed that the random carrier-grouping gains more in the lower traffic case. As traffic gets higher, the gain is less. It can be predicted that both schemes will perform the same if the system is fully loaded. There is no obvious change in spectral efficiency for either scheme. As shown in Figure 13.8, when each user uses three carriers, the comparison of fixed carrier-grouping and random carrier-grouping is done with two and five subcarrier groups per cell. It is observed that random carrier-grouping outperforms the fixed one, due to its interference averaging advantage. As the number of subcarrier groups increases from two to five, user outage probability gets lower, which could be explained as more interference diversity gain is obtained while more subcarrier groups are available in each cell. In this way, good system fairness can be achieved.

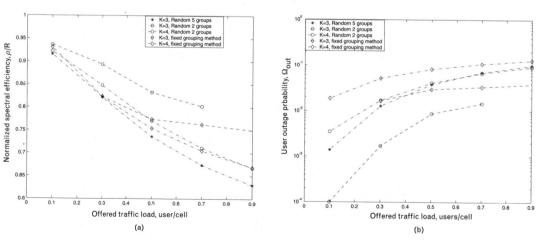

FIGURE 13.8   (a) Average throughput of carrier-grouping schemes. (b) User outage probability of carrier grouping schemes.

13.8 Concluding Remarks

This chapter presents a tutorial overview on downlink RRM of an MCA-based system in wireless IP networks. The wireless technologies that enable IP packets to be transmitted over radio at a high speed with high spectral efficiency and good coverage are discussed. MCA and non-MCA systems are reviewed under the context of multidimensional definition of a wireless channel. The impact of MCA-based system on wireless IP networks is described. Then investigations on downlink performance are presented. A multicarrier approach gives more flexibility on offering high spectral efficiency and variable data rate. It is a good candidate for wireless IP networks. We are aware that cochannel interference is the main barrier to combat. To achieve a truly wireless Internet, the investigation of radio resource management properties of MCA systems is of great importance. Hence, different techniques including frequency diversity, power control, and carrier-grouping are depicted on the network level. Wireless IP is currently a cutting edge technology overlaying wireless and computer networks. Further knowledge on this topic can be found in recent books [19–21].

Acknowledgments

The author thanks the editors and reviewers for their valuable comments that improved this chapter significantly. Thanks are also due to Dr. Slimane for the discussions on detailed investigations in Sections 13.5, 13.6, and 13.7. The author especially appreciates Professor Jens Zander for his comments and discussions on resource management and wireless IP.

REFERENCES

[1] Dixit, S., et al., "Resource Management and Quality of Service in Third Generation Wireless Networks," *IEEE Communications Magazine*, February 2001, pp. 125–133.

[2] http://www.3gpp.org.

[3] http://www.ietf.org.

[4] http://www.britannica.com.

[5] Umehira, M., et. al. "Wireless and IP Integrated System Architectures for Broadband Mobile Multimedia Services," *WCNC 1999*, Vol. 2, pp. 593–597.

[6] Nanda, S., et al., "Adaptation Techniques in Wireless Packet Data Services," *IEEE Communications Magazine*, January 2000, pp. 54–64.

[7] Lei, Z., and S. B. Slimane, "A Multicarrier Allocation (MCA) Scheme for 3G Wireless Systems," *IEEE Communications Magazine*, June 2000, pp. 86–91.

[8] Hara, S., and R. Prasad, "Overview of Multicarrier CDMA," *IEEE Communications Magazine*, Vol. 35, No. 12, December 1997, pp. 126–133.

[9]   Chuang, J., and N. Sollenberger, "Beyond 3G: Wideband Wireless Data Access Based on OFDM and Dynamic Packet Assignment," *IEEE Communications Magazine*, July 2000, pp. 78–87.

[10] Milstein, L. B., "A Conceptual Overview of Wideband Code Division Multiple Access," *2000 IEEE Loth Intl. Symp. on Spread Spectrum Techniques and Applications,* Vol. 1, 2000, pp.226–229.

[11] Hara, S., and R. Prasad, "Overview of Multicarrier CDMA," *IEEE Communications Magazine*, December 1997, pp. 126–134.

[12] Ahlin, L., and J. Zander, *Principles of Wireless Communications*, Lund, Sweden: Student litteratur, 1998.

[13] Cai, J., and D. J. Goodman, "General Packet Radio Service in GSM," *IEEE Communications Magazine*, October 1997, pp. 122–131.

[14] Kumagai, T., et al., "A Maximal Ratio Combining Frequency Diversity ARQ Scheme for OFDM Signals," *9th Intl. Symp. on Personal, Indoor, and Mobile Radio Communications,* Vol. 2, September 8–11, 1998, pp. 528–531.

[15] Lei, Z., "On the Downlink Performance of Multicarrier Allocation (MCA)-Based System for Wireless IP Networks," *4th Intl. Symp. on Wireless Personal Multimedia Communications (WMPC'01),* Aalborg, Denmark, September 9–12, 2001.

[16] Zander, J., "Performance of Optimum Transmitter Power Control in Cellular Radio Systems," *IEEE Transactions on Vehicular Technology*, Vol. 41, No. 1, February 1992, pp. 305–311.

[17] Grandhi, S. A., R. Vijayan, and D. J. Goodman, "Distributed Power Control in Cellular Radio Systems," *IEEE Transactions on Communications*, Vol. 42, No. 2/3/4, February/March/April 1994, pp. 226–228.

[18] Grandhi, S. A., R. D. Yate, and D. J. Goodman, "Resource Allocation for Cellular Radio Systems," *IEEE Transactions on Vehicular Technology*, Vol. 46, No. 3, August 1997, pp. 581–587.

[19] Ojanperä, T., and R. Prasad, Eds., *WCDMA: Towards IP Mobility and Mobile Internet*, Norwood, MA: Artech House, 2001.

[20] Prasad, R., *Towards a Global 3G System: Advanced Mobile Communications in Europe, Volume 1 and Volume 2*, Norwood, MA: Artech House, 2001.

[21] Zander, J., and S.-L. Kim, *Radio Resource Management for Wireless Networks*, Norwood, MA: Artech House, 2001.

Part III:
TCP/IP in Wireless IP Networks

CHAPTER 14

TCP/IP over Next-Generation Broadband Wireless Access Networks

Ivana Stojanovic, Manish Airy, David Gesbert, and Huzur Saran

14.1 Introduction

The demand for broadband Internet access is booming. Delays in the deployment of 3G high-speed wireless networks, such as WCDMA and cdma2000, as well as the slow progress in satisfying demand for wired solutions such as Digital Subscriber Line (x-DSL) and cable modems, place high expectations on alternative access technologies such as fixed wireless. The fixed wireless technologies of focus here address the lower frequency bands (MMDS, 3.5 GHz) dedicated to mass market—residential, *small offices home office* (SoHo)—use. Current Internet *broadband wireless access* (BWA) technologies are based on the existence of a line-of-sight link between the subscriber's unit and the access point. This assumption allows the avoidance of multipath fading and equalization; however, it puts very stringent limits on the scalability and ubiquity of the technology, preventing low-cost mass deployment and also preventing any future evolution to support mobility.

Figure 14.1 illustrates the architecture and the main network components of BWA systems. These systems are deployed as cellular systems where one access point, a BTS, serves multiple subscriber units (SUs) located in its proximity area, defined as a cell. This architecture assumes that all SU nodes talk to and hear from the base station and that all communication must go through BTS node. In other words, two SUs within the same cell cannot communicate directly. Multiple cells, sufficiently spaced apart, can use the same frequency lowering down the requirement for available spectrum in a given area. Frequency reuse is enabled by natural phenomena of signal strength loss with transmission distance and can further be improved by

FIGURE 14.1
*Illustrating BWA
network components.*

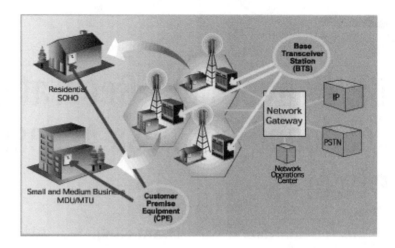

different techniques that enable interference mitigation from neighboring cells. Ideally, each cell reuses the same frequency channel, translating into highly desirable frequency reuse 1 system. In practice, however, this reuse is very difficult to achieve.

The goal of the next-generation (2G-BWA) design is to demonstrate feasibility of Internet access over very unreliable channels with satisfactory TCP/IP quality. This chapter presents a state-of-the-art overview for Internet access over 2G-BWA networks. The challenge for these, mostly fixed access technologies, stems from *non-line-of-sight* (NLOS) deployment with high coverage requirements and very efficient spectrum use, while maintaining low latency TCP/IP link performance. This performance has to be at least comparable to that of competing wired technologies like digital *subscriber line* (DSL) and cable modem. At the same time, the system has to maintain user friendliness, especially during installation, and low cost of operation. Key features to a successful 2G-BWA design lay at the physical (PHY) layer and MAC layer. Typical overall figure-of-merit requirements for these upcoming systems are shown in Table 14.1.

TABLE 14.1   REQUIREMENTS FOR 2G-BWA

FIGURE OF MERIT	REQUIREMENT
Aggregate rates	4–7 Mbps
Spectrum efficiency	2 bits/Hz/BTS
Coverage	4–6 miles (90% area)
Latency	Comparable to DSL
Link reliability	0.999

Non-line-of-sight broadband wireless channels are prone to frequency-selective and time-selective fading [1]. In addition, because of the quasi-stationary nature of the subscriber unit, the Doppler spread in the channel is low (less than 1 Hz), causing fades to typically last longer than wireless mobile applications (up to a few seconds). To provide good link performance at the TCP/IP level over such channels, fade mitigation at either the PHY or MAC layers becomes absolutely critical.

For this emerging industry branch, standardization efforts have already been made within the IEEE 802.16 Working Group on Broadband Wireless Access. This group is creating the IEEE 802.16 family of WirelessMAN standards for *wireless metropolitan area networks* (WMANs). The IEEE Standard Board formally approved IEEE Standard 802.16. This standard sets the platform for global deployment of 10- to 66-GHz metropolitan area networks that support voice and data application at the quality level customers demand, while making highly efficient use of available bandwidth. A standardization branch for 2- to 10-GHz air-interface is still under development and a closure is expected shortly. Also, under this effort, the IEEE Standard Board adopted Channel Models [2] proposed jointly by Stanford University and Sprint. These models provide fixed wireless service providers and equipment vendors with the definition of NLOS, the deciding factor whether the fixed broadband wireless technology can deliver over unreliable wireless channel.

A number of emerging technologies that satisfy this need have been investigated for use in the BWA systems. Note that only generic or standard versions of the system features are mentioned here. This chapter first presents a tutorial background on the windowing and loss recovery mechanisms of TCP and address difficulties faced by TCP over unreliable and lossy wireless links. Then, a selection of technologies, both at the MAC and PHY, that are suited for NLOS BWA are presented. Next, end-to-end link performance (including TCP/IP) for a typical 2G-BWA system exposed to time/frequency fading channels is presented. In particular, the focus is on the respective impact of key PHY and MAC layer features on end-to-end TCP/IP performance over these channels. The conclusion carries design recommendations for 2G-BWA.

14.2 TCP Background

In this section, the TCP windowing and loss recovery mechanisms are recalled and differences between several widely used versions, Tahoe and Reno through TCP evolution process, are emphasized.

TCP is a dynamic window-based flow control protocol. The window size, denoted by $W(t)$, varies in response to packet *acknowledgments* (ACKs) and loss detection. While the receive processes are the same for all TCP versions, their transmit processes and window adjustment algorithm in response to loss detection are different [3].

At any given time t, the TCP transmitter maintains several variables for each connection: lower window edge, congestion window, and slow start threshold. Lower window edge $X(t)$, denotes that all data packets numbered up to $X(t) - 1$ have been transmitted and acknowledged. Each ACK receipt causes $X(t)$ to advance by the amount of acknowledged data, and thus, lower edge window evolution is a monotonically nondecreasing time function. The transmitter's congestion window $W(t)$ defines the maximum amount of data packets the transmitter is permitted to send without an acknowledgment, 'starting from $X(t)$. Control mechanism upon ACK reception allows $W(t)$ to increase or decrease within W_{max} boundaries, whose value is advertised by the receiver during connection setup. Slow start threshold $W_{th}(t)$ controls the increments in $W(t)$. If $W(t) \leq W_{th}(t)$ transmission is assumed to be in slow start phase, where each ACK causes $W(t)$ to be incremented by 1. This rapid growth switches to slow increase once the window crosses slow start threshold. This phase, called the congestion avoidance phase, aims to cautiously probe for extra bandwidth. Each ACK reception increments window size $W(t)$ by $1/W(t)$.

On the receiver side, packets are accepted out of order, but delivered only in sequence to an intended TCP user. The destination returns *cumulative* acknowledgment for every correctly received packet. Each ACK carries the next in-order packet sequence number expected at the receiver. Thus, a single packet loss can be detected if the same "next expected" number is received in all consecutive acknowledgments of packets successfully transmitted after the lost packet. Reaction on packet loss detection depends upon the TCP version. In the simplest case, TCP OldTahoe, the TCP transmitter continuously sends packets till the congestion window is exhausted and then waits for timer expiry. The sender maintains a timer only for the last transmitted packet; that is, each time a new packet is sent, the transmitter resets the already running transmission timer. The timer is set for a round trip timeout (*rto*) value that is derived from a smoothed round-trip time (*rtt*) estimator applying the low pass filter over the last estimate and the current measurement of the *rtt*. The timeout values are set only in multiples of timer granularity: in a typical implementation the timer granularity is 500 ms. According to the basic recovery algorithm, upon timeout, window parameters are adjusted: $W_{th}(t+)$ is set to $W(t)/2$ and $W(t)$ is reset to 1, and retransmission is initiated from the first missing packet.

More sophisticated TCP protocols such as Tahoe and Reno implement *fast retransmit* and *loss recovery* procedures. The fast retransmit procedure takes advantage of additional information that duplicate acknowledgments carry. The transmitter waits for several duplicate ACKs to arrive (typically three) to account for possible network delay and packet reordering. Then, the missing packet is retransmitted before the timer expiry. In subsequent loss recovery, the TCP Tahoe transmitter behaves as if a timeout has occurred and resorts to the basic recovery algorithm.

In the case of Reno, only the lost packet is fast retransmitted and loss recovery procedure is handled differently. Congestion window recorded at this time is referred to as *loss window*. After Nth duplicate ACK at time *t*, the sender adjusts slow start threshold $W_{th}(t^+)$ to $W(t)/2$ and $W(t^+)$ to $W_{th}(t^+)$ + N. Congestion window $W(t)$ is additionally incremented by N to account for packets that have successfully left the network and produced duplicate ACKs. While waiting for the ACK for the first retransmission, any successfully received outstanding packet produces a duplicate ACK, which further increments $W(t)$ by 1. If a single packet were lost, the packet retransmission would complete loss recovery. At this time, congestion window $W(t)$ is set to $W_{th}(t)$, and the transmission resumes according to the normal window control algorithm. However, the algorithm can experience a possible stall time if multiple packet losses occur within the loss window. In this case, the congestion window might close before a sufficient number of duplicate ACKs has been generated to trigger multiple fast retransmits. In that case, algorithm waits for timer expiry. The Reno recovery algorithm outperforms the Tahoe version, but it is pessimistic and suffers performance degradation in case of multiple losses within the loss window [3]. In this chapter, the focus is mainly on TCP Reno.

14.2.1 Performance of TCP on Wireless Links

From the discussion above it is clear that TCP is not designed for lossy links. The TCP protocol has been developed for wired networks to deal with network congestion, rather than nonnegligible random losses of wireless links that raise the packet error probability several orders of magnitude. The TCP sending rate drops upon loss detection and increases only gradually afterwards. Congestion window evolution, clocked by arrival of acknowledgments, largely depends on round trip time. If round trip time is large and the congestion window is decreased frequently due to multiple losses, the TCP connection may not be able to fully utilize available channel bandwidth. Although some of the previous studies show that different versions of the TCP protocol perform better or worse in response to wireless lossy link with correlated errors, none of them gives satisfactory performance. This calls for new techniques that will allow reliable transmission in heterogeneous networks comprising both wired and wireless hops. Comparative analysis of several higher layer schemes proposed to alleviate the effects of non–congestion-related losses has been conducted in [4]. The authors classify the schemes in two categories according to the approach they take to improve TCP performance in lossy systems. The first category includes link–layer protocols, split connection approaches, and TCP aware link layer schemes. These protocols try to locally solve the problem and hide wireless losses from the TCP sender. As a result, wireless hops appear as high-quality links with reduced effective bandwidth. The second class of technique attempts to make the sender aware of the existence of non-congestion-related losses. The idea is that the sender restricts itself from invoking the congestion avoidance algorithm in response to such events. Second class schemes, however, require undesirable change in TCP end-to-end algorithm. Comparison analysis shows that local protocols outperform the second class techniques [4]. Moreover, reliable link layer schemes offer higher throughput than the split connection approach, and at the same time preserve end-to-end protocol semantics.

14.3 Design Features for 2G-BWA Networks

In this section, the focus is on BWA system design features that attempt to overcome unreliability introduced by time and frequency selective fading and interference-prone channels. Both PHY and MAC layer features are presented, while TCP fixes that have been proposed to deal with wireless channel behavior are ignored. The focus here is on a set of selected features relevant to good network economics and performance in terms of coverage, spectrum efficiency, and satisfactory TCP/IP quality. The MAC layer features ensure a low resultant error rate seen by TCP. The PHY layer features improve link quality and enable higher data rates.

14.3.1 MAC Layer

14.3.1.1 Weighted Round Robin Scheduling

In order to facilitate the QoS requirement of both links, centralized MAC protocol with arbitration and complexity moved to the BTS is suitable for packet access networks such as cellular. In this protocol type, the BTS has explicit control on the access of the medium. Downlink transmission schedule is easy to compute since all data transmission requests arrive at the BTS. On the other hand, uplink channel needs a specific protocol that enables SUs to transmit according to application quality attributes. Delay-sensitive real-time applications are hard to achieve required delay bounds with random access [5] protocol where each packet has to contend for transmission. On the other hand, guaranteed access [6] protocol featuring polling mechanisms put additional overhead in the network, especially in the case of bursty data and do not guarantee bounded delays either. QoS guarantees can be fulfilled with demand assignment protocols where the request and transmission channels are separated. In this case, the central node (BTS) collects bandwidth requests and schedules transmission matching requirements against available resources. The request channel is typically random access where only short control messages carrying bandwidth requests contend for access according to slotted ALOHA-based algorithm. Additional requests, instead of going through contention, can be piggybacked on data transmission.

Then, at the beginning of each frame, the BTS transmits a *MAC assignment protocol* (MAP) message that allows a particular SU to determine which timeslots it should decode or transmit in and what modulation/coding to use.

The scheduling policy used to compute the uplink and downlink transmission schedules at the BTS has to meet the following requirements and constraints:

- Ensure that *constant bit rate* (CBR) service flows get allocated a constant bandwidth with acceptable delay jitter;

- Ensure that *unspecified bit rate* (UBR) service flows share available bandwidth in a fair manner;

- Support channel dependent scheduling—that is, compute transmission schedule based on the knowledge of modulation/coding being used over current channel state;

- Provide ability to temporarily shut off transmission to an SU with a bad link.

14.3.1.2   Automatic Retransmission/Fragmentation

Automatic retransmission/fragmentation (ARQ/F) is widely considered to be a key tool for dealing with errors occurring over wireless channels. A low-latency acknowledgment and retransmission mechanism is implemented at the MAC layer between the SU and the BTS. The MAC layer fragments IP packets into *atomic data units* (ADUs). A technique complementary to coding, ARQ/F only introduces redundancy during the fraction of time when data gets corrupted. A retransmission of only erroneous ADUs is efficient in dealing with fades at the cost of only moderate additional link latency. ARQ/F-based systems can be designed to operate at high BER levels while still providing satisfactory TCP/IP performance, as shown later in the results section.

At a transmitter, *protocol data units* (PDUs) received for transmission from higher protocol layers are fragmented into ADUs and transmitted sequentially, with each outgoing ADU being uniquely identified by its sequence number. At a receiver, uncorrupted incoming ADUs are stored in a reassembly buffer. Whenever all of the ADUs comprising a PDU are received, the reassembled PDU is passed up to the higher layer protocols. To accommodate situations where several ADUs comprising a PDU are received in error, the receiver periodically generates an acknowledgment message (ACK) notifying the transmitter about sequence numbers of corrupted ADUs. When the transmitter receives an ACK, it retransmits the "missing" ADUs from its transmit buffers. These retransmitted ADUs are given higher priority over regular transmissions. ADUs that have been successfully received at the receiver are freed from the transmit buffer.

The ARQ protocol is based on a sliding window mechanism—that is, each flow is allowed to transmit only if it has no more than a window size (W) worth of unacknowledged ADUs. More formally, the following inequality must always hold:

$$T(N) - T(U) <= W \quad if \quad T(N) > = T(U) \tag{14.1}$$

$$MAX\_SEQ\_NUM - T(U) + T(N) <= W \quad if \quad T(N) < T(U) \tag{14.2}$$

where MAX_SEQ_NUM denotes the size of the sequence number space, $T(N)$ denotes the sequence number of the next in-sequence ADU eligible to be transmitted at a transmitter, and $T(U)$ denotes the sequence number of the "oldest" ADU that has not been positively acknowledged by the receiver. The window size W shall be configurable at the time of link initialization, but it cannot be larger than $MAX\_SEQ\_NUM/2$ ADUs. This limit arises from the size of the sequence number space and its uniqueness requirement.

14.3.1.3   Adaptive Modulation

Adaptive modulation (AM) and coding let a user adapt its data rate as a function of channel conditions (e.g., SINR). The technique has been popularized in the EDGE cellular standard [7]. AM allows a several times rate improvement by exploiting all SINR margins available at any time/location. The idea behind location adaptive modulation is depicted in Figure 14.2. By comparison, a nonadaptive system must be deployed with most conservative modulation/coding for all users in order to preserve good coverage and frequency reuse. However, location self-adaptivity to even slowly changing conditions, is a key requirement in extracting significant capacity gains. Further improvement is obtained by faster adapting. For TCP/IP, AM will result in larger pipe size and additional statistical multiplexing.

To detect an appropriate modulation/coding level, the MAC layer uses a set of *packet error rate* (PER) and SINR statistics obtained from the PHY layer. An algorithm assigns each user a modulation and coding level that is a

FIGURE 14.2 *AM.*

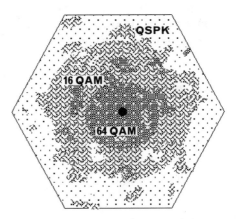

function of individual channel conditions, while satisfying target error rate constraint [8].

14.3.2 Physical Layer

14.3.2.1 Spatial Diversity

Spatial diversity (SD) is obtained through the use of multielement antennas at either the BTS and/or the SU. Antenna combining is used to deal with fading (Figure 14.3) and allows advanced interference canceling algorithms. The basic idea of diversity combining is that several uncorrelated fading elements are much less likely to fade simultaneously than a single element. Use of SD can reduce the SINR requirement by 10 dB to 15 dB with no loss of TCP/IP performance. This is exploited to extend the coverage, to increase data rates, and most importantly to allow a tight frequency reuse that will work in arbitrary terrain. Recent measurement campaigns [9] for BWA

FIGURE 14.3
Multiple antennas provide a robust source of diversity in NLOS channels.

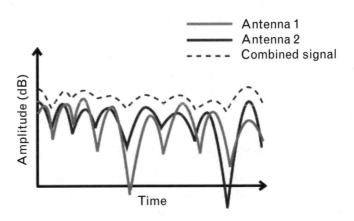

suggest that decorrelation is achieved with 1 to 2 wavelengths of element separation; less spacing is achieved with dual-polarized antenna elements.

The signal processing techniques used to combine the antennas can vary. MRC and minimum mean square error (MMSE) combining are typical examples of combining algorithms when the channel coefficients are known (e.g., at the receiver) [10]. Upon transmission, where the exact channel coefficients cannot be known unless a bandwidth consuming feedback link is used, a blind transmit spatial diversity must be used. Examples are space-time coding algorithms recently developed [11]. Delay-diversity is used for transmitting and MRC is used for receiving.

14.3.2.2   MIMO and Spatial Multiplexing

In the case of multielement antennas at both ends (*multiple-input multiple-output*, or MIMO), an additional data rate increase can be obtained by exploiting simultaneous transmission over the eigenmodes of the channel, as seen in [10, 12]. A simple *spatial multiplexing* (SM) algorithm works as follows (more advanced variations combined with coding are typically used in practice): a high rate signal is demultiplexed into a set of independent bit streams, which are independently transmitted by the multiple antennas. The signals are mixed with each other in the channel since they occupy the same time and frequency resource. At the receiver, the multiple antennas are used to first learn the spatial signature (channel) of each stream using training sequences, then combined so as to separate the different bit streams from each other. The streams are finally multiplexed to offer the original high rate signal. This operation consumes only the bandwidth normally needed to transmit one single bit stream. The increase in data rate is achieved at the expense of increased interference. The procedure is illustrated in Figure 14.4.

When there is only one usable eigenmode due to rank deficiency of the MIMO channel, the streams are not separable and the system must fall back to a diversity scheme [12]. This technology is particularly well suited for BWA because, unlike handsets, compact multielement subscriber units can be designed with two or three elements.

The gain in data rate is proportional to the minimum of the number of transmit and receive antennas. The impact of this technology on TCP/IP performance is examined in the results section.

14.3.2.3   Frequency Diversity

Broadband transmission over multipath channels introduces frequency selective fading, as illustrated in Figure 14.5. Mechanisms that spread information bits over the entire signal band will mitigate the effect of fading occurring at certain frequencies. An example of such multipath-friendly

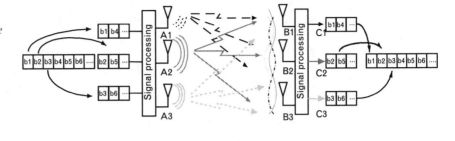

FIGURE 14.4
Simplified SM scheme with three antennas. This gives three-fold data rate improvement.

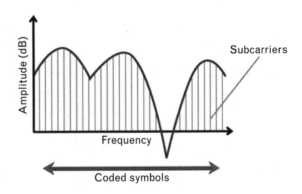

FIGURE 14.5
OFDM with coding transforms multipath into a source of anti-fading diversity.

mechanisms is frequency-coded multicarrier (e.g., OFDM) modulation [13]. This technique requires lower SINR to achieve the same spectral efficiency as compared to single carrier system.

14.4 End-to-End Performance

14.4.1 Simulation Methodology

The main objective of end-to-end simulation is to capture all aspects of system design and examine their impact on TCP/IP. System evaluation approach organizes simulations into several hierarchical layers where each module sees the abstracted behavior of a layer below. The simulation is constrained from the bottom by BWA channel models and from the top by TCP/IP usage models. The three main design layers with several degrees of abstraction and implementation detail need to be captured—namely, transport, MAC, and PHY layers. Different tools are used to implement each layer.

Network simulator (NS) [14] closely mimics TCP-based transport layer and is used to capture user behavior, create network nodes, and apply different traffic models to source nodes. Detailed proprietary MAC layer functionality is developed in C/C++ and interfaced to NS. PHY layer behavior is abstracted to a form suitable to the MAC layer, through error and data rate traces, that reflect performance of coding, modulation, and multiple antennae processing techniques over realistic fixed broadband wireless channels. Realistic channel models are developed from fixed MIMO channel measurements and standardized in IEEE 802.16 body.

The NS module developed for fixed broadband wireless system evaluation models one target multiuser TDMA-based cell. The interference caused by other cells, however, is also captured through the data and error rate traces, assuming a 2 × 3 frequency reuse. The 2 × 3 frequency reuse assumes that each cell is divided in three sectors and that the set of three frequencies is repeated every second cell. The wired backbone network is represented as a set of Web servers (traffic sources) connected to the BTS through high bandwidth, high propagation delay, and low BER links. The wired links converge to the BTS (Figure 14.6), which maintains separate queues for each SU. Multiple TCP connections originating from different servers can get routed to the same queue, modeling a business customer or residential users with multiple open connections. Packets are transmitted

FIGURE 14.6
*Simulation
environment.*

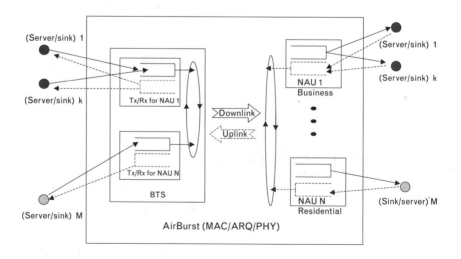

over the air and are subject to errors as predicted by PHY error traces. A connection from the BTS to any particular SU is represented with a link whose average quality is dictated by long-term channel condition specific for each subscriber location.

14.4.2 MAC Model

The MAC layer model captures detailed design features such as segmentation and reassembly of IP packets, scheduling, ARQ, UL access via contention, and link adaptation.

IP packets are fragmented into ADUs and are delivered in order to higher (TCP) layers according to the ARQ/F protocol. The instantaneous channel state and transmit mode determine ADU loss probability. Lost ADUs are identified and marked for retransmission. The centralized scheduler enforces a weighted round robin policy. Allocations are made for a number of time-frequency blocks considering transmit mode and service attributes of a particular user. On the reverse link, from an SU to the BTS, initial access is obtained over a contention channel. Once data transfer starts, additional bandwidth requests are piggybacked on the data ADUs. If enabled, both UL and DL are subject to transmit mode adaptation based on SINR measurements made at the receiver.

Many specific channel parameters are programmed by MAC layer and broadcasted in downlink messages. However, several parameters are left unspecified and configurable during the SU registration process in order to optimize performance for a particular deployment scenario.

14.4.3 PHY Model

The PHY performance is summarized through estimation of modem setpoints for each transmit mode. A transmit mode is comprised of a coding mode (choice of MQAM and coding rate) and antenna mode (diversity or spatial multiplexing). The modem setpoint for a particular mode, as defined by the modulation, coding, and number of transmit antennas, is defined as the SNR required to achieve a desired pre-ARQ operating error rate. Here, two cases can be distinguished. If the mode is static over time, a modem setpoint represents the average SINR level over a long period of time attained at one receive antenna element. In the second case, link adaptation tracks channel variations, allowing maximization of data rate whenever possible through selection of "optimal" transmit mode over a time window of interest. In this case both packet error rate trace, as well as corresponding mode trace are generated and supplied as input to the end-to-end simulator.

14.4.4 Wireless Channel Model

Physical properties of wireless communication channels exhibit lots of impairments unknown to their wired counterparts. In a wireless system, signal propagation is subject to scattering, reflection, refraction, or diffraction of surrounding objects, such as buildings, hills, streets, trees, and moving mobiles, as shown in Figure 14.7. The received signal is attenuated due to wireless propagation impairments such as path loss, shadowing, and fast fading.

Path loss is the drop in radiated signal power due to signal propagation from the transmitter to the receiver and environment characteristics, such as water, foliage, and so forth. Ideal free space propagation follows inverse square law spreading with the distance, whereas in a real cellular environment, the path loss exponent varies from 2.5 to 5.

Shadowing, or slow fading, comes from signal blocking by large obstructions such as buildings and hills that are positioned between base station and access terminal. Multipath fading plays a central role in determining the nature of wireless channel. Signal interactions with numerous physical objects in its path to the receiver result in a presence of many signal components, or multipath signals, at the receiver. Multipath arrives from different directions, with different attenuation and propagation delays resulting in received signal amplitude that varies in many dimensions—time, frequency, space.

Time selectivity, measured through Doppler spread, comes from the fact that a pure tone is frequency dispersed due to environment change in the receiver's vicinity. In the mobile channel, received signal envelope has Rayleigh distribution in the time domain, which translates into Doppler power spectrum of the received signal that has a "horn" shape. On the other hand, fading characteristics of the fixed wireless channels are very different from the mobile case and result in a "rounded" Doppler spectrum [2]. Techniques that exploit this property are adaptive modulation, which tracks channel quality over time, and radio link protocols, like ARQ/F, which attempt to fix fades by retransmission of lost packets in good channel states.

Frequency selectivity or multipath delay spread is caused by several distinct and delayed versions of a transmitted signal. It is characterized by coherence bandwidth representing maximum frequency separation for which channel response remains correlated. Signal components delayed by even a small fraction of the symbol period give rise to intersymbol interference. OFDM for wireless is well suited to combat frequency selectivity by coding over subcarriers.

Space selective fading, or angle spread, refers to distinct multipath angle of arrival that translates into signal amplitude that depends on antenna location. This property allows for space and MIMO diversity techniques in the systems with antenna arrays.

FIGURE 14.7
Radio wave propagation with diffraction, refraction, and multiple reflection.

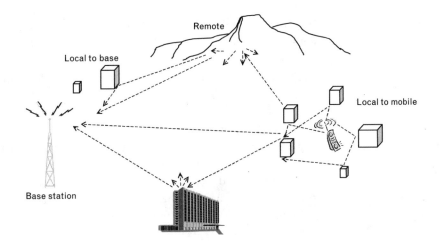

The channel models, adopted by IEEE 802.16, reflect specific broadband wireless channels developed from measurements and published by Sprint and Stanford University [2]. These models capture radio link impairments mentioned above. Here, the emphasis is on the use of NLOS models with low Ricean K factors in order to design systems with high reliability requirements. The use of multiple transmit and receive antennas calls for antenna correlation modeling. The system under study assumes cross-polarized antennas. Cross-polarization gives rise to attenuation due to polarization mismatch (XPD) between transmit and differently polarized receive antenna. For LOS components, 10-dB attenuation is recommended, but for Raleigh component XPD loss goes as low as 4 dB. This effect should be included in PHY modeling. Key parameters for the different channel models, SU1 to SU6, are listed in Table 14.2.

14.4.5 Traffic Model

The exponential growth of the WWW has caused the traffic originating from Web applications to dominate the Internet both in terms of volume and active users. This trend, likely to continue in upcoming years, causes

TABLE 14.2   KEY PARAMETERS FOR CHANNEL MODELS

	SU1	SU2	SU3	SU4	SU5	SU6
rms delay spread (μs)	0.1	0.2	0.3	1.3	3	5.2
Overall K (linear)	10	5	0	0	0	0
Antenna correlation	0.7	0.5	0.25	0.25	0.25	0.25

downstream and upstream link asymmetry. The majority of the Internet applications today, such as Web browsing and file downloads, require larger bandwidth pipes on the downstream. However, there is an increasing demand for streaming data (audio and video) applications, video conferencing, and high-QoS services that will require networks to support more symmetric traffic.

Accurate traffic modeling has high relevance in system evaluation and capacity planning. In particular, traffic model requests need to mimic a certain population of real users. The literature shows that such a model has been difficult to develop due to a number of unusual Web workload characteristics [15]. Empirical studies of operating servers show that their workload is highly variable, both in volume and number of open connections. This feature indicates that proper attention should be paid to properties of Web streams such as request interarrivals and file sizes. Another important Web traffic characteristic is self-similarity—that is, significant variability over a wide range of time scales. Self-similarity negatively impacts network performance and becomes important to address.

In the system evaluation a focus should be on a traffic model that attempts to closely imitate HTTP requests originating from a population of Web users. Traffic models can be characterized at two levels: macroscopic and microscopic. The macroscopic description determines the business-to-residential user penetration ratio, the number of active connections per business SU, and SU activity factors across the subscriber population. A good approximation assumes an active set composed of 20% business SUs with an average of five active connections each.

The microscopic description describes individual user behavior over time. A user can request data or act as a server to a remote client. From a global perspective, data transfer volume can be characterized as long-term throughput average or workload, with an assumption of a 20-Kbps downlink and 8-Kbps uplink per connection workload. Introduction of uplink traffic diminishes link asymmetry and allows for more symmetric loads that account for future traffic development. In addition to these data workloads, control traffic in terms of MAC/ARQ or TCP acknowledgments is inherently generated. The workload is averaged over several WWW sessions, including "think" time.

According to the standardized SURGE WWW usage model [15], each user starts several sessions during an on-line period. User behavior during a session is described as a superposition of *active downloads* and *think times*. Tails of the download (or request) sizes are described by a Pareto distribution, with the body of download sizes following a log-normal distribution. The average request size used in simulations is 27 KB. The think time is Pareto distributed with an average of 3.5 seconds. The model parameters are shown in Table 14.3 and illustrated in Figure 14.8.

TABLE 14.3   PARAMETERS OF WWW TRAFFIC USAGE PATTERN AND FILE SIZES

REQUEST BODY LOG-NORMAL	REQUEST TAIL PARETO	ACTIVE OFF PARETO	SESSION LENGTH INVERSE GAUSSIAN
$\mu = 7.881$	K = 34	K = 1	$\mu = 3.86$
$\sigma = 1.339$	$\alpha = 1.177$	$\alpha = 1.4$	$\sigma = 6.08$

FIGURE 14.8
User activity model.

14.5   Results

In this section, performance benefits of key system design aspects that enable efficient broadband wireless operation are presented. This section specifically addresses AM, SD, and MIMO techniques, as well as operating error rate.

AM and coding allow the system to adapt transmit modes across users in a cell as a function of channel conditions at each location. The highest AM allowable rate is 12 times that of nonadaptive system deployed with the most conservative mode to preserve required coverage and frequency reuse. Twelve available data rates at the physical layer in the adaptive system form a set of 1.1, 2.2, …, 13.6 Mbps. Diversity of transmit modes is obtained through MIMO and SD features of the physical layer. In Figure 14.9, performance of both systems as a function of offered load is presented. A nonadaptive system sustains several times lower workloads with even worse request service delays. In other words, good coverage of a nonadaptive system is bounded by quality of service requirement, which in return greatly reduces the number of users the system can support.

Figure 14.9 also illustrates how an unequal data rate of users with disparate SINR levels reduces the system's offered troughput. Active users are separated into N classes according to their SINR level and corresponding data rate. A user in a class n receives data at rate R_n bps. Now, suppose that all users request equal traffic amounts making a relative frequency of packets of class n, denoted by P_n, proportional to the number of users in the class.

FIGURE 14.9
AM offers several-fold improvement in capacity, efficiency, and QoS. Note that the offered load does not include ~16% of upper layer protocol overhead.

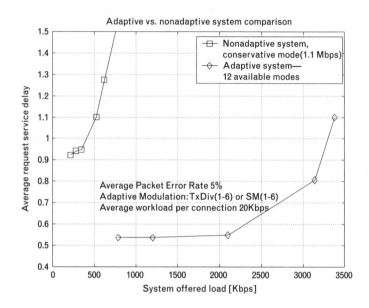

Assume infinite traffic load of each class and suppose a scheduling scheme where slots are assigned one at a time succesively to each user.

As a result, the bandwidth share of a given class is proportional to the class size. The system's avarage data rate, or throughput, is

$$R_{av} = \sum_{i=1}^{N} P_n R_n \text{ bps} \tag{14.3}$$

This means that fixed size packets of lower data rate users will have proportionally higher latencies. That is, if the latency of a B-bit packet is considered, the number of slots consumed by each user in class n will be B/R_n, and hence the latency L_n is inversely proportional to the rate R_n.

On the other hand, in practice there is a requirement is that all users have essentially the same latency irrespective of the R_n they can support and that a finite amount of traffic is requested by each user within a certain time window. Then, each class will consume bandwidth directly proportional to its size and inversly proportional to its rate. Round robin scheduler assumption and QoS constraint on bounded delays require the system to operate below its full capacity. Achieved system throughput differs from the average data rate, presented in (14.3), due to multiplexing of users with disparate SINR levels. Assume that each user requests B bits and that users of nth class require $S_n = k/R_n$ (k being a constant) slots to trasmit those bits. In this case, effective throughput is defined as

$$R_{av} = \frac{\sum_{n=1}^{N} P_n R_n S_n}{\sum_{n=1}^{N} P_n S_n} = \frac{1}{\sum_{n=1}^{N} P_n / R_n} bps \qquad (14.4)$$

For example, SINR distribution across the cell with 5% average ADU error rate yields an average throughput of 6 Mbps which translates into 4 Mbps of effective throughput. Figure 14.9 shows that after including 16% of upper layer protocol overhead, the system offered load does not go beyond 4 Mbps at full link utilization.

The above analysis indicates cost paid in throughput for latency equalization. Less bandwith "unfair" allocation can be made if delay garantees are relaxed for the user with worse link conditions.

MIMO and SD are key to enhancing system performance. Both techniques take advantage of the rich fading structure of the wireless channel, allowing multiple transmit modes to be deployed. Physical layer simulations estimate that a user in a single antenna system (SISO) requires a 12-dB additional SINR margin in comparison to MIMO system with two transmit and three receive antenna elements at the same transmit power and the same operating target error rate. System analysis shows that attainable cell size of SISO deployment in a macrocell environment shrinks down 2.5 times. To preserve the same large cell size of a MIMO system, a cell edge user in SISO system would need to operate at 50% PER, unacceptably degrading post-ARQ delays. Furthermore, adaptive modulation can not exploit higher modes causing average downlink peak data rate of SISO system to drop almost three times. As a result, SISO systems are highly inefficient in extracting available capacity both in coverage and data rate.

Given the above-mentioned underlying physical layer differences, TCP level quality of service is compared on Figure 14.10. The cumulative distribution function of throughput during file download is presented for files above 64 KB. "Small" files are excluded from the plot, since the TCP, rather than PHY and MAC layers, gates measured throughput for "small" files. In particular, the connection round-trip time and the slow start phase of congestion avoidance algorithm dictate the rate at which packets arrive at the wireless link. "Small" files are never able to fully utilize available bandwidth because the transmit window does not advance into congestion avoidance phase. Also, large round-trip times cause large delays between consecutive congestion window increments, due to the window advancement mechanism that is clocked by ACK arrivals. This effect is also present during a "big" file download, but only for a fraction of the actual download time. Thus, the achievable throughput during file downloads asymptotically approaches effective throughput because of TCP.

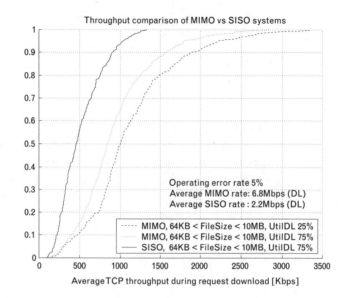

FIGURE 14.10
*MIMO system offers
gain in coverage,
thoughput, and in-
creases the load the
system can support.*

The number of users corresponding to link utilization of 75% in SISO, only build up to 25% link utilization in MIMO deployments. This corresponds to a three-fold difference in average data rate in these systems. In other words, to bring utilization back to 75%, the MIMO system can support up to three times more active users per cell site, at the same time offering better quality of service.

Operating ADU error rate plays a pivotal role in user perceived quality of service and area coverage. Table 14.4 lists the peak and effective data rates averaged across the cell, as well as coverage gain for several target error rates under study. The results presented are for a macrocell environment defined by cell radius of 6 miles, rooftop SUs, and a BTS antenna height of 30m. Two interesting trends can be derived from the data in Table 14.4. First, the average gross rate improves as the ADU loss probability increases, since

TABLE 14.4   OPERATING ERROR RATE VERSUS THROUGHPUT AND COVERAGE

		TARGET PER	1%	5%	10%
DL	Avg. peak data rate [Mbps]		5.7	6.3	6.3
	Peak data rate (1-error rate)		Ref	+6%	+0.5%
	Effective data rate [Mbps]		3.7	4.0	4.0
UL	Avg. peak data rate [Mbps]		2.8	3.1	3.1
	Peak data rate (1-error rate)		Ref	+6%	+0.7%
	Effective data rate [Mbps]		1.9	2.1	2.1
	Coverage		86%	90%	92%

relaxed SINR requirements of all modes allow users with poorer link quality to transmit at higher data rates. The rationale is to let ARQ recover from deep fades by retransmission, but to allow faster communication link during "good" channel states. On the other hand, the nonlinear slope of SINR curves does not allow for significant rate gain as target error rate moves up to 10%.

However, net throughput, defined a post-channel data rate does not monotonically increase. Compared to the reference error rate of 1%, there is a 6% gain in throughput for 5% PER and only 0.5% gain for error rate of 10%. Higher PERs improve coverage since more users with poorer channel conditions are accepted into system.

Request service delays closely relate to wireless link quality and delay bounds determine the system operating error rate. The average request service delay as a function of link utilization is plotted in Figure 14.11 for the three average loss probabilities of interest (listed in Table 14.3). As expected, a 10% ADU error rate and extended fades do not allow ARQ to recover quickly. The other two cases have very close performance, with 1% PER being marginally better across the whole utilization range. The figure shows that the system can be safely loaded up to 60% of available bandwidth without any cost paid in service quality. Also, request service delays do not exhibit a sharp knee at any utilization point, showing that the system is stable for the whole loading range.

From the above follows the conclusion that TCP throughput and latency requirements for mainstream Web applications can easily be obtained with up to a 5% ADU loss over the air, due to the ARQ/F mechanism. Emphasis of the conclusion is on ADU error rate, since without the ARQ/F mechanism, unfragmented higher layer packets would experience unacceptable losses, forcing TCP to decrease its window often and retransmit excessively. Furthermore, the ARQ/F mechanism permits high spectral efficiency with acceptable delays. In conclusion, link layer ARQ/F recovers from channel errors efficiently and prevents excessive TCP timeouts—that is, link layer ARQ/F permits a TCP-friendly design.

14.6 Concluding Remarks

Several promising technologies, which play a critical role in providing high-performance broadband wireless TCP/IP access over (typical) unreliable channels, have been discussed. Results show that these wireless link technologies have a significant impact on QoS provisioning, and therefore careful design of PHY and MAC layers is critical for acceptable network performance. AM allows aggressive system design that takes the most advantage of each link condition. While coverage remains guaranteed, AM

FIGURE 14.11
Link layer design permits operating error rate as high as 5%.

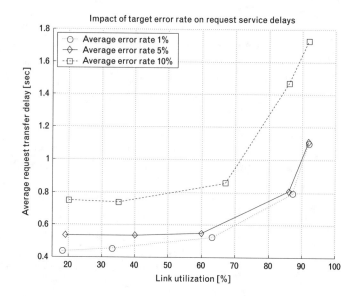

FIGURE 14.11
Link layer design permits operating error rate as high as 5%.

greatly improves user-perceived service quality, measured through average request delay, as well as the number of users the system can support. Underlying technology that enables finer granularity of AM is multiantenna signal processing at both ends. Results indicate that MIMO and SD offer additional gain in coverage, almost 2.5 times improvement over SISO system in attainable cell size and data rates. The design of link layer, ARQ/F mechanism, and MAC scheduling in particular, allow the network layer to perform well over the wireless channel with ADU loss rates as high as 5%. ARQ/F mechanism hides wireless link losses from TCP, preventing TCP control window restriction and successful recovery from channel impairments.

Other concepts that enhance spectral efficiency of 2G-BWA systems that were not presented in this chapter include interference mitigation either at the PHY or MAC, aggressive frequency reuse to ease deployment strategies, and channel-dependent scheduling to enhance system capacity.

Interference cancellation can be employed on both links to reduce modem setpoints. Array processing techniques at PHY or intelligent scheduling that coordinates transmissions across interfering BTSs lead to a system with aggressive frequency reuse. Additional capacity gain can be achieved with multiuser channel-dependent scheduling that allocates bandwidth to a user with the best channel quality while supporting desired QoS.

REFERENCES

[1]   Erceg, V., et al., "A Model for the Multipath Delay Profile of Fixed Wireless Channels," *IEEE Journal of Selected Areas in Communications*, Vol. 17, No. 3, March 1999, pp. 399–410.

[2]   Hary, K. V. S., and K. Sheikh, "Interim Channel Models for G2 MMDS Fixed Wireless Applications," *IEEE 802.16 Broadband Wireless Access Working Group*, http://ieee802.org/16.

[3]   Kumar, A., "Comparative Analysis of Versions of TCP in a Local Network with Lossy Link," *IEEE/ACM Transactions on Networking*, August 1998.

[4]   Balakrishnan, H., et al., "A Comparison of Mechanisms for Improving TCP Performance over Wireless Links," *IEEE/ACM Transactions on Networking*, June 1997, pp. 336–350.

[5]   Chen, K. C., and C. H. Lee, "RAP—A Novel Medium Access Protocol for Wireless Data Networks," *Proc. IEEE GLOBECOM*, 1993, pp. 1714–1717.

[6]   Znagh, Z., and A. S. Acampora, "Performance of a Modified Polling Strategy for Broadband Wireless LANs in Harsh Fading Environment," *Proc. IEEE GLOBECOM '91*, 1991, pp. 1141–1146.

[7]   Furuskar, A., et al., "System Performance of EDGE, a Proposal for Enhanced Data Rates in Existing Digital Cellular Systems," *Vehicular Technology Conf.*, 1998, Vol. 2, pp. 1284–1289.

[8]   Catreux, S., et al., "Adaptive Modulation and MIMO Coding for Braodband Wireless Data Networks," *IEEE Communications Magazine*, Vol. 40, No. 6, June 2002, pp. 108–115.

[9]   Baum, D. S., et al., "Measurement and Characterization of Broadband MIMO Fixed Wireless Channels at 2.5 GHz," *2000 IEEE International Conference on Personal Wireless Communications*, 2000, pp. 203–206.

[10]   Paulraj, A. J., and T. Kailath, "Increasing System Capacity in Wireless Broadcast Systems Using Distributed Transmission/Directional Reception (DTDR)," U.S. Patent No. 5,345,599, http://www.uspto.gov/.

[11]   Tarokh, V., N. Seshadri, and A. R. Ralderbank, "Space-Time Codes for High Data Rate Wireless Communication: Performance Criterion and Code Construction," *IEEE Transactions on Information Theory*, Vol. 44, No. 2, March 1998, pp. 744–764.

[12]   Foschini, G. J., and M. J. Gans, "On Limits of Wireless Communication in a Fading Environment When Using Multiple Antennas," *Wireless Personal Communications*, Vol. 6, 1998, pp. 311–335.

[13]   Keller, T., and L. Hanzo, "Adaptive Modulation Techniques for Duplex OFDM Transmission," *IEEE Transactions on Vehicular Technology*, Vol. 49, No. 5, September 2000, pp. 1893–1906.

[14]   http://www-nrg.ee.lbl.gov/ns/.

[15]   Barford, P., and M. E. Crovella, "Generating Representative Web Workloads for Network and Server Performance Evaluation," *Proceedings of Performance '98/ACM SIGMETRICS '98*, Madison, WS, pp. 151–160.

CHAPTER 15

Reliable Multicast Congestion Control for TCP/IP in Hybrid Networks

Melody Moh, Len Wesley, and Hong Li

15.1 Introduction

Wireless Internet access, such as e-mail, Web browsing, mobile computing, and so forth, have gained increased attention due to rapid advances in both wireless communication and Internet technology. There is an increasing number of Internet applications that require robust and efficient multicast services. Examples include teleconferencing, distributed games, software and file distribution, video distribution, and replicated database updates. Wireless users will need these Internet applications in the near future.

The TCP is an end-to-end connection-oriented transport protocol in the TCP/IP protocol suite. Many research studies have shown that TCP works well in the wired network where the bit error rate is very low, but suffers considerably when some end hosts are connected with wireless, mobile connections. It has also been widely recognized that TCP does not support fairness among heterogeneous connections.

There have been significant efforts to enhance TCP for mobile, wireless networks; few of them, however, address the needs of multicast applications. Significant research has recently focused on the development of multicast congestion control algorithms, but little work has considered hybrid networks where network links and receivers have very different characteristics such as capacity, bit error rate, and mobility.

To meet the demand of multicasting over hybrid wired, wireless networks [1–3], this chapter describes the development of a reliable multicast congestion control algorithm over IP-centric, hybrid wired/wireless, mobile networks. Based on the framework for rate-based *reliable multicast congestion control* (RMCC) proposed by Whetten [4], our new algorithm, RMCC-Mobile, which is described here, is specifically designed for hybrid networks.

RMCC-Mobile addresses the challenges of supporting wireless, mobile links/receivers—for example, dealing with nondeterministic factors such as the receiver locations, link bandwidth, propagation delay, and packet loss rates. Furthermore, RMCC-Mobile also ensures that multicast streams are well controlled and coexist fairly with unicast TCP streams.

15.1.1 Major Challenges

TCP/IP Transparent
Most existing Internet end hosts (whether sources or destinations) and applications use the TCP/IP protocol suite. TCP, however, was designed originally for a low loss rate network medium, in which it is assumed that an extended RTT or a duplicated ACK is caused by congestion in the network. This assumption is no longer true in hybrid networks. An extended RTT may be due to roaming or hand-off users; duplicated ACK may be caused by unreliable wireless links. Furthermore, when extending the TCP window-based mechanism to multicast connections, TCP suffers *drop-to-zero* throughput and feedback implosion problems [4].

TCP Friendly and TCP Fairness
One way to deal with the above problems is to use TCP-friendly rate-based multicast congestion control [5]. This, however, requires accurate feedback including estimation of packet loss rate and RTT [6] to mimic the throughput control of TCP. This is challenging especially in mobile networks where feedback from wireless mobile receivers may be inaccurate, incomplete, or even conflicting. In addition, a RMCC scheme needs to provide fairness both among receivers of the same multicast session (intrasession fairness) and among simultaneous multicast and unicast TCP sessions (intersession fairness). This is especially challenging since TCP is biased against distant connections, and also since there is still a lack of a generally accepted definition of "fairness" among unicast and multicast receivers in different sessions.

Scalable
With the rapid increase of Internet and mobile users, a RMCC scheme needs to be scalable. It should support a large variety and number of

receivers. A large number of receivers present a well-known problem in multicasting congestion control—acknowledgment implosion. It also presents challenges to intermediate routers (or base stations that support wireless, mobile users), which need to intelligently consolidate ACK from downstream receivers before conveying a proper ACK to its upstream node.

15.2   Background and Related Studies

15.2.1   TCP Congestion Control and the TCP Formula

By far the majority of traffic on the Internet is transported by TCP, which uses the congestion avoidance and control algorithm proposed by Jacobson. In setting a guideline for requirements of multicast congestion control, the IETF required that any standard multicast congestion control must be TCP-friendly [7].

Various studies have modeled and analyzed the theoretical throughput behavior of the TCP congestion control algorithm. They derived a relationship between the throughput (data rate) of TCP flow as a function of packet loss rate and RTT. This became known as the *TCP formula*. The notion of TCP-friendliness, therefore, can be loosely described as the behavior of a congestion control algorithm that yields similar (and not more than the) throughput as the TCP formula.

Among pioneering work reported in the literature [8], there is a response model that relates the throughput of a TCP under low loss rate (P is below 0.01) and normal operating conditions where RTT is the round-trip time in seconds and C is a constant between 0.9 and 1.3 [8]

$$\begin{aligned} Throughput &= F_1\left(PacketSize, RTT, P\right) \\ &= \frac{C \cdot PacketSize}{RTT \cdot \sqrt{P}} \end{aligned} \tag{15.1}$$

A more complete analysis of the TCP response function has been presented in [9]. This model incorporated the TCP timeout *Tzero*, and described the throughput under conditions in which the loss rate P may be high. In the throughput formula shown below, C and C' are constants given by $C = \sqrt{2b/3}$ and $C' = \sqrt{27b/8}$ where b is the number of packets acknowledged by a received ACK.

$$Throughput = F_2 \left(PacketSize, RTT, T0, P \right)$$

$$= \frac{PacketSize}{C \cdot RTT \cdot \sqrt{P} + T0 \cdot Min(1, C' \cdot \sqrt{P}) \cdot P \cdot (1 + 32 \cdot P^2)} \tag{15.2}$$

15.2.2   Rate-Based Multicast Congestion Control

Much effort has been undertaken to design TCP-friendly, rate-based multicast congestion control algorithms. One of the first efforts was described in [10]. This work pioneered many key concepts including one of using *exponentially weighted moving average* (EWMA) for reporting packet loss rates. This scheme is adopted in this work and will be discussed in greater detail in the next section.

On rate-based congestion control, Whetten studied the response functions representing the theoretical TCP throughput and illustrated safety and fairness properties in both unicast and multicast environments [4]. They presented a modified version of the complex TCP formula for reliable multicasting, by setting *T0* to be equal to *T*RTT* and adding a scaling factor *QoS*.

$$T_{RM} = T_{RM} \left(PacketSize, RTT, T', P \right)$$

$$= \frac{PacketSize}{RTT} \cdot \frac{1}{(C / QoS)\sqrt{P} + T \cdot C' \cdot P^{1.5} \cdot (1 + 32 \cdot P^2)} \tag{15.3}$$

A QoS value less than 1 makes a reliable multicast stream more responsive than TCP. Setting a QoS value larger than 1 allows a network manager to give a particular stream higher priority than other streams. Whetten specified that $b = 1$, and therefore $C = 0.82$ and $C' = 1.83$. When QoS = 1, (15.3) is equivalent to (15.2).

Based on the above extended response function, Whetten proposed a generic reliable multicast congestion control mechanism for a flat-structured network in which no intermediate routers will provide congestion control. They further proposed the *Reliable Multicast Transport Protocol* (RMTP)-II, a reliable multicast congestion control scheme that works on hierarchical networks [4]. The generic mechanism is adopted in this work as the framework for developing reliable multicast congestion control schemes. Details are given in the next section.

Recently, Moh and Zhang presented a rate-based reliable multicast congestion control scheme for flat-structured networks [3]. The work addressed two fundamental challenges: TCP-friendliness and scalable feedback. They also proposed four scalable feedback schemes for reliable multicasting over heterogeneous wired/wireless/mobile networks. Using sampling techniques, these feedback schemes adapt their feedback generation frequency according to network dynamics including congestion status, end-host mobility, and packet losses.

On the hybrid wired/wireless ATM networks, Moh and Chen have proposed a multicasting flow control for both point-to-multipoint and multipoint-to-point ABR data traffic [1]. Later, Moh and Mei proposed a multicasting scheme for *guaranteed frame rate* (GFR) service supporting TCP/IP traffic [2].

15.2.3 Window-Based Multicast Congestion Control

On window-based congestion control, Rhee, Balaguru, and Rouskas proposed *multicast TCP* (MTCP), a scalable TCP-like congestion control for reliable multicast [11]. Its features include hierarchical congestion status reports, a TCP-like congestion window, and a retransmission window that regulates flows of repair packets to prevent local recovery from causing congestion.

More recently, Rizzo reported a single-rate, window-based multicast congestion control scheme [12]. A representative receiver is selected from the multicast group. This receiver and the sender then run a TCP-like algorithm to control packet transmission. The major challenge of this approach is to select the right representative—one uses the most congested resources in the network. If the point of congestion moves as the multicast session continues, the algorithm needs to continuously find new representatives.

Golestani and Sabnani presented some fundamental observations on multicast congestion control [6]. They studied and established a three-way relationship among regulation parameter choices (such as rate or window size), RTT estimation requirement, and fairness types. They showed that window-based regulation using a common window size for the whole session leads to unnecessary restrictions on the throughput. They thus proposed a multicast window scheme using a distinct window size for each receiver.

15.3 Reliable Multicast Congestion Control Schemes

15.3.1 A Generic RMCC Framework

This section describes a generic, hierarchical, rate control framework for reliable multicast. The framework is extended from a study done by Whetten [4], initially for a flat-structured (instead of hierarchical) network where no assistance for congestion control is assumed from intermediate routers.

15.3.1.1   TCP Response Function

As mentioned in Section 15.2.2, Whetten presented a modified version of the complex TCP equation for reliable multicasting, which is shown in (15.3). The equation will be used at each of the multicast receivers to calculate throughput, which will be sent back in the ACK to the sender.

15.3.1.2   Receiver: RTT and Loss Rate (p) Estimation

In the feedback scheme, each ACK contains information on estimated RTT, loss rate p, and estimated throughput T—which is the target rate calculated using the formulae given in (15.1) and (15.3).

For a multicast rate control mechanism to work well, it is vital that RTT and p be accurately estimated. Montgomery has suggested to use EWMA for reporting network feedback (including packet loss rates and RTT values) [10]. It is easier to measure and more readily adapts to network changes. Whetten later improved the measuring by adapting the weight to be half of the value of current measurement.

For a series of n measurements, rtt_i, $1 \leq i \leq n$, it reports to the source the current value, RTT_n as

$$
RTT_n = \alpha \cdot rtt_n + (1 - \alpha) \cdot RTT_{n-1}
$$
$$
\alpha = 1/2
$$

(15.4)

Accurate measurement of packet loss rate P is also important, especially since the value of the TCP response function greatly depends on P [referring to (15.1) to (15.3)] above. On one hand, if receivers report transient p values, this might lead to an unstable condition or cause drop-to-zero effect. On the other hand, if linear averaged values were reported, the source would not respond promptly or appropriately. Studies have shown that a value of an average of 4 to 8 measurements appears to result in the best trade-off between minimizing the drop-to-zero effect and providing the highest responsiveness to changes in network conditions [4, 13].

Loss rate may be estimated using the EWMA method discussed above. Another way to estimate loss rate is by using a weighted average of eight samples, such that the first four are weighted equally while the last four have a linearly decaying representation. That is, for eight sample measurements $[X_1, X_2, ..., X_8]$, the weights on each are [1, 1, 1, 1, 0.8, 0.6, 0.4, 0.2]. This method has been examined using detailed network simulation of a rate controller for unicast traffic and has been shown to be effective in avoiding drop-to-zero effect resulting from fractal nature of network traffic [13].

15.3.1.3 Sender: Rate Adjustments

One important goal of RMCC is to provide TCP-fair rate allocation for multicast connections. While receiving ACKs from various receivers, the RMCC source uses the *worst-path* method, which takes the minimum estimated throughput value among all the receivers as its rate estimation. Alternatively, the source may use a *double-worst* policy, which takes the longest *RTT* and the largest p to calculate the rate estimation for receivers. This policy, however, suffers the most from the drop-to-zero problem, even though it is provably fair to all competing TCP connections.

The *target rate*, which is the worst-path throughput reported among all the receivers, will serve as a reference for rate adjustment. However, to avoid rate oscillation and to prove TCP fairness, RMCC does not simply adjust the rate to be exactly equal to the worst-path throughput. Instead, the following steps are executed:

1. *Slow start*: At the beginning of a connection, RMCC requires the sender to start sending at an initial minimum rate, and increases this rate exponentially as ACKs are received, similar to TCP slow start. This slow start phase continues until one of the following two conditions occurs: (1) the target rate reported is smaller than the current sender throughput, or (2) a set of slow start threshold losses is reported from any receiver.

   The lower bound of the rate during this period follows the equation given below [10]:

   $$T = (NumAcks * PacketSize) / InitRTT$$
   $$+ Max (MinimumThroughput, PacketSize / InitRTT)$$

   where *InitRTT* is an initial, conservative worst-case estimation of group *RTT*.

2. *Normal mode:* Once a connection experiences its first decrease, the maximum increase in a sender's rate is linear. That is, the maximum increase is one packet per *RTT*, where *RTT* is the maximum reported at receivers' ACKs. The maximum increased rate T_{max} is

   $$T_{max}(t + 1) = T(t) + PacketSize / RTT$$

The maximum decrease rate T_{min} in a sender's rate is multiplicative:

$$T_{min}(t + 1) = T(t) / (2^{Delta})$$

where $Delta = 1 / RTT$.

15.3.2   Enhanced Mechanisms for Hierarchical RMCC over Hybrid Networks

Section 15.3.1 outlined major parts of an RMCC framework for flat-structured networks. This section presents additional mechanisms both at receivers and at the sender (or BSs) for building an efficient RMCC on hierarchical, tree-based networks.

15.3.2.1   Receiver Mechanism

Rate Advertisement. Since last-mile links usually have very diverse characteristics in a hybrid network, it is essential for receivers to advertise their maximum receiving rates. The TCP response functions [(15.1) to (15.3)] suggest a very aggressive rate when packet loss rate p is small. Without an advertised rate that fits an individual receiver's capacity, it is likely to cause RMCC to be more aggressive than TCP (as experienced in the simulation runs) in low data rates. In high data rates, this might cause unnecessary packet losses.

15.3.2.2   Sender/BS Mechanisms

Rate-Update Time. In a fixed, wired network, the time to update the sender's data rate may be triggered by a fixed control interval. In a hybrid network with wired, wireless, and mobile receivers, this time should be dynamically adjusted to reflect the current network traffic situation.

Responsiveness Improvement. In a wired network, a sender would wait until it receives ACK from all the downstream nodes/branches to decide upon and update the data rate. In a wireless/mobile networks, however, some receivers may experience a very unstable wireless link, or may be moving from one area to another. If a sender or a BS were to wait until receiving all the ACKs, the algorithm would not be responsive enough. To avoid unnecessary delay, a sender or a BS should respond after a major portion, say 90% or 75%, of ACKs have been received.

15.3.3   RMCC-Preliminary

In this and the next two sections, three RMCC algorithms are described. A detailed comparison is given in Table 15.1 at the end of the section. In the following, we describe RMCC-Preliminary, a basic version of RMCC that is included here primarily for a base-line reference. Detailed features of RMCC-Preliminary are first described, followed by a high level description of the algorithm.

- The RTT measurement follows the EWMA method, as given in (15.4), using = 0.875. This follows the RTT calculation used by TCP simulated in the Network Simulator 2 (NS2).

- *Loss rate* is measured using the weighted average of eight samples [13], as described in Section 15.3.1.2. Each sample is calculated by monitoring a sliding window of 300 received packets. The sliding window method used for calculating loss rate has been used in the simulator presented for reliable multicasting using active networks [14].

RMCC-Preliminary

I.   *Receiver Mechanisms*

I.a.   *Target–Rate Calculation*

The target rate is calculated following the TCP response function given in (15.3).

I.b.   *RTT Estimation*

Follows the EWMA method, as given in (15.4), using $\alpha = 0.875$.

I.c.   *Loss Rate Estimation*

Uses the weighted average of eight samples, as described above.

I.d.   *Rate Advertisement*

No rate advertisement is implemented.

II.   *Sender Mechanisms*

II.a.   *Slow-Start Mode Rate Adjustment*

No slow-start mode is implemented.

II.b.   *Normal Mode Rate Adjustment*

The sender takes the worst (smallest) target rate reported from all the receivers as the new sending rate.

II.c.   *Rate-Update Time*

The sender uses a fixed control period for rate adjustment.

II.d.   *Responsiveness Improvement*

No mechanism is used to improve sender responsiveness.

II.e.   *Policy with Multiple BSs*

Sender uses the worst-path policy (taking the smallest target rate) as described in Section 15.3.1.3.

15.3.4   RMCC-Whetten-Enhanced

As implied from its name, this algorithm is enhanced from the one proposed by Whetten [4].

Detailed enhancement features are given here, followed by a high-level description of the algorithm:

- *Loss-rate estimation* uses EWMA with $\alpha = 0.5$ in (15.4) in Section 15.3.1.2. This has been verified by simulation to be more accurate than the weighted average of eight samples, and was also discussed in Whetten's work [4].

- *Rate advertisement* at the receiver uses the bandwidth of the last link (usually a wireless link). At the beginning of the multicast session, each receiver informs the sender its last-mile bandwidth. The minimum of these advertised rates would serve as an upper bound for sender's data rate.

- At the sender, *rate adjustments* are divided into slow-start mode and normal mode, as discussed in Section 15.3.1.3. Slow-start mode terminates when a target rate lower than the current sending rate is received at the sender. Rate adjustment mechanisms at the two modes are the same as described in Section 15.3.1.3.

- Finally, the *control period for rate adjustment* is dynamically adjusted according to RTT, and extra control messages are triggered by packet loss occurrences, which in turn causes rate adjustment. This extra mechanism helps the sender to adapt to packet losses.

RMCC-Whetten-Enhanced

I.   *Receiver Mechanisms*

I.a.   *Target-Rate Calculation*

The target rate is calculated following the TCP response function given in (15.3).

I.b.   *RTT Estimation*

Using the EWMA method, as given in (15.4), using $\alpha = 0.875$.

I.c.   *Loss Rate Estimation*

Follows EWMA method, as given in (15.4), using $\alpha = 0.5$.

I.d.   *Rate Advertisement*

Uses the link bandwidth of the last hop as the advertised rate.

II.   *Sender Mechanisms*

II.a.   *Slow-Start Mode Rate Adjustment*

Follows Section 15.3.1.3; terminates when a received target rate is smaller than the current rate.

II.b.   *Normal Mode Rate Adjustment*

Follows Section 15.3.1.3.

II.c.   *Rate-Update Time*

Control period dynamically adjusted according to RTT; extra control periods triggered by extra ACK caused by packet losses.

II.d.   *Responsiveness Improvement*

No mechanism is used to improve sender responsiveness.

II.e.   *Policy with Multiple BS*

Uses worst-path policy (taking the smallest target rate) as described in Section 15.3.1.3.

15.3.5   RMCC-Mobile

This algorithm is newly proposed with an aim to provide congestion control for reliable multicasting over a hybrid networks with wired, wireless, and mobile receivers. Important features are designed specifically for effective multicasting among hybrid receivers. These features, added over RMCC-Whetten-Enhanced, are presented below and followed by the algorithm description.

- *Precise loss-rate estimation* is especially important for multicasting over a hybrid network. As mentioned before, TCP response function is very sensitive to loss-rate estimation P—referring to (15.1) to (15.3). For each loss-rate estimation reported to the sender, an average of six moving (sliding) samples is used in which each sample is averaged over six loss periods. (Each loss period is the period during which one packet loss occurred.) This would ensure that it is an average over sufficiently long period of time with sufficient packet loss occurrences. This is important since packet loss rate fluctuates rapidly in a wireless/mobile link, and this rate differs from one receiver to another. Furthermore, there is a relatively long period of time before an ACK reaches the sender and before the sender's response in rate change would affect the receiver (approximately, and slightly longer than, the RTT). It is therefore important to report an accurate, significant indication of packet loss rate to the sender.

- *Slow-start mode* at the sender is terminated by either the first packet loss reported, or a reported target rate lower than the current rate. At the slow-start mode (referring to Section 15.3.1.3), rate increases exponentially. This usually causes a rapid rate increase in multicast traffic, more rapidly than TCP unicast traffic due to initial small packet loss. Packet-loss indication is therefore added to terminate slow-start mode, this improves TCP-friendliness at the initial phase.

- *Responsiveness improvement* is done by speeding up responses at the sender for rate adjustment. In most multicasting algorithms, the sender (or an intermediate router as in the hierarchical structure) would respond to an ACK and adjust rate only when ACK have been received from all the downstream nodes. This period might be long due to wireless and mobile receivers. The sender or a base station (or intermediate router) will respond for the first time when 90% or more ACK are received, and subsequently when 75% or more ACK are received. For each response period, the sender (or BS) would take the minimum value of all the target rates reported, even though some might belong to an ACK of a previous (but not current) packet. In this case, delayed ACKs are also considered without penalizing responsiveness of the algorithm.

RMCC–Mobile

I.   *Receiver Mechanisms*

I.a.   *Target-Rate Calculation*

Follows the TCP response function given in (15.3).

I.b.   *RTT Estimation*

Uses EWMA method, as given in (15.4), using $\alpha = 0.875$.

I.c.   *Loss-Rate Estimation*

Uses an average of six moving (sliding) samples, each sample is averaged over six loss periods.

I.d.   *Rate Advertisement*

Uses the link bandwidth of the last hop as the advertised rate.

II.   *Sender Mechanisms*

II.a.   *Slow-Start Mode Rate Adjustment*

Follows Section 15.3.1.3; terminates when the first packet loss is reported, or a received target rate is smaller than the current rate.

II.b.   *Normal Mode Rate Adjustment*

Follows Section 15.3.1.3.

II.c.   *Rate-Update Time*

Control period dynamically adjusted according to RTT; extra control periods triggered by extra ACK caused by packet losses.

II.d.   *Responsiveness Improvement*

There is 90% or more ACK received from downstream nodes for the first rate adjustment; 75% or more ACK needed for all subsequent rate adjustments.

II.e.   *Policy with Multiple BS*

Uses worst–path policy (taking the smallest target rate) as described in Section 15.3.1.3.

TABLE 15.1   COMPARISON OF RMCC ALGORITHM

	RMCC-PRELIMINARY	RMCC-WHETTEN-ENHANCED	RMCC-MOBILE
Receiver: target-rate calculation	Response function	Response function	Response runction
Receiver: RTT estimation	EWMA with weight factor $\alpha = 0.875$	EWMA with weight factor $\alpha = 0.875$	EWMA with weight factor $\alpha = 0.875$
Receiver: loss-rate estimation	Eight weighted samples; each sample calculated by a sliding window of 300 received packets	EWMA with weight factor $\alpha = 0.5$	*Average of six samples, each sample is an average of six loss periods*
Receiver: rate advertisement	None	*Link bandwidth of the last hop*	*Link bandwidth of the last hop*
Sender: slow-start mode	Not applicable	Exponential rate increase. Terminates when reported target rate is less than current rate	Exponential rate increase. Terminates when reported target rate is less than current rate *or when first packet loss reported*
Sender: normal mode	Take the worst rate of all the receivers	If current rate < target rate, increase 1 Pkt per RTT	If current rate < target rate, increase 1 Pkt per RTT
		If current rate > target rate, drop to half	If current rate > target rate, drop to half
Sender: rate update time	Fixed control period	Control period dynamically adjusted according to RTT, and *extra control messages are triggered by loss*	Control period dynamically adjusted according to RTT, and *extra control messages are triggered by loss*
Sender: responsiveness improvement	None	None	*First time requires 90% of receivers' ACKs; subsequently requires 75% of receivers' ACKs to update the sending rate*
Sender policy for multiple BSs	Worst path	Worst path	Worst path

15.4   Performance Evaluation

15.4.1   Simulation Objectives and Settings

One important goal of the simulation study is to investigate the effectiveness of the proposed RMCC schemes, and to see how well they coexist fairly with TCP. For this purpose, four network and traffic configurations have been designed for our performance study. They use different network topologies, link characteristics, and traffic scenarios. The four network configurations are summarized in Table 15.2.

Each of the four configurations is used in each of the following four sections. In each section, sending rate, receiving rate, and packet loss rate of the RMCC schemes are measured and plotted against those of TCP. The last section evaluates fairness of the proposed RMCC schemes with TCP and presents the TCP fairness index.

TABLE 15.2   SUMMARY OF FOUR SIMULATED CONFIGURATION

	Topology description	Major features	Simulation objectives
Configuration 1 (Figure 15.1)	One sender, one BS, and six end nodes as multicast receivers	Vary traffic load of multicast and unicast traffic from 200 Kbps to 800 Kbps	Evaluate performance in a congested, hybrid wired/ wireless network
	Three of the six end nodes are also TCP unicast receivers		
	The downstream links are 0.75 Mbps, lower than the upstream link (1 Mbps) to create congestion at the router		
Configuration 2	Same as Configuration 1, except that the six end nodes are connected with wireless links	Vary the packet loss rate of the six wireless links, ranging from 0.01% to 10%	Evaluate performance in a congested, hybrid network with lossy wireless links.
Configuration 3	Twice as big as Configuration 1, with one sender, 2 BSs, and 11 multicast receivers	Same as in Configuration 1	Evaluate performance in a large, congested, hybrid wired/wireless network
	Six of the 11 end nodes are also TCP unicast receivers		
Configuration 4	Same as in Configuration 3	Same as in Configuration 1, with one receiver (node $n2$) moves from one BS to the other	Evaluate performance in a large, congested, hybrid wired/wireless/mobile network.

Table 15.3 summarizes major parameters used throughout the simulation.

15.4.2 Configuration 1—A Simple Network

Configuration 1 is shown in Figure 15.1 (refer to Table 15.2 for more discussion on this simulation experiment). It consists of one BS (or an intermediate router) connected to the sender with a wired link of 1 Mbps and 10-ms propagation delay. The BS connects with six nodes with links of 0.75 Mbps and 20-ms propagation delay. All six nodes are receivers of the multicast session; in addition, nodes $n2$, $n4$, and $n6$ are also receivers of unicast TCP flows 1, 2, and 3, respectively. By having three TCP unicast flows sharing with the multicast flow the link between node $n0$ and BS, a congestion (and thereby packet loss) scenario is created. To ensure that each RMCC scheme performs effectively and yet fairly with TCP, variables in (15.3) need to be fine-tuned—the results are shown in Table 15.4.

TABLE 15.3   SIMULATION PARAMETERS

PARAMETER	VALUE
Simulator	NS-2
Traffic type	Exponential; burst-time = 2 seconds, idle-time = 0.1 second
Traffic load	Varies, 200–800 Kbps
Averaged packet size	1,000 bytes
Link bandwidth	1 Mbps (sender–BS)
	0.75 Mbps (BS–receivers)
Link loss rate	0.0 (sender–BS)
	65535.1 (BS–receivers)
	Varies 0.1–0.0001 (BS–receivers, Configuration 2)
Per-link propagation time	10 ms (upstream)
	20 ms (downstream)
Initial RTT	0.2 second
Minimum throughput	40 Kbps
Buffer size at BS	256 KB
T in (15.3)	4
B, C, C', QoS in (15.3)	Varies; see each configuration for values
Simulation duration	60 seconds
Router queuing scheme	Random early decision (RED)

FIGURE 15.1
Configuration 1.

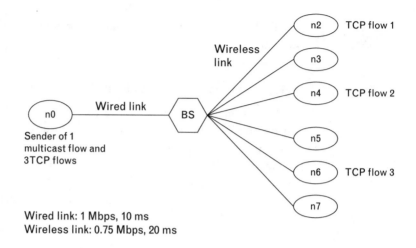

Wired link: 1 Mbps, 10 ms
Wireless link: 0.75 Mbps, 20 ms

TABLE 15.4   EQUATION (15.3) VARIABLES FOR CONFIGURATIONS 1 AND 2

SCHEME	EQUATION (15.3) VARIABLES FOR CONFIGURATION 1 AND 2
RMCC-Preliminary	$B = 1$, $C = 0.82$, $C' = 1.83$, $QoS = 1.4$
RMCC-Whetten-Enhanced	$B = 1$, $C = 0.82$, $C' = 1.83$, $QoS = 1.4$
RMCC-Mobile	$B = 2$, $C = 1.15$, $C' = 2.6$, $QoS = 1.4$

Again referring to Table 15.2, experiments are carried out using the three RMCC schemes for congestion control of multicast traffic. Figure 15.2 shows—with initial traffic load of 400 Kbps for all flows and for the first 35 seconds—sending rate versus time for the three unicast TCP flows and the multicast flow using RMCC-Mobile. It is clear that RMCC-Mobile is able to control the sending rate to be smooth among, and slightly below, all TCP flows. Figure 15.3 shows the same characteristics of the multicast flow controlled by all three RMCC schemes, and of a TCP flow. It can be seen from both figures that TCP fluctuates the most; RMCC-Preliminary suffered from very low sending rate; RMCC-Whetten-Enhanced and RMCC-Mobile both performed well, with RMCC-Mobile maintains a most stable sending rate.

Next, the initial traffic load of both unicast and multicast flows are varied ranging from 200 Kbps to 800 Kbps. Results on averaged receiving rate, and packet loss rate measured at receivers are shown in Figures 15.4 and 15.5, respectively (results on averaged sending rate are not shown, as it was very similar to those of receiving rate). It is clear that RMCC-Mobile has the highest receiving rate among the three RMCC schemes yet maintains a rate slightly lower than TCP. All control schemes are able to maintain a very low packet loss rate (below 4%), except RMCC-Preliminary; this is due to

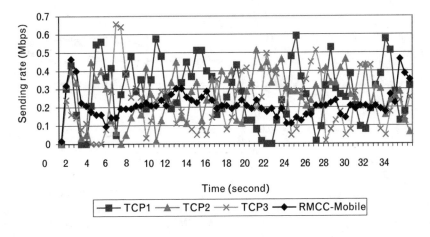

FIGURE 15.2
Configuration 1: sending rate characteristics (three TCP flows and one RMCC-Mobile flow).

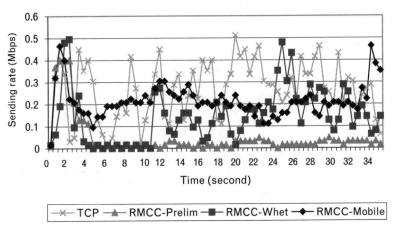

FIGURE 15.3
Configuration 1: sending rate characteristics (one TCP flow, one RMCC-Prelim flow, one RMCC-Whet flow, and one RMCC-Mobile flow).

the absence of slow–start rate control at the sender that causes an initial large packet loss.

15.4.3 Configuration 2—A Wired/Wireless Network with Wireless Links of Different Loss Rates

Configuration 2 is the same as Configuration 1, except that the packet loss rate on wireless links connecting BS to receivers are varied from very low to very high to simulate nondeterministic, lossy characteristics of wireless links. First, with a high loss rate of 0.1 and the initial traffic load fixed at 400 Kbps, the sending rate characteristics are plotted on Figure 15.6. Figure 15.6 shows one simulation run. It is clear that both RMCC–Preliminary and RMCC–Whetten–Enhanced suffer from very low throughput (drop-to-zero problem).

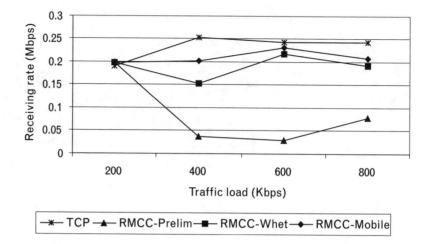

FIGURE 15.4
Configuration 1: comparison of receiving rate.

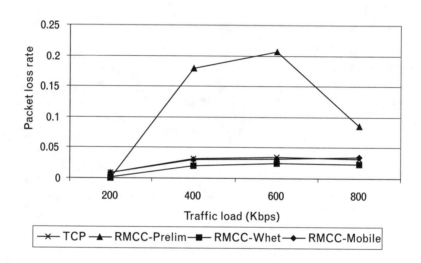

FIGURE 15.5
Configuration 1: comparison of loss rate.

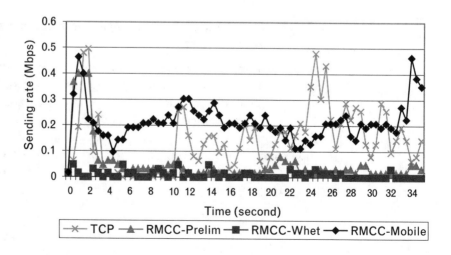

FIGURE 15.6
Configuration 2: sending rate characteristics (one TCP flow, one RMCC-Prelim flow, one RMCC-Whet flow, and one RMCC-Mobile flow).

Next, the loss rate at wireless links varies from 0.0001 to 0.1, and initial traffic load is fixed at 400 Kbps—the results are shown in Figures 15.7 and Figure 15.8. One can see that in very low packet loss rate (0.0001), both RMCC-Whetten-Enhanced and RMCC-Mobile have a throughput higher than TCP. This is likely because (15.3) does not represent well TCP throughput at very low packet loss rate; (15.1), which is for low loss rate networks, should be used instead. When packet loss rate is in mid range (i.e., 0.001 and 0.01), both RMCC-Whetten-Enhanced and RMCC-Mobile performed well. When packet loss rate is very high (i.e., 0.1), most schemes dropped their throughput to very low values. RMCC-Mobile, on the other hand, maintained a relatively stable throughput in the lossy-link environment. Figure 15.8 shows that all schemes except RMCC-Preliminary keep a low loss rate. When the link is very lossy (0.1 loss rate), RMCC-Mobile still maintains a loss rate lower than TCP even while keeping a higher throughput. This set of experiments verified that RMCC-Mobile performs well in lossy links; it maintains a stable throughput and a low packet loss rate.

15.4.4 Configuration 3—A Large Network with Two Base Stations

Configuration 3, shown in Figure 15.9, consists of one sender that sends one multicast flow and six TCP unicast flows. There are two BSs, each connecting with six and five receivers, respectively. End nodes $n2$, $n4$, $n6$, $n10$, $n12$, and $n14$ are also TCP unicast flow receivers. Parameter values used in (15.3) for this and the next configurations are shown in Table 15.5.

Similar experiments of Configuration 1 are carried out for Configuration 3. Results are shown in Figures 15.10 and 15.11. Compared with those of Configuration 1, their results are very similar. RMCC-Mobile has a relatively high receiving rate, but lower than TCP; all RMCC schemes (except

FIGURE 15.7
*Configuration 2:
comparison of receiving
rate.*

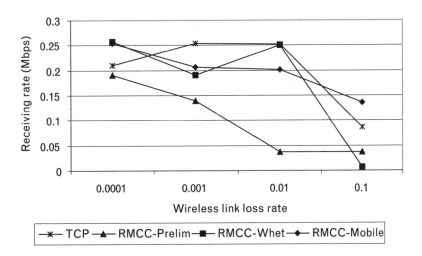

FIGURE 15.8
Configuration 2: comparison of loss rate.

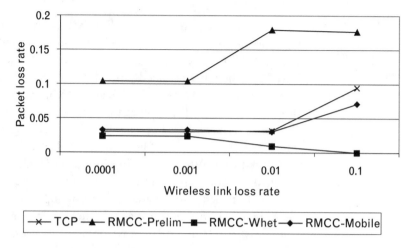

FIGURE 15.9
Configuration 3.

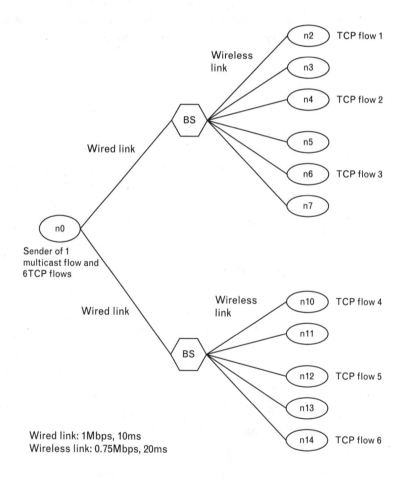

Wired link: 1Mbps, 10ms
Wireless link: 0.75Mbps, 20ms

RMCC-Preliminary), along with TCP are able to maintain a low packet loss rate (again below 4%).

TABLE 15.5 EQUATION (15.3) VARIABLES FOR CONFIGURATION 3 AND 4

SCHEME	EQUATION (15.3) VARIABLES FOR CONFIGURATION 3 AND 4
RMCC-Preliminary	$B = 1$, $C = 0.82$, $C' = 1.83$, $QoS = 2.0$
RMCC-Whetten-Ehanced	$B = 1$, $C = 0.82$, $C' = 1.83$, $QoS = 2.0$
RMCC-Mobile	$B = 2$, $C = 1.15$, $C' = 2.6$, $QoS = 2.0$

FIGURE 15.10
Configuration 3: comparison of receiving rate.

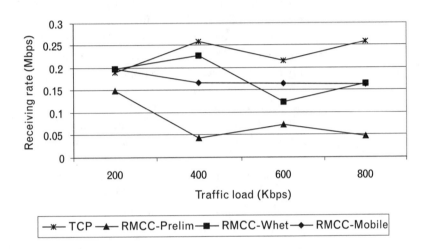

FIGURE 15.11
Configuration 3: comparison of loss rate.

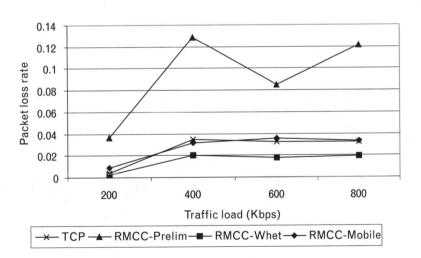

15.4.5   Configuration 4—A Wired/Wireless/Mobile Network with Mobile Nodes

Configuration 4 is identical to Configuration 3, with the addition of having one mobile node (*n2*) moving from the top BS to the bottom BS. It is used to evaluate the RMCC schemes in a mobile environment. The same set of experiments (as in Configurations 1 and 3) is carried out.

Figure 15.12 shows the characteristics of sending rate versus time for three TCP unicast flows and the multicast flow, at 400 Kbps initial traffic load for all flows and for the first 35 seconds. Since the mobile node *n2* moves at the 30th second, it is most interesting to examine the figure at the 30th second and beyond. Note that since *n2* is the receiver of both multicast flow and TCP flow 1, the movement of *n2* causes traffic load at the two wired links to change. Initially, the upper and the lower wired links both carry three TCP flows and one multicast flow. After *n2* moves from the upper to the lower BS, the upper wired link then carries only two TCP flows (plus the multicast flow), while the lower wired link needs to carry one extra TCP flow (added to become a total of four TCP flows plus the multicast flow). This causes greater congestion in the lower wired link. Examining Figure 15.12 one can see that in responding to the mobile node's movement, the multicast flow of RMCC-Mobile dropped its sending rate at 30th second and beyond. This verifies that RMCC-Mobile is able to handle mobility. On the other hand, TCP flows still fluctuate greatly in the 30th second and beyond; this should not be a surprise since the TCP scheme was not designed to handle mobile nodes.

Figures 15.13 and 15.14 show the receiving rates and the packet loss rate of all the schemes while the initial traffic load of all flows are varied from 200 Kbps to 800 Kbps. Again, the results are consistent with those obtained from Configurations 1 and 3: RMCC-Mobile achieves high, yet slightly lower

FIGURE 15.12
Configuration 4:
sending rate character-
istics (three TCP
flows and one
RMCC-Mobile
flow).

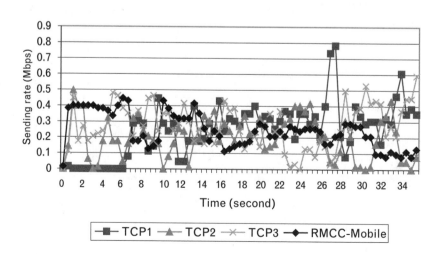

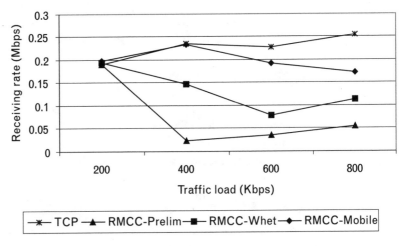

FIGURE 15.13
*Configuration 4:
comparison of receiving
rate.*

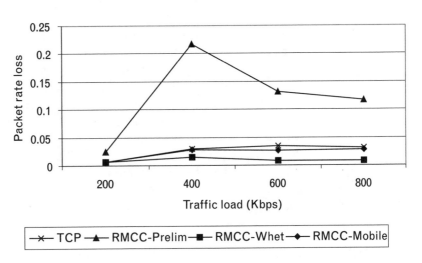

FIGURE 15.14
*Configuration 4: com-
parison of loss rate.*

than TCP, throughput. The average packet loss rate measured at receivers for all schemes, except for RMCC-Preliminary, is still low (around 4%).

15.4.6   TCP Fairness Index

This section evaluates the TCP fairness of the proposed RMCC schemes. The following *fairness index* (FI) is used:

$$FI = \frac{\left(\sum_{x=0}^{N-1} T(x)\right)^2}{N\sum_{x=0}^{N-1} T(x)^2} \tag{15.5}$$

where $T(x)$ is the throughput of the xth protocol. In this chapter, each $T(x)$ is the averaged RMCC or TCP throughput received by receiver x. Tables 15.6 and 15.7 summarize fairness index values of Configurations 2 and 3, respectively (results for Configurations 1 and 4 are omitted due to length limitation).

Table 15.6 shows the FI values. The second row shows FI among all RMCC-Mobile receivers (intrasession fairness for multicast receivers); the third row shows FI among all TCP unicast flow receivers (intersession fairness for unicast flows); and the fourth row shows FI among all RMCC-Mobile and TCP receivers (intersession fairness for multicast and unicast receivers). Specifically, the FI values are calculated using (15.5), where the $T(x)$ values are the receiving rates of (1) RMCC-Mobile receivers (second row), (2) TCP receivers (third row), and (3) both RMCC-Mobile and TCP receivers (fourth row).

From the two tables, it is clear that FI of RMCC-Mobile receivers are all above 0.99. This shows that RMCC-Mobile achieves excellent intrasession fairness. For TCP flows, there are a few exceptions when FI goes below 0.99, including the case in Configuration 2 when packet loss rate is 0.1. This, as discussed above, may be due to the fact that TCP scheme was not designed for high loss-rate links or for mobile networks; it therefore does not perform well in those network environments. For both RMCC-Mobile and TCP receivers (intersession), FI is not as high as that of intrasession, especially when traffic load reaches 800 Kbps or the loss rate on wireless links is 0.1. But for intersession, even the value of 0.947846 is already a good FI. This shows that RMCC-Mobile and TCP flows share the bandwidth fairly.

TABLE 15.6   FAIRNESS INDEX OF CONFIGURATION 2

Loss rate	0.0001	0.001	0.01	0.1
RMCC-Mobile	0.999919	0.999999	0.999991	0.99995
TCP	0.994339	0.997615	0.996367	0.979185
RMCC-M & TCP	0.989376	0.987596	0.985413	0.947846

TABLE 15.7   FAIRNESS INDEX OF CONFIGURATION 3

Traffic load	200K	400K	600K	800K
RMCC-Mobile	0.999989	0.999987	0.999655	0.999943
TCP	0.999974	0.987212	0.992138	0.997216
RMCC-M & TCP	0.99955	0.987064	0.977287	0.949622

15.5   Summary and Future Enhancement: Use of Belief Functions

The most important challenge of multicasting congestion control over hybrid networks is that the receivers and their connected links have nondeterministic, dynamic, and uncertain characteristics. The *belief function* (BF) calculus is a mathematical tool that deals with incomplete, imprecise, inaccurate information. It may be used for future extension of RMCC schemes. This section presents a brief overview of BF and an example that shows how it could be used in RMCC enhancement. In particular, none of the RMCC schemes presented above considers buffer occupancy at BSs for rate adjustment (the BF example presented below makes use of buffer occupancy rate).

15.5.1   BF Calculus: An Overview

The belief function calculus can be viewed as a mathematically sound extension to classical probability theory and statistical inference. It is a mature calculus that has been shown to be well suited for representing and reasoning from information that is incomplete, imprecise, and inaccurate to varying degrees (commonly called evidential information) [15]. Some motivating factors for using the BF calculus are as follows:

- Its ability to capture and differentiate, in a single representation, the degree to which a network feedback (propositional sentence) is true or false, and the degree ignorance about the truth or falsity of the feedback.

- Its ability to readily identify which parts of network feedback contributed to the truth, falsity, and ignorance associated with the suggested rate change.

15.5.2   An Abstract Example

Having these capabilities is important for advancing the ability of congestion control algorithms to adjust rates based on the situational dynamics of a network. Indeed, it must be anticipated that the highly dynamic nature of communication networks will induce a varying degree of uncertainty about the current or projected state of the network.

An attractive aspect of the BF calculus is that the computations for integrating evidential information and drawing inferences can be carried out in as parallel a manner as can be supported by the host platform. This, in addition to other performance enhancing implementations, allows a system to readily assess the situation of a network and to make a more informed decision about rate changes than is possible with current congestion control algorithms.

Within the BF calculus, possible answers to questions like "How should the current rate change?" are represented as a set of interrelated propositional statements, called a *frame of discernment* (FOD) or simply a frame. Elements in a FOD represent a mutually exclusive and exhaustive set of propositions that can be viewed as possible answers to some question of interest such as the one above. For example, a "TARGET RATE" FOD Θ = (*Don't change rate, Decrease rate to 90% of current rate., Decrease rate to 80% of current rate., Decrease rate to 70% of current rate., ..., Increase current rate by 10%, Increase current rate by 20%, Increase current rate by 30%, ...*), might be used to represent possible answers to the above question. We could have represented our example frame as a continuous frame from—for instance, [–90, +20], where each value (say, +10) in the interval corresponds to a propositional sentence of the form "The current rate should be adjusted by x%." The choice between using a continuous or discrete frame depends on the form of the answer that is desired by the user. For simplicity, we will use the discrete frame above to introduce the BF calculus without loss of generality.

To discern which proposition in the frame is the one and only true proposition involves acquiring and combining distinct opinions about the truthfulness or falsity of subsets of Θ. Sources of information can range from statistical data about the percentage of receivers (or mobile hosts) that are having a particular loss rate, the current overall loss rate at all receivers, through the buffer occupancy at BS. Such information can be viewed as metric information that can be used to infer if, how, and when to adjust transmission rates at the sender.

In practice, a single body of information (feedback) will not always directly suggest the truth or falsity of propositions (rate adjustment) in the TARGET RATE (TR) frame. Rather, input information (called an opinion) is often expressed in terms of propositions in a frame Θ' that is distinct and indirectly related to possibilities in Θ. To arrive at a consensus over the TR frame, opinions that are conveyed over indirectly related frames must be first translated to the TR frame before a consensus can be formed about the true proposition in the TR frame. The relationships between propositions in different frames are represented as a *compatibility relation* (*C*) that is defined to be

$$C{:}2^{\Theta} \; a \, 2^{\Theta'}$$

where every mapping from $p \in 2^{\Theta}$ to $q \in 2^{\Theta'}$ means that if p is true then q can be simultaneously true, and vise versa.

In practice, separate frames corresponding to each MobileHosts (MH), LossRate (LR), and BufferOccupancy (BO) information would typically be constructed and related to the TargetRate (TR) as illustrated in Figure 15.15. The figure also shows an intermediate cross-product frame MHLRBO that captured dependencies between the MH, LR, and BO frames. The need for MHLRBO follows from the fact that an adjustment to a rate depends on particular combinations of values in MH, LR, and BO, rather than these values being related to TR independent of each other.

As shown in Figure 15.15, the values in the MH frame show abbreviated notation for propositions such as "<10% of the mobile hosts have the maximum loss rate and the percentage of mobile hosts with this loss rate is decreasing," "Between 10% and 30% of mobile hosts have the maximum loss rate and the percentage of mobile hosts with this loss rate is decreasing," through " >50% of mobile hosts have the maximum loss rate and the percentage of mobile hosts with this loss rate is increasing." The intuition behind the "…and the percentage of mobile hosts with this loss rate is decreasing/increasing" part of each proposition is that a rate change should increase or decrease depending on whether the values in each frame is increasing or decreasing. Similar "propositional" interpretations can be given for the values in the respective LR and BO frames. The cross-product frame MHLRBO contains propositions that are logical conjunctions of a subset of propositions in MH, LR, and BO. It is worth noting that the number of propositions in the MHLRBO is not equal to the product of the number of propositions in each of the MH, LR, and BO frames. There are some combinations that are not logically possible, and hence are omitted from MHLRBO. The frames and compatibility relations in Figure 15.15

FIGURE 15.15
Example BF frames and compatibility relations C(…) that relate values between frames.

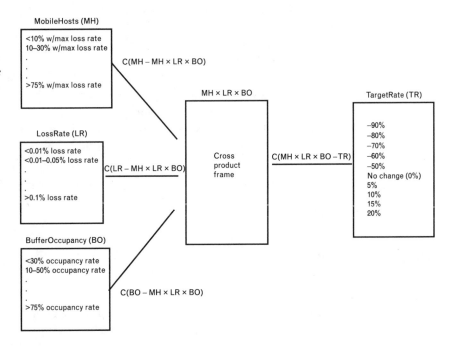

represent just what is possible, not which proposition in TR is probably true.

15.5.3   Future Directions

We have constructed an example to evaluate the role of a BF model in enhancing RMCC by reasoning about changes to the sender's transmission rate. Currently, we have started designing RMCC-BF, a BF-assisted RMCC. We continue to explore the following two attractive features of the BF calculus and how they may be applied to rate control in hybrid networks. First, it allows us to quantify our ignorance and provides a rich representation of what is and is not know so that a network has at least the option of investigating the current situation and refine its feedback to discern the most appropriate rate change at the multicast sender. Second, it is able to extract information about which portion of all the feedback received has the greatest or least impact on the current suggested rate change. This information could be useful when trying to identify which feedback information to focus refinement efforts.

15.6   Summary

This chapter presents a pioneering work on congestion control for reliable multicast over hybrid, IP-centric networks. The major challenges are identified, and a set of sender and receiver mechanisms are presented that are significant for rate-based, TCP-friendly control schemes over networks of combined wired, wireless, mobile nodes. With careful developing of enhanced features, RMCC-Mobile, a scheme designed specifically for hybrid, wired, wireless/mobile Internet access, is proposed. RMCC-Mobile is evaluated and compared with two other RMCC schemes using extensive simulation experiments. It is found that while TCP and other RMCC schemes suffered drop-to-zero throughput problem in lossy, mobile networks, RMCC-Mobile continued to perform well in those environments, while maintaining stable throughput, low packet loss, and TCP-fairness. Finally, BF calculus is presented as a potential tool for future enhancement of RMCC over hybrid networks.

Acknowledgments

The authors appreciate helpful discussions with Dr. Brian Whetten (via e-mail). Melody Moh is supported in part by KDDI R&D Labs, Japan, and NSF grant NCR-9714700 of the United States; Len Wesley is supported in part by NSF grant 21-1307-2045; Hong Li is supported in part by SJSU Graduate Student Research Assistantship.

REFERENCES

[1] Moh, W. M., and Y. Chen, "Multicasting Flow Control over Hybrid Wired/Wireless ATM Networks," *Performance and Evaluation Journal*, Vol. 40, March 2000, pp. 161–194.

[2] Moh, W. M., and H. Mei, "Design and Evaluation of Multicast GFR for Supporting TCP/IP over Hybrid Wired/Wireless ATM," *ACM/Baltzer Journal on Wireless Networks (WINET)*, Vol. 6, No. 5, November 2000, pp. 401–410.

[3] Moh, W. M., and S. Zhang, "A Study on Rate-Based, Reliable Multicast Congestion Control," invited paper, *Proceedings of SPIE Workshop on Scalability and Traffic Control in IP Networks*, Vol. 4526, 2001, pp. 369–380.

[4] Whetten, B., "Reliable Multicast Congestion Control and Tree Based Reliable Multicast," Ph.D. dissertation, Department of EECS, University of California, Berkeley, Fall 2000.

[5] Handley, M., S. Floyd, and B. Whetten, "Strawman Specification for TCP Friendly (Reliable) Multicast Congestion Control (TFMCC)," Technical Report presented in The Reliable Multicast Research Group Meeting, Arlington, VA, December 1998.

[6] Golestani, S. J., and K. K. Sabnani, "Fundamental Observations on Multicast Congestion Control in the Internet," *Proc. of IEEE INFOCOM*, March 1999.

[7] Mankin, A., et al., "IETF Criteria for Evaluating Reliable Multicast Transport and Application Protocols," RFC 2357, June 1998.

[8] Mathis, M., et al., "The Macroscopic Behavior of the TCP Congestion Avoidance Algorithm," *ACM Computer Communication Review*, Vol. 27, July 1997, pp. 67–82.

[9] Padhye, J., et al., *Modeling TCP Throughput: A Simple Model and Its Empirical Validation*, Technical Report CMPSCI TR 98-008, Universtiy of Massachusetts, 1998.

[10] Montgomery, T., *A Loss Tolerant Rate Controller for Reliable Multicast*, Technical Report NASA-IVV-97-011, West Virginia University, August 1997.

[11] Rhee, I., N. Balaguru, and G. Rouskas, "MTCP: Scalable TCP-Like Congestion Control for Reliable Multicast," *Proc. of IEEE INFOCOM*, March 1999.

[12] Rizzo, L., "A TCP-Friendly Single-Rate Multicast Congestion Control Scheme," *ACM SIGCOMM'00*, August 2000.

[13] Floyd, S., et al., "Equation Based Congestion Control for Unicast Applications," *Proc. of SIGCOMM*, August 2000.

[14] Kasera, S., et al., "Scalable Fair Reliable Multicast Using Active Services," *IEEE Networks*, January/February 2000.

[15] Wesley, L. P., "Evidential-Based Control in Knowledge-Based Systems," Ph.D. dissertation, Department of Computer and Information Science, University of Massachusetts, Amherst, MA, June 1988.

Part IV:
Handoff, Mobility, and Signaling

CHAPTER 16

Mobile IP: A Challenge in the Mobile World

Christina A. Nika, Dimitrios D. Vergados, and Michael Theologou

16.1 Introduction

Over the past few years, the explosive growth of the Internet and the increasing popularity of notebook computers, which are both portable and extremely powerful, have been noticed [1]. People want to work remotely from home and office, almost everywhere [2]. Workers must be able to log in to their companies' networks at any moment during the day, regardless of where they are working. All of this means a boom in mobile computing [3, 4]. Consider how cellular phones have given people new freedom in carrying out their work.

Mobile IP is an Internet standards-track protocol that enhances the existing IP to accommodate mobility. Mobile IP in wireless networks is intended to be a direct extension of the existing fixed/wireline networks with uniform end-to-end QoS guarantees.

Mobile IP is changing the way that everybody works. Anyone with access to the Internet will be able to build a computing environment wherever one goes. In addition, seamless roaming and access to applications will make users more productive, and they will be able to access applications on the fly, perhaps giving them an edge on the competition. In today's networks, there is one layer in the protocol stack, which is becoming the most ubiquitous and an obvious piece to the overall puzzle [5]. That piece is the Internet Protocol of the TCP/IP protocol suite. The Internet of today lacks mechanisms for the support of users travelling through the world. IP is the common base for thousands of applications and runs over dozens of different networks. This is the reason for supporting mobility at the layer IP.

In this chapter a review of the challenge in the use of Mobile IP in wireless networks is presented. In the following section, the need for Mobile IP together with the advantages and the requirements are discussed in detail. In Section 16.3, an overview of Mobile IP is given. The procedures and the major components are also discussed together with a description of the routing datagrams in a mobile IP network. In Sections 16.4 and 16.5, a description is presented of problems that are in Mobile IPv4 and as well as an outline of the differences between Mobile IPv4 and IPv6. Sections 16.6 and 16.7 describe some of the suggested solutions for the micro-mobility problem in Mobile IPv4, like Cellular IP and HAWAII. The chapter's results are presented analytically in the last section.

16.2   The Need for Mobile IP

It has been foreseen that mobile computing devices will become more pervasive, more useful, and more powerful in the future. The power and usefulness will come from being able to extend and integrate the functionality of all types of communication such as Web browsing, e-mail, phone calls, information retrieval, and perhaps even video transmission. For Mobile IP computing to become as pervasive as stationary IP networks of the world, an ubiquitous protocol for the integration of voice, video, and data must be developed. The most widely researched and developed protocol is Mobile IP.

The advantages of using Mobile IP can be summarized as follows [6]:

- It allows fast, continuous low-cost access to corporate networks in remote areas where there is no public telephone system or cellular coverage.

- It supports a wide range of applications from Internet access and e-mail to e-commerce.

- Users can be permanently connected to their Internet provider and charged only for the data packets that are sent and received.

- Lower equipment and utilization costs for those requiring reliable high-speed data connections in remote locations worldwide.

- A user can take a palmtop or laptop computer anywhere without losing the connection to the home network.

- Mobile IP finds local IP routers and connects automatically. It is phone-jack and wire-free.

- Other than mobile nodes/routers, the remaining routers and hosts will still use current IP. Mobile IP leaves transport and higher protocols unaffected.

- Authentication is performed to ensure that rights are being protected.

- Mobile IP can move from one type of medium to another without losing connectivity. It is unique in its ability to accommodate heterogeneous mobility in addition to homogenous mobility.

The Mobile IP network is also characterized by some disadvantages as well [7]:

- There is a routing inefficiency problem caused by the "triangle routing" formed by the home agent, correspondent host, and the foreign agent. It is hoped that Mobile IPv6 can solve this problem.

- Security risks are the most important problem facing Mobile IP. Besides the traditional security risks with IP, one has to worry about faked care-of addresses. By obtaining a mobile host's care-of address and rerouting the data to itself, an attacker can obtain unauthorized information. Yet another issue related to the security is how to make Mobile IP coexist with the security features coming in use within the Internet.

The characteristics that should be considered as baseline requirements to be satisfied by any candidate for a Mobile IP are the following [8]:

- *Compatibility:* A new standard cannot require changes for applications or network protocols already in use. Mobile IP has to remain compatible to all lower layers used for the standard nonmobile IP. It must not require special media or protocol.

- *Transparency:* Mobility should remain "invisible" for many higher layer protocols and applications. Besides maybe noticing a lower bandwidth and some interruption in service, higher layers should continue to work even if the mobile changed its point of attachment to the network.

- *Scalability and efficiency:* Introducing a new mechanism into the Internet must not degrade the efficiency of the network. Due to the growth rates of mobile communication, it is clear that Mobile IP must be scalable over a large number of participants in the whole Internet.

- *Security:* All messages used to transmit information to another node about the location of a mobile node must be authenticated to protect against remote redirection attacks.

The requirements of Mobile IP may be summarized as follows [7]:

- A mobile node must be able to communicate with other nodes after changing its link-layer point of attachment to the Internet, yet without changing its IP address.

- Application programs must be able to operate continuously over a single session while the network attachment point of the mobile host changes.

- A mobile node must be able to communicate with other nodes that do not implement these mobility functions.

- All messages used to update another node with the location of a mobile node must be authenticated in order to protect against redirection attacks.

16.3   The Mobile IP in Wireless Networks: A Simple Architecture

Mobile IP is a new proposed standard of the IETF designed to support mobile users [5]. It is also a new Internet standard for the Web and private networks. There are currently two standards, one to support the current IPv4, and one for the upcoming IPv6. One of the big requirements on the new standard is that it should support both ordinary and wireless networks [7]. The big issue is to solve the overlapping between different networks—a user should be able to move between different networks without packet losses.

The solution proposed by a working group within the IETF suggests that the mobile node should use two different IP addresses: a fixed home address and a care-of address, that changes at each point of attachment. The solution requires two additional components, the *home agent* (HA), and the *foreign agent* (FA). The mobile host is then able to move between different networks, while keeping the same IP address. The HA is placed on the user's local network, while the FA is placed on the network that the host currently is visiting. If the mobile host were able to handle the things that the FA is doing, then the FA would be superfluous.

The protocol works as follows [6]:

- *Mobile agents* (MAs) advertise their presence by sending agent advertisement messages.

- A mobile host may solicit MAs by sending agent solicitation messages.

- A mobile host uses the MA advertisements to determine if it is on the home or the foreign network.

- When the host is on the home network, it acts independently of the HA.

- When a mobile host returns from a foreign network, it must deregister with the HA through Registration Request and Registration Reply messages.

- When a mobile host finds it has moved to a new, foreign network, it obtains a COA from the FA or from other means such as DHCP.

- When a mobile host on the foreign network obtains its COA, it registers the new care-of address with the HA using a Registration Request and Registration Reply.

- Datagrams sent to the home network are received by the HA. They are encapsulated in a new datagram that contains the care-of address and are sent to the FA or to the mobile host if it is acting without the aid of the FA.

- Datagrams sent by the mobile host on the foreign network need not be returned to the HA, but could be sent directly to the destination.

This procedure can be described by the three major components of Mobile IP:

1. Agent discovery;

2. Registration;

3. Tunneling.

16.3.1 Agent Discovery

The agent discovery procedure used in Mobile IP is based on the *Internet Control Message Protocol* (ICMP) Router Advertisement standard protocol. Agent advertisements are typically broadcast at regular intervals (e.g., once a second, or once every few seconds) and in a random fashion, by home agents and foreign agents.

When a mobile node is away from home, it wants to find agents so that it does not lose access to the Internet. There are two ways of finding agents. The first is by selecting an agent from among those periodically advertised, and the second is by sending out a periodic solicitation until it receives a response from a mobility agent. The mobile node thus gets its COA, which may be dynamically assigned or associated with its FA.

A mobility agent advertising its services on a link transmits agent advertisements. Mobile nodes use these advertisements to determine their current point of attachment to the Internet.

16.3.2   Registration

A mobile node registers whenever it detects that its point-of-attachment to the network has changed from one link to another. The mobile node registers its COA with its HA in order to obtain service. The registration process can be performed directly from the mobile node, or relayed by the FA to the HA, depending on whether the COA was dynamically assigned or associated with its FA. Note that simultaneous registration with multiple COAs is possible.

In Mobile IP this can be accomplished by using the registration procedure (see Figure 16.1). The mobile node sends a registration request (using the UDP) with the COA information. This information is received by the HA and, if the request is approved, adds the necessary information to its routing table and sends a registration reply back to the mobile node.

In the registration process the mobile node either obtains its COA from an FA or registers itself to another protocol, say, a DHCP server. The mobile node should send a registration request to the FA or directly send a request to its HA by using collocated COA as the source IP address of the request.

In case the mobile node cannot contact its predefined HA, it is possible that this mobile node will register to another unknown HA on its home network. This method, called automatic HA discovery, works by using a directed broadcast IP address that reaches IP nodes on the home network, instead of the HA's IP address. The IP nodes in the home network that can operate as HA will receive the directed broadcast IP packet and will send a rejection to the mobile node. This rejected message will, among others, contain the IP address of its source node. The mobile node will then be able to use this IP address in a new attempted registration message. During the registration procedure, there is a need to authenticate the registration information.

FIGURE 16.1
Registration in
Mobile IPv4.

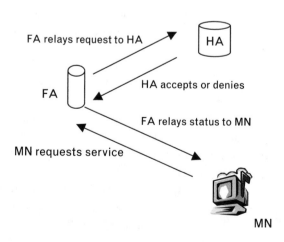

Registration in Mobile IP must be made secure so that fraudulent registrations can be detected and rejected. Otherwise, any malicious user on the Internet could disrupt communications between the HA and the mobile node by the simple expedient of supplying a registration request containing a bogus COA. This would effectively disrupt all traffic destined for the mobile node.

There are three authentication extensions defined for use with Mobile IP [9]:

1. The mobile-home authentication extension (which is required by all registration requests and replies);

2. The mobile-foreign authentication extension;

3. The foreign-home authentication extension.

Each extension includes a *security parameter index* (SPI) that indicates the mobility security associate that contains the secret and the other information needed to compute the authenticator. Exactly one mobile-home authentication extension is required to be present in all registration requests and registration replies. The location of the extension also marks the end of the data authenticated by the mobile node.

16.3.3 Tunneling

In the most general tunneling, illustrated in Figure 16.2, the source, encapsulator, decapsulator, and destination are separate nodes. The encapsulator node is considered the entry point of the tunnel, and the decapsulator node is considered the exit point of the tunnel. Multiple source-destination pairs can use the same tunnel between the encapsulator and decapsulator.

The HA, after a successful registration, will begin to attract datagrams destined for the mobile node and tunnel each one to the mobile node at its COA. The tunneling can be done by one of several encapsulation algorithms, but the default algorithm that must always be supported is simple IP-within-IP encapsulation [6]. Encapsulation is the mechanism of taking a packet consisting of packet header and data and putting it into the data part of a new packet. Encapsulation is a very general technique, used for many

FIGURE 16.2
General tunneling in Mobile IPv4.

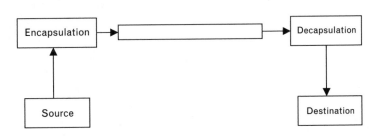

different reasons including multicast, multiprotocol operations, authentication, privacy, defeating traffic analysis, and general policy routing.

Other types of encapsulation mechanisms can also perform the tunneling procedure. These mechanisms are included in different encapsulation protocols such as the *Minimal Encapsulation Protocol* (MEP) and the *Generic Routing Encapsulation* (GRE) Protocol. In the GRE encapsulation protocol a *source route entry* (SRE) is provided in the tunnel header. By using the SRE, an IP source route, which includes the intermediate destinations, can be specified. In the MEP the information from the tunnel header is combined with the information in the inner minimal encapsulation header to reconstruct the original IP header. In this manner the header overhead is reduced, but the processing of the header is slightly more complicated.

Decapsulation is the reverse process of encapsulation. During service time (after the registration process and before the service time expiration), the mobile node gets forwarded packets from its FA, which were originally sent from the mobile node's HA.

Tunneling is the method used to forward the message from HA to FA and finally to the mobile node. After the mobile node returns home, it deregisters with its HA to drop its registered COA. In other words, it sets its COA back to its home address. The mobile node achieves this by sending a registration request directly to its HA with the lifetime set to zero. There it has no need to deregister with the FA because the service expires automatically when the service time expires.

16.3.4   Mobile IP Datagram

Figure 16.3 illustrates the routing of datagrams to and from a mobile node away from home, once the mobile node has registered with its HA.

The mobile node is presumed to be using a care-of address node provided by the FA:

FIGURE 16.3
The routing of datagrams to and from a mobile node.

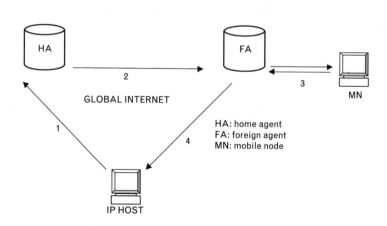

1. A datagram to the mobile node arrives on the home network via standard IP routing.

2. The datagram is received by the HA and is tunneled to the COA.

3. The datagram is detunneled and delivered to the mobile node.

4. For datagrams sent by the mobile node, standard IP routing delivers each to its destination.

This datagram gives a rough outline of operation of the Mobile IP:

• Mobile agents make themselves known by sending agent advertisement messages. An impatient mobile node may optionally solicit an agent advertisement message.

• After receiving an agent advertisement, a mobile node determines whether it is on its home network or a foreign network. A mobile node basically works like any other node on its home network when it is at home.

• When a mobile node moves away from its home network, it obtains a COA on the foreign network, for instance, by soliciting or listening for agent advertisements or DHCP or PPP.

• While away from home, the mobile node registers each new COA with its HA, possibly by way of an FA.

• Datagrams sent to the mobile node's home address are received by its HA, tunneled by its home agent to the COA, received at the tunnel endpoint, and finally delivered to the mobile node.

• In the reverse direction, datagrams sent by the mobile node are generally delivered to their destination using standard IP routing mechanisms, not necessarily passing through the HA.

16.4   Open Issues in Mobile IPv4

The Mobile IPv4 can provide mobility support to portable devices that are roaming through different wireless subnetworks, but there are still several issues that have to be solved.

In interdomain mobility (macro-mobility), which defines the movement of a mobile node from one subnetwork to another subnetwork, there are the following open issues [6, 7]:

• *Triangle routing:* This refers to the path followed by a packet from the correspondent host to a mobile node, which must first be routed, via the mobile node's HA. In Figure 16.1, the packet sent by the correspondent host follows path 1, 2, and 3, while the packet sent by the mobile node will follow routes 3 and 4.

- *Inefficient direct routing:* The process of selecting a next hop and an outgoing interface over which an IP packet should be forwarded in Mobile IPv4, measured in number of hops or end-to-end delay, independent of the triangle routing situation, is inefficient.

- *Inefficient HA notification:* In Mobile IPv4 the process by which a mobile node informs its HA and various correspondent hosts of its current COA is inefficient, since the HA has to be notified during each interdomain handover.

- *Inefficient binding deregistration:* If a mobile node moves to another (new) FA, then the previous FA could release the resources used by the mobile node. In Mobile IPv4 this is not possible due to the fact that the previous FA waits until a binding registration lifetime expires.

In intradomain mobility (micro-mobility), which defines the movement of a mobile node within a subnetwork (i.e., handover from cell to cell), the Mobile IPv4 does not provide solutions for:

1. *Local management of micro-mobility events:* Due to the fact that the micro-mobility events can happen with relatively high frequency, they should be managed as much as possible locally.

2. *Seamless intradomain handover:* After intradomain handover, which provides mobility and handover support for frequently moving hosts, the IP data stored into the previous entities (e.g., BSs) should be transferred to the new BS.

3. *Mobility routing crossing in a intranet:* During intradomain handover, when the mobile node moves and approaches a new BS and redirects its packets from the old to the new BS, the routing crossings should be avoided as much as possible.

The QoS provided by Mobile IPv4 does not specify usable features or capabilities that guarantee the transport of real-time data in the wireless Internet. However, Mobile IPv4 should be able to assist QoS algorithms or mechanisms. The open issues are as follows:

1. *Efficient Mobile IP aware reservation mechanisms:* Definition of reservation mechanisms that can be used to assist for fast routing during Mobile IPv4 intradomain handovers.

2. *RSVP operation over IP tunnels:* Due to the tunneling operation the Mobile IPv4 cannot interwork with the RSVP.

3. *RSVP reservation on Mobile IP triangle route situations:* Due to the triangle route operation the Mobile IPv4 cannot interwork with the RSVP.

Simultaneous bindings refer to the possibility of a host to register more than one COA at the same time. This is an optional feature in Mobile IPv4—for example, a host can register to different COAs that identify with two neighboring subnetworks. There is a following open issue:

1. *Inefficient maintenance of simultaneous bindings:* The problem occurs when a FA receives a tunneled packet destined to a mobile node, and this mobile node is not in a visitor list entry of the foreign agent, then this tunneled packet should be dropped. The foreign agent cannot recognize the home address of the mobile node. This situation will have negative consequences on the higher-level layer associations.

Security means that the Mobile IP protocol permits mobile internetworking to be done on the network layer. Generally, security refers the science of protecting computers, network resources, and information against unauthorized access, modification, and destruction. Mobility introduces the need for enhanced security. The open issues in security are as follows [7, 10]:

1. *Ingress filtering:* In an ISP any border router may discard packets that contain a source IP address that is not being configured for one of the ISP's internal networks. The ISPs begin filtering IP packets in their routers to make sure that the IP source address of a packet is legitimate before it is forwarded. Ingress filtering can mitigate these attacks but it cannot solve them. This is because the flooding attack can still be waged from spoofed addresses so long as those IP addresses appear to emanate from the appropriate point in the network.

2. *Minimize the number of required trusted entities:* If the number of the required trusted entities (for example, HA and FA) is decreased, the security may be enhanced.

3. *Authentication:* Authentication is the process of someone or something's claimed identify. Authentication involves challenging a person to prove that he or she has physical possession of something or that he or she has knowledge of something. Authentication protocols define the message flows by which this challenge and response are sent and received by the parties being authenticated. The Mobile IPv4 authentication techniques between mobile nodes and FAs are not reliable enough.

4. *Authorization:* One of the purposes of the network is to provide a way to share resources among many users. These resources must be pro-

tected against unauthorized access. An organization that owns and/or operates a network would need to decide who may attach to this network and what network resources the attaching node might use.

5. *Nonrepudiation:* Proving that a source of some data did in fact send data that user might later deny sending. In the future wireless Internet, a recipient of a message should have the opportunity to prove that a message has been originated by a sender.

6. *Encryption key distribution:* Using some forms of cryptography can provide authentication, integrity, and nonrepudiation. Cryptography is the science of transforming data in seemingly bizarre ways to accomplish surprisingly useful things. A cryptography system consists of two components, algorithm and keys. Message senders and receivers require the distribution of encryption key information.

7. *Location privacy:* A sender of a message should be able to control which, if any, receivers know the location of the sender's current physical attachment to the network. Location privacy is concerned with hiding the location of a mobile node from correspondent hosts.

8. *Use one single subscription for all service types: Network access identify* (NAI) is used to identify ISP subscribers during roaming operations. Cellular service providers are evolving their home cellular networks to provide 3G cellular services using NAIs with a single subscription.

9. *Firewall support in Mobile IP:* Firewall is a device that protects a private network against intrusion from nodes on a public network. The firewalls are configured to interrupt traffic to and from interloper nodes. If a mobile node has to enter a private Internet network that is securely protected by a firewall, then Mobile IP aware support at this firewall is required.

16.5   Mobile IPv6

Mobility support in IPv6 follows the design for Mobile IPv4. The basic idea, that a mobile node is reachable by sending packets to its home network and that the HA sends packets from a home network to the mobile node's current COA, remains the same. Also, similar to the method used before (IPv4), the HA encapsulates the packets for delivery from the home network to the COA. While discovery of a care-of address is still required, the mobile node can configure its COA by using address autoconfiguration and neighbor discovery [6].

In Mobile IPv6 there are no FAs, but there is still the need to provide a central point to assist with Mobile IP handoffs. The major differences between Mobile IPv4 and Mobile IPv6 are:

1. The size of the addresses, 128 bits in IPv6 versus 32 bits in IPv4, which means several benefits of such large addresses:
   - There is no real chance of running out of addresses during the lifetime of IPv6;
   - Allows aggregation of many network-prefix routes into a single network-prefix route;
   - Allows nodes to autoconfigure using very simple mechanisms.

2. Support for route optimization is the fundamental part of Mobile IPv6, and it solves the triangle routing prblem existing in Mobile IPv4. Packets sent by a correspondent to a mobile node connected to a foreign link are routed first to the mobile node's HA and then tunneled to the mobile node's COA. However, packets sent by the mobile node are routed directly to the correspondent, as shown in Figure 16.4.

   This optimized routing is potentially more efficient in terms of delay and resource consumption than triangle routing because the

FIGURE 16.4
*(a) Triangle routing
and (b) optimized
routing.*

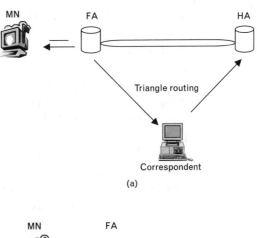

MN   FA   HA

Triangle routing

Correspondent

(a)

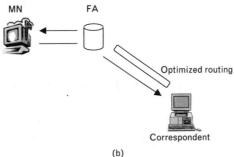

MN   FA

Optimized routing

Correspondent

(b)

packets will have to traverse fewer links on their way to their destination. In Mobile IPv4 the route optimization feature is being added on as an optional set of extensions that may not be supported by all IP nodes.

As mobile nodes move from one point of attachment to the next within the Internet, it would be nice if the handoffs were as smooth as possible. This could be a problem if datagrams heading toward one point of attachment were dropped because the mobile node had just left to attach somewhere else nearby. With route optimization such problems will almost certainly arise because there is no way that all correspondent nodes can instantaneously receive updated binding reflecting the node's movement.

It is important to deliver datagrams correctly even though they may arrive at the wrong COA. Route optimization enables the solution to this problem by allowing previous FAs to maintain a binding for their former mobile visitors, showing a current COA for each. With such information, a previous FA can reencapsulate a datagram with the right COA and send it along to the mobile node.

The main obstacle to route optimization relates to security. For a correspondent to tunnel directly to a mobile node, the correspondent must be informed of the mobile node's current COA.

3. A new feature is specified in Mobile IPv6 that allows mobile nodes and Mobile IP to coexist efficiently with routers that perform ingress filtering. Many border routers discard packets coming from within the enterprise if the packets do not contain a source IP address configured for one of the enterprise's internal networks. Because mobile nodes would otherwise use their home address as the source IP address of the packets they transmit, this present difficulty. Solutions to this problem in Mobile IPv4 typically involve tunneling outgoing packets from the COA, but the difficulty is how to find a suitable target for the tunneled packet from the mobile node. The use of reserved tunnels to the HA can overcome the restriction imposed by ingress filtering. A reserved tunnel is a tunnel that is established starting from the mobile node's COA (not from the mobile mode's home address as in Mobile IPv4) up to the HA. When the mobile node moves to a foreign network, it detects FAs that are supporting reserve tunneling by listening advertisements. The mobile node selects such a foreign agent by registering through it.

4. Mobile IPv6, unlike Mobile IPv4, uses IPsec for all security requirements such as sender authentication, data integrity protection, and replay protection for binding updates, which serve the role of both registration and route optimization in Mobile IPv4. IPv6 nodes are expected to implement strong authentication and encryption fea-

tures to improve Internet security. This affords a major simplification for IPv6 mobility support since all authentication procedures can be assumed to exist when needed and do not have to be specified in Mobile IPv6.

5. In Mobile IPv6 the functionality of the FAs can be accomplished by IPv6 enhanced features such as neighbor discovery and address autoconfiguration. Address autoconfiguration is a very powerful mechanism for nomadic nodes to acquire an address and begin communicating on their current link. Address autoconfiguration, however, does not change the nature of network-prefix routing or the need to devise special ways to route packets to mobile nodes that move from link to link while retaining any ongoing communications.

   Neighbor discovery in IPv6 is the set of functions that relate to how nodes discover and interact with neighboring nodes on their current link. The packets, which arrive at the home network and are destined for a mobile node that is away from home, are received by the mobile node's HA using IPv6 neighbor discovery rather than ARP, as is used in Mobile IPv4.

6. There are three types of addresses in IPv6 [6, 7]:
   - *Unicast address:* A packet is sent to a unicast address and then it is delivered to exactly one node, which has the unicast address, assigned to one of its interfaces.
   - *Multicast address:* A packet is sent to a multicast address and then it is delivered to every node that is a member of the group identified by the multicast address.
   - *Anycast address:* A packet is sent to an anycast address and then it is delivered to "any" one of several possible nodes that are identified by that address, typically the node that is closest to the sender of the packet.

   The mobile mechanism is more efficient and more reliable. This is because only one packet needs to be replied to the mobile mode.

7. In Mobile IPv6 the routing to mobile nodes is done with tunneling and source routing instead of only tunneling in IPv4. In Mobile IPv4 an ICMP error that is created due to a failure in delivering an IP packet to the COA will be returned to the home network without containing the IP address of the original source of the tunneled IP packet. This is solved in the HA by storing the tunneling information.

16.6   Possible Solutions for Micro-Mobility in Mobile IP Networks

The Mobile IP has been developed for computers that are moving. With the Mobile IP, the mobile nodes are able to send and receive data despite their current point of attachment to the Internet. In the standard Mobile IP, mobile nodes have to report their every movement in the foreign network to their home networks. This causes a huge amount of signaling traffic and disturbing latency during handoffs. Because of these problems, several protocol proposals have been defined to solve this so-called micro-mobility problem. In all of these solutions the home network does not have to know the exact location of the mobile node. Instead, the home network only has to know in which visited network the mobile node is located, and the local micro-mobility is managed inside the visited network. The mobile node's packet loss during and after handoffs may be disturbingly high because the delay of informing the HA, possibly on the other side of the world, may be several seconds in the current Internet. Also, for mobiles that are using QoS, acquiring a new COA on every handoff would trigger the establishment of new QoS reservations between the HA and the FA. However, most of the path from mobile node to a correspondent node would be the same before and after the handoff.

Proposals for the micro-mobility include HMIP, Cellular IP, and HAWAII. None of these proposals is trying to replace Mobile IP. Instead, they are enhancements to the Mobile IP.

16.6.1   Cellular IP

Mobile users expect the same level of service quality as wireline users. That translates to high-speed access with seamless mobility, which it is defined as the ability of the network to support fast handoff between base stations with low delay and minimum or zero packet loss. As the number of mobile subscribers grows, so does the need to provide efficient location tracking in support of idle users and paging in support of active communications. In order to achieve scalable location management, the wireless Internet needs to handle the location tracking of active and idle hosts independently. Support for passive connectivity balances a number of important design considerations. Mobile IP does not support the notion of seamless mobility, passive connectivity, or paging. The future wireless Internet will need to support these requirements in order to deliver service quality, minimize signaling, and scale to support hundreds of millions of subscribers.

16.6.1.1   General

Cellular IP is a micro-mobility protocol that provides seamless mobility support in limited geographical area. Cellular IP incorporates a number of important design features present in cellular networks but remains firmly based on IP design principles. The protocol is specifically designed to support seamless mobility, passive connectivity, and paging. Also, it is specified to provide handover support for frequently moving hosts and is intended to be applied on a local level (e.g., on a campus or in a metropolitan area network). Cellular IP can interwork with Mobile IP to support wide area mobility (mobility between Cellular IP networks). Cellular IP access networks require minimal configuration, thereby easing the deployment and management of wireless access networks. An important concept in Cellular IP design is simplicity and the minimal use of explicit signaling, enabling low-cost implementation of the protocol. Cellular IP employs per-mobile-host states and hop-by-hop routing to achieve fast handoff control [6, 8].

16.6.1.2   A Cellular IP Access Network

The universal component of Cellular IP access networks is the BS, which serves as a wireless access point and router of IP packets. The BS is built on a regular IP forwarding engine with the exception that IP routing is replaced by Cellular IP routing and location management. Cellular IP access networks are connected to the Internet via gateway routers. Mobile hosts attached to an access network use the IP address of the gateway as their Mobile IP COA.

Figure 16.5 illustrates a view of multiple Cellular IP networks that have access to the Mobile IP–enabled Internet. Packets are first routed to the host's HA of the Internet backbone and then tunneled to the gateway. The gateway detunnels packets and forwards them toward a BS. The Cellular IP routing protocol ensures that packets are delivered to the host's actual location. Packets transmitted by mobile hosts are first routed toward the gateway and from there on to the Internet.

Periodically, each BS transmits beacon signals. These signals provide statistical information related to signal strength that can be used by mobile nodes to locate the nearest BS. All IP packets transmitted by a mobile node are routed from the BS to the gateway by hop-by-hop shortest path routing regardless of the destination address. The nodes used in the Cellular IP network are called Cellular IP nodes. These nodes can route IP packets inside the Cellular IP network and communicate with mobile nodes via wireless interface. A Cellular IP node that has a wireless interface can also be called and BS.

FIGURE 16.5
(a) Multiple Cellular IP access network, and (b) network model of HAWAII.

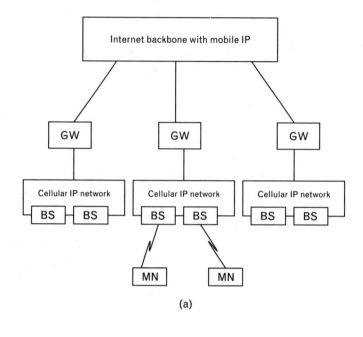

(a)

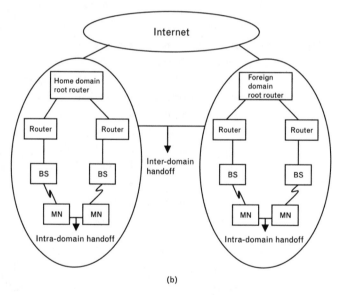

(b)

16.6.2 HAWAII

HAWAII is one of the extension protocols of Mobile IP in which host-based forwarding entries are installed in specific routers to support intradomain micro-mobility and paging functionality. The installation of these entries is accomplished using specialized path setup schemes. In general, by using these entries the performance is enhanced. This is due to the reduction of the mobility-related disruptions to user applications and to the reduction of the number of mobility related updates. Furthermore, in HAWAII mobile nodes retain their network address while moving within the domain, simplifying QoS support.

Figure 16.5(b) shows a network model of HAWAII that is divided into hierarchies based on domains. In general, the HAWAII nodes are IP routers. Each domain has a gateway that is called domain root router and each host has an IP address and a home domain. When a mobile node moves within its home domain, its IP address is retained. The packets that are destined to the mobile node can reach the domain root router based on the subnetwork address of the domain. The root router decapsulates and again encapsulates the packet to forward it to either the intermediate router or BS, which decapsulates the packet and finally delivers it to the mobile node. The received packets are then forwarded to the mobile node by using special dynamically established paths.

In order to maintain IP routing between the domain root router and the mobile host, HAWAII establishs special paths as the mobile host moves. The algorithm used to update selected routers to establish or maintain connectivity with a mobile host is termed a path setup scheme.

There are different path setup schemes, which can be used to dynamically establish paths. The path setup is done by update messages that the mobile node sends to the gateway in order to create entries (used during the reserve path) in the intermediate nodes.

Power-Up

This path setup scheme is active during power-up. The mobile host is assigned a dynamic address through DHCP during power-up. The use of a dynamic address for mobile hosts is similar to the dial-up model of service provided by Internet service providers to fixed hosts. The difference is that users in wireless networks are mobile and the home domain is determined by where the host is powered up rather than which modem access number is dialed. Apart from requiring fewer IP addresses than static allocation of IP addresses, this also results in optional routing with no tunneling as long as the user does not move out of a domain while powered up.

Micro-Mobility

This path setup scheme is active during handoff. This happens when the mobile host is handed off from one BS to other. The mobile host maintains its IP address since this movement is within a HAWAII domain. When the mobile node can only receive/transmit from/to one BS at a time, this path is called the forwarding path setup scheme in HAWAII. This path is useful in networks like TDMA networks where the mobile host cannot listen to two BSs simultaneously.

In the case of networks such as CDMA, where the mobile host could listen to multiple BSs simultaneously, this leads to a different algorithm for updating the routers and is called a *nonforwarding path setup scheme* in HAWAII.

The advantage of custom tailoring these path setup schemes for different wireless networks is that disruption to user traffic can be minimized. This is especially critical in next-generation wireless data networks where VoIP and other multimedia traffic will likely be carried.

16.6.3 Hierarchical Mobile IP

There are several protocol suggestions that have the same basic idea of hierarchical structure of visited (foreign) networks. In all these proposals, the mobile node does not have to inform its HA of every movement it performs inside the visited network. Instead, there is a network element that takes care of the mobile node's registrations.

One of proposals, Mobile IP regional tunnel management, was recently proposed in the IETF (Figure 16.6). The proposal provides a scheme for performing registrations locally in the visited (foreign) domain, thereby reducing the number of signaling messages forwarded to the home network as well as lowering the signaling latency that occurs when an mobile node moves from one FA to another. If the foreign network supports regional tunnel management, there is a special kind of FA called a *gateway foreign agent* (GFA). The mobile node uses the GFA's IP address as its COA when it registers to the HA. This COA does not change when the mobile node moves between the foreign agents that are located under the same GFA. After first registration, the mobile node makes its registrations with the GFA. Registrations are not done with the home agent as long as it is moving under the same GFA. If the mobile node changes GFA, within or between visited domains, it must again register with the home agent [3, 8].

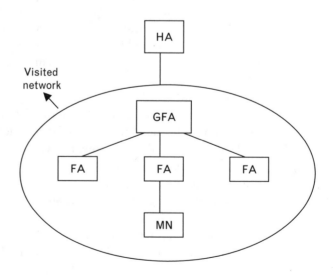

FIGURE 16.6
Network model of hierarchical Mobile IP.

There is one extension for Mobile IPv6 that will reduce the amount of signaling to CNs and the HA and may also improve the performance of Mobile IPv6 in terms of handoff speed. The proposed hierarchical mobility management for Mobile IPv6 is extended with a new function which is *mobility anchor point* (MAP). It simply provides an optional mobility management function that can be located at any level in the hierarchy starting from the *access router* (AR). MAP allows for fast handoffs, which will minimize the packet losses due to handoffs and consequently improve the throughput of the best-effort services and performance of real-time data services over the radio interface.

The aim of the hierarchical mobility management model in Mobile IPv6 is to enhance the network performance while minimizing the impact on Mobile IPv6 or other IPv6 protocols. This is achieved by using the Mobile IPv6 protocol combined with layer 2 features to manage both IP micro- and macro-mobility, leading to rationalization and less complex implementations in the mobile node and other network nodes.

16.7 Conclusions and Future Directions

Mobile IP will give users a great degree of mobility across private and public networks. Mobile IP is a newly defined protocol that supports mobile users but also is compatible with the current IP. It is also a new recommended Internet protocol designed to support the mobility of a user (host). Host mobility is becoming important because of the recent blossoming of laptop computers, PDAs, digital cellular communication, and the high desire to have continuous network connectivity anywhere the host happens to be [10]. It is still in the process of being standardized, and there are still many items that need to be worked on and enhanced, such as the security issue and the routing issue. It allows a node to change its point of attachment from one link to another while using the same IP address.

All of the micro-mobility solutions described in this chapter point to the fact that the Mobile IP is able to handle the macro-mobility between the networks, but they all define a different micro-mobility support protocol to be used inside networks. In the worst case this could lead to a situation where the mobile nodes could not roam to a network that is using an unknown micro-mobility solution. There are basically two ways of handling this problem. The first is to define a single micro-mobility protocol that is used everywhere in the Internet. The second is to define how different micro-mobility solutions can coexist simultaneously in the Internet. The most important thing is not which one of these micro-mobility solution technologies is the best, but rather, that there has to be a solution to make the coexistence of these micro-mobility protocols in different parts of the Internet possible. If coexistence is possible—for example, HAWAII, Cellular IP, and so forth—then every network administrator can choose whichever micro-mobility solution it wants to use inside the network, but this choice will not affect mobile nodes, HA, or the micro-mobility solutions in other parts of the Internet. This would lead to a situation where the best technologies would probably win the race and become the most popular.

Security needs are getting active attention and will take advantage of the deployment efforts underway. The IETF standardization process requires the working group to rigorously demonstrate interoperability among various independent implementations before the protocol can advance. Test results have given added confidence that the Mobile IP specification is sound, implementable, and of diverse interest throughout the Internet community. It is possible that the deployment phase of Mobile IP will track that of IPv6, or that the requirements for supporting mobility in IPv6 nodes will give additional impetus to the deployment of both IPv6 and mobile networking. The increased user convenience and the reduced need for application awareness of mobility can be a major driving force for adoption. Both IPv6 and Mobile IP have little direct effect on the operating systems of mobile computers outside of the network layer of the protocol stack, application. Mobile computing with Mobile IP is a topic that is surely going to receive more attention. With so little in the way of research and test networks today, it remains to be seen how far and how quickly Mobile IP can take us. With the popularity of IP-based applications, there are sure to be plenty forward-thinkers and early investors willing to try and make it happen, even though numerous technical problems still remain to be solved. Finally, much effort should be concentrated on the solidity of security solutions.

REFERENCES

[1] Fernandes, L., "Developing a System Concept and Technologies for Mobile Broadband Communications," *IEEE Personal Communications,* February 1995.

[2] Wong, P., and D. Britland, *Mobile Data Communication*, Norwood, MA: Artech House, 1993.

[3] Soldatos J., et al., "Applying Mobile IP over Wireless ATM Communication Networks," *IEEE 54th Vehicular Technology Conference (2001 VTC - Fall VTS)*, IEEE, Atlantic City, NJ, October 7–11, 2001.

[4] Nika, C., et al., "The Challenge of Mobile IP in Wireless Networks," *4th Wireless Personal Multimedia Communication (WPMC 2001) International Conference*, Demark, p. 1601.

[5] Perkins, C., "IP Mobility Support," IETF RFC 2002, October 1996.

[6] Perkins, C., "Mobile IP," *IEEE Comm.*, Vol. 35, No. 5, 1997.

[7] Perkins, C., *Mobile IP: Design Principles and Practice*, Reading, MA: Addison-Wesley Longman, 1998.

[8]   Agrawal, P., et al., "SWAN: A Mobile Multimedia Wireless Network," *IEEE Personal Communications,* April 1996.

[9]   Geiger, R., et al., "Wireless Network Extension Using Mobile IP," *IEEE Micro.,* Vol. 17, No. 6, 1997.

[10] Vergados, D. D., et al., "Mobile IP in Wireless ATM Networks," *IEEE 4th International Conference on ATM (ICATM 2001)*, IEEE, Seoul, Korea, 2001.

CHAPTER 17

IP Micro-Mobility Management Using Host-Based Routing*

K. Daniel Wong, Hung-Yu Wei, Ashutosh Dutta, Kenneth Young, and Henning Schulzrinne

17.1 Introduction and Background

Global IP mobility solutions using Mobile IP or SIP are not optimized to handle micro-mobility management. For micro-mobility situations, low-latency handoffs are essential to reduce performance degradation. *Host-based routing* (HBR) schemes such as HAWAII, Cellular IP, and *Micro-Mobility Protocol* (MMP) are one of two main classes of IP micro-mobility management schemes; the other is hierarchical Mobile IP–derived schemes. This chapter discusses HBR schemes and examines their performance. Various simulation results and prototype system measurements demonstrate the superiority of HBR schemes over both MIP and hierarchical MIP-derived micro-mobility schemes in terms of fewer packets dropped per handoff for UDP traffic and better TCP throughput in various scenarios.

17.1.1 Chapter Overview

An overview of Mobile IP has been provided in Chapter 16. An overview of SIP for macro-mobility is included in Section 17.1.2. We do not claim that either Mobile IP or SIP is better for macro-mobility, but merely present them as alternatives. Both schemes are more suitable for macro-mobility than micro-mobility, though. Micro-mobility schemes are introduced in Section 17.1.3, although the introduction of schemes for micro-mobility based on HBR is deferred to Section 17.2 for more detailed coverage.

* This material is based upon work supported by DARPA under Contract MDA972-00-9-0009 (sub-contract RK3105 from BAE).

Following that discussion, performance-related issues are explored in Section 17.3. Section 17.4 contains selected performance results from our simulations and prototype test bed that illustrate the performance of HBR schemes. This is followed by conclusions in Section 17.5. Practical issues related to the integration of macro-mobility and micro-mobility protocols are beyond the scope of this chapter, but the reader is referred to [1].

17.1.2 The Application-Layer Macro-Mobility Management Alternative

A strength of Mobile IP, it being a network layer mobility solution, is that it is transparent to the applications above it. On the other hand, if the mobility solution were to be implemented at a higher layer (e.g., separately by each application), it might arguably be inefficient. However, this argument may not be as strong if a widely used application layer protocol were to be able to handle mobility.

Indeed, SIP [2] is rapidly gaining widespread acceptance (e.g., in IETF and 3GPP) as the signaling protocol of choice for Internet multimedia and telephony services. It fits into a possible future IP multimedia stack (Figure 17.1). SIP allows multiple participants to establish sessions consisting of multiple media streams. SIP components [i.e., user agents, servers (proxy and redirect), and registrars] provide an application layer mobility solution while interacting with other network protocols like DNS and DHCP. While SIP supports personal mobility (see Section 17.1.2.1) as part of its signaling mechanism, it can also be extended to provide support for terminal, service, and session mobility (Figure 17.2). Handoff, registration,

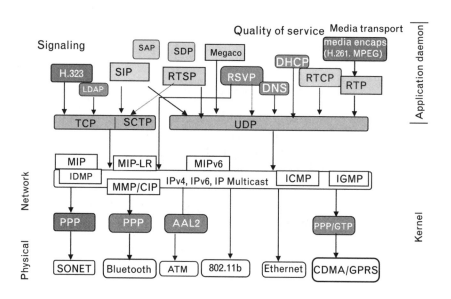

FIGURE 17.1
A possible future IP multimedia stack.

FIGURE 17.2
*SIP terminal mobility
illustrated.*

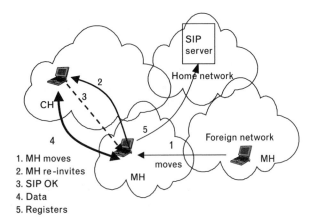

configuration, dynamic address binding, and location management are key requirements for an SIP-based mobility scheme [3].

Mobility management in the wireless Internet may involve terminal, session, service, and/or personal mobility. Mobile IP and its derivatives, variations, and auxiliary schemes are network layer solutions that provide continuous media support when nodes move around, handling the terminal mobility problem. However, Mobile IP and related schemes by themselves do not provide means of device-independent personal, session, or service mobility. For delay-sensitive real-time applications, Mobile IP-based solutions suffer from limitations like triangular routing/registration, encapsulation overhead, and need for an HA in the home network. *Mobile IP with route optimization* (MIP-RO) alleviates the triangular routing problem but also tunnels binding updates to the HA. It requires changes in the operating system of the end hosts. MIPv6 is similar to SIP-based terminal mobility, updating the IP address on the *correspondent host* (CH) directly. However, it needs a 16-byte home address destination option.

Multimedia traffic is real-time or non-real-time, depending on delay and loss characteristics. Different transport mechanisms may be used for each type of traffic. Most real-time traffic should be carried over RTP (Real Time Transport Protocol)/UDP, whereas non-real-time traffic has traditionally been carried over TCP. SIP-based terminal mobility provides subnet and domain handoff while a session is in progress. The SIP-based scheme provides a different approach for achieving terminal mobility by means of application layer signaling. This scheme does not rely on the mechanism of the underlying network components in the network core. Instead, proxy servers instituted by any third party service providers can provide mobility support.

When the mobile station moves from one subnet to another within the same administrative domain, SIP would support subnet handoff during the session as described below:

- The *mobile host* (MH) obtains a new temporary IP address through a protocol like DHCP.

- The MH reinvites the CH to its new temporary address. The identifier of the outbound proxy in the visited network is inserted in the SIP INVITE header.

- In case of domain handoff, a complete registration takes place.

A complete handoff procedure for a SIP session would consist of SIP signaling between the corresponding entities and actual media delivery. Delay associated with handoff would consist of several factors such as delay due to layer-2 detection, IP address acquisition by the mobile, activating the SIP signaling with the new address parameters, and actual delivery of media.

If the MH and CH are situated wide apart, then it may take some time for the reinvite to reach the CH. It has been proposed [4] that an RTP translator can be affiliated with a SIP proxy server that would intercept the traffic and would send the media to the current location of the mobile host. Thus, RTP translators reduce the end-to-end handoff delay (due to traversal of the INVITE request) to a one-way delay between the MH and the SIP proxy. In cases when both communicating hosts move during a session, each side would have to issue INVITE requests through their respective home proxy servers, where the MHs register their new location address after the movement.

While the RTP translator concept may reduce the micro-mobility problem somewhat, SIP does not in itself provide an optimized, targeted solution to the micro-mobility problem. Like Mobile IP, it is optimized for macro-mobility. Based on this brief examination of Mobile IP–based and SIP-based macro-mobility management, it can be deduced that a highly desirable property for a micro-mobility scheme is flexibility to work with a variety of macro-mobility schemes, and not just Mobile IP–based macro-mobility.

17.1.2.1 SIP Support for Other Types of Mobility

In addition to terminal mobility, SIP also supports other mobility concepts—namely, personal mobility, service mobility, and session mobility. Arguably, SIP offers a more unified macro-mobility management scheme than Mobile IP and its variations, which are more limited.

Personal mobility is the ability of users to originate and receive calls and access the subscribed network services on any terminal in any location in a transparent manner, and the ability of the network to identify end users as they move across administrative domains. SIP's Uniform Resource Indentifier (URI) scheme and registration mechanism are some of the main components used in providing personal mobility. A roaming subscriber is accessible independent of the device the subscriber uses. Service mobility refers to the subscriber's ability to maintain ongoing sessions and obtain services in a transparent manner regardless of the subscriber's point of attachment. Session mobility allows a user to maintain a media session even while changing terminals such as transferring a session that began on a mobile device to a desktop PC after entering an office.

17.1.3 Micro-Mobility Management

The requirement that Mobile IP registration (or SIP reinvites) be performed every time an MH moves between subnets may cause high handoff latency. Various solutions have been proposed. The proposals generally implicitly or explicitly use a concept of micro-mobility regions where these regions comprise numerous subnets, and registrations with the HA are not necessary for movement of the MH within these regions. Registration with the HA is still necessary for movement between micro-mobility regions. Typically, Mobile IP handles the macro-mobility (mobility between micro-mobility regions), while a micro-mobility scheme handles micro-mobility (mobility within micro-mobility regions). Micro-mobility management schemes reduce the high handoff latency of Mobile IP by handling mobility within micro-mobility regions with low-latency local signaling.

17.1.3.1 Hierarchical Mobility Agent Schemes

Micro-mobility solutions using a hierarchy of mobility agents include Mobile IP *with regional registration* (MIP-RR) [5] and TeleMIP/*Intradomain Mobility Management Protocol* (IDMP) [6].

MIP-RR involves few modifications to Mobile IP. In a foreign network, the two level mobility hierarchy contains the upper-layer GFA and several lower-layer *regional foreign agents* (RFAs). All MHs under the GFA share the same COA.

Suppose an MH moves between subnets under a GFA with which it is already registered. As shown in Figure 17.3(b), the MH initiates its registration with FA2. Then the registration request is sent to GFA1. Since MN is already registered with GFA1, GFA1 does not initiate a home registration to HA, but just sends the registration reply to the MH through FA2. Since the HA does not need to be contacted in this scenario, MIP-RR reduces the handoff latency.

If the MH changes its GFA, it needs to register with its HA. As shown in Figure 17.3(a), the MH moves from FA3 to FA2, and its GFA is no longer GFA2. The MH sends a registration request to its new RFA, which is FA2, and then GFA1. Because GFA1 is a new GFA, it has to register with the HA. The HA sends the registration reply all the way through GFA1 and FA2 to the MH.

FIGURE 17.3
*Mobile IP regional
registration: (a) move-
ment between regions;
(b) movement within a
region.*

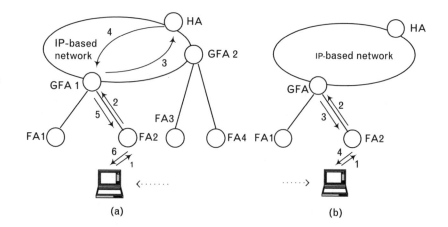

(a)   (b)

If the MH moves frequently within a GFA domain, it does not need to perform the time-consuming registration procedure with its HA. Therefore, the average handoff latency is reduced.

Like MIP-RR, TeleMIP is a hierarchical IP-based architecture that provides lower handoff latency and signaling overhead than Mobile IP. TeleMIP uses Mobile IP as a macro-mobility management protocol and can interwork with SIP-based mobility. IDMP is the micro-mobility protocol used with TeleMIP (Figure 17.4). IDMP uses multiple COAs that are taken care of by SAs and the MA at the subnet and domain level, respectively. SAs are FAs or DHCP servers at the subnet level that provide an MH with a

FIGURE 17.4
TeleMIP's IDMP.

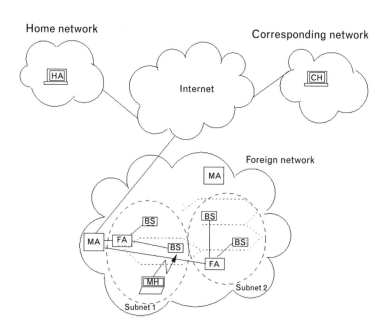

locally scoped address, and they are analogous to MIP-RR RFAs. The locally scoped address provided by an SA identifies the MH's location within the domain. An IDMP MA is similar to a MIP-RR GFA and acts as a domain-wide point for packet redirection.

The serving MA provides an MH with a global COA that stays constant as the MH moves within the domain. Unlike MIP-RR, multiple MAs can be provisioned for load balancing and redundancy within the domain. All packets from the global Internet are tunneled to the MA (or one of the MAs, in the case of multiple MAs). The serving MA forwards packets to the MH using regular IP routing, with the local COA (colocated or FA) as the destination. It does this by un-encapsulating the packet and then performing a second encapsulation of the IP packet, with the local COA as the destination address.

On subsequent movement within the domain, the MH only obtains a new local COA. At that point, the MH needs only to update its MA with its new local COA. By limiting intradomain location updates to the MA, it reduces the latency associated with intradomain mobility. In addition, IDMP also provides the added advantage (over MIP-RR) of dynamic load balancing, within a domain.

17.2 HBR Overview

A class of micro-mobility management schemes is that which employs HBR, including Cellular IP [7], HAWAII [8], and MMP [9]. HBR schemes for micro-mobility could be considered a class of auxiliary schemes that deal with the handoff latency problem of Mobile IP. However, they have grown beyond being just a class of auxiliary schemes (e.g., MMP is designed to be usable with SIP mobility for real-time traffic, and with a Mobile IP variant for nonreal-time traffic). This flexibility is an advantage over micro-mobility schemes based on a hierarchical Mobile IP structure (e.g., MIP-RR or TeleMIP). A second major advantage of HBR schemes is that they offer the lowest latency networking rerouting solution for micro-mobility. This is because hierarchical Mobile IP–derived schemes like MIP-RR and IDMP only reduce the latency problem inherent in Mobile IP registration. The GFAs or MAs still need to cover a large area to be scalable and cost-effective. Location updates still need to reach them in these schemes. On the other hand, with HBR schemes, updates take an optimal path to the closest node that should handle the location/route update, namely the crossover node, as will be explained shortly. The performance results in Section 17.4 support these assertions with numerical results.

The distinctive characteristics of HBR schemes for micro-mobility are that (1) HBR is used within the micro-mobility regions, (2) very low-latency handoffs are possible since the update message only needs to propagate to the crossover node, and (3) one or more special nodes (known as gateways or root nodes) are used as the demarcation point between each micro-mobility region and the rest of the Internet.

With HBR schemes, forwarding behavior is specified separately for each host. For example, nodes may route packets according to tables or caches indexed by unique host identifiers (e.g., their IP address). MMP, HAWAII, and Cellular IP are examples of HBR where the indexing is by host IP addresses. HBR differs from group-based routing schemes, where forwarding behavior is specified for groups of hosts. For example, for group-based routing, nodes may route packets according to tables or caches indexed by group identifiers (e.g., IP address prefix and netmask)—that is, packets with different destination addresses, but where the destination addresses match the prefix and netmask, will be routed in the same way.

A critical advantage of HBR schemes is that location management and routing can be integrated. With Mobile IP, location management is handled by registrations, while routing is overlaid on the existing IP-based network routing. The location management requires possibly long-latency registrations to a potentially distant HA whenever subnet boundaries are crossed, while the use of overlay routing over standard IP routing creates problems like triangular routes and encapsulation overhead. HBR, on the other hand, gives the power to update routes simultaneously and precisely with location management, precisely because the routes are host specific and do not affect routes to/from other MHs when changed. For the lowest latency handoffs, the intuition would be to update the routing information for an MH at the closest intersection with the old route (also known as the crossover node), whenever it performed a handoff. And this is indeed what happens with the HBR schemes for micro-mobility management.

Rather than present three separate descriptions of the three HBR schemes (MMP, HAWAII, Cellular IP), a generic, bare-bones HBR scheme will be introduced in Section 17.2.1, and then differences between MMP, CIP, and HAWAII will be discussed in Section 17.2.2. This approach brings out the essence of the HBR schemes before explaining the minor differences between them. A comparison with ad hoc routing protocols that use HBR-like routing follows in Section 17.2.3.

17.2.1 A Generic HBR Solution for IP Micro-Mobility

A generic HBR scheme is now presented to illustrate the essential workings of HBR in facilitating low-latency handoffs for IP micro-mobility. Actual protocols differ from this generic scheme in some aspects. For example, actual protocols use various enhancements to provide more seamless handoffs. There may be multiple nonoverlapping micro-mobility domains in a given network. Movement between micro-mobility domains may be handled by a macro-mobility protocol like Mobile IP or SIP.

In the generic HBR scheme, each HBR micro-mobility domain has one root router that serves as the interface between the micro-mobility domain and the rest of the network. The other infrastructure nodes in the HBR micro-mobility domain are arranged in a strict inverse tree structure beneath the root router. In other words, every node is a child of one and only one other node (possibly the root router), and may be a parent of zero, one, or more other nodes. The inverse tree structure can be seen in Figure 17.5, which shows an abstraction of a generic HBR scheme for IP micro-mobility. Some of the nodes at the bottom of the hierarchy are BSs, with wireless interfaces, and whose coverage areas are called cells. Each infrastructure node other than the root router has one upstream interface and zero, one, or more downstream interfaces. The upstream interface is the interface towards the root router, while the downstream interface(s) is/are towards the BSs and MHs. How nodes know which are the upstream interfaces and which are the downstream interfaces is unspecified. However, one way this may happen is by listening to beacon messages sent periodically down by the root router (the interface through which the beacon arrives is recorded as the upstream interface and is used as the next hop for routing of any packet to the root router; the rest are considered downstream interfaces). Another way might be if the HBR scheme is implemented as an overlay over standard IP routers, as with HAWAII. Alternatively, they may be preconfigured.

When an MH first enters an HBR micro-mobility domain, the network would use access control mechanisms, such as authentication of the MH. Some form of macro-mobility signaling would be initiated, so that macro-mobility can be handled by the relevant protocol. The MH may use its home IP address, or it may obtain a temporary address for use in the HBR domain. The temporary address, if applicable, is called a COA, although it is

FIGURE 17.5
Abstraction of generic HBR scheme for IP micro-mobility management.

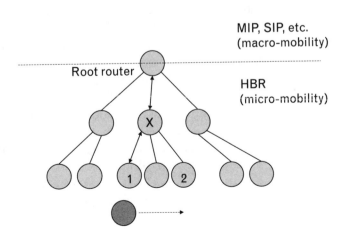

MIP, SIP, etc.
(macro-mobility)

Root router

HBR
(micro-mobility)

not necessarily exactly like a Mobile IP COA. It may be obtained through various means, and may be a globally routable address of the root router, or be colocated. In any case, the MH uses one IP address as long as it is in the same HBR domain, even when it moves from cell to cell.

Upstream routing of packets (from MH to gateway) is simple: once packets from a particular MH are admitted into the micro-mobility domain infrastructure, each node (BSs included, root router excluded) merely forwards upstream packets to the root router through their upstream interface. Therefore, the root router eventually gets upstream packets. It then routes them normally through the IP-based network, or it sends them back through its micro-mobility domain infrastructure, depending on the destination address.

Downstream routing of packets depends on routing caches maintained in each node. By definition of HBR, there are separate cache entries for each MH. These entries are set whenever upstream data passes through a node. The node reads the source IP address to identify the MH, and then binds it in the cache to the interface address (plus MAC address, if necessary) of the incoming packet, with the assumption that it is the right interface (and MAC address) to use for downstream packets destined to that MH. In order to facilitate simpler handoffs, the cache entries have a soft state and need to be periodically refreshed. Since it cannot be assumed that upstream data is always being transmitted, periodic control packets called route updates are transmitted by each MH to refresh the cache entries after a period of no upstream transmissions from that MH.

To explain how handoffs are handled, Figure 17.5 shows an MH (the unattached node at the bottom of the figure) moving from the cell covered by the BS labeled 1 (BS1) to the cell covered by the base station labeled 2 (BS2). The node marked "X" is the crossover node for this handoff. The crossover node is defined as the lowest node that is in the upstream of the paths from both BSs to the root router. In some cases, it may be the root router itself, but often the crossover node is below the root router. The MH initiates handoff by sending a route update through the new BS, BS2 in this case. As this route update propagates up to the root router, for each node below the crossover node, the routing cache may not have any binding for the MH, and the appropriate entry is added. At the crossover node, the critical switch occurs as the binding for the MH is updated to point towards the interface heading towards BS2. As the route update continues up to the root router from the crossover node, the routing cache entry at each node is updated as normal, as though no handoff had occurred.

HBR may not require any signaling with the old BS, BS1, nor any of the nodes between BS1 and the crossover node (some signaling occurs through these nodes in HAWAII, though, in one of the enhancements to provide seamless handoffs). Since the routing caches contain soft-state entries, the entries will naturally time out and be removed without needing additional signaling. This is one of the reasons why soft-state routing cache entries are used. Another reason is that the MH does not need to perform any deregistration when it leaves the HBR domain, which it would need to do to update the routing caches if they had hard-state entries. Furthermore, if for any reason the MH loses connectivity or crashes, the hard-state entries would not be removed.

The generic HBR solution could also have some kind of paging scheme to provide paging gains over Mobile IP (see Section 17.3.1.1). One way this could be done would be by using paging caches, similar but parallel to the routing caches. The paging caches would have longer expiry times, so that they need not be refreshed as often as routing caches. When an MH is idle, it only needs to send paging updates occasionally.

17.2.2 Comparison Between HBR Schemes

In this section, specific differences between HBR schemes like HAWAII, Cellular IP, and MMP are discussed. Although each of these protocols has much in common with the generic HBR scheme just described, their design goals are not all the same. For HAWAII, the design goals [8] are to limit disruption to user traffic, enable efficient use of access network resources, enhance scalability by reducing updates to the HA, provide intrinsic support for QoS, and enhance reliability. For Cellular IP, the design principles [7] are to use universal building blocks as nodes, be a plug-and-play solution, be scalable, minimize the burden on the MH, and support passive connectivity. For MMP, the design principles are to limit disruption to user traffic, be robust, reliable, and survivable, enable efficient use of low-bandwidth access network resources, be a plug-and-play solution, be scalable, minimize the burden on the MH, support passive connectivity, and facilitate QoS support.

The MH in HAWAII is a Mobile IP client. The HAWAII domain looks like an FA to the MH, and its root router looks like an FA to the HA of the MH. On the other hand, the MH in Cellular IP needs to use Cellular IP signaling (e.g., route updates and the root router takes care of the Mobile IP signaling). As for MMP, the MH needs to use MMP signaling, and the root router acts on behalf of the MH to perform signaling for the macro-mobility scheme, whether it is Mobile IP or SIP. The advantage of having the MH be an ordinary Mobile IP client is that no changes are needed to MHs that already have Mobile IP client software. However, this choice is not made in Cellular IP, since that would go against the "universal building block" principle because then the BSs would have extra IP-level functionality in addition to Cellular IP node functionality (i.e., they would have to send updates to the root router on behalf of the MH). Furthermore, this would reduce the ability to put together a plug-and-play wireless IP access network. This second reason applies to MMP as well. Additionally, however, MMP MHs cannot be Mobile IP MHs, since MMP is designed to work with several macro-mobility schemes, not just Mobile IP.

Although both Cellular IP and HAWAII are designed to work with Mobile IP as the macro-mobility management scheme, each works with a different mode of Mobile IP. In HAWAII, the MH acts as a Mobile IP MH in colocated COA mode, while in Cellular IP, the gateway acts as an enhanced FA that takes care of Mobile IP registration on behalf of the MH. Another major choice is whether the HBR scheme is implemented as an overlay over standard IP routers running in intradomain routing protocol (e.g., RIP, OSPF), as with HAWAII, or whether the HBR routing is the only routing that the HBR nodes are capable of, as with Cellular IP. The overlay approach allows more sophisticated seamless handoff techniques, as it allows HBR nodes to communicate with one another to forward buffered packets, and so on. It also may be able to rely on some of the reliability mechanisms of the underlying intradomain routing protocol, and may be easier to implement using existing equipment. However, it is not as lightweight and scalable over a range of wireless access environments as the nonoverlay approach, since full-fledged IP routers are used.

One of the design goals of MMP is to be robust, reliable, and survivable. Because of the efficiency in signaling and distribution of the routing information in the generic HBR scheme, only the nodes along the path between the serving BS and the root router know how to route to the MH. This could result in disruption in communications if a node or link fails, or if the root router fails. The problem is shared by Cellular IP, and to some extent, by HAWAII. With MMP, however, there is the option to use multiple root routers, and there is also the option for some or all the nodes to have more than one parent node. This provides for robustness and survivability, at the cost of a little more complexity and signaling traffic within the HBR domain.

Another difference between the schemes is in how handoffs are treated. The basic HBR handoff scheme is equivalent to the hard handoff scheme of MMP and Cellular IP. Although HBR schemes can achieve fast, low-latency handoffs by using only their very fast local updates, much faster than Mobile IP, packets could still be dropped. Therefore, various seamless handoff schemes have been proposed to improve performance even further. In HAWAII, there are four schemes for setting up the new path when handoffs occur. With the *multiple stream forwarding* (MSF) and *single stream forwarding* (SSF) schemes, the old BS receives the initial handoff message and signals with the new BS to set up a path with the new BS to forward packets there that have been buffered at the old BS. Using MSF could result in multiple out-of-order streams arriving at the BS through the new BS, as some earlier packets being forwarded from the old BS to the new BS may arrive later than newer packets from the crossover node to the new BS for very short periods of time. SSF is more sophisticated, and results in a single stream of packets being forwarded to the new BS. The crossover node in this case needs to be informed by the old BS when it has cleared its buffers of packets for the MH, and only then would the crossover node switch packets for the MH over to the new BS. With the other two handoff schemes in HAWAII —*unicast nonforwarding* (UNF) and *multicast nonforwarding* (MNF) —it is the new BS that receives the initial handoff message. UNF, in the network, is like the basic hard handoff except that the old BS is informed through signaling from the new BS and it sends an acknowledgment back to the MH. The difference from simple hard handoff is that the MH is assumed to be able to communicate with both BSs during the handoff period, to reduce packet losses associated with hard handoff. As for MNF, it is like UNF except it uses a special dual-cast scheme (from crossover node to both BSs) for a short period of time from the time the crossover node receives the handoff message so the MH does not have to talk to two BSs simultaneously. On the other hand, the only seamless handoff scheme with Cellular IP, semi-soft handoff, is somewhat like MNF. The difference is that the MH switches back to the old BS after sending the initial handoff message, to reduce packet loss while the message is traveling to the crossover node.

There are also other differences between the actual protocols, such as paging schemes, but for more details, the reader is referred to the source documents (e.g., [10]).

17.2.3 Comparison with Ad Hoc Mobility Schemes

Another class of routing problems where various nonhierarchical, HBR-like schemes have been proposed for handling mobility is that of ad hoc routing in *mobile ad hoc networks* (MANET) [11]. In ad hoc networks, the nodes are not arranged in a fixed infrastructure, typically because they are mobile and are constantly changing positions with respect to one another. The distinction between the fixed infrastructure and mobile hosts may vanish, and every node may well be a mobile router. The lack of a fixed infrastructure makes the route discovery problem more difficult than in the case of the HBR schemes for micro-mobility management.

A variety of hierarchical routing protocols have been proposed, using concepts of ad hoc clusters of nodes based on factors like proximity and geographical location. However, such protocols may depend on assistance from the GPS or depend on the existence of a core of "backbone" nodes or use some heuristics for selecting cluster heads. Other ad hoc routing protocols are nonhierarchical but flat, which is closer to the routing within HBR micro-mobility domains. Among the flat ad hoc routing protocols, some, like AODV routing, are reactive. These differ from the HBR schemes in that the routes are computed in an on-demand manner, as needed.

The flat, proactive ad hoc routing protocols such as DSDV routing may be closer to the HBR schemes. In DSDV, all nodes maintain a routing table that contains separate entries for all the possible destinations, which are periodically refreshed. Two differences between DSDV and HBR for micro-mobility are noticeable. While in DSDV, all nodes maintain a routing table for all the destinations; in HBR only the infrastructure nodes maintain these tables, and then only for MHs being served by one of their children or descendent nodes. The other difference is that the refresh problem is an order of magnitude more challenging in DSDV, since the nodes all are moving. Infrequent transmissions of full dumps are needed, where full dumps contain all routing information, and periodic transmissions of incremental packets are used to relay information on changes occurring since the last full dump. On the other hand, for HBR, route updates are always incremental and specific to mobile hosts, very quickly providing up-to-date information on MH location after movement occurs. Even the periodic refreshes are meant to be for optimizing network resource utilization more than to handle significant network topology changes.

Despite these advantages of HBR, ad hoc schemes must still be used in cases where the mobility situation demands it. However, whenever HBR can be used, it should be—for example, in less mobile or semi mobile networks (e.g., a tactical network where some "infrastructure" nodes move infrequently and remain on the same hierarchy even after moving)—since it also has additional advantages over ad hoc schemes. The root router provides a natural transition point between the micro-mobility region and the macro network. Moreover, MANETs run into a scalability problem with as few as 50 to 100 nodes because of all the updating and exchange of route information that goes on, whereas HBR domains can handle thousands of nodes.

An interesting and open area of research is where the crossover between an HBR network and a MANET might occur. For situations where there is a relatively stable infrastructure, HBR makes sense, whereas ad hoc routing protocols would need to be used in more mobile, fluid networks. A key question is how to qualify and quantify what is meant by a relatively stable infrastructure with core nodes, in which the network can take advantage of the core nodes to reduce signaling overhead. Ad hoc routing protocols like CEDAR that assume that core, high-bandwidth backbone nodes can be found, are closer to HBR in a sense. One could imagine a self-organizing protocol where nodes perform some self-discovery of the network and switch into an HBR mode or an ad hoc routing mode depending on certain conditions. It could periodically check if the conditions have changed, and switch modes if necessary.

....

17.3   Performance Issues

Performance is examined qualitatively in Section 17.3.1. Previously reported quantitative results are discussed in Section 17.3.2, as a prelude to discussing our performance results in Section 17.4.

17.3.1   A Qualitative Perspective

The major goal of micro-mobility schemes using HBR is providing fast, low-latency handoffs. This may result in fewer packets dropped during handoffs, leading to less disruption of UDP traffic and better TCP through-put performance.

While obtaining significant reductions in handoff latency compared to MIP is a major accomplishment of micro-mobility management schemes, other advantages have been claimed. These include:

• Reduction in signaling overhead through paging concepts;

• Reduction in packet header overhead in the low-bandwidth radio access network through not using encapsulation;

• Easier integration with QoS provisioning;

• Better use of scarce IP address resources by using fewer IP addresses than Mobile IP or hierarchical Mobile IP derivatives

These advantages will be discussed, qualitatively, in Sections 17.3.1.1 to 17.3.1.4. Other issues, such as scalability, will be discussed in Sections 17.3.1.5 to 17.3.1.6.

17.3.1.1   Paging Gains

The idea behind paging gains is that Mobile IP does not differentiate between active and idle MHs. It requires that MHs go through the registration process whenever MHs move between subnets, regardless of the activity level of the MH. There are two problems with this. First, the signaling overhead is high, even if the MH is idle (communications-wise) but moving around rapidly. Second, each of the registrations with movement between subnets would consume power from the MH's battery. The first of these problems is largely dealt within an HBR domain even without paging, because macro-mobility signaling would not need to be invoked upon every handoff occurring. The second problem, however, can be dealt with using an idea (paging) borrowed from traditional wireless cellular networks.

Cellular mobile networks have long differentiated between active and idle states of an MH. When an MH is idle, it registers less often with the network, the trade-off being that the network knows the MH's position with less precision and needs to page the MH to reach it when it needs to communicate with it. The less often the MH registers, the less precisely the network will know its position and the larger the paging area. The HBR schemes implement variants of the paging concept. However, this is not a fundamental flaw of Mobile IP, nor a fundamental advantage of HBR schemes. Moreover, a paging extension to Mobile IP has been proposed [12].

17.3.1.2   Reduction in Packet Header Overhead

Mobile IP adds at least 8 to 12 bytes per packet for minimal encapsulation, and more for alternative encapsulation schemes. With small packets, this can make a significant difference compared to a scheme without encapsulation overhead [13]. Since micro-mobility regions are often in wireless access networks where bandwidth efficiency may be at a premium, it is an advantage of HBR schemes that there is no encapsulation overhead. This advantage is over Mobile IP, and also over other non-HBR micro-mobility schemes that use encapsulation, like TeleMIP/IDMP and MIP-RR.

17.3.1.3   QoS

In the IntServ model for providing QoS, resource reservation protocols like RSVP are used to reserve network resources. The resource reservation, however, assumes that the endpoints have unchanging IP addresses. When an endpoint changes IP address (e.g., an MH obtains a new COA), the old reservations cannot be used, and new reservations need to be made. With HBR schemes, MHs keep the same IP address within an HBR domain, providing a more stable endpoint for RSVP than Mobile IP does.

17.3.1.4   Use of IP Addresses

There is a shortage of IP addresses in the IPv4 address space. Using Mobile IP for micro-mobility would require a pool of IP addresses to be set aside for use as COAs in every subnet. Since MHs with HBR can keep one IP address as they move within an HBR domain, a pool of COAs can be set aside for the entire HBR domain, if necessary, resulting in a more efficient use of IP addresses.

17.3.1.5   Scalability

HBR has a potential scalability problem in that the forwarding cache grows linearly with the number of hosts. To deal with the scalability problem, one solution is to use a group-based routing scheme. The Internet is an example of a network with group-based routing. Furthermore, the groups are hierarchical, with smaller groups as subsets of larger groups, providing a very efficient and flexible way to specify routing behavior. At the minimum, a routing table may contain a default route that specifies how to route all packets.

The Internet uses a hierarchical routing scheme because HBR does not scale. An alternative way to deal with the scalability issue is to restrict the number of hosts involved, for example, to just the hosts roaming within a certain region. HBR works for micro-mobility because it is confined to a definite region, with a gateway between the micro-mobility region and the rest of the Internet.

17.3.1.6   Communications Between Two MHs in the Same HBR Domain

For cases where the CH is not in the same HBR domain as the MH, the routing within the HBR domain is optimal in both directions. Uplink packets go straight to the root router as they should, and downlink packets go straight to the correct BS, as they should. However, what happens when the CH is also an MH in the same HBR domain? The way that CIP, HAWAII, and MMP have been specified at this time, the route would be through the root router and down to the other BS. Even if the two MHs had a crossover node that was the direct parent of their respective BSs, packets would still go to the root router. The reason is that the HBR nodes along the path simply forward any uplink packets to the root router regardless of final destination.

Should this routing "inefficiency" be removed? A straightforward solution might be to check the destination address of every uplink packet with the contents of the routing cache to see if it should be forwarded down rather than to the root router. The problem, however, is that the uplink forwarding would become less efficient. It is unclear if the trade-off is worthwhile, given that the percentage of traffic from one MH to another in the same domain might be very low. Even if that percentage is not that low, the "inefficient" route through the root router is not a serious inefficiency because the added latency would not be very high and would be naturally bounded by the size of the HBR domain. Nevertheless, a solution has been proposed that introduces the concepts of optimizing Cellular IP node, proxy route-update packet and optimizing teardown packet [14].

17.3.2   Quantifying Performance

Attempts have also been made to quantify the performance of HBR schemes. Reference [7] describes an experimental prototype and measurements made thereupon, which show that TCP throughput decreases as the handoff rate increases. Measurements also show how semi soft handoff (also known as advanced binding handoff) performs better than hard handoffs, experiencing less of a TCP throughput decrease as the handoff rate increases. Reference [8] provides performance results using a novel network simulator developed at Harvard University. It compares HAWAII with Mobile IP in terms of average number of dropped packets (UDP case) per handoff. TCP throughput of HAWAII is compared with that of Mobile IP as handoff frequency varies from 0.5 to 4 times per second.

While the first-order performance improvements are based on much reduced handoff latency (over Mobile IP) using HBR, second-order performance improvements may be available through various additional optimizations. For example, attempts at seamless (or almost seamless) hand-offs, including semi-soft, MNF, UNF, MSF, SSF, can further reduce packet losses, at the cost of additional complexity. In this chapter, we focus more on the performance improvement related to reduced handoff latency over Mobile IP. Our simulation results extend, as well as complement, the performance results of [7, 8].

....

17.4   Performance Results

In this section, performance results (mostly from computer simulations using NS2, but also including some analytical and laboratory prototype results) are discussed. Section 17.4.1 introduces the simulation environment. Various results are discussed in Section 17.4.2, which is divided into two main types of simulations: UDP simulations and TCP simulations. Finally, Section 17.4.3 describes some measurement results from the laboratory prototyping.

17.4.1 Simulation Environment

The base simulation setup for our simulations of HBR schemes for micro-mobility management is illustrated in Figure 17.6. A simple wireless model is used that assumes perfect overlapping coverage, no propagation delay, and no transmission errors. Furthermore, handoffs are smooth and instantaneous at layer 2 and below. The link latency for the plain (straight-line) links in the micro-mobility domain are 2 ms, whereas the link latency of the dash-dot links in the Internet are 10 ms. The link bandwidths are 375 Kbps within the micro-mobility region, and 1.544 Mbps in the "Internet." Routing and paging cache entries need to be refreshed, and the route update and paging update intervals used are 3 and 60 seconds, respectively. The size of each update packet was 100 bytes. Lightly loaded network conditions were simulated, with only one MH and one CH. TCP Tahoe was used as the transport protocol, with a flat application data rate of 200 Kbps. The application is assumed to not be delay sensitive (i.e., it is equally acceptable for the

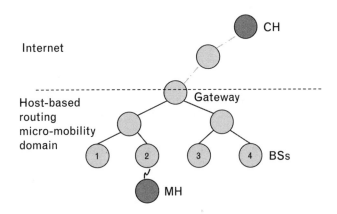

FIGURE 17.6
Base simulation setup for HBR micro-mobility simulations.

instantaneous throughput to fluctuate a lot as it is for the instantaneous throughput to be relatively constant, provided that the average throughput is the same). The handoff rate is once every 5 seconds on average, with exponentially distributed interhandoff intervals, and the handoffs are back and forth between the two base stations labeled BS1 and BS2. Each simulation was run at least as long as needed for 200 handoffs to occur.

Simulations were run comparing performance in terms of (1) average number of packets dropped per handoff for UDP traffic and (2) the TCP throughput. In different simulations, the effects of varying link latencies, handoff frequencies, application data rate, link bandwidths and other parameters were investigated. While TCP Tahoe was the default TCP used, TCP Reno was also simulated for comparison. Since there have been various seamless handoff schemes proposed, it has been decided that for this chapter only the basic HBR hard handoff and one representative seamless handoff scheme, the semi-soft handoff scheme, be simulated. The reason for including basic HBR hard handoff is to bring out the performance gains of HBR schemes resulting simply from the low-latency route updates even without any auxiliary schemes for seamless handoffs. The reason for including one representative seamless handoff scheme is merely to confirm and illustrate that such schemes do indeed help further improve performance. In this, the choice of semi soft handoffs is somewhat arbitrary, partly because its performance is expected to be somewhat moderate compared with the other seamless handoff schemes.

The performance of HBR schemes has been compared with the performance of Mobile IP where the base simulation setup for Mobile IP simulations is shown in Figure 17.7. The wireless model, handoff model, and other parameters are almost identical to those for the HBR simulations, and the topology is similar to that in Figure 17.6, in order to allow for meaningful comparisons. Each BS now becomes a different subnet and has either an FA or an HA. The Mobile IP registration request and reply messages are each 100 bytes long. Minimal encapsulation is simulated, adding only 12 bytes to the unencapsulated packets. MIP periodic reregistrations are not simulated in this model because the intervals typically are long, in the order of many minutes, and so would hardly impact the simulation results. As for the dashed link, the link latency on that link was varied (from the original 2 ms to 10 ms to 100 ms), to simulate the real possibility that the HA is further away. It should be noted that the case where the dashed link has only 2 ms of latency should be useful to indicate the performance of either (1) MIP where the HA is very close to the FA, or (2) micro-mobility management by MIP-RR or TeleMIP/IDMP (without seamless handoff enhancements). In the case of MIP-RR or TeleMIP/IDMP, the HA in Figure 17.7 would be analogous to the GFA or the MA, in that it is very close to the FA (which would be the RFA or SA for MIP-RR or IDMP, respectively).

17.4.2 Simulation Results

17.4.2.1 CBR UDP Traffic

CBR UDP traffic was applied from CH to MH to investigate HBR performance compared with Mobile IP, in terms of number of packets dropped per handoff. The simulation environment was as described in Section 17.4.1. It could be expected that the results would be little impacted by varying the handoff rates, for reasonable handoff rates (not too large). Indeed, simulations verified this assumption. Recall that in the base case for both HBR and Mobile IP simulations (described in Section 17.4.1), the application data rate is 200 Kbps. For the base UDP simulations, this rate is accomplished by sending packets of 1,000 bytes every 40 ms. This can be described as a "heavy traffic" scenario, since the 375-Kbps links are more than 50% loaded. A "light traffic" scenario will also be investigated next, where the application data rate is 20 Kbps.

Some results for the simulations in the heavy traffic scenario are shown in Figure 17.8. As can be expected, increasing the link bandwidth would

FIGURE 17.7
Base simulation setup for Mobile IP micro-mobility simulations.

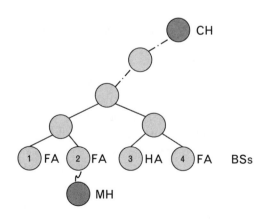

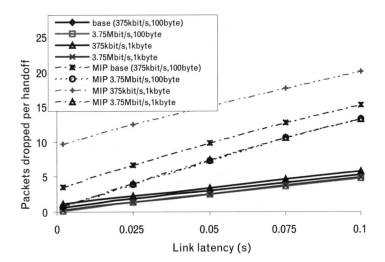

FIGURE 17.8
*Packets dropped per
handoff in a heavy-
traffic scenario.*

decrease the number of packets dropped per handoff, because the packets spend less time in transit. Similarly, increasing the size of the update packets would increase the number of packets dropped per handoff, as the update packets will take longer to arrive at their destination. These two effects were investigated by simulating micro-mobility domains with 10 times larger bandwidths, as well as cases of 10 times larger update packets. The plots marked "3.75 Mbps" have micro-mobility domains with link bandwidths of 3.75 Mbps, while the plots marked "1 kbyte" use large update packets 1 KB long. Curves whose labels are prefixed by "MIP" are those in which Mobile IP was run, and the rest use HBR by default (this statement applies to the rest of the performance results as well, not just this figure). The *x*-axis shows the link latency, which is the transmission latency of the straight-line links in the simulation setup. The *y*-axis shows the packets dropped per handoff.

Since this is a heavy traffic scenario, the number of packets dropped per handoff can be quite significant. However, it is the relative performance of HBR and Mobile IP that is of interest. With Mobile IP, the performance is worse (more packets dropped per handoff), even for this best-case Mobile IP scenario where the FA and HA are close together. Also noteworthy is the spread between the performance of the different cases when the link band-widths are modified and/or the update packet size is modified. With HBR, the spread is very slight, from the best case (large link bandwidths and regu-lar size update packets) to the worst case (regular size bandwidths and large update packets). With Mobile IP, the spread is much more pronounced, showing that it is much more sensitive to the settings of such variables.

Looking next at a light traffic scenario, two ways of reducing the traffic load are compared. First, reducing the size of the UDP packets 10-fold but keeping the rate of the packets at one every 40 ms can reduce traffic load from 200 Kbps to 20 Kbps. Second, keeping the same packet size (1,000 bytes) but reducing the rate of the packets 10-fold can also reduce traffic load to 20 Kbps. As would be expected, the number of packets dropped per handoff in the first case would be significantly higher than in the second case. Indeed, this is the case, as illustrated in Figure 17.9. In this figure, "light" refers to the light traffic scenario, "sp" refers to short packets, and "lr" refers to low rate (the two ways to reduce the traffic load). It can be seen that for both HBR and Mobile IP, sp has more dropped packets per handoff. However, as in the heavy traffic scenario, the variation is greater for Mobile IP, in addition to the actual numbers being worse.

Having had a flavor of some simulation results, it is appropriate to address the issue of how generally the results can be interpreted. One question that arises is whether the results would be limited only to the unlikely case that an MH just moves back and forth between two BSs, rather than more realistic movement (e.g., randomly moving between all four BSs in the base simulation setup). This is a valid question, but the "to-and-fro" movement results are more generally useful because they can be extended to more general movement patterns according to the following methodology:

- Suppose the HBR domain is an n-tier domain, so the crossover node could be one level above the BSs, or up to n levels above the BSs.

- For each level from $i = 1$ to n, to-and-fro handoffs are simulated between any two BSs whose crossover node is i levels above the BSs. Let λ_i be the number of packets dropped per handoff.

- Given the HBR domain (or subregion within it) and mobility pattern, compute $E[h_i]$, the expected number of ith-tier handoffs, for each $i = 1$ to n, and define $h = \sum_{i=1}^{n} E[h_i]$.

- The overall expected number of packets dropped per handoff is then computed as the weighted average

$$\lambda = \sum_{i=1}^{n} \frac{E[h_i]}{h} \lambda_i \tag{17.1}$$

For example, for our base simulation setup, $n = 2$. For a handoff distribution that is uniform over the other three BSs, the crossover node would be two levels up for two of them and one level up for the other BS. Hence, the result λ_2 should be weighted by 2/3 and λ_1 by 1/3. It would be expected that the resulting value would be similar to what could be obtained by actually simulating handoffs under such conditions of random motion. Figure 17.10 shows the results. The curve labeled "1-up" is for to-and-from handoffs

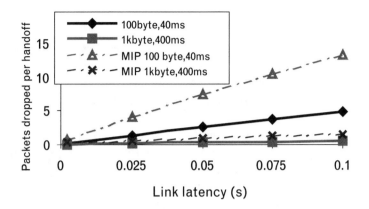

FIGURE 17.9
Packets dropped per handoff in a light-traffic scenario.

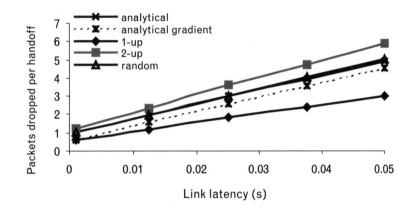

FIGURE 17.10
Comparing simulations of random handoffs with analytical model.

when the crossover node is one level above the BSs. The curve labeled "2-up" is similarly for to-and-fro handoffs when the crossover node is two levels above the BSs. The curve labeled "analytical" is where the performance in the more general random case is computed analytically according to (17.1). The curve labeled "random" is for the same case, but with results from actual simulations. It can be seen that "analytical" and "random" are almost the same curve, demonstrating that the analytical methodology works. One other curve can be seen in the figure, labeled "analytical gradient." This was obtained for the case where the 2-up simulation results were not used, but a 2-up scenario was emulated by simulating a 1-up scenario with double the link latency (up to 0.1 second instead of 0.05 second). The reason this underestimates the actual number of packets dropped per handoff for the random case is that it excludes the time for store-and-forward, and processing, at the intermediate nodes (just one such node in this case). Since this time is relatively constant, the offset of the resulting estimate from the real values is also roughly constant. Nevertheless, it at least provides the gradient of the correct curve and so is labeled "analytical gradient."

17.4.2.2   TCP Traffic

TCP throughput versus link latency is shown in Figure 17.11. "Base HBR" and "base MIP" are the results for the base simulations already described. "HBR semisoft" is for the case where the semi soft handoff scheme is used to further reduce handoff latencies. The figure shows the TCP throughput as link latency (of the straight-line links in Figure 17.6 and Figure 17.7) varies. HBR with semi-soft handoffs tolerates the most link latency and provides the highest throughput for any given link latency. Base Mobile IP shows a sharp deterioration in TCP throughput, which gets worse as the HA moves further away (not shown in this figure), which is more realistic.

The remaining curve in Figure 17.11 is labeled "flood MM." In this case, the root router floods the whole HBR domain with downstream packets, and upstream packets can arrive from any BS to the root router. This is not an HBR scheme, but is included to act as a performance bound. It is expected that no packets would be lost, since every BS is receiving the same stream, except for packets actually in the middle of being transmitted over the air when a handoff occurs. Notice that the throughput of flood MM starts to decrease at a link latency of just over 0.15 second. It may be conjecture that the reason is because that is where it runs into the so-called bandwidth-delay product bound. The TCP window size in the simulations is 20 KB, so the TCP "pipe" from sender to receiver can only hold that much data. Let link latency (in the HBR domain) be x seconds, and recall that the link latency of the "Internet" dash-dot links is 0.01 second. There are two of each type of link between the CH and MH. Therefore, the bandwidth-delay product from sender to receiver is (in kilobits)

$$\beta = 2 \times 0.01 \times 1,544 + 2x \times 375 \tag{17.2}$$

In order to be within the window size constraint, it is necessary that $\beta < 160$, and so $x < 0.1722$. It would be expected, however, that the

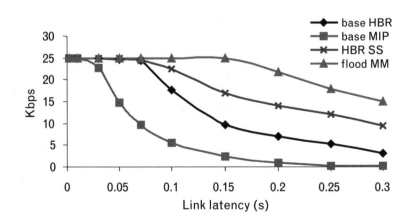

FIGURE 17.11
*TCP throughput
versus link latency.*

throughput would start dropping for link latencies even a little less than 0.1722 second because the sender needs to allow time for acknowledgments from the receiver (ideally, therefore, it would need the bandwidth-delay product to be slightly less than the window size to allow room for it to keep sending without interruptions while waiting for the ACKs). Therefore, the performance of "flood MM" is reasonable and shows the bounds in performance for this scenario. It can be seen that HBR SS gets the closest to this bound. Furthermore, it can be expected that "flood MM" will not perform as well when there are multiple MHs, as flooding will affect other MHs the most, whereas HBR SS would still provide good results.

As in Section 17.4.2.1 for UDP, the results of the to-and-fro handoffs are also compared with that of actually simulating random handoffs, where the MH hands off from any BS with a uniform distribution to any of the other three. In Figure 17.12, the results are plotted for both HBR (with hard handoffs) and HBR SS (with semisoft handoffs). For "base HBR random" and "HBR SS random," the link latency is as given on the x-axis. For "base HBR to-and-fro" and "HBR SS to-and-fro," the actual link latencies used for the points on the curve are 5/3 the values on the x-axis.

This is because for the random handoff cases, the crossover node is expected to be two levels above the BSs two-thirds of the time and one level above the BSs one-third of the time; and through a similar reasoning process as for the UDP simulations, the 5/3 weighting factor for the to-and-fro simulations can be derived. Unlike the case of the UDP simulations, it is not expected that the curves will line up so well. The reason is that the number of packets dropped per handoff in the UDP simulations is linearly related to the handoff latencies, whereas the relationship for the TCP case is nonlinear. The results confirm this. Therefore, for TCP simulations, to-and-fro simulations may be useful to get an approximate understanding of performance for specific handoff situations like random handoffs, but actual simulations of the actual handoff situations are necessary to get specific performance results for specific scenarios.

FIGURE 17.12
Comparing to-and-fro with random handoffs.

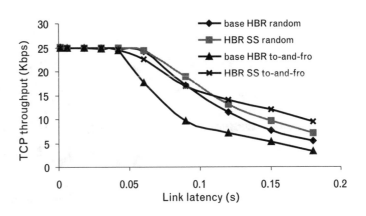

Although packets dropped per handoff is a useful performance indicator for UDP traffic, packet retransmission behavior is arguably more useful for TCP traffic. This is because throughput is more closely related to the packet transmission and retransmission behavior. Slight differences in number of packets dropped may result in timeouts occurring in one situation and not another, resulting in rather significant differences in packet retransmission behavior, and hence in throughput performance. The packet retransmission behavior for the different schemes is illustrated in Figure 17.13. For each of the three schemes (base HBR, base Mobile IP, and HBR SS), both the retransmit ratio and the number of retransmissions per handoff are plotted. The retransmit ratio refers to the ratio of retransmitted packets to total transmitted packets, and it is plotted as vertical bars with the values on the left y-axis. The values for the number of retransmissions per handoff, on the other hand, are plotted as regular curves, with the values on the right y-axis. The results for "flood MM" were also computed, but not plotted in the figure, because zero retransmissions occur throughout. Even the throughput decline for higher latencies is due to the bandwidth-delay product being constrained by the window size, not packet losses.

Looking at the retransmissions per handoff, it can be seen that all three curves follow the same basic pattern of increasing first, and then decreasing. The reason is that as the link latency increases, TCP is able to keep up with the increasing number of dropped packets by increasing the retransmissions and varying the instantaneous throughput so that can exceed 25 Kbps, allowing the average throughput to be still about 25 Kbps. There is a point, however, where TCP cannot "catch up" because of too many dropped packets, and the (average) throughput drops as a result. Since the average rate of packets is decreasing in this phase, the number of dropped packets per handoff, and retransmitted packets per handoff, also declines. It is interesting to note that Mobile IP has the worse performance in that the peak retransmission per handoff is the highest, and it occurs with the smallest link

FIGURE 17.13
Packet retransmission
behavior.

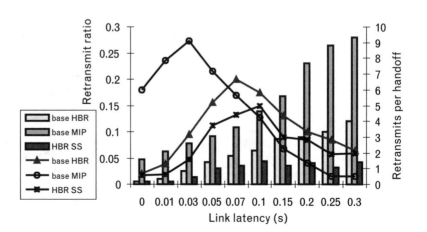

latency before entering the declining throughput phase. Similarly, HBR SS performs the best in having the lowest peak occurring with the largest link latency. As for the retransmit ratios, these tend to increase as link latency increases. In the worst case, with Mobile IP, somewhere between a link latency of 0.2 second and 0.25 second, over one-fourth of the packets arriving at the MH are retransmitted packets.

In order to see how the TCP throughput is affected by the network latency between the HA and FA in Mobile IP, the latency on the dashed link in Figure 17.7 is varied from the base value of 2 ms to 100 ms to 500 ms. The resulting TCP throughput is shown in Figure 17.14, where the dashed link has a latency of 100 ms and 500 ms for Mobile IP 0.1 and MIP 0.5, respectively. These are not unreasonable values for Mobile IP, where HA and FA could be very far apart. The resulting degradation on performance is evident. The performance of "base HBR" is also included in the figure for reference. As expected, in this region where the link latency is 0.05 second or less, HBR can achieve 25 Kbps throughput, unlike Mobile IP with the various latencies.

In the base simulations, the handoff rate is one handoff every 5 seconds on the average. To investigate the impact of different handoff rates, simulations were run (for HBR, HBR SS, and Mobile IP) with three average handoff intervals: 1 second, 5 seconds, and 10 seconds. The results are plotted in Figure 17.15. The solid lines show the performance of HBR. The dashed lines show the performance of HBR SS. The dotted lines show the performance of MIP. For each of these cases, the lines marked with crosses show the performance with handoff interval of 10 seconds, the lines with circles show the performance with handoff interval of 5 seconds, and the lines with triangles show the performance with handoff interval of 1 second. By looking at the three curves with crosses, it can be seen that the best performance is with the longer handoff intervals; whereas by looking at the three curves with triangles, it can be seen that the worst performance is with the shorter handoff intervals. This is expected because the more frequent the

FIGURE 17.14
Performance degrada-tion as latency between HA and FA increases.

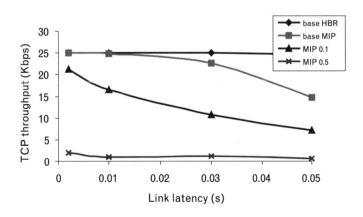

FIGURE 17.15
*Effect of varying
handoff rate.*

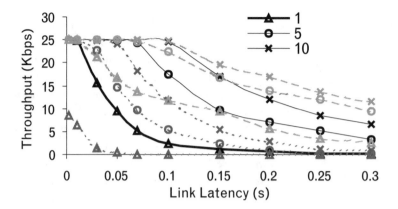

handoffs, the more frequent the dropping of packets during handoff, result-ing in more frequent transition of TCP into fast retransmit and slow start. It should also be noticed from the figure that for any given handoff rate, HBR SS performs best, followed by HBR, and then by Mobile IP.

Simulations were also run with TCP Reno. TCP Reno improves on the performance of TCP Tahoe when single packets are dropped, because of how it deals with congestion. In both cases, lost packets would result in duplicate ACKs being sent back to the transmitter, resulting in a retransmis-sion of the lost packet (the fast retransmit algorithm). However, this is fol-lowed by slow start with TCP Tahoe, which can have a big impact on the TCP throughput. With TCP Reno instead, the fast recovery algorithm is used, going back to congestion avoidance instead of slow start. Despite per-forming better with single dropped packets, it has been previously found that TCP Reno performs poorly in general when multiple consecutive packets are dropped [15]. It was verified that this is the case for TCP Reno over HBR and Mobile IP as well.

Various other simulations were also run. These include cases where the basic simulation topology is modified, through vertical expansion (adding more layers of hierarchy) and horizontal expansion (adding more leaf nodes to each parent node). Also, cases of uneven link latencies have been explored. The results are similar to those with the basic simulation environment.

17.4.3   Results from Laboratory Prototype

The laboratory prototype used is shown in Figure 17.16. The purpose of the prototype was to confirm the simulation results with actual measurements performed on a prototype implementation. The nodes labeled "root node," "HBR BS1," and "HBR BS2" run the Linux (kernel 2.2.14) operating sys-tem, as do the MH and CH. The router is a standard Cisco router. The HBR prototype is based on the CIP version 1.1 software from Columbia

FIGURE 17.16
Laboratory prototype configuration.

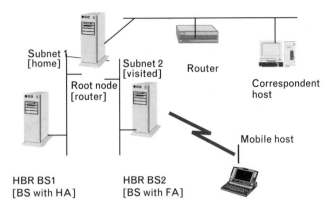

TCP throughput between MH and CH was measured for both HBR
and Mobile IP, using ttcp. Some of these results are shown in Figure 17.17.

University (http://comet.ctr.columbia.edu/cellularip), while the Mobile
IP prototype is based on the Sun Laboratories Mobile IP prototype
(http://playground.sun.com/pub/mobile-ip/sunlabs).

If the HBR measurements are performed with one set of platforms and
the Mobile IP measurements with another, then some irrelevant differences
might creep in—for example, differences in processor speeds, cross-traffic,
and so forth—that would reduce the accuracy of the comparisons. In order
to reduce irrelevant differences, both the HBR scheme and Mobile IP were
therefore run on the same platforms and with the same hardware configuration and network connections. Switching back and forth between HBR
and Mobile IP is a matter of typing a few commands to change a few interface configurations and routing table entries, and turn IP forwarding off (for
HBR, the routing/paging caches are used instead) or on (for Mobile IP).
Since the same hardware was used, both setups (HBR and Mobile IP) are
shown on the same diagram (Figure 17.16). Where the functionality differs,
the Mobile IP functionality is shown in square brackets beneath the HBR
functionality (e.g., HBR BS1 becomes the BS with HA in the Mobile
IP case).

TCP throughput between MH and CH was measured for both HBR
and Mobile IP, using ttcp. Some of these results are shown in Figure 17.17.
The values are averages over several measurements made at the ttcp receiving process, for CH to MH communications. The throughput of MH to
CH traffic has also been measured and shows similar behavior. Throughput
with Mobile IP for more than six handoffs per minute has not been included
because the results become unstable in that region.

Measurements of TCP throughput were also made for traffic from the
MH to the CH (Figure 17.18). The throughput degradations using Mobile
IP with an increasing number of handoffs per minute are similar in this case
as the previous. However, HBR performs better in this upstream direction.
This is because even before the crossover node is aware of the handoffs, data
packets following the handoff message are already taking the right path up to

FIGURE 17.17
TCP throughput for data transfer from the CH to the MH.

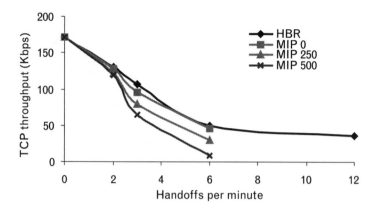

FIGURE 17.18
TCP throughput for data transfer from the MH to the CH.

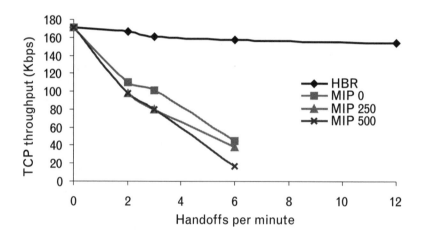

the gateway (TCP acknowledgments may be lost during this time, though, accounting for the slight degradation in throughput as the handoff rate increases). Typically, the studies on HBR schemes focus on the downstream to the MH, because that is more critical for many applications (e.g., MH obtaining streaming video from a network server). It should be noted, however, that HBR schemes can perform even better in the upstream (this assertion is based on the assumption that data packets can immediately follow a route update packet). If security measures are in place that do not allow that, then upstream performance will be affected.

17.5   Conclusions and Future Directions

This chapter deals with HBR schemes, and in particular on how they are designed for reducing IP-level handoff latency to possibly orders of magnitude less than what may be experienced with pure macro-mobility schemes based on Mobile IP and its variants, or SIP. This minimizes IP-level handoff disruptions, resulting in significantly fewer dropped packets and higher throughput. This chapter has provided an overview of the macro-mobility schemes and their shortcomings, especially where the shortcomings relate to handoff latency problems and the fact that they are not optimized to handle micro-mobility. An overview of two classes of micro-mobility schemes has been provided. The first class is the hierarchical Mobile IP-derived schemes like Mobile IP-RR and TeleMIP/IDMP. The second class is that of the HBR schemes. First, a generic HBR scheme is described that contains the essential features of the HBR schemes. Second, differences between actual HBR schemes like MMP, Cellular IP and HAWAII are explained. Some of the differences arise from differences in the design objectives, and the range of implementations of HBR schemes should provide network architects with sufficient flexibility to choose a micro-mobility solution that best fits a particular network. Third, comparisons are made between HBR domains and MANET. It was concluded that for certain types of very mobile networks, MANET ad hoc routing protocols must be used, but where a certain degree of fixed infrastructure exists that can support the HBR micro-mobility schemes, HBR would be preferred, because of reasons such as scalability and a more efficient distribution of routing information.

Next, performance issues of HBR micro-mobility schemes are discussed. In addition to providing the lowest latency handoffs of all the protocols discussed, the HBR schemes also (1) reduce signaling overhead through paging concepts, (2) improve bandwidth efficiency in the low-bandwidth radio access network by not using encapsulation in the radio access network, (3) facilitate QoS reservations by using an unchanging IP address while moving within the HBR domain, and (4) use IP address resources more efficiently than other schemes like Mobile IP with colocated COA. Results of simulations and laboratory prototype measurements are also reported. The goal of the performance studies reported in this chapter is to demonstrate the improvements of using HBR over Mobile IP for micro-mobility management. It is also explained how the Mobile IP results could be applied to hierarchical Mobile IP–derived schemes like MIP-RR, demonstrating that HBR performs better because of the lowest latency handoffs it provides. One implication of the simulation results is that using TCP over an IP micro-mobility management scheme magnifies the differences in handoff latencies between the schemes, because of the workings of the congestion control mechanisms like slow start.

Close to the top of the list of further simulations to do are simulations where there are multiple MHs in an HBR domain, where the traffic to/from other MHs would impact the performance of each MH. One expectation is that "flooding" micro-mobility management might not perform as well as it did in the results in this chapter (which is the best-case scenario for using flooding), because it would have the most impact on other

MHs, so it would not be practical. Further study is needed on transport protocols other than TCP, such as RTP and SCTP. Several modifications of TCP itself have been proposed for the problem of TCP interpreting errors and delays on wireless links as congestion [16]. The performance of such schemes over HBR micro-mobility schemes might be worth investigating.

This chapter has focused on the routing aspects of HBR schemes. However, it is very important to consider the implications of using HBR schemes together with schemes for QoS and security. Additionally, network management issues related to the HBR schemes should be investigated. HBR schemes may be more challenging to implement commercially than some other micro-mobility schemes, but they can be implemented slowly, in steps. Additional practical considerations are beyond the scope of this chapter. It is hoped, however, that the good performance of HBR schemes for IP micro-mobility management would serve as an incentive for further investigation in standards bodies like the IETF, and for engineers to work out the implementation issues.

References

[1]   Wong, K. D., "Architecture Alternatives for Integrating Cellular IP and Mobile IP," *Proceedings of IEEE IPCCC 2002*, Phoenix, AZ, April 2002, pp. 197–204.

[2]   Rosenberg, J., et al., "SIP: Session Initiation Protocol," RFC 3261, IETF, June 2002.

[3]   Dutta, A., et al., "Application Layer Mobility Management Scheme for Wireless Internet," *3G Wireless International Conference*, San Francisco, CA, May 2001.

[4]   Schulzrinne, H., and E. Wedlund, "Application Layer Mobility Support Using SIP," *ACM Mobile Computing and Communications Review*, Vol. 4, No. 3, July 2000, pp. 47–57.

[5]   Gustafsson, E., A. Jonsson, and C. Perkins, "Mobile IP Regional Registration," Internet Draft, IETF, September 2001, work in progress.

[6]   Das, S., et al., "TeleMIP: Telecommunications Enhanced Mobile IP Architecture for Fast Intra-Domain Mobility," *IEEE Personal Communications Magazine*, Vol. 7, August 2000, pp. 50–58.

[7]   Valko, A., "Design and Analysis of Cellular Mobile Data Networks," Ph.D. Dissertation, Technical University of Budapest, 1999.

[8]   Ramjee, R., et al., "HAWAII: A Domain-Based Approach for Supporting Mobility in Wide-Area Wireless Networks," available at http://www.bell-labs.com/user/ramjee/papers/hawaii.ps.gz.

[9]   Dutta, A., et al., "Multilayered Mobility Management for Survivable Network," *Proceedings of MILCOM*, October 2001.

[10]   Ramjee, R., T. La Porta, and L. Li, "Paging Support for IP Mobility Using HAWAII," Internet Draft, IETF, June 1999.

[11]   Royer, E. M., and C.-K. Toh, "A Review of Current Routing Protocols for Ad Hoc Mobile Wireless Networks," *IEEE Personal Communications*, April 1999, pp. 46–55.

[12]   Zhang, X., et al., "P-MIP: Minimal Paging Extensions for Mobile IP," Internet Draft, IETF, July 2000, work in progress.

[13] Joa-Ng, M., and K. D. Wong, "IP Mobility Management for the ACN Platform," *Proceedings of MILCOM*, October 2000, pp. 465–469.

[14] Shelby, Z., et al., "Cellular IP Route Optimization," Internet Draft, IETF, June 2001, work in progress.

[15] Fall, K., and S. Floyd, "Simulation-Based Comparisons of Tahoe, Reno and SACK TCP," *Computer Communications Review*, Vol. 26, No. 3, July 1996, pp. 5–21.

[16] Balakrishnan, H., et al., "A Comparison of Mechanisms for Improving TCP Performance over Wireless Links," *IEEE/ACM Transactions on Networking*, December 1997, pp. 756–769.

CHAPTER 18

Handoff Initiation in Mobile IPv6

Torben W. Andersen, Anders Lildballe, and Brian Nielsen

18.1 Introduction

The IP is expected to become the main carrier of traffic to mobile and wireless nodes. This includes ordinary data traffic like HTTP, FTP, and e-mail, as well as voice, video, and other time-sensitive data. To support mobile users, the basic Internet protocols have been extended with protocols (Mobile IP) for intercepting and forwarding packets to a mobile and possibly roaming node. Seamless roaming requires that users and applications do not experience loss of connectivity or any noticeable hiccups in traffic. This is not only important for time-sensitive traffic, but also for TCP-based traffic, as TCP performance is highly sensitive to packet loss and reordering.

It is therefore imperative that a handoff is initiated in such a way that network connectivity is maintained for the longest possible period of time, and that the handoff latency and packet loss is minimized. However, little is known about the performance of the Mobile IPs in an actual network. In particular, it is not understood how different handoff initiation algorithms influence essential performance metrics like the packet loss and the duration of a handoff.

To improve this situation this chapter studies the performance of two basic handoff initiation algorithms: *eager cell switching* (ECS) and *lazy cell switching* (LCS) [1]. The study uses both a theoretical approach that derives a mathematical model for handoff latency, and an empirical approach that includes experiments in a Mobile IPv6 test bed and an office building. The chapter also compares ECS and LCS with a novel proactive strategy, *parametric cell switching* (PCS), that considers link layer information about signal quality.

18.1.1   Mobile IPv6 Operation

A Mobile IP node is associated with a permanent home network and is assigned a static IP home address of the home network. The home address identifies the node globally. When the node is attached to a foreign network, it selects one router as its default router, and obtains an additional IP address, the COA, which identifies the current location of the mobile node. The network to which the mobile node is currently attached is called its primary network. In the basic mode of operation, a correspondent node sending packets to the mobile node addresses these to its home address. A router serving as HA must be present at the home network. The HA is responsible for tunneling IP packets sent to the home address to the mobile nodes current COA. The result is the triangular routing depicted in Figure 18.1.

A handoff in Mobile IP occurs when the mobile node switches from one (foreign) network to another, and thus obtains a new COA. It then registers the new COA at the HA by sending it a binding update message. The HA adds this to its binding cache and replies with a binding acknowledgment. Thus, packets tunneled after the mobile node has changed networks but before the binding update has reached the HA may be lost. The mobile node may also decide to send binding updates to the correspondent node(s), allowing it to address the mobile node directly and thereby bypass the HA. This avoids triangular routing, and is referred to as route optimization.

To discover new networks, the mobile node listens for router advertisements broadcast periodically from access routers. A router advertisement contains the network prefix of the advertised network and a lifetime denoting how long this prefix can be considered valid. The mobile node may then use stateful or stateless autoconfiguration to generate its new COA. In stateless autoconfiguration a host generates its own IP address based on the network prefix learned through router advertisements and the IEEE 802 address of its network interface. Stateful autoconfiguration involves additional communication with a DHCP server, but also allows configuration of other network services such as DNS.

FIGURE 18.1
Basic Mobile IP
operation.

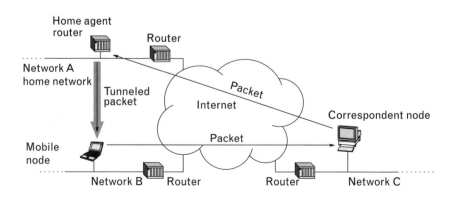

A handoff algorithm has three major responsibilities: (1) detecting and quality assessing available networks, (2) deciding whether to perform a handoff, and (3) executing the handoff. Handoff initiation consists of the first two activities. A seamless handoff requires that no packets are lost as a consequence of the handoff. In general, it is also desirable that packets are not reordered, duplicated, or extraordinarily delayed.

18.1.2 Handoff Initiation

The Mobile IPv6 specification [2] contains only a weak specification of handoff initiation algorithms. Two conceptually simple handoff initiation algorithms that have gained considerable interest are ECS and LCS [1]. Both operate at the network layer without requiring information from the lower (link) layers.

First consider the scenarios depicted in Figure 18.2. Here the ranges of two wireless networks (1 and 2) are depicted as circles. A mobile user moves from point A to point B. In the situation shown in Figure 18.2(a), where the networks do not overlap, no Mobile IP handoff initiation algorithm could avoid losing packets (one might imagine a very elaborate infrastructure where packets were multicast to all possible handoff targets and that packets could be stored there until the mobile node arrives, but even then a long period without network access would most likely be noted by the user). In contrast, if the networks overlap sufficiently as shown in Figure 18.2(b), seamless handoff is possible. Figure 18.2(c) shows that there are situations where a handoff is possible, but not desirable.

ECS proactively initiates a handoff every time a new network prefix is learned in a router advertisement. Conversely, LCS acts reactively by not initiating a handoff before the primary network is confirmed to be unreachable. When the lifetime of the primary network expires, LCS probes the current default router to see if it is still reachable. If not, a handoff to another network is initiated.

Consider what happens when ECS and LCS are subjected to the movement in Figure 18.2(b). Figure 18.3 illustrates a timeline for LCS where

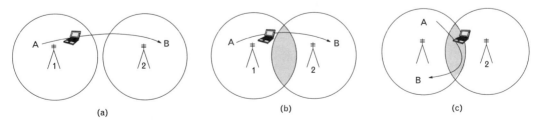

(a) (b) (c)

FIGURE 18.2   *A node moves from point A to B: (a) no seamless handoff is possible, (b) seamless handoff theoretically possible, and (c) seamless handoff possible, but should not be performed.*

FIGURE 18.3
LCS.

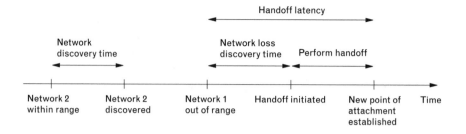

significant events have been pointed out. The first event is that network 2 gets within range. However, the mobile node cannot, in general, observe this before it receives a router advertisement from network 2. This results in a network discovery delay. LCS does not yet perform a handoff. Next, network 1 gets out of range. This cannot, in general, be detected immediately, as this requires active communication with the base station, and it gives rise to a network loss discovery delay. LCS declares the network unreachable when the lifetime of the last received router advertisement has expired and the following probing is unsuccessful. LCS then hands off to one of the alternative networks known to it through router advertisements—in this case network 2—and thus establishes a new point of attachment. Thus, LCS will lose packets in the handoff latency interval.

The behavior of ECS is illustrated in Figure 18.4. ECS hands off immediately when a new network is discovered. If the mobile node has an interface that is capable of receiving from the old network while attaching to the new, a seamless handoff can be performed provided a sufficient network overlap.

The performance of ECS thus depends on the frequency with which access routers are broadcasting router advertisements. Similarly, LCS also depends on the frequency of broadcasted router advertisements, but additionally depends on the lifetime of network prefixes and probing time. The theory and data needed to decide what handoff initiation algorithms to use in what circumstances, how to tune protocol parameters, and where to put optimization efforts, are missing.

Both ECS and LCS are very simple-minded approaches. PCS is our proposal for a more intelligent handoff initiation algorithm that also considers measured signal-to-noise ratio and round-trip time to access routers.

FIGURE 18.4
ECS.

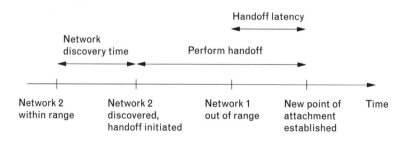

Periodically (currently every 0.5 second) the algorithm sends an echo request to the default router at all available networks. The default routers are expected to reply to the echo. These echo requests are sent for three reasons:

1. It is only possible to measure the SNR of a link to a network if there is traffic at the network.

2. It can be determined faster that a network has become unreachable than by monitoring the lifetime of network prefixes.

3. The round-trip time is an indication of the capacity of a network.

Parametric cell switching only performs handoff when a significantly better network is available [3]. The cost of using the PCS algorithm is a slightly increased network load.

18.1.3   Handoff Performance

There are a number of important metrics that should be considered when evaluating the performance of a handoff initiation strategy as experienced by a mobile node:

- *Handoff latency:* The handoff latency is the period of time where the mobile node is potentially unreachable. In general, it is caused by the time used to discover a new network, obtain and validate a new COA, obtain authorization to access the new network, make the decision that a handoff should be initiated, and, finally, execute the handoff, which involves notifying the home agent of the new COA and awaiting the acknowledgment from the HA.

- *Number of performed handoffs:* The more handoffs a given strategy will perform in a given scenario, the more likely it is that the user will observe them, and the more the network is loaded by signaling messages.

- *User value:* When several networks are candidates as target for a handoff, the one most optimal from the user's perspective should be chosen. This may be the network that offers the most bandwidth, cheapest price, the most stable connection, and so on.

18.1.4   Comparison with Other IP Mobility Schemes

The basic mobility concept of IPv6 is very similar to that of IPv4. However, the required support for mobility is better embedded in the core IPv6 protocols. Besides expanding the address space, IPv6 provides stateless autoconfiguration, (proxy) neighbor discovery, and more flexible header extensions and options. For instance, the home address option is used in a packet sent by a mobile host to inform the receiver about the mobile node's home address, which is the real source address of the message. The mobile node cannot use its home address as source because a router performing ingress filtering will drop packets with a source address that contains a network prefix different from the prefix of the router's interface at which the packet arrived. The binding update option is used to inform the home agent, and possibly also corresponding nodes (for route optimization), about its current COA. The routing header is used to specify a set of intermediate nodes the packet must traverse on its path to its destination. It reduces the tunneling overhead and simplifies treatment of forwarded ICMP packets.

The main architectural difference is that the foreign agent that assigns COAs and forwards tunneled messages received from the HA to the mobile node in Mobile IPv4 is absent in IPv6.

Currently, most work on handoff in Mobile IP is concerned with handoff execution using micro-mobility schemes such as HAWAII [4] and Cellular IP [5]. Micro-mobility aims at reducing the network load and handoff latency in environments of limited geographical span but with many frequently migrating nodes by minimizing and localizing the propagation of binding updates and binding acknowledgments.

For example, in Cellular IP a wireless access network consisting of a collection of interconnected Cellular IP nodes functioning as simple routers and base stations provides Internet access to wireless nodes via a gateway. The required signaling for an intra-access network handoff is purely local. Each Cellular IP node maintains a routing cache that maps the home address of the mobile node to the last of its neighbors that forwarded a packet sent by the mobile node (i.e., the cellular IP nodes maintains a soft-state path from the mobile node to the gateway, and uses the reverse path as route to the mobile node). Active mobile nodes refresh the path by periodically sending real or dummy packets containing a route update option. After handoff, the new path is established automatically by the transmitted route update packets. The locations of idle nodes are not tracked accurately. Rather, idle nodes are paged when needed. This reduces the network load. When handoff to another access network is needed, Mobile IPv6 is used.

In a hard handoff the packets transmitted via the old path are lost until the route update reaches the crossover point of the new and old path. Cellular IP also provides a semisoft handoff mechanism that can be used to reduce packet loss during a handoff. The mobile node can request in its route update packet that packets are to be forwarded along both the old and new route (i.e., normal message forwarding is turned into a multicast operation).

Like Mobile IP, handoffs are initiated by the mobile node based on its knowledge of available networks that it has learned from beacon signals similar to Mobile IP router advertisements emitted periodically by base stations. No particular handoff initiation strategy seems to be assumed or proposed.

While reducing handoff execution time is important, this is insufficient to obtain seamless handoffs, especially when the network topology is not controllable, as is the case in the Internet. It seems beneficial to take a proactive approach to handoff initiation that has the potential of reducing the packet loss to zero. This strategy has also proven successful in existing wireless networks such as GSM [6].

18.2   Mathematical Models

Because the focus of the models is the performance of handoff initiation algorithms, they do not include the propagation delay of binding updates and binding acknowledgments used in the complete handoff procedure. The handoff (initiation) latency is the time from the current primary network gets out of range until an event occurs at the mobile node that triggers it to perform a handoff. To obtain the total handoff latency, the round-trip time to the HA must be added to the numbers stated in this chapter. A typical number used for the continental United States is approximately 200 ms.

The goal is to predict the variation in handoff latency and its average as a function of the primary protocol parameters. The variation in latency is important to real-time sensitive traffic because it also indicates the best and worst case amount of time the mobile node risks being unreachable. The idea, therefore, is to derive a density function from which average and variation of handoff latency can be computed.

18.2.1   Basic Definitions

The mathematical models assume perfect cell boundaries (i.e., getting within range of a new network coincides with leaving the range of the primary network). According to [7], a router must pick a random delay between each broadcast of an unsolicited router advertisement in order to avoid routers synchronizing. The minimum possible time between two consecutive router advertisements is denoted R_{min} and the maximum time between two consecutive router advertisements R_{max}, meaning that the period between any two router advertisements must be found in the interval $[R_{min}, R_{max}]$. An additional simplifying assumption is that all access routers broadcast with the same frequency range. The time at which a mobile node enters the range of a new network is denoted by C_{time}.

The lifetime of broadcasted network prefixes is denoted by T_1. It is assumed that the time it takes to probe a default router is distributed uniformly within the interval $[Q_{min}, Q_{max}]$.

18.2.2   Eager Cell Switching

For ECS the theoretical handoff latency L_{eager} is the period from getting out of range of the primary network until the reception of a router advertisement from a new network. This is illustrated in Figure 18.5(a).

The goal is a density function for L_{eager} that can be used to derive the average value of L_{eager}. Because the primary variables C_{time} and R are random but dependent, their joint density function $P_{C_{time},R}(c_{time},r)$ must first be computed. It can be expressed as

$$P_{C_{time},R}(c_{time},r) = P_{C_{time}|R}(c_{time}|r) \cdot P_R(r) \tag{18.1}$$

where $P_{C_{time}|R}(c_{time}|r)$ is the probability distribution for C_{time} given R. As C_{time} is evenly distributed in the interval $[0, R]$, the probability distribution $P_{C_{time}|R}(c_{time}|r)$ can be calculated as

$$P_{C_{time}|R}(c_{time}|r) = \frac{1}{r} \cdot 1_{c_{time} \in [0,r]} \tag{18.2}$$

where $1_{c_{time} \in [0,r]}$ is an indicator function with a value of 1 when $c_{time} \in [0,r]$ and 0 otherwise.

The density function $P_R(r)$ expresses the probability that C_{time} should occur in an interval of size $R = r$. Intuitively, the probability of C_{time} occurring in an interval is proportional to the size of the interval. When the interval size is given as r, the function $f(r) = r \cdot 1_{r \in [R_{min}, R_{max}]}$ exhibits this intuitive property. The density function $P_R(r)$ is obtained as $f(r)$ divided with the area of $f(r)$. This yields

$$P_R(r) = \frac{r}{\int_{R_{min}}^{R_{max}} r\,dr} \cdot 1_{r \in [R_{min}, R_{max}]}$$

$$= \frac{2r}{R_{max}^2 - R_{min}^2} \cdot 1_{r \in [R_{min}, R_{max}]} \tag{18.3}$$

FIGURE 18.5
*Model for computing
(a) L_{eager} and (b) density function $P_{L_{eager}}(l)$
for ECS.*

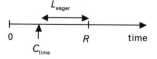

$$L_{eager}$$

0   C_{time} R time

0 : Router advertisement sent prior to R
R : First router advertisement received after C_{time}
$C_{time} \in [0,R]$
$L_{eager} = R - C_{time}$

(a)

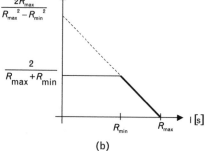

(b)

The joint density function $P_{C_{\text{time}},R}(c_{\text{time}},r)$ obtained by combining these results becomes

$$
P_{C_{\text{time}},R}(c_{\text{time}},r) = P_{C_{\text{time}}|R}(c_{\text{time}}|r) \cdot P_R(r)
$$

$$
= \frac{2}{R_{\text{max}}^2 - R_{\text{min}}^2} \cdot 1_{c_{\text{time}} \in [0,r]} \cdot 1_{r \in [R_{\text{max}},R_{\text{min}}]}
$$

(18.4)

The density function $P_{L_{\text{eager}}}(l)$ for L_{eager} can be obtained by integrating over the joint density function $P_{C_{\text{time}},R}(c_{\text{time}},r)$ for all possible values of c_{time} and r. As r can be expressed as $r = c_{\text{time}} + l$ (which follows from $l = r - c_{\text{time}}$), we have

$$
P_{L_{\text{eager}}}(l) = \int_{-\infty}^{\infty} P_{C_{\text{time}},R}(c_{\text{time}}, c_{\text{time}} + l) dc_{\text{time}}
$$

$$
= \frac{2}{R_{\text{max}}^2 - R_{\text{min}}^2} \cdot 1_{l \in [0,R_{\text{max}}]} [c_{\text{time}}]_{\max(0,R_{\text{min}}-l)}^{R_{\text{max}}-l}
$$

(18.5)

$$
= \begin{cases} \dfrac{2}{R_{\text{max}} + R_{\text{min}}}, & \text{if } 0 \le l \le R_{\text{min}} \\ \dfrac{2(R_{\text{max}} - l)}{R_{\text{max}}^2 - R_{\text{min}}^2}, & \text{if } R_{\text{min}} \le l \le R_{\text{max}} \\ 0, & \text{otherwise} \end{cases}
$$

In Figure 18.5(b) the density function $P_{L_{\text{eager}}}(l)$ from (18.5) is plotted. Observe that when using ECS, the handoff latency is bounded by the value of R_{max} and that there is the highest probability of obtaining handoff latencies in the range $[0, R_{\text{min}}]$.

Given the density function $P_{L_{\text{eager}}}(l)$ it is easy to obtain the average value of L_{eager}, denoted by $\overline{L}_{\text{eager}}$, by integrating over the product of $L_{\text{eager}} = l$ and $P_{L_{\text{eager}}}(l)$ for all possible values of L_{eager}:

$$
\overline{L}_{\text{eager}} = \int_0^{R_{\text{max}}} l \cdot P_{L_{\text{eager}}}(l) dl
$$

$$
= \frac{R_{\text{max}}^3 - R_{\text{min}}^3}{3(R_{\text{max}}^2 - R_{\text{min}}^2)}
$$

(18.6)

18.2.3 Lazy Cell Switching

For LCS the theoretical handoff latency L_{lazy} is the time from leaving the range of the primary network until concluding that the primary network is unreachable. This occurs when the lifetime of the primary network has expired and probing it has failed. Figure 18.6 depicts the used model where the last router advertisement from the primary network was received at time 0. The goal is a density function for L_{lazy} that can be used to calculate the average value of L_{lazy}.

Intuitively, the handoff latency consists of the remaining lifetime of the primary network plus the probing time Q used to determine that the primary network is unreachable. The remaining lifetime is the lifetime T_1 of the last router advertisement minus the time the network was reachable, C_{time}. The handoff latency L_{lazy} can thus be expressed as

$$L_{lazy} = \textit{Lifetime remaining of primary network} + Q$$
$$= T_1 - C_{time} + Q \tag{18.7}$$

Because Q is assumed to be uniformly distributed within the interval $[Q_{min}, Q_{max}]$ the density function for Q is

$$P_Q(q) = \frac{1}{Q_{max} - Q_{min}} \cdot 1_{q \in [Q_{min}, Q_{max}]} \tag{18.8}$$

Using a similar line of reasoning and method of calculation as outlined for ECS, the density function for LCS denoted $P_{L_{lazy}}(l)$ can be computed. That is, the joint density function of the involved parameters is integrated for all possible values. However, the intermediate calculations are more involved, and only the result is stated in (18.9). Further details on its derivation can be found in [8].

$$P_{L_{lazy}}(l) = \int_{-\infty}^{\infty} P_Q(l - l_{np}) \cdot P_{C_{time}}(T_1 - l_{np}) dl_{np},$$
$$\textit{where } P_{C_{time}}(l) = P_{L_{eager}}(l)$$
$$\textit{and } l_{np} = T_1 - c_{time} \tag{18.9}$$

FIGURE 18.6
Model for computing
L_{lazy}.

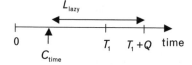

0 : Last router advertisement received from primary network
C_{time} : Leaving the range of the primary network
T_1 : Lifetime of network prefix
Q : Probing time
$L_{lazy} = T_1 - C_{time} + Q$

The average handoff latency $\overline{L}_{lazy}$ can be computed by integrating over the density function, or more simply, directly from (18.7) that states that $L_{lazy} = T_1 - C_{time} + Q$. The average latency $\overline{L}_{lazy}$ can be computed as the average remaining lifetime plus the average probing time $\overline{Q}$:

$$\overline{L}_{lazy} = T_1 - \overline{C}_{time} + \overline{Q}$$
$$= T_1 - \frac{R_{max}^3 - R_{min}^3}{3(R_{max}^2 - R_{min}^2)} + \frac{1}{2}(Q_{max} - Q_{min})$$

(18.10)

The average remaining lifetime of the primary network is the lifetime of the last received router advertisement minus the average amount of time that the primary network was reachable. On average, the primary network is reachable for an amount of time corresponding to the time used to discover a new network. This corresponds exactly to the average handoff latency for ECS. Thus, $\overline{C}_{time}$ equals $\overline{L}_{eager}$ calculated in (18.6).

18.3   Experimental Results in Test Bed

Section 18.2 presented the mathematical models needed to compute the handoff latency as a function of essential protocol parameters. This section presents the design of a Mobile IPv6 testbed and compares the theoretically predicted handoff latency with the handoff latency experienced by a mobile node in the Mobile IPv6 test bed.

18.3.1   The Mobile IPv6 Test Bed

The test bed is depicted in Figure 18.7 and consist of four nodes: three routers and one host. The three routers (iridium, platin, and nikkel) are assigned an IPv6 prefix for each network device. The mobile node (lantan) is manually assigned an IPv6 address at the fec0:0:0:1::/64 network, its home network. When lantan is not at its home network, it uses stateless autoconfiguration to obtain an IPv6 address as its COA. The home agent is located at iridium, which also hosts an application corresponding with an application at the mobile node. The mobile node can roam between the two ARs platin and nikkel. The link media used in the experiments reported here are standard 802.3 10-Mbps Ethernet devices. However, the test bed also runs 802.11b 11-Mbps WLAN connections. The connection between the ARs platin and nikkel allows them to coordinate on whom should offer access to the mobile node. It also allows experiments with route optimizations because it offers an alternative path to the mobile node.

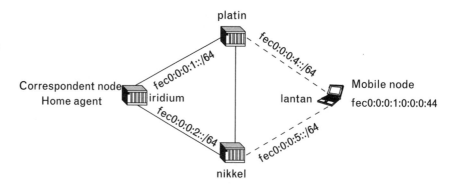

All nodes run FreeBSD version 4.1 [9]. On top of FreeBSD, the KAME package [10] is installed. The KAME package includes Mobile IPv6 support and IPsec support. The KAME package installed is the weekly snap–release of 25/9-2000. A snap-release is the newest version of the package and may include functionality that is still under development and is not fully tested. The Mobile IPv6 code supplied with KAME is an example of such functionality. The Mobile IPv6 implementation included in KAME can be configured to use either the ECS or the LCS handoff initiation algorithm.

18.3.2 Experimental Approach

Two different scenarios have been emulated in the testbed. In the *no network overlap scenario*, perfect cell boundaries are assumed. A mobile node moving out of the range of one network therefore coincides with the mobile node moving within range of another network. This scenario corresponds to the one for which mathematical models were derived in Section 18.2. In the *network overlap scenario*, cell boundaries are overlapping between the two networks and the mobile node is always able to reach at least one network. This scenario was applied to investigate whether ECS was able to avoid packet loss when able to receive and send packets via two networks at the same time. LCS behaves identically in both scenarios.

The mobile node moving in and out of the range of a network has been emulated by operating a firewall at the ARs. When the firewall is enabled, the mobile node is not able to receive or send any traffic through that particular AR. Accordingly, when the firewall is disabled, all traffic is allowed to pass to and from the mobile node.

To determine the handoff latency in an experiment we applied the following method:

- UDP packets are sent from the correspondent node to the mobile node. Each packet contains a send timestamp and a sequence number. The UDP packets are sent with a random interval between 95 and 105 ms. The interval is randomized to make sure the network does not adjust itself to any particular sending frequency.

- The UDP packets are received and timestamped at the mobile node. The sequence number and the send and receive timestamp are stored upon reception of a packet as an entry in a log file.

- A handoff is registered from packets missing in the log file. We compute the measured handoff latency by multiplying the number of lost packets with the average period between sending packets (0.1 second). The precision of the measured latency is thus 0.1 second. If a handoff is performed without losing packets, it will not be registered. Both the average and the frequency distribution of handoff latencies can be computed by inspecting the log.

By reducing the interval between UDP packets (increasing frequency), a higher accuracy will be obtained and the measured latencies will approach the theoretical latencies defined in Section 18.2. The interval of 95 to 105 ms was chosen to avoid too many UDP packets being sent. Due to a memory leakage in the KAME Mobile IPv6 software, only a limited number of packets can be sent from a correspondent node before it crashes. In the experiments presented in this chapter it was possible to perform 300 to 400 handoffs in sequence before the correspondent node crashed. Further confidence in the mathematical models has been obtained by implementing a simulator in JAVA. This simulator has confirmed the theoretically predicted density functions for a range of configurations for both ECS and LCS.

18.3.3 Overview of Performed Experiments

The Mobile IPv6 test bed has been used to perform the following experiments:

- *Default configuration:* In this experiment the router advertisement interval and network prefix lifetime is set as recommended in [2]. This means a router advertisement interval randomly chosen between 0.5 and 1.5 second and a lifetime of 4 second. The purpose of this experiment is to reveal handoff latency using the default configuration. In the latest versions of the Mobile IPv6 specification [3] the minimum time between router advertisements has been reduced from 0.5 seconds to 0.05 second. However, this change has little impact on the results of this chapter.

- *Latency as a function of router advertisement interval:* Handoff performance is measured for different router advertisements interval, but with an identical network prefix lifetime. The purpose of this experiment is to investigate how the interval between sending router advertisements affects the handoff latency.

- *Latency as a function of network prefix lifetime:* In this experiment the handoff latency is measured for different network prefix lifetimes, but with a fixed range for the intervals between sending router advertisements. The purpose is to investigate how the lifetime of router advertisements affects the handoff latency.

All experiments have been performed using both the network overlap and the no network overlap scenario. Only a selection of the empirically obtained results can be presented here. The full set of results is given in [8]. All plots also show the theoretically predicted handoff latency such that the theoretical and empirical results can easily be compared.

18.3.4 Default Settings

First, the theoretically predicted probability distributions are compared with the measured frequency distributions using the default configuration of access routers in the no overlap scenario. Next, their performance is compared numerically.

The histogram in Figure 18.8(a) depicts the frequency distribution of the experimentally measured handoff latencies for ECS. The continuous line shows the theoretically predicted density function. The observed handoff latencies lie in the range 0.1 to 1.5 seconds. Also note that no handoff latencies in the interval [0,0.1] are present. This is caused by the experimental setup in which the precision is limited by the frequency of packets from

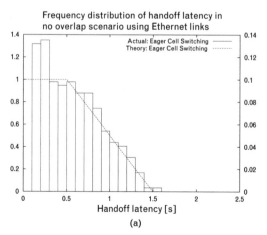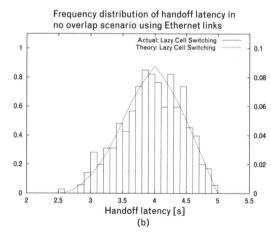

FIGURE 18.8   *Frequency distribution for handoff latency (a) using ECS and default configuration in the no network overlap scenario, and (b) using LCS and default configuration in the no network overlap scenario.*

the corresponding node (i.e., latencies below 0.1 second cannot be observed). Instead, these handoff latencies are recorded in the [0.1,0.2] interval. In conclusion, the experimental results for ECS conforms well to those predicted by the mathematical models.

Figure 18.8(b) depicts the histogram obtained for LCS. Here the observed latencies range from 2.5 seconds to 5 seconds with most values centered around 4 seconds. It can be seen that LCS is generally unable to avoid packet loss as it does not initiate a handoff until after the primary network has become unavailable. In conclusion, the empirically obtained frequency distribution conforms well to the density function predicted by the mathematical model. The comparison of ECS and LCS is summarized in Table 18.1.

When no overlap between network ranges exists, ECS yields an average handoff latency of 0.54 second. This corresponds to the time it takes to discover the new network. LCS yields a much worse latency with an average of 3.96 seconds. Also, best- and worst-case values are higher. With respect to handoff latency, ECS outperforms LCS. Furthermore, the experiments with overlapping networks show that ECS is able to avoid packet loss altogether during a handoff, provided that a sufficient overlap between network ranges exists.

A disadvantage of ECS, however, is that it always performs a handoff when discovering a new network, whether or not this offers stable connectivity. Consequently, ECS will likely perform many unnecessary handoffs resulting in an increased network load and loss of connectivity. In conclusion, both ECS and LCS have serious performance deficiencies, but the performance of ECS indicates that proactive handoff initiation has the potential to avoid packet loss.

18.3.5 Varying Advertisement Frequency

One of the primary protocol parameters is the frequency of router advertisements. Its effect on handoff latency is shown in Figure 18.9, which plots

TABLE 18.1 SUMMARY OF EXPECTED AND ACTUAL RESULTS FOR THE NO NETWORK OVERLAP SCENARIO USING DEFAULT ROUTER CONFIGURATION

HANDOFF STRATEGY	L [s] (THEORY)			LATENCY [s]		
	Avg.	Min.	Max.	Avg.	Min.	Max.
Eager	0.54	0	1.5	0.54	0.10	1.52
Lazy	3.96	2.5	5.0	3.97	2.54	4.97

Note: The probing time Q for LCS is assumed to be in the interval [0,1].

average handoff latency as a function of the interval between broadcasting router advertisements.

ECS behaves like what would be expected intuitively—that is, a higher frequency implies that networks are discovered sooner, which again implies faster handoffs. Surprisingly, however, observe that the LCS latency is actually decreasing when the interval between broadcasting router advertisements is increased. The explanation for this is that the lifetime is fixed at a constant value of 5 seconds in this experiment. This result thus indicates that the handoff latency for LCS can be minimized by configuring ARs with a prefix lifetime very close to the maximum interval between broadcasting router advertisements.

18.4   Optimizing Protocol Configuration

This section demonstrates that the default configuration of access routers proposed in [2] does not result in optimal handoff performance for either ECS or LCS. In [2] it is suggested that a router should broadcast unsolicited router advertisements distanced by a random period chosen from the interval [0.5,1.5]. This gives an average network load of one router advertisement every second. In later versions of the Mobile IPv6 specification [3], the default configuration has changed the router advertisement period to [0.05,1.5] seconds. This gives a slight increase in average network load to 1.3 router advertisements per second. Using these values, our theoretical models predict that the average handoff initiation latency for ECS is 0.5 second and for LCS is 4 seconds (i.e., a small improvement for ECS and a small drawback for LCS). As demonstrated in the following, however, the models can be used to find a new set of parameters that reduce handoff latency *without* increasing the network load. Similarly, the models can be used to find the optimal settings should an increased network load be accepted.

Using the same network load as the suggested rate of an average of one advertisement per second, the average ECS latency $\overline{L}_{eager}$ can be minimized by adjusting R_{min} and R_{max} in (18.6) subject to the constraint that the sum of R_{min} and R_{max} must equal 2. Close inspection reveals that ECS performs best when R_{min} and R_{max} are configured with values as close together as possible. Optimal performance for ECS can therefore be obtained when both R_{min} and R_{max} are set to a value of 1.

The same method applied to LCS reveals that LCS performs better when R_{min} and R_{max} are configured with values far from each other. For LCS, then, optimal performance can be obtained by configuring R_{min} to a value of 0 and R_{max} to a value of 2. This is in direct contradiction to the optimal configuration for ECS. However, for LCS the dominating factor for the handoff latency is the lifetime of broadcasted network prefixes. As this

FIGURE 18.9
*The average handoff
latency as a function of
router advertisement
interval for ECS and
LCS in the no net-
work overlap scenario.*

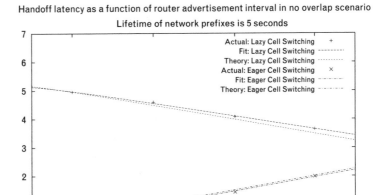

Handoff latency as a function of router advertisement interval in no overlap scenario
Lifetime of network prefixes is 5 seconds

lifetime cannot be configured to be lower than the value of R_{max}, a reduction of R_{max} (which is the case when R_{min} and R_{max} is configured to have values close to each other) can also benefit the performance of LCS, if the lifetime is configured close to the value of R_{max}.

In Table 18.2 theoretical values of handoff latency for three different configurations of access routers are shown. The probing time for LCS is assumed to be in the interval $[0,1]$.

Observe that both ECS and LCS perform better with the proposed configuration of access routers with $R_{min} = 0.9$ and $R_{max} = 1.1$. For ECS the average handoff latency is reduced from 0.54 second to 0.50 second, and the worst-case handoff latency is reduced from 1.5 seconds to 1.1 seconds. Similarly, for LCS the average handoff latency is reduced from 3.96 seconds to 1.1 seconds and the worst-case handoff latency is reduced from 5.0 to 2.1

TABLE 18.2 THE MATHEMATICALLY PREDICTED HANDOFF LATENCY IN THE NO NETWORK OVERLAP SCENARIO FOR DIFFERENT CONFIGURATIONS

HANDOFF STRATEGY	ROUTER CONFIGURATION			L [s] (THEORY)		
	R_{min}	R_{max}	T_1	Avg.	Min.	Max.
Eager	0.5	1.5	4	0.54	0	1.5
Lazy	0.5	1.5	4	3.96	2.5	5.0
Eager	0.9	1.1	1.1	0.5	0	1.1
Lazy	0.9	1.1	1.1	1.1	0	2.1
Lazy	0	2	2	1.83	0	3.0

seconds. The new proposed settings thus simultaneously improve on average, best, and worst case for both ECS and LCS.

Observe from the last row in Table 18.2 that for LCS the advantage of configuring R_{min} and R_{max} far from each other is outweighed by the fact that the lifetime T_1 has to be configured at a higher value.

The performance of the new settings has been tried out in the test bed. The experiment confirmed the theoretically predicted values [8].

An alternative to reducing the lifetime of router advertisement messages is to exploit the *advertisement interval option* in router advertisements proposed in [2]. This option contains the maximum time (R_{max}) between router advertisements that mobile nodes should expect. This would allow a mobile node to probe its default router if no router advertisement has been received for a period corresponding to the value of R_{max}. This, in effect, forces LCS to become more proactive.

18.5   Building Wide Experiment

The following simple experiment compares the handoff performance of ECS, LCS, and parametric cell switching handoff initiation algorithms in a more realistic scenario than the Mobile IPv6 test bed.

18.5.1   Experimental Setup

Three ARs (nikkel, blue, and vismut) are all installed with 802.11b WLAN network devices and are configured with the improved router configuration: R_{min} = 0.9 second, R_{max} = 1.1 second, and T_1 = 2 seconds (the value permitted in the test bed nearest the desired 1.1 seconds). These three access routers have been deployed at Departments of Computer Science in three different locations. The mobile node *lat11* is carried at normal walking velocity along an approximately 2-minute walk. A single walk was performed for each of the handoff initiation algorithms.

During the experiment, a program at iridium that acts as home agent and correspondent node sends UDP packets with intervals randomized between 95 and 105 ms. A program at the mobile node *lat11* receives the packets and counts the number of dropped packets and performed handoffs. In addition, it measures the signal-to-noise ratio for each of the three networks once every second during the walk. Figure 18.10 plots the signal-to-noise ratios measured along the path.

18.5.2   Results

Table 18.3 summarizes the results. Surprisingly, it can be observed that ECS shows the poorest performance. ECS drops many packets presumably

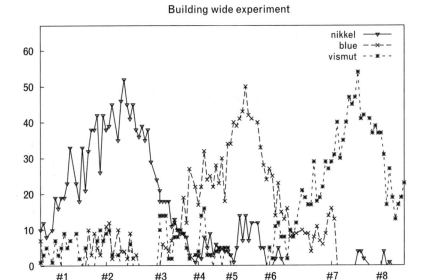

FIGURE 18.10
*Measured signal-to-
noise ratios from the
three access routers
when the mobile node
is moved along the
walking path. The
numbers along the x-
axis corresponds to dif-
ferent locations along
the walking path.*

TABLE 18.3   SUMMARY OF RESULTS FOR BUILDING WIDE EXPERIMENT

HANDOFF STRATEGY	NUMBER OF ATTEMPTED HANDOFFS	NUMBER OF LOST PACKETS	AVERAGE LATENCY [s]	MAXIMUM LATENCY [s]
Eager	24	48	0.14	0.41
Lazy	3	10	0.10	0.10
Param	2	0	0.00	0.00

because it initiates many handoffs as each network disappears and reappears several times along the path. Often it tries to perform handoffs to networks that are only sporadically available. LCS performs much better. It has reduced the number of handoffs significantly and lost only 10 packets. These were found to be single packets dropped when approaching the maximum range of the primary network.

PCS shows excellent performance. It keeps the number of handoffs at only two handoffs in this scenario. The number of handoffs is optimal because PCS only initiates a handoff when a network is present that has significantly better signal-to-noise ratio than the primary network and has a stable low round-trip time to the access router. In this scenario the SNR behaves in a way that gives PCS opportunity to hand off to a new network before the connection to the primary network becomes unstable.

More importantly, the results show that it is possible to achieve very good handoff performance using Mobile IP, but also that this requires a more intelligent (proactive) initiation algorithms than the proposed eager and lazy cell switching.

18.6   Conclusions and Future Directions

The mathematical models for the ECS and the LCS handoff initiation algorithms were found to be able to predict handoff performance. Using a testbed installed with FreeBSD 4.1 and the KAME Mobile IPv6 software, these models were shown to accurately reflect the handoff latency experienced by an actual roaming node. The mathematical models were also used to optimize Mobile IPv6 protocol configuration to reduce the handoff latency without increasing network load due to router advertisements.

The mathematical models have a perspective beyond academic satisfaction and protocol optimization. They can, for instance, be used to calculate how much cells should overlap. If seamless handoff should be possible they must overlap by at least the duration it takes to initiate and perform a handoff. Translating the amount of overlap measured in time to one measured in geographical distance requires assumptions about the speed with which users move. The models offer a way of relating protocol parameters with assumptions for movement speed and requirements for cell overlap.

Both existing handoff initiation strategies, ECS and LCS, have serious performance lacks, but ECS has the potential to avoid packet loss. But as initial results from the building-wide experiment indicate, ECS does not perform well in an actual wireless network because sporadic router advertisements from new but unstable networks barely within reach cause ECS to handoff to the new network. A more intelligent algorithm is needed. Parametric cell switching, which takes signal quality into account as well as the throughput and price of a link, indicates that good performance is possible using Mobile IP. It would be interesting to investigate if the technique can be further improved by additional parameters such as measured bit error rate or media type. One might even consider (BS) topology information that could be broadcasted from access routers or made available upon request.

Further work should include a more complete study than presented here to determine the number of performed handoffs in various cell configurations and movement patterns, possibly performed as a simulation study. Also the behavior in a wide area setting should be investigated more thoroughly.

More work should also be done on how AAA can be done such that a limited overhead is added to a handoff. This may involve using techniques like obtaining authorization concurrently with the home agent binding update, using preauthorization, allowing time limited access while authorization is ongoing, or using smart card technology.

REFERENCES

[1] Perkins, C. E., *Mobile IP Design Principles and Practices*, Reading, MA: Addison-Wesley, 1998.

[2] Johnson, D. B., and C. Perkins, "Mobility Support in IPv6," Internet draft, draft-ietf-mobileip-ipv6-13.txt, November 2000, work in progress.

[3] Johnson, D. B., and C. Perkins, "Mobility Support in IPv6," Internet draft, draft-ietf-mobileip-ipv6-16.txt, March 2002.

[4] Ramjee, R., et al., "HAWAII: A Domain-Based Approach for Supporting Mobility in Wide-Area Wireless Networks," *Seventh Annual International Conference on Network Protocols*, October 31–November 3, 1999, Toronto, Canada, p. 283.

[5] Campbell, A. T., et al., "Design, Implementation and Evaluation of Cellular IP," *IEEE Personal Communications*, Special Issue on IP-Based Mobile Telecommunications Networks, June/July 2000.

[6] Mouly, M., and M.-B. Pautet, *The GSM System for Mobile Communications*, 1992, Palaiseau, France: M. Mouly et M.-B. Pautet, F-91120, 1992.

[7] Deering, S., "ICMP Router Discovery Messages," RFC 1256, September 1991.

[8] Andersen, T. W., and A. Lildballe, "Seamless Handoff in Mobile IPv6," Masters thesis, Department of Computer Science, Aalborg University, Denmark, 2001.

[9] FreeBSD, http://www.freebsd.org.

[10] KAME, http://www.kame.net.

CHAPTER 19

Location Independent Network Architecture and Mobility Handling in IPv6

Masahiro Ishiyama, Mitsunobu Kunishi, and Fumio Teraoka

19.1   Introduction

Mobile computing devices are evolving very rapidly, and nowadays a great number of people access the Internet through various types of mobile terminals. The number of Internet access points is expected to increase, and wireless access links such as IEEE 802.11 and Bluetooth are expected to become more widely deployed. This is paving the way for a real mobile computing environment in which portable devices can access the Internet and communicate with each other while on the move. It would be too restrictive, however, to adopt a usage model in which only the receiving terminals are mobile and the sources of information (e.g., Web servers) are fixed. Therefore, demand for a protocol that enables bidirectional peer-to-peer communications is emerging. IPv6 [1] is the protocol that enables such communication; however, IPv6 does not include a mobility support mechanism—it needs an additional protocol to support mobile nodes. To support mobility in IPv6, Mobile IPv6 [2] has been proposed and is being discussed within the IETF. However, Mobile IPv6 has several fundamental problems.

For example, it is very difficult to make Mobile IPv6 fault tolerant. MIPv6 needs a dedicated router, an HA, that is in charge of forwarding packets to a mobile node for mobile communications. The home agent is a single point of failure in Mobile IPv6. While managing of HAs in a distributed manner is a prerequisite, it is very difficult in the current Mobile IPv6 specification to scale them in a distributed fashion.

MIPv6 does not contain the concept of a node identifier. MIPv6 introduces the home address that is a fixed address independent of the location of a mobile node. Although it may work as a means to identify a node in a limited context, it is not a node identifier. The home address depends of the location of its HA, and if the network to which the HA belongs is renumbered, the home address must also be renumbered.

Mobile IPv6 has a packet length overhead. It uses optional headers for all packets that a mobile node sends, so it consumes more bandwidth and the transmission delay is greater then IPv6.

These problems arise from the fact that MIPv6 was not developed based on architectural considerations on providing mobility, but rather it was built as a simple extension of IPv6 from the viewpoint of routing.

In this chapter "mobility" is defined as the functionality that satisfies the following two capabilities: (1) communication with nodes regardless of their location; (2) continuation of communication even if a correspondent node changes its location. From the mobility viewpoint, the fundamental problem of conventional network architectures is in the duality of the network address. For example, in both IPv4 and IPv6, an IP address has two meanings: the node identifier and the interface locator. The IP address specifies not only the identity of the node but also the point of attachment to the Internet. The IP address of the node changes if the node moves to another subnet. Consequently, the identity of the node is no longer preserved. This problem inherent in conventional network architectures is caused by assigning an address to the network interface of a node, not to the node itself. The locator of the network interface is regarded as the identifier of the node. In other words, there is no notion of a node identifier in the current Internet architecture.

This chapter introduces the *location independent network architecture* (LINA) [3], which employs separation of identifier and locator to support node mobility. There have been some network protocols based on the separation of the node identifier and the interface locator, such as Xerox Internet Datagram [4], VIP [5, 6], and the GSE proposal for IPv6. However, an entire network architecture covering the layers from network to application and considering node identity has not appeared yet. The primary goal of LINA is to build an encompassing network architecture that is applicable to the design of practical protocols. This chapter also introduces a new mobility support protocol called location independent networking for IPv6 (LIN6), which was designed by applying LINA to IPv6. LIN6 provides mobility to IPv6 without impacting the existing IPv6 infrastructure and maintains compatibility with traditional IPv6. LIN6 has several advantages compared to Mobile IPv6 in terms of performance and system stability.

19.2 LINA

This section introduces a new network architecture called LINA. LINA is based on the idea of separating the identifier and the locator of a node. In the application layer, a target node can be specified by its identifier in addition to the conventional model in which the target node is specified by the locator. The network layer is divided into two sublayers: the identification sublayer and the delivery sublayer. When the application specifies a target node, the identification sublayer maps the identifier to the corresponding locator, and then the delivery sublayer "embeds" the identifier in the locator. This embedding of identifiers into locators is done based on the new addressing model called embedded addressing.

19.2.1 Basic Concept

As described above, in conventional network architectures including IPv4/IPv6, the network address of a node denotes its identity as well as its location. This is a critical problem for supporting mobility at the network layer because there is no location independent identity for a mobile node. To solve this problem, LINA introduces two entities in the network layer to support node mobility.

- *Interface locator:* uniquely identifies the node's current point of attachment to the network. It is assigned to the network interface of a node and is used to route a packet to the network interface.

- *Node identifier:* signifies the identity of the node. It is assigned to the node itself and does not change even if the point of attachment to the network changes and a new interface locator is assigned to the network interface of the node.

Figure 19.1 depicts the basic communication model based on LINA. The network layer is divided into two sublayers: the identification sublayer and the delivery sublayer. The identification sublayer converts the node identifier and the interface locator mutually while the delivery sublayer delivers the packet destined to the interface locator. An application program can specify a target node by either a node identifier [Figure 19.1(a)] or an interface locator [Figure 19.1(b)].

In case (a), an application wants to communicate with a node specified by the node identifier. This means that an application wants to communicate with a particular node regardless of its location. The transport layer maintains the connection with the pair of node identifiers. The identification sublayer in the network layer converts the node identifier to the appropriate interface locator, and the delivery sublayer delivers the packet. In this

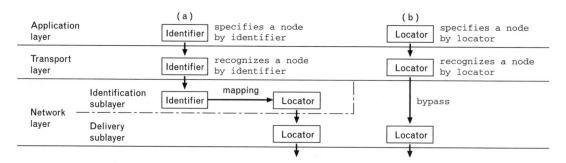

FIGURE 19.1   *Communication model: The network layer is divided into two sublayers to separate the notion of locator and identifier. An application can specify either (a) a locator or (b) an identifier to specify a correspondent node.*

case, mobility is supported because the node identifier never changes even when the node moves.

In case (b), an application wants to communicate with a node located at the point specified by the interface locator. This means that an application wants to communicate with a node at the particular location of the network regardless of its identity. The transport layer maintains the connection between the pair of interface locators. The identification sublayer is bypassed, and the delivery sublayer delivers the packet. In this case mobility is not supported between nodes and the transport connection will reset if one of the nodes moves.

Specifying an interface locator at the application layer has several advantages. This feature can be used in the following cases:

- An application wants to directly specify an interface. For example, a target node has several interfaces, and the application wants to communicate through a specific interface of the target node.

- An application wants to communicate with a particular node without identity. For example, a node wants to communicate with the node that is present at a particular location regardless of its identity.

19.2.2   Embedded Addressing Model

On the basis of the layering model described above, the network layer header in a packet should be divided into two headers in general protocol layering. This is not an efficient method, however, since adding a new protocol header results in a greater overhead compared to existing architectures that use the interface locator as the node identifier. To solve this problem, LINA embeds the node identifier in the interface locator. This addressing model is called the embedded addressing model.

The interface locator that follows this addressing model is called the ID-embedded locator. Although an ID-embedded locator is information for the delivery sublayer, it also implies a node identifier that is information for the identification sublayer. Thus, the identification sublayer header can be integrated into the delivery sublayer header by using ID-embedded locators in the network layer header, and the overhead issue is avoided accordingly. Detailed properties of the ID-embedded locator are described in the following section. However, a node still needs to determine the interface locator from the node identifier. In this model, a node simply refers to an association of the two, which is managed outside of the packet header. Details of this mechanism are described in Section 19.2.5.

19.2.3 Embedding and Extraction

Figure 19.2 shows entities and operations that are used in the embedded addressing model. LINA assumes that a node has one or more node identifiers that are assigned by an authority. It also has one or more interface locators when a node is connected to the network. Such locators are called current locators.

Embedment is an operation that determines the ID-embedded locator from the current locator and the node identifier. An ID-embedded locator is also assigned to the interface of the node. Thus, the node is assigned not only current locators but also ID-embedded locators that are determined by performing embedment. The ID-embedded locator satisfies the following conditions:

1. An ID-embedded locator uniquely determines a node identifier without referring to other information.

2. It is possible to distinguish between an ID-embedded locator and a current locator.

3. An ID-embedded locator is a valid interface locator. That is, the format and the functions of the ID-embedded locator are equivalent to

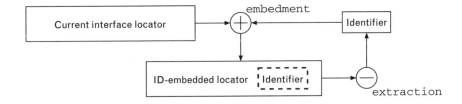

FIGURE 19.2 *Operations of the embedded addressing model: embedment determines ID-embedded locator from the current locator and the node identifier, and extraction determines the node identifier from the ID-embedded locator.*

those of the interface locator and an intermediate node can deliver a packet even if the destination is specified by an ID-embedded locator.

LINA also introduces the concept of extraction, which is the inverse operation of embedment. Extraction is an operation that determines a node identifier from an ID-embedded locator.

19.2.4   Generalized ID

In the concept described in Figure 19.1, an application can specify not only a node identifier but also an interface locator. In this case, it is required that both the application layer and the transport layer handle both node identifier and interface locator. Thus, from an engineering standpoint, it would be easier to process the node identifier and the interface locator in the same format.

For that reason, LINA introduces the concept of dedicated embedment. LINA introduces the dedicated locator that is the dedicated interface locator for dedicated embedment. The dedicated locator is a predefined well-known fixed value. The dedicated locator does not determine any physical point of the network, whereas the ID-embedded locator and the current locator uniquely determine a physical point of attachment to the network.

In dedicated embedment, the dedicated locator is used for the current locator. The dedicated locator is a fixed value; thus, the result of dedicated embedment has a one-to-one correspondence to a given node identifier. That is, the result can be used as an identifier for the node. The result of dedicated embedment is called the generalized identifier. Since the generalized identifier conforms to the format of an interface locator, an application layer and a transport layer does not need to discriminate between them, and hence the above issue is resolved.

19.2.5   Mapping: Resolving Interface Locator from Node Identifier

When a node performs embedment, the node needs to associate of the node identifier with the current locator of the node. This association is called mapping.

LINA introduces a function called a mapping agent that maintains this mapping. Designated mapping agents, which are the mapping agents that maintain the mapping of a particular node identifier, are introduced. That is, "designated mapping agents of the node A" means that those mapping agents maintain the mapping of node A.

A node registers its mapping periodically with its designated mapping agents. It also registers a new mapping when the node changes its location on the network.

When a node performs embedment to determine an ID-embedded locator of a target node, the node first determines the designated mapping agents of the target node and queries one of the designated mapping agents to acquire the mapping of the target node. Then the node can determine the current locator of the target node, and the node can perform embedment.

19.2.6 LINA Communication Model

This section describes the detailed process of sending and receiving a packet in LINA. Figure 19.3 shows the communication model of LINA, which is based on an application of the embedded addressing model to the basic communication concept shown in Figure 19.1.

Upon sending a packet, an application specifies the destination with a generalized identifier for the target node. An identification sublayer examines whether the given destination is a generalized identifier or an interface locator. If the destination is a generalized identifier, the identification sublayer first performs extraction to obtain the node identifier, following which it determines designated mapping agents of the node identifier and queries the mapping of the node. Since it can derive the current locator of the target node when it obtains the mapping, it performs embedment and derives the ID-embedded locator of the target node. Then it passes the packet to the delivery sublayer with the ID-embedded locator. The delivery sublayer transmits the packet that is destined for the ID-embedded locator. If the destination is not a generalized identifier, the identification sublayer is bypassed.

When a packet is received, the delivery sublayer receives the packet and examines whether the specified source locator in the packet is an ID-embedded locator or not. If the source is an ID-embedded locator, the

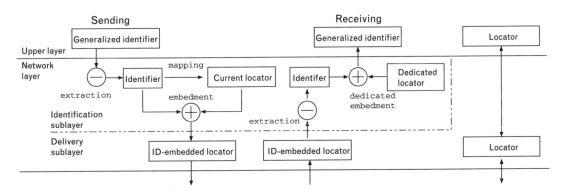

FIGURE 19.3   *Communication model of LINA. Upper layers identify a correspondent node with generalized identifier. The ID-embedded locator derived from the result of embedment is used for a packet delivery. Embedment is not performed when a locator is specified in upper layers.*

delivery sublayer passes the packet to the identification sublayer. The identification sublayer performs extraction and obtains the node identifier of the source node, following which it performs dedicated embedment to derive the generalized identifier of the source node. The identification sublayer informs the upper layer that the source locator of the packet is the generalized identifier. If the source is not an ID-embedded locator, the identification sublayer is bypassed.

19.3   LIN6: An Application of LINA to IPv6

This section presents LIN6, a protocol that supports mobility by applying LINA to IPv6. For practical purposes, LIN6 is carefully designed to maintain compatibility with conventional IPv6 so that there is minimal impact on the existing IPv6 infrastructure. This section shows a method for applying the concept of embedment of LINA to IPv6, followed by a description of a mechanism to determine a designated mapping agent.

19.3.1   Embedded Addressing in LIN6

The concept of the ID-embedded locator of LINA shall now be applied to IPv6. In this context, the locator is called a LIN6 address. Currently in IPv6, addresses are assigned according to the *aggregateable global unicast address (AGUA)* [7] format, whose structure is shown in Figure 19.4. In AGUA, the upper 64 bits of 128-bit IPv6 address indicate the network prefix to which the address belongs; and the lower 64 bits represent an interface ID. The interface ID is not required to be unique in the Internet but only in the subnet. Also, it is required to conform to the IEEE EUI-64 format [8].

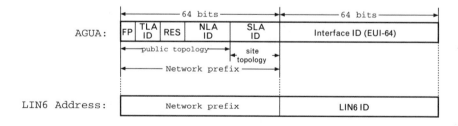

FIGURE 19.4   *The format of AGUA and LIN6 address. In AGUA, the upper 64 bits represent the location of a subnetwork and the lower 64 bits represent the identifier of an interface, not of the node. In LIN6 address, although the upper 64 bits are the same as for AGUA, the lower 64 bits represent the node identifier, LIN6 ID.*

Since this means that practical IPv6 subnetworks have a 64-bit network prefix length, the basic strategy in applying embedment to IPv6 is to use the lower 64 bits of the IPv6 address for the node identifier. That is, in LIN6, the address space of the node identifier is 64 bits. This 64-bit node identifier is called the LIN6 ID. Although a 64-bit address space is far smaller than the IPv6 128-bit address space, 64 bits can accommodate approximately 10^{19} nodes, which is considered sufficiently large to support LIN6 nodes.

This strategy satisfies condition (1) of the ID-embedded locator that is described in Section 19.2.3. To satisfy conditions (2) and (3), LIN6 forms a LIN6 ID so that it can be identified as a LIN6 address—that is, LIN6 uses part of the LIN6 ID to examine whether it is a LIN6 address or not. In this method, a LIN6 address can coexist with AGUA—that is, LIN6 can use the same prefix as in AGUA on a foreign network for the upper 64 bits of a LIN6 address, and use a specially formed LIN6 ID for the lower 64 bits. The following method is an example of how this can be realized. In AGUA, the lower 64 bits are required to be constructed in EUI-64 format. In EUI-64, the upper 24 bits denote the organizationally unique identifier (OUI) assigned by IEEE, and the lower 40 bits are the value that is assigned by an administrator who is assigned an OUI. If an OUI is assigned for LIN6, a LIN6 address can be identified by examining the OUI part of the lower 64 bits of a given IPv6 address. This satisfies both condition (2) and condition (3) because a LIN6 address is a valid AGUA since a LIN6 ID completely follows the EUI-64 format in this method. Although the real address space of LIN6 ID decreases to 40 bits, this method does not have impact on existing IPv6 networks.

Since a LIN6 ID includes the specific OUI assigned for LIN6, strictly speaking, the LIN6 ID is in reality just the lower 40 bits, excluding the 24 bits assigned to the OUI. However, for convenience in following discussions, the entire 64-bit field including the OUI is simply called the LIN6 ID.

19.3.2 Embedment in LIN6

In LIN6, the current locator of an LIN6 node is the IPv6 address assigned to the interface of the node. This address is generally an AGUA, and the mapping is an association between the LIN6 ID of the node and an AGUA assigned to the node at that time.

The operation of embedment in LIN6 is as follows. The upper 64 bits of a given IPv6 address, which are the current interface address of a target LIN6 node, are concatenated with the LIN6 ID of the target LIN6 node. The extraction operation simply draws out the lower 64 bits from a given LIN6 address. These operations are shown in Figure 19.5.

FIGURE 19.5
Embedment in LIN6.
A LIN6 address is
derived by concatenat-
ing the prefix of the
current IPv6 address
and a node identifier.

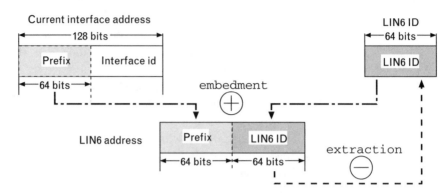

19.3.3 Generalized Identifier in LIN6

LIN6 assumes that a 64-bit-length network prefix for dedicated embedment, which is called the LIN6 prefix, has been allocated. The LIN6 prefix is a predefined, fixed value, and is expected that it is well known to all LIN6 nodes in advance. This is the same assumption that all IPv6 nodes know link-local prefix, local-loopback address, and so forth. It does not identify any physically existing subnetwork and is only used for dedicated embedment in LIN6, and thus LIN6 does not need routing information for it. The operation of dedicated embedment in LIN6 is simply to concatenate the LIN6 prefix to the LIN6 ID. The result of dedicated embedment is called LIN6 generalized ID, which corresponds to a generalized identifier in LINA. The LIN6 generalized ID completely conforms to the IPv6 address format, and existing IPv6 applications can specify this ID without any modifications as the destination of a correspondent node. Table 19.1 summarizes the correspondence between LINA and LIN6.

TABLE 19.1   CORRESPONDENCE BETWEEN LINA AND LIN6

LINA	LIN6
Node identifier	LIN6 ID, 64 bits but includes a specific OUI
Current locator	An IPv6 address that is generally an AGUA assigned to one of the interfaces of a LIN6 node
ID-embedded locator	LIN6 address completely follows AGUA format.
Dedicated locator	LIN6 prefix, a 64-bit-length predefined prefix assumed to be assigned
Generalized identifier	LIN6 generalized ID, a concatenation of the LIN6 prefix and LIN6 ID

19.3.4 Finding Designated Mapping Agents

LIN6 needs mapping information to perform embedment. The address space of the LIN6 ID is 64 bits, and it might be very difficult to maintain mappings in a centralized single database.

In the Internet, the *Domain Name System* (DNS) resolves a similar issue. DNS is a distributed database that maps a *fully qualified domain name* (FQDN) to an IP address and vice versa, and provides other types of information related to an IP address and an FQDN. For example, consider the case in which one wants to know an FQDN associated with a particular IPv6 address. In this case, one sends a query with an IPv6 address as a key to any DNS server and obtains the FQDN associated with the requested IPv6 address. This feature of DNS, called reverse lookup, works fine in the Internet.

DNS can be used to determine the mapping associated with a particular node identifier—that is, DNS servers can be used as mapping agents. Using DNS servers as mapping agents, however, presents a serious problem since DNS is designed for handling "static" data. DNS assumes that the association between an IPv6 address and FQDN does not change frequently, hence DNS servers can cache data for load balancing. Since DNS is a very large-scale distributed database, its success is mainly due to this caching mechanism. However, the content managed by mapping agents may change frequently, and thus DNS is inappropriate for mapping agents.

Consequently, LIN6 does not use DNS for the mapping agent, but introduces a new dedicated server. The node on which the server runs is called a mapping agent. A network administrator of a LIN6 node decides the designated mapping agent of the node, and then the administrator registers this relationship with DNS. That is, an association between the LIN6 generalized ID of the node and each IPv6 address of designated mapping agents of the node is registered in appropriate DNS servers. Consequently, a node can acquire the IPv6 addresses of designated mapping agents for the LIN6 node by querying DNS, specifying the LIN6 node's LIN6 generalized ID as the key, as in the case of reverse lookup. The DNS server program must be modified so as to be able to handle information on mapping agents. However, this modification is only required for the master and the slave servers managing the LIN6 generalized ID. LIN6 does not need any modifications to root and intermediate DNS servers, since intermediate servers are only interested in a query name, and the query name is the LIN6 generalized ID that is compatible with an address in IPv6 address format. An association between a LIN6 node and its designated mapping agents will not change frequently, and consequently, this information is suitable for handling in DNS.

To acquire the mapping of a target LIN6 node, a node first sends a query to DNS to obtain information on designated mapping agents, and then sends a query to any of these mapping agents to acquire the mapping. An obtained mapping is cached in the node, and the caching area of mapping is called the mapping table. A mapping has a lifetime and is discarded when the lifetime expires. Also, addresses of mapping agents may be cached.

When an LIN6 node moves, it sends its mapping to one of the designated mapping agents of the node, and the designated mapping agents maintain consistency of the mapping by notifying each other of the received mapping. Since a LIN6 node can obtain the addresses of its designated mapping agents by querying DNS, LIN6 nodes do not need to know its designated mapping agents in advance. The communication mechanism of LIN6 is summarized in Figure 19.6.

19.3.5   Handoff of a Mobile Node

In LIN6, the current locator of a node is stored in the mapping entry as a cache on each correspondent node. Thus, if the node moves, a correspondent node needs to update the mapping to get hold of the new location of the node.

Mapping can be updated in one of two ways: (1) notification from the node that has moved, and (2) autonomous update by the correspondent node that is communicating with the node that has moved.

When a LIN6 node moves to a new location, it sends the new mapping to one of its designated mapping agents. In addition, the mobile node may send the new mapping to all correspondent nodes. This operation is called the mapping update. The mobile node can find its correspondent nodes by inspecting its mapping table.

If a mobile node sends the mapping update, some security mechanism such as IPsec [9] is needed to protect against spoofing attack. However, the mobile node might not be able to use IPsec with the correspondent node. In

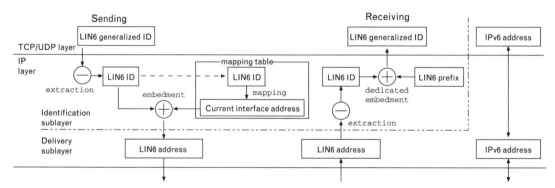

FIGURE 19.6   *Communication mechanism of LIN6. On sending, the LIN6 generalized ID is transferred to the LIN6 address by embedment with referring to the mapping table. On receiving, the LIN6 address is transferred to the LIN6 generalized ID by dedicated embedment. The LIN6 generalized ID is used in upper layers, while the LIN6 address is used to route a packet.*

this case, the mobile node can send the mapping refresh request instead of the mapping update to the correspondent node. The mapping refresh request message does not include the mapping. It only requests the correspondent node to update its current mapping from the mapping agent. Thus, this message is free of any spoofing attack. When a node receives a mapping refresh request, it queries the mapping to the mapping agent and obtains the new mapping.

Correspondent nodes might not receive the mapping update for some reason (e.g., the mapping update packet is lost in the intermediate network, or a mapping of the correspondent node is deleted from the mapping table). LIN6 deals with this problem by using the ICMP *destination unreachable message* (Dst Unreach) [10]. A Dst Unreach message is sent to the packet sender from an intermediate router if reachability to the destination is lost. If a LIN6 node receives a Dst Unreach message when communicating with another LIN6 node, it queries the designated mapping agents about the new mapping. If the obtained mapping contains a new current interface address, it can derive the new appropriate LIN6 address of the correspondent LIN6 node and can continue to communicate. If a Dst Unreach message is received, it can be assumed that the correspondent LIN6 node is temporarily disconnected from the network and is just in motion. If the content of the obtained mapping is the same as the cached one, or the mapping cannot be obtained, the Dst Unreach message is notified to the upper layer. This situation signifies that the correspondent LIN6 node may be off-line for a long time.

19.3.6 Compatibility with Traditional IPv6 Nodes

Let us consider the case when an LIN6 node wants to communicate with a traditional IPv6 node. In this case, an application specifies a traditional IPv6 address, not an LIN6 generalized ID, as the destination, a situation that corresponds to communicating via an interface locator in LINA. The identification sublayer is consequently bypassed—that is, no LIN6 specific operation such as embedment is performed. As a result, this situation is completely equivalent to traditional IPv6 communication. Thus, LIN6 nodes are able to communicate with traditional IPv6 nodes; however, when this happens, mobility cannot be supported since interface locators are used instead of node identifiers.

19.4 Communication Example of LIN6

This section describes LIN6 through an example scenario depicted in Figure 19.7. MN1 and MN2 are LIN6 nodes. G1, G2, and G3 represent

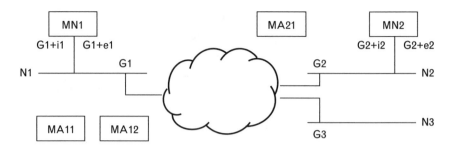

network prefixes advertised by a router in each network. i1 and i2 represent the LIN6 IDs of MN1 and MN2, and e1 and e2 represent interface IDs of MN1 and MN2 for AGUA. "+" is used as the concatenation operator; that is, "G1+e1" represents an IPv6 address that is a result of concatenation of network prefix G1 and interface ID e1. "O" represents a LIN6 prefix; that is, "O+i1" represents a LIN6 generalized ID of MN1. MA11 and MA12 are the designated mapping agents of MN1, and MA21 is the designated mapping agent of MN2. It is assumed that both MN1 and MN2 know their LIN6 identifiers and that they can obtain the addresses of a DNS server.

19.4.1   Bootstrap

Let us consider that MN1 boots up and connects to network N1. MN1 registers the current mapping with one of its designated mapping agents as follows (see Figure 19.8):

1. MN1 receives the router advertisement messages from a router on N1. At this point, MN1 acquires at least two global addresses: one is G1+e1 (a traditional AGUA), and the other one is G1+i1 (a LIN6 address).

2. MN1 sends a query to a DNS server to obtain the addresses of its designated mapping agents.

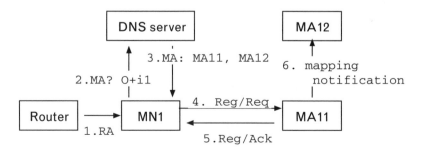

3. The DNS server replies with the addresses of the designated mapping agents of MN1 (MA11 and MA12). Recursive query of DNS is abbreviated from the figure.

4. MN1 selects a designated mapping agent at will. Assuming that it chooses MA11, MN1 sends a registration packet of the mapping to MA11.

5. MA11 sends an acknowledgment of the registration back to MN1.

6. MA11 obtains the designated mapping agents of MN1 in a similar way and finds that there is one more designated mapping agent of MN1 (MA12). Then, MA11 notifies MA12 of the mapping to keep consistency of mapping of MN1.

19.4.2 Communication Between LIN6 Nodes

Let us consider that MN1 communicates with MN2 that is on network N2. It is assumed that MN2 has already registered its mapping and G2+i2 as its LIN6 address. When an application on MN1 wants to communicate with MN2, it specifies LIN6 generalized ID of MN2, which is O+i2, as the destination address. Communication between MN1 and MN2 takes place as follows (see Figure 19.9):

1. To acquire the current mapping of MN2, MN1 asks a DNS server about the designated mapping agents of MN2. As a result, MN1 finds MA21 is the designated mapping agent of MN2.

2. MN1 sends a query to MA21 and acquires the current mapping for MN2. The mapping contains G2+e2, which is MN2's current interface address.

3. MN1 performs extraction and obtains i2, the LIN6 ID of MN2, and then it performs embedment and derives G2+i2, the LIN6 address for MN2.

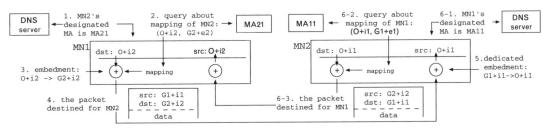

FIGURE 19.9 *Example of LIN6 communication: upper layers of MN1 and MN2 identify each other with LIN6 IDs, and LIN6 addresses are used as source and destination addresses in packets that flow in the Internet.*

4. As a result, the destination address of the packet transmitted by MN1 is G2+i2, and the application recognizes the destination as O+i2. Similarly, the source address of the packet is G1+i1, the LIN6 address of MN1.

5. MN2 receives this packet and examines the source address field. Since the source address is a LIN6 address, MN2 performs extraction and obtains i1, the LIN6 ID of MN1 that is the source node of the packet. Then MN2 performs dedicated embedment and obtains O+i1, the LIN6 generalized ID of MN1, and then informs the application that O+i1 is the source address of the packet.

6. When MN2 sends a packet to O+i1, similar procedures are followed where the destination address in the packet MN2 transmits is G1+i1 and the source is G2+i2.

Thus, between MN1 and MN2 in the intermediate network, communication is perceived to be between G1+i1 and G2+i2, whereas on the node MN1, it is perceived to be between O+i1 and O+i2.

19.4.3   Handoff of a LIN6 Node

Let us now consider the case when MN1 moves to network N3 while communicating with MN2. When using mapping refresh request, an update of MN1's mapping is performed as follows (see Figure 19.10):

1. MN1 sends a *router solicitation* (RS) message, and it receives a router advertisement message from the router on N3. At this point, MN1 acquires G3+e1 as the AGUA and G3+i1 as the LIN6 address.

2. MN1 registers a new mapping (i1, G3+e1) with MA11.

3. MN1 inspects its mapping table to find current correspondent nodes. Since MN1 has cached MN2's mapping (i2, G2+e2), MN1 recognizes that MN2 is one of them. MN1 sends a mapping refresh request

FIGURE 19.10
Handoff sequence using mapping refresh request: MN1 registers a new mapping with its designated mapping agent, then sends a mapping refresh request to let MN2 update its cached copy of MN1's mapping.

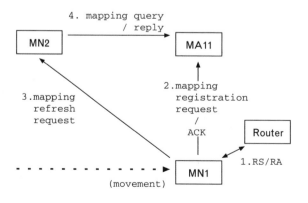

message to MN2 to let MN2 update its cached copy of MN1's mapping.

4. When MN2 receives the mapping refresh request, MN2 sends a mapping request to MA11 to update the mapping (it is assumed that a relationship between MN1 and MA11 is already cached at this point), and receives a new mapping for MN1.

19.5 Implementation

19.5.1 Implementation Status

A prototype of LIN6 was implemented on NetBSD/i386 1.5 with the KAME IPv6 stack. Operations such as embedment, dedicated embedment, and extraction, which are packet header manipulations, are implemented in the kernel, along with the mapping table. Functions such as registration and acquisition of mapping are implemented in user space. This program is called the mapping resolver. It uses a new socket (called the LIN6 socket) to communicate with the kernel's mapping table management functions. The motion detector is an application program to detect the movement of the node and its new location. The reason for separating the mapping resolver from the motion detector is that the motion detector can be easily modified in order to try several motion detection mechanisms that might have different performance. The mapping agent is also implemented as an application program.

19.5.2 Motion Detection Mechanism with IPv6

In the current prototype system, it uses information from layer 3 (i.e., IPv6 layer) to detect the movement of a mobile node, and it considers any change of the default router in the routing table of the kernel to be a movement of the mobile node. The motion detector watches changes of the default router. The discovery and selection of a default router is done using the neighbor discovery mechanism [11], which is part of the basic specification of IPv6. All IPv6 routers basically advertise their existence by a router advertisement message. Although neighbor discovery specification states that router advertisement be periodically sent to a link to which the router is connected, a node may request routers on the same link to send RA messages in order to avoid waiting for the next period. This request message is called router solicitation. When the default router changes, it reads the current interface to be used and the new network location of the node, then notifies the mapping resolver of the movement through the LIN6 socket. The selection and update of the default router is performed in the kernel and is a native function of the KAME IPv6 stack. When the mapping resolver receives a movement notification from the motion detector, it generates a current locator (i.e., an LIN6 address) and assigns the address to the current interface of the mobile node, then performs a mapping registration.

19.6   Protocol Evaluation of Mobility Handling

This section evaluates the mobility handling mechanism of a prototype LIN6 implementation. A handoff of a mobile node was performed on an experimental network, and packets on the network were observed by using tcpdump.

19.6.1   Experimental Network

Figure 19.11 shows the configuration of the experiment network. R1, R2, and R3 represent IPv6 routers. MN is an LIN6 mobile node, and CN is the correspondent node of MN. MA is a designated mapping agent of MN and CN. CN sends VoIP packets to MN, and MN moves from network N2 to network N3 while MN is receiving VoIP packets from CN. The intervals between each VoIP packet is set to 20 ms. All nodes are connected by 10BASE-T.

Packets observed at both network N1 and network N3 were logged. H represents the host who logs packets on the points Pa and Pb. From these logs, the behavior and performance of handoffs in LIN6 were analyzed.

19.6.2   Experimental Results and Considerations

Figure 19.12 shows the results of the experiment of a handoff using mapping update. The number printed after the line number on each line indicates the time in seconds when the packet was received by tcpdump. The number in parentheses is the time difference from the previous line. In the

FIGURE 19.11
The configuration of the experiment network. MN performs a handoff from N2 to N3 while receiving VoIP packets from CN.

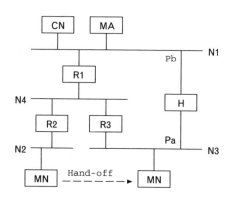

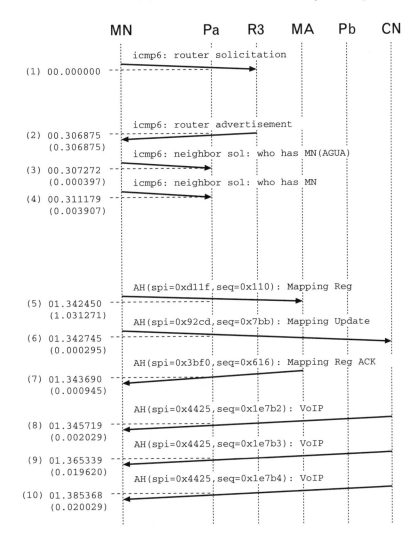

FIGURE 19.12
Experiment 1: observing the handoff with DAD count 1, enabled random delay of router advertisement, using the mapping update.

experiments, the IPv6 address of the mapping agent is already cached in MN and CN; thus, DNS query packets are not observed.

When MN detects a connection to an another link N3, MN sends an RS message to the link (1). R3 receives this RS message and sends a router advertisement message to MN (2).

MN receives the RA message, then it generates its new LIN6 address and assigns the address to its interface. Numbers (3) and (4) denote neighbor solicitation packets for *duplicate address detection* (DAD) [12]. The specification states that DAD is performed on unicast addresses prior to assigning them to an interface in order to verify their uniqueness on the link to avoid address conflicts. Number (5) denotes a mapping registration with mapping agent from MN and (7) denotes acknowledgment of this from mapping agent. These packets use IPsec AH header [13] to protect against spoofing attack. Number (6) denotes a mapping update to CN from MN. This packet also uses the AH header. At this point, LIN6 handoff has completed. Numbers (8), (9), and (10) represent VoIP packets from CN. It shows that communication between CN and MN has been reestablished. Thus, the overhead of LIN6 mobility handling can be approximated from the elapsed time from sending an RS message (1) to receiving the first VoIP packet (8).

In this case, theoretically, the elapsed time to complete motion handling is expected to depend mainly on the round-trip time between MN and CN. However, the elapsed time from (1) to (8) is about 1,345 ms, whereas the round-trip time that is estimated from the elapsed time from sending mapping registration (5) to receiving ACK (7) is less than 10 ms.

In Figure 19.12, it takes a very long time from (2) to (5), about 1,036 ms. This is a majority of the time that elapsed during handoff. This is the time required for the DAD procedure. The IPv6 specification states that an address whose uniqueness on a link is being verified is not allowed to be used until the procedure is finished. The current KAME IPv6 stack only sends one neighbor solicitation packet for DAD in default. After MN sends the neighbor solicitation packet, it waits 1,000 ms, a value that is specified in the IPv6 specification as the default, for replying to the solicitation. The mapping resolver must wait for this procedure to complete. The mobile node is not allocated a valid global IPv6 address and thus no packets can be sent to it until this procedure ends. This is the reason for the long elapsed time from (2) to (5). This problem is not unique to LIN6 but also applies to other protocols that need to assign a new IPv6 address by the method that conforms to the IPv6 specification such as MIPv6.

The time elapsed between sending an RS message (1) and receiving a router advertisement message (2), takes 307 ms. The reason for this time is that the router intentionally delays sending the router advertisement message in response to the RS message. The IPv6 specification states that when an IPv6 router sends a router advertisement message, the router must delay the transmission for a random amount of time in order to prevent routers on the same link from sending router advertisement messages at exactly the same time. The default maximum time of delay is 300 ms, and the KAME IPv6 stacks conforms to this value. This is the reason for the elapsed time until the RA message from the router is received. This problem is also confronted by all protocols that use RA messages (e.g., Mobile IPv6 in the case of using address autoconfiguration).

It is acceptable, however, to decrease this random delay if the network has few routers on the same link, and it is possible to design such a network in a real environment. Practically, address conflicts will not occur, especially in LIN6 because LIN6 addresses use LIN6 IDs in the interface ID part of IPv6 address format and the LIN6 ID is required to be unique on the Internet. It is possible that DAD be disabled on networks to support high-speed handoffs on certain networks.

Next, the same experiment was performed as shown in Figure 19.12 except for the following two points: (1) the random delay was disabled when R3 sends a router advertisement message in response to a RS message; and (2) DAD count was set to zero on MN. This means that DAD is disabled. The result of this experiment is shown in Figure 19.13.

In this case, the elapsed time from sending the router advertisement message (1) to receiving the first VoIP packet (6) is about 46 ms, and it can be estimated that MN drops two VoIP packets. Although this evaluation does not include the time needed to disconnect from the previous link, the extent of this overhead is acceptable enough since the packets lost can easily be recovered by error correction mechanisms such as FEC. The average measured time of the handoff was about 46.4 ms with 10 tries.

The case of using mapping refresh request was also tested. The result is shown in Figure 19.14, with random delay and DAD disabled.

Numbers (1) and (2) are RS and router advertisement messages, and (3) and (4) are mapping registration and its ACK, respectively. Number (5) is the mapping refresh request to CN from MN. The mapping refresh request is sent after receiving the ACK of registration from the mapping agent (4), whereas MN does not wait for the ACK when the mapping update is sent. The reason that MN waits for the ACK is to prevent CN from querying

FIGURE 19.13
Experiment 2: observing handoff with DAD count 0, disabled random delay of RA, using the mapping update.

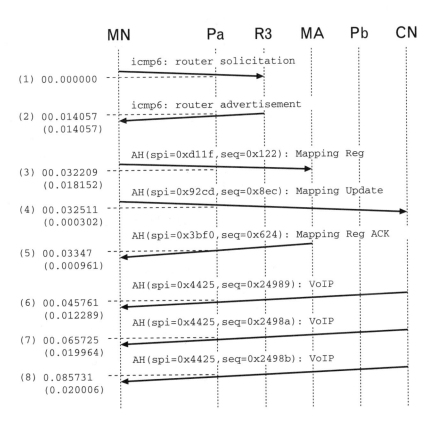

FIGURE 19.14
Experiment 3: observing handoff with DAD count 0, disabled random delay of router advertisement, using the mapping refresh.

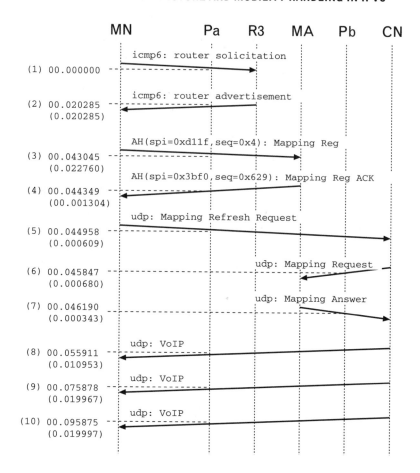

before the mapping agent learns the new mapping of MN and to ensure that CN can receive the new mapping of MN from the mapping agent. Notice that it does not use an authentication header since it does not require IPsec, as mentioned in Section 19.3.5. When the CN receives the mapping refresh request from MN, it sends the mapping query message to the mapping agent (6), and the mapping agent replies to it (7). At this point, the CN now has the new mapping of the MN that includes the new location of the MN. In this case, the overhead of the LIN6 motion handling procedure mainly depends on three RTTs: MN-MA (mapping agent), MN-CN, and CN-MA. The result shows that the overhead is about 56 ms, and the average is 49.7 ms with 10 tries.

The results show that mapping refresh requests have performance implications over mapping updates, as the RTT between the CN and the mapping agent increases. However, since mapping refresh requests do not require IPsec mechanisms, they can be used when CNs do not support IPsec and prove to be useful.

19.7   Comparison of LIN6 and Mobile IPv6

19.7.1   Single Point of Failure: Mapping Agent and HA

An LIN6 node needs a mapping agent to acquire mapping. Consequently, the mapping agent is the single point of failure in LIN6. On the other hand, an HA is the single point of failure in Mobile IPv6.

In order to enhance fault tolerance, managing of such agents in a distributed manner is a prerequisite. For Mobile IPv6, however, the location of the HA is dependent on the home address of a mobile node—that is, the HA must be placed on the subnetwork of the home address. It is possible to place multiple HAs in the same subnetwork, but placing them on a more distributed scale is very difficult. Thus, enhancing fault tolerance is difficult in Mobile IPv6.

On the other hand, the location of MAs is completely independent of the node identifier; hence, designated MAs can be placed at any point in the network. Thus, LIN6 is more fault-tolerant than Mobile IPv6.

19.7.2   Overhead of Packet Header Length

Mobile IPv6 needs at least one extra destination option, a home address destination option, to inform the recipient of that packet of the mobile node's home address. In addition, to route a packet to the mobile node through the optimal route, which means the packet is not routed through the HA, Mobile IPv6 uses the routing header, which is another IPv6 extension header. If two Mobile IPv6 nodes communicate with each other through an optimal route, both the routing header and the home address option must be added to all of the packets that are exchanged between the nodes. The header length of the home address option is at least 20 bytes and the routing header is at least 24 bytes. A 44-byte packet header overhead for every packet is incurred when Mobile IPv6 nodes communicate with each other.

In recent years, VoIP, a new Internet application, has been attracting a great deal of attention. Since VoIP requires exceedingly low delay, it uses small packets, and much work [14] has been done on compressing not only the payload of packet but also the headers to reduce delay. The overhead of packet header length that Mobile IPv6 bears could be a serious problem in such applications.

In contrast to Mobile IPv6, LIN6 does not incur any overhead on protocol header length. An LIN6 address contains not only a node identifier but also an interface locator. As a result, LIN6 supports the concept of separation of identifier and locator, but needs no extra protocol headers or options.

19.7.3 End-to-End Communication

In Mobile IPv6, if a node does not have the binding of a target correspondent mobile node, the packet destined for the mobile node is routed to the home address and the HA intercepts this packet and forwards it to the mobile node through IPv6-in-IPv6 tunneling. Thus, Mobile IPv6 does not always guarantee end-to-end communication.

On the other hand, in LIN6, mapping between the LIN6 generalized ID and the LIN6 address is performed only at the end nodes, once the mapping is obtained from the mapping agent. Thus, LIN6 does not require additional processing by intermediate nodes for packet delivery and always guarantees end-to-end communication without using tunnels.

When packets are not transferred end-to-end, issues such as decline in reliability arise. In Mobile IPv6, if the optimal route is not used, three paths of communication are required for it to function correctly. That is, a path from the mobile node to the correspondent node, from the correspondent node to the HA, and from the HA to the mobile node. When end-to-end communication is guaranteed, a single path between the mobile node to the correspondent node is all that is required.

19.8 Summary

This chapter introduced a new network architecture called LINA. LINA employs the notion of separating the identifier and locator in order to support mobility and integrates the conceptually separated sublayers into a single packet header based on the embedded addressing model. This embedded addressing model enables LINA to avoid packet length overhead. This chapter also introduced LIN6, which is a realization of LINA on IPv6 that provides node mobility to IPv6. LIN6 is compatible with traditional IPv6 and does not affect existent IPv6 infrastructure. This chapter also showed that LIN6 has several advantages when compared to Mobile IPv6 in terms of header length overhead and system stability.

Mobility handling of LIN6 was analyzed using the prototype implementation in an experimental network. LIN6 provides two distinct handoff mechanisms, mapping update and mapping refresh request, to accommodate various security needs, and both mechanisms were evaluated. The results showed that the mapping update and mapping refresh request can support handoff on average at 46.4 ms and 49.7 ms, respectively. This is acceptable to many applications such as VoIP if suitable error correction mechanisms are used. The experiments also showed that parts of current IPv6 specifications such as DAD and the random delay in routing advertisement do not adequately support fast mobility.

Topics for future work include performance comparisons of Mobile IPv6 and LIN6 and experiments in wide-area commodity networks with practical applications such as videoconferencing.

REFERENCES

[1]   Deering, S., and R. Hinden, *Internet Protocol, Version 6 (IPv6) Specification*, RFC 2460, December 1998.

[2]   Johnson, D. B., C. Perkins, and J. Arkko, *Mobility Support in IPv6*, Internet-draft, June 2002.

[3]   Kunishi, M., et al., "LIN6: A New Approach to Mobility Support in IPv6," *Proceedings of the Third International Symposium on Wireless Personal Multimedia Communications*, November 2000.

[4]   Xerox, *Internet Transport Protocols. XSIS 028112*, 1981.

[5]   Teraoka, F., et al., "VIP: A Protocol Providing Host Mobility," *Communications of the ACM*, Vol. 37, No. 8, August 1994, pp. 67–75.

[6]   Teraoka, F., "Mobility Support with Authentic Firewall Traversal in IPv6," *IEICE Transaction on Communications*, Vol. E80-B, No. 8, August 1997.

[7]   Hinden, R., M. O'Dell, and S. Deering, *An IPv6 Aggregatable Global Unicast Address Format*, RFC 2374, July 1998.

[8]   IEEE, "Guidelines for 64-Bit Global Identifier (EUI-64) Registration Authority," 1997.

[9]   Kent, S., and R. Atkinson, *Security Architecture for the Internet Protocol*, RFC 2401, November 1998.

[10] Conta, A., and S. Deering, *Internet Control Message Protocol (ICMPv6) for the Internet Protocol Version 6 (IPv6) Specification*, RFC 2463, December 1998.

[11] Narten, T., E. Nordmark, and W. Simpson, *Neighbor Discovery for IP Version 6 (IPv6)*, RFC 2461, December 1998.

[12] Thomson, S., and T. Narten, *IPv6 Stateless Address Autoconfiguration*, RFC 2462, December 1998.

[13] Kent, S., and R. Atkinson, *IP Authentication Header*, RFC 2402, November 1998.

[14] Douskalis, B., *IP Telephony*, Upper Saddle River, NJ: Prentice Hall, 2000.

CHAPTER 20

Distributed Signaling and Routing Protocols in iCAR*

Hongyi Wu, Chunming Qiao, and Sudhir S. Dixit

......

20.1   Introduction

The cellular concept was introduced for wireless communication to address the problem of having scarce frequency resources. It is based on the subdivision of geographical area to be covered by the network into a number of smaller areas called cells, and the use of a low-power transmitter to serve each cell. The limited transmission range of the transceivers allows the frequency to be reused in cells far away from each other, thus increasing the system's capacity. In order to avoid potential channel interference resulting from frequency reuse, however, it creates cell boundaries that prevent the channel resource of a cell from being accessible to users in another cell. Thus, a *mobile host* (MH) in a system can use only the *cellular bandwidth* (CBW) of the BTS located in the same cell, which is a subset of the CBW available in the system. There is no access to CBW in other cells by the MH, which limits the bandwidth usage and consequently the system capacity. More specifically, when a connection request arrives in a congested cell that does not have enough free CBW, this request will be rejected even though free CBW is available in other cells in the system.

Qiao, Wu, and Tonguz [1] introduced the *Integrated Cellular and Ad Hoc Relaying* (iCAR) System to address the congestion problem in cellular systems and the scalability problem in ad hoc network, and to provide interoperability for heterogeneous networks. In iCAR, an MH is allowed to use the bandwidth available in a nearby cell by relaying through *ad hoc relaying stations* (ARSs), which are placed at strategic locations in a system. By using ARSs, it is possible to divert traffic from one (possibly congested) cell

* This research is in part supported by NSF under the contract ANIR-ITR 0082916 and Nokia Research Center.

to another (noncongested) cell. This helps to circumvent congestion and makes it possible to maintain (or handoff) connections involving MHs that are moving into a congested cell, or to accept new requests for connections involving MHs that are in a congested cell. Both the analysis and simulation [1–3] have shown that iCAR can effectively balance the traffic load among cells, which, in turn, breaks the cell boundaries and leads to significant increase in system capacity over a conventional cellular system.

The explosive growth of the Internet, and in particular, the introduction of IPv6 resulting in a huge address space and a phenomenal increase in the number of mobile users and wireless nodes that all have their own globally unique IP addresses, has stimulated the interest in the development of packet switching data services in existing and future cellular systems. While GPRS and the 3G wireless systems can support packet access in addition to conventional voice traffic, it is desirable to have a seamlessly converged next-generation system that is based on IP techniques and supports connection-oriented services at the same time. This is because IP is a connectionless protocol, and as such it is difficult for it to meet the QoS requirements of real-time traffic.

In order to introduce IP into wireless mobile networks, carriers and infrastructure providers face a major challenge in meeting the increased bandwidth demand of mobile Internet users, and the bursty and unbalanced IP traffic will exacerbate this problem of limited capacity in existing cellular systems. It is expected that congestion will occur in some cells during peak usage hours or when providing emergency rescue services at a disaster site. iCAR, with its ability to leverage both the cellular and ad hoc relaying techniques to increase a system's capacity, is a promising evolution path for the cellular systems. Nevertheless, in order for iCAR to support real-time IP-based applications in a wireless mobile environment, efficient signaling and routing protocols are needed to set up a relaying path with reserved bandwidth, so as to guarantee the required QoS.

Although several connection-oriented signaling protocols, such as RSVP and MPLS, have been proposed for wired data networks, very little research has been done on QoS-capable connection-oriented packet switching in the wireless networks, especially for integrated networks with heterogeneous technologies such as cellular and ad hoc relaying. This chapter presents a distributed signaling and routing protocol for establishing QoS (i.e., bandwidth)-guaranteed connections in iCAR. When using such a distributed protocol, each ARS serves as a router and is responsible for relaying route selection. The *packet switching controller* (PSC)—which may be the BSC in GPRS or the radio network controller RNC in 3G systems—only maintains and provides the current bandwidth usage information of the BTSs, and is not involved in routing. Intuitively, this allows the protocol to scale well to a large number of ARSs and MHs, requiring only minor modifications to the components of the existing cellular system. The performance of iCAR using the proposed protocol is evaluated via simulations. The results show that iCAR can effectively improve the system performance in terms of request rejection rate over a conventional cellular system, albeit with only a limited increase in signaling overhead.

The rest of this chapter is organized as follows. Section 20.2 reviews the basic operations and key benefits of the iCAR system. Section 20.3 describes the proposed distributed signaling protocol, including connection setup and release as well as the ARS routing. Section 20.4 discusses the simulation and the performance results of the protocols in terms of request rejection rate and signaling overhead. Finally, Section 20.5 concludes the chapter and discusses future work.

20.2   An Overview of the iCAR System

This section describes the basic operations and main benefits of iCAR (see [1] for more details). For simplicity, the following presentation will focus on cellular systems where each BTS is controlled by a PSC. Major differences between BTSs and the proposed ARSs are as follows. Once a BTS is installed, its location is fixed. An ARS, on the other hand, is a wireless communication device deployed by a network operator. It may, under the control of a PSC, have limited mobility in order to adapt to varying traffic patterns[1] and communicate *directly* with an MH, a BTS, or another ARS via the appropriate air interfaces. In addition, it has a much lower complexity and fewer functionalities as compared to a BTS.

20.2.1   Basic Operations

An example of relaying is illustrated in Figure 20.1 where MH X in cell B (congested) communicates with the BTS in cell A (or BTS A, which is not congested) through two ARSs (there will be at least one ARS along which a relaying route is set up). Note that each ARS has two air interfaces: the C (cellular) interface for communicating with a BTS and the R (relaying) interface for communicating with an MH or another ARS. Also, MHs should have two air interfaces: the C interface for communicating with a BTS, and the R interface for communicating with an ARS. For example, the C interface may operate at or around 1,900 MHz in the PCS, and the R

---

1   This chapter, however, only considers static ARSs. The benefit of ARSs with limited mobility will be examined in future work.

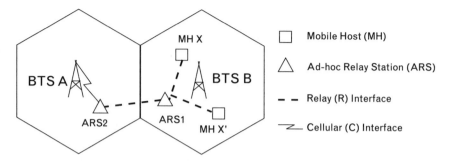

interface at 2.4 GHz in the unlicensed ISM band.[2] The R interface and the MAC protocol used are similar to that used in WLANs or ad hoc networks (see, for example, [4–6]). Note that because multiple ARSs can be used for relaying, the transmission range of each ARS using its R interface can be much shorter than that of a BTS, which implies that an ARS can be much smaller and less costly than a BTS. At the same time, it is possible for ARSs to communicate with each other and with BTSs at a higher data rate than MHs can, due to limited mobility of ARSs and specialized hardware (and power source).

The rest of this section will review two basic relaying operations in iCAR: primary and secondary relaying.

20.2.1.1   Primary Relaying

In an existing cellular system, if MH X is involved in a new connection request (as a sender or receiver) but it is in a congested cell B, the new request will be rejected. In iCAR, the request may not have to be rejected. More specifically, MH X, which is in the congested cell B, can switch over to the R interface to communicate with an ARS in cell A, possibly through other ARSs in cell B (see Figure 20.1). We call this strategy primary relaying.

With primary relaying, MH X can communicate with BTS A, albeit indirectly (i.e., through relaying). Hereafter, the process of changing from the C interface to the R interface (or vice versa) is referred to as switchover, which is similar to (but different from) frequency-hopping [7, 8]. Of course, MH X may also be relayed to another nearby noncongested cell other than cell A. A relaying route between MH X and its corresponding (i.e., sender or receiver) MH X′ may also be established (in which case, both MHs need to switch over from their C interfaces to their R interfaces), even though the probability that this occurs is typically very low.

---

2   It is possible, however, to set aside a (subset of) cellular-band data channel(s) for relaying in order to avoid the need to equip an existing MH with the additional R interface.

20.2.1.2   Secondary Relaying

If primary relaying is not possible, because (e.g., in Figure 20.1) ARS 1 is not close enough to MH X to be a proxy (and there are no other nearby ARSs), then one may resort to secondary relaying so as to free up the amount of requested bandwidth from BTS B for use by MH X. Two basic cases are illustrated in Figure 20.2, where MH Y denotes any MH in cell B that is currently involved in an active connection. More specifically, as shown in Figure 20.2(a), one may establish a relaying route between MH Y and BTS A (or a BTS in any other cell). In this way, after MH Y switches over, the bandwidth used by MH Y can now be used by MH X. Similarly, as shown in Figure 20.2(b), one may establish a relaying route between MH Y and its corresponding MH Y′ in cell B or in cell C, depending on whether MH Y is involved in an intracell connection or an intercell connection. Note that congestion in cell B implies that there are a lot of ongoing connections (involving candidates like MH Y); hence, the likelihood of secondary relaying Figure 20.2 should be better than that of primary relaying (Figure 20.1). In addition, although the concept of having an MH-to-MH connection via ARSs only (i.e., when no BTSs are involved) is similar to that in ad hoc networking, a distinct feature (and advantage) of the proposed integrated system is that a PSC can perform (or at least assist in performing) critical connection management functions such as authentication, billing, and locating the two MHs and finding and/or establishing a relaying route between them. Such a feature is also important to ensure that switchover of the two MHs (this concept is not applicable to ad hoc networks) is completed fast enough so as not to disconnect the ongoing connection involving the two MHs or cause severe QoS degradation (even though the two MHs may experience a glitch or jitter).

FIGURE 20.2
Secondary relaying to free up bandwidth for MH X: (a) MH Y to BTS A, and (b) MH Y to MH Y′.

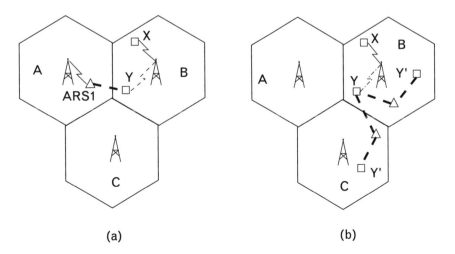

(a)   (b)

20.3   Signaling and Routing Protocol

This section describes the proposed distributed signaling and routing protocol for iCAR to establish and release bandwidth-guaranteed connections possibly involving ARS relaying. Such a protocol aims to address the QoS need of IP-based real-time applications. Since iCAR integrates the cellular system and ad hoc networks, the signaling protocol is a hybrid of those two signaling systems, requiring a novel design to search for relaying routes, for example, in order to achieve a high efficiency.

20.3.1   Connection Setup

Figures 20.3 and 20.4 illustrate the protocol for connection setup. When an MH (e.g., MH_0) needs to set up a QoS-guaranteed connection to the core network, it sends a CBW request message *CBW_REQ* to the BTS located in the same cell, called home BTS and denoted by H_BTS (see step 1 in Figure 20.3). H_BTS will forward the message to PSC (see step 2 in Figure 20.3), which is responsible for admission control and bandwidth allocation. If there is enough CBW available, PSC responds with a CBW allocation message *CBW_AL* to H_BTS, which in turn generates a positive acknowledge *CBW_ACK* to MH_0 (see steps 3a and 4a in Figure 20.3), and the connection request is satisfied. Otherwise, MH_0 will receive a negative acknowledge *NAK*. So far, this is the same as the process in a conventional cellular system.

20.3.1.1   Signaling for Primary Relaying

What is different in iCAR is that when there is not enough CBW available, MH_0 will also receive the CBW information of other BTSs in the system from the PSC (see steps 3b and 4b in Figure 20.3) and try primary relaying by switching to the R interface. More specifically, it will contact the nearby

FIGURE 20.3
Distributed signaling protocol for connection request via primary relay.

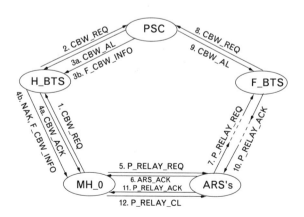

FIGURE 20.4
*Distributed signaling
protocol for connection
request via secondary
relay.*

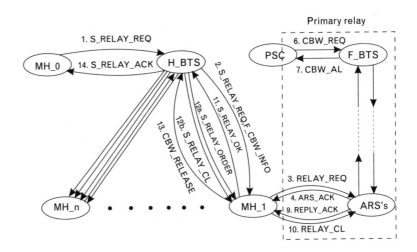

ARSs by broadcasting a primary relaying request message P_RELAY_REQ, which includes the addresses of a set of nearby BTSs with available CBW, and start a timer T1 at the same time (see step 5 in Figure 20.3). If the MH is not covered by any ARSs, it will not receive any acknowledgment before T1 times out, and thus, primary relaying will fail. In this case, the request will be rejected unless secondary relaying is tried and succeeds (to be discussed later). When a nearby ARS receives the relaying request, it will look up its routing table and respond with a positive ARS_ACK if at least one of the desirable BTSs is reachable (see step 6 in Figure 20.3). After receiving an ARS_ACK, MH_0 will clear T1 and start another timer T2, which has longer timeout interval. Meanwhile, the ARSs that have responded positively try to relay the request message to one of the nearby (or foreign) BTSs and reserve bandwidth along the relaying path, using a routing protocol to be described in Section 20.3.2. In the event of any resource shortage, however, primary relaying will be aborted. For example, between MH_0 and a nearby ARS, or between two ARSs along the relaying path, a certain amount of *relaying bandwidth* (RBW) needs to be reserved. When there is not enough RBW at an ARS, no relaying traffic could pass through it. In addition, since each ARS has limited CBW, primary relaying will also fail if the ARS serving as the last hop of the relaying path runs out of available CBW and thus cannot communicate with an intended foreign BTS.

If the relaying request message is eventually relayed to a foreign BTS (denoted by F_BTS in Figure 20.3) that has free CBW, F_BTS will reserve the CBW and send back a positive P_REPLY_ACK (see steps 7–11 in Figure 20.3), which may contain bandwidth and hop count information on the path. If MH_0 receives more than one P_REPLY_ACK message before T2 times-out, it will choose one route to transmit data, and broadcast a relay cancel message P_RELAY_CL to all other routes to release the reserved bandwidth.

Several approaches for route selection by MH_0 involving trade-offs between bandwidth efficiency and load balancing can be considered:

- *Shortest path/minimum distance-bandwidth usage:* The distance-bandwidth is defined to be the product of the requested bandwidth and the number of hops of the relaying route. This approach minimizes the relaying bandwidth used for data transmission by selecting a relaying route with a minimum number of hops (or ARSs).

- *Fastest path/minimum delay:* This approach minimizes the propagation delay on the relaying path (as well as the latency in selecting a route) by accepting the first relaying route for which MH_0 receives the positive *P_RELAY_ACK*.

- *Widest bottleneck:* The bottleneck value of a relaying path is defined to be the minimum free bandwidth (CBW or RBW) over all hops along the relaying path. In this approach, the path with the largest bottleneck value will be chosen for relaying.

- *Widest path:* This approach is based on the inverse of the width of a relaying path, which is defined to be $\sum_{i=0}^{N} 1/Free\_BW$ in which $Free\_BW$ is the free bandwidth (CBW, RBW) at each hop, and N is the total number of hops. The lower the value, the wider the path.

20.3.1.2 Signaling for Secondary Relaying

If primary relaying fails, MH_0 will try secondary relaying by sending a *S_RELAY_REQ* message to H_BTS (see step 1 in Figure 20.4), which keeps the addresses of the active MHs within its cell coverage in a table MH_TB. After receiving *S_RELAY_REQ*, H_BTS will contact the active MHs by broadcasting a *S_RELAY_REQ* message that includes the addresses of a set of BTSs with enough free CBW (see step 2 in Figure 20.4) and will start a timer T3. Whenever a MH (e.g., MH_1) receives the *S_RELAY_REQ* message, it will try to set up a relaying path in the same way as that used for primary relaying (see steps 3–10 in Figure 20.4). After it receives a positive *REPLY_ACK*, the MH will send an acknowledge *S_RELAY_OK* to H_BTS (see step 11 in Figure 20.4), which may receive multiple such acknowledges before T3 times out, and will send *S_RELAY_ORDER* to the best one to confirm the usage of that relaying path (see step 12a in Figure 20.4). In addition, it will broadcast a *S_RELAY_CL* message to all other MHs and ARSs to cancel the reserved bandwidth (see step 12b in Figure 20.4). After receiving a *S_RELAY_ORDER* message, the MH will switch to the relaying path and release its CBW, which may be in turn used by MH_0. If none of the MHs in MH_TB can do a successful relaying, H_BTS sends a *NAK* to MH_0 and this request will be blocked.

So far we have discussed the scenario where a MH is the sender. If the MH is a receiver, the request message will be routed from PSC to the corresponding BTS (e.g., BTS_Y) in which the MH is located. When BTS_Y is congested, it will request that the MH find a relaying path to another BTS that has free CBW. If either primary or secondary relaying is successful, BTS_Y will inform PSC to reroute the incoming data (either for this MH if primary relay is applied, or for another MH if secondary relay is applied) to the new BTS.

20.3.2 ARS Routing

This section discusses multihop relaying among ARSs (Figure 20.5). The ARSs in iCAR form a special type of ad hoc networks, which have the following properties. First, a network of ARSs is usually small in size (dozens of nodes), and there are multiple such networks separated from each other covering different cells. Only when the ARSs cover the entire cellular system does the ARS network become a big, single ad hoc network. Second, ARSs have low (or no) mobility and are controlled by the system, while most of current research work about ad hoc networks focus either on large systems with hundreds and thousands of nodes (e.g., sensors) or small systems with highly unpredictable end user–controlled mobility. Due to these special properties, ARS routing in iCAR will be different from routing in pure mobile ad hoc networks (Figure 20.5).

In addition, none of the existing well-known wireless ad hoc network and wired network routing protocols can be used in iCAR without any modifications. More specifically, the reactive protocols for ad hoc networks (e.g., AODV [5], DSR [6], and so on) flood the route discovery message, which may waste system resources by reserving bandwidth in many BTSs

FIGURE 20.5
A example of ARS routing, where two relaying routes are established.

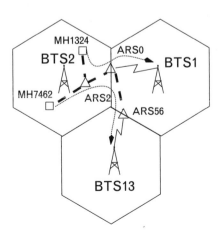

and along many relaying routes. On the other hand, most proactive routing protocols only maintain a shortest or most stable path [4]. But in iCAR, a blocked MH wants to find relaying paths to any nearby BTS, and hence multiple choices are desirable for traffic engineering purposes (see the route selection approaches described in Section 20.3.1). Here, we propose a modified distance vector protocol for ARS routing. The main motivations behind choosing distance vector-based routing are that each ARS only needs to exchange information with its immediate neighbors and the ARS network is fairly stable so the slow-convergence of distance vector-based routing is not an serious issue. Nevertheless, route metrics need be augmented as discussed below.

Each ARS maintains a routing table, which includes the addresses of reachable BTSs, the next hops to reach the BTSs (e.g., the addresses of other ARSs or the BTSs themselves), and the total number of hops to reach those BTSs. An example of an ARS routing table is shown in Figure 20.6(a). Since the number of BTSs—especially the reachable BTSs by an ARS via relaying in a system—is not usually large, it is feasible to include all reachable BTSs in a single table. In addition to the routing table, an ARS also maintains the free bandwidth information (*RBW*, *CBW*) of itself and the neighboring ARSs [see Figure 20.6(b)].

When an ARS powers on, it detects the current reachable BTSs and initializes the routing table by putting their addresses into it [see the first two entries of the routing table in Figure 20.6(a)]. Then, the ARS periodically broadcasts update information including its routing table entries [in Figure 20.6(a)] as well as the bandwidth information [Figure 20.6(b)] to neighboring ARSs. Whenever an ARS receives an update message from a neighbor, it adds the reachable BTSs of the neighboring ARS into its routing table and records the neighboring ARS as the next hop. If two entries have the same BTS address and next hop, only the one with the shorter distance in terms of the number of hops will be kept. If there are N cells in the system and an ARS has M ARS neighbors, the maximum number of entries in its routing

Reachable BTS	Next Hop	Distance
BTS_1	BTS_1	1
BTS_2	BTS_2	1
BTS_2	ARS_2	2
BTS_13	ARS_56	2
......	......	......

(a)

ARS_No	CBW(Units)	RBW(Units)
......	......	......
ARS_56	2	9
......	......	......
ARS_2	3	8
......	......	......

(b)

Conn_ID	Next_Hop	Previous_Hop
......	......	......
MH_7462	ARS_56	ARS_2
......	......	......
MH_1324	BTS_1	ARS_2
......	......	......

(c)

FIGURE 20.6   *(a) An example of ARS routing table; (b) an example of a bandwidth information table assuming that initially there are 10 units of RBW and 3 units of CBW; and (c) an example of ARS routing buffer; all maintained at ARS_0, which is assumed to be along the border of two cells in which BTS1 and BTS2 are located (see Figure 20.5).*

table is N $(M + 1)$. An ARS will also update the bandwidth information of neighbors. The frequency of sending update messages can be adjusted according to the signaling traffic load, and it is usually low because the ARSs have low (or no) mobility.

Whenever an ARS receives a relay request message *RELAY_REQ*, which includes the source MH address and a set of foreign BTS addresses, it looks up the routing table. If a foreign BTS is found to be reachable and free CBW (if the next hop is BTS) or RBW (if the next hop is ARS) is available, it reserves RBW/CBW, stores the routing information into what we call the routing buffer [as shown in Figure 20.6(c)], and relays the message to the next hop. The relaying path in the routing buffer is identified by the connection ID, which is equal to the address of the requesting MH in this example. If there is more than one choice of next hop, the one with the shorter path is chosen. If more than one shortest path exists, the ARS will relay the message to the next hop with the most free bandwidth. When the destination BTS is reached, an acknowledgment message *ACK* containing the total number of hops, the bottleneck link bandwidth, and/or the width of the relaying path (as defined in Section 20.3.1) will be sent back to the source MH. Here, the routing algorithm finds a shortest path and as a tie-breaker, chooses a wider next hop. The source MH will select either a shortest, fastest, widest bottleneck, or widest path later on. Otherwise, if the message cannot be relayed further, the ARS sends a *NAK* to the previous hop. When an ARS receives the *NAK*, it removes the corresponding entry in the routing buffer, releases the reserved RBW/CBW, and looks up the routing table again and tries to find other possible routes. If the route is already in the routing buffer, the signaling message or data packets will be relayed immediately (see further discussion in the next section about route caching).

20.3.3 Connection Release

When data transmission is completed, the MH sends a connection release message to either the BTS (if without relaying) or the ARS (if with relaying) (Figure 20.7). The ARSs on the relaying path release the reserved RBW/CBW and forward the release message to the BTS, which provides CBW, and the latter will release the bandwidth and update the bandwidth information to the PSC. Although the bandwidth is released, the routes in the relaying buffer will be kept and attached with a timer, which is reset

FIGURE 20.7
Signaling protocol for connection release.

whenever the route becomes active. When the timer times-out, the route will be flushed out from the routing buffer.

20.4   Simulation Results and Discussion

To evaluate the system performance in terms of request rejection probability and signaling overhead, a simulation model has been developed using PARSEC language [9] and GloMoSim simulator [10]. The simulated system includes a cell X and its neighboring cells (see Figure 20.8), which are controlled by a PSC. The cells are modeled as hexagons with the center to vertex distance of 2 km. For simplicity, each connection is assumed to require 1 unit bandwidth, while each BTS has 50 units of CBW. An ARS covers an area whose radius is 500m, and is randomly placed in the boundary region of cell X, which is bounded by two dashed circles as shown in Figure 20.8. Figure 20.8(a) shows the case where 30 ARSs are deployed in the area between two dashed circles whose radii are 2,500m and 1,500m, respectively. Figure 20.8(b) shows that there are 60 ARSs between the two dashed circles whose radii are 2,500m and 1,000m, respectively. Note that since the ARSs are randomly distributed, not all of them result in effective coverage. In particular, some ARSs cannot relay traffic from one cell to another either directly or through other ARSs. The signaling protocol discussed in Section 20.3 has been implemented with various parameters, such as the number of ARSs, the amount of CBW and RBW of each ARS, and the traffic intensity in cell X. Here, the traffic intensity is in Erlangs, which is the product of request arrival rate (Poisson distributed) and the holding time

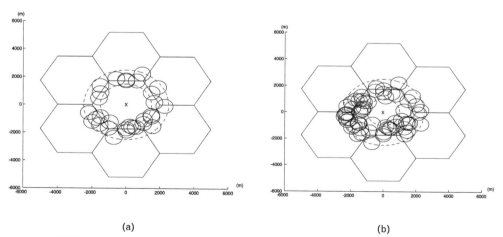

           (a)   (b)

FIGURE 20.8   *ARS placement in the simulated system. The cell in the center is the cell being considered (cell X): (a) 30 ARSs and (b) 60 ARSs.*

(exponentially distributed). In order to obtain converged statistical results, 1,600 MHs are simulated and randomly placed in the system. The simulation runs for 100 hours for each traffic intensity before the results are collected.

Although the results are not shown here, the path selection schemes discussed in Section 20.3 have only little impact on the performance, and none of them performs definitely better than the others. Nevertheless, in most situations, the shortest path/minimum bandwidth approach achieves good results. Therefore, in the following discussion, it is adopted as the default approach. The request rejection rates for primary and secondary relaying are shown in Figure 20.9. The performance improvement of an iCAR system in terms of the request rejection rate is due to the relaying ability of a system, which in turn depends on the effective ARS coverage and the amount of relaying bandwidth. With the increase in the number and coverage of ARSs, the rejection rate of primary and secondary relaying is reduced significantly, as expected. In addition, increasing the bandwidth will also help reduce the request rejection rate. Specifically, increasing CBW of ARSs from 1 unit to 3 units may reduce the request rejection rate by about 50% in secondary relaying. Nevertheless, ARS coverage has a stronger impact on the system performance than relaying bandwidth.

The signaling overhead of PSC is shown in Figure 20.10. (The results for ARSs, BTSs, and MHs show similar trends and are thus omitted.) As can be seen, the number of signaling messages increases with the traffic intensity since a higher traffic intensity results in more relaying requests, and therefore more signaling overhead. It is interesting to note that the primary relaying increases the average overhead by a little but not much, because only the MHs associated with the blocking requests will generate signaling messages. On the other hand, secondary relaying results in an exponential increase in

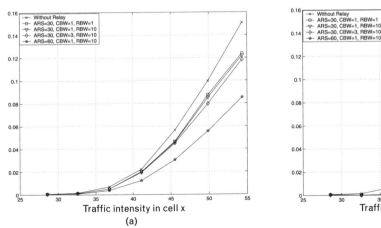
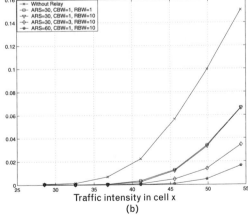

FIGURE 20.9 *Request rejection rate in cell X: (a) primary relaying, and (b) secondary relaying.*

signaling overhead since many MHs and ARSs are involved in each secondary relaying request.

The number of ARSs is another important factor that affects the signaling overhead. As can be seen from Figure 20.10, the number of signaling messages increases with the number of ARSs because, with more ARSs, more MHs whose requests have been rejected initially by PSC will try primary or secondary relaying. In addition, increasing the number of ARSs will not only increase the ARS coverage but also (in this simulation) increase the density of the ad hoc network (see Figure 20.8), and consequently increase the connectivity of ARSs. Accordingly, ARSs will search for many more possible relaying routes for secondary relaying in a densely connected network (e.g., with 60 ARSs) than in a sparsely connected network (e.g., with 30 ARSs), although on the positive side, this will also result in a much higher possibility of successfully finding a relaying route.

Although separate results for connection setup and release are not shown here, many of the extra signaling messages in iCAR are introduced by the search for relaying routes, especially in secondary relaying. Accordingly, the connection release requests result in only a small portion of total signaling overhead. In addition, it should be noted that most of the signaling messages are short and therefore may be squeezed into existing signaling or data packets without significantly increasing additional bandwidth requirement for signaling.

FIGURE 20.10
*Signaling overhead in
PSC per successful
connection.*

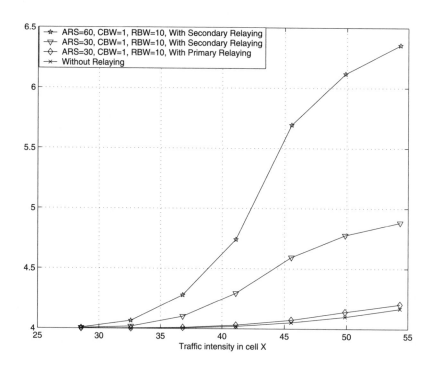

20.5 Conclusion and Future Work

This chapter has presented distributed signaling and routing protocols for establishing QoS (e.g., bandwidth)-guaranteed connections in the next-generation wireless system called iCAR. In particular, it has discussed the distributed signaling and routing protocol for establishing relaying routes from a MH to a BTS in a nearby cell via ARSs. The performance of the proposed protocol has been evaluated in terms of request rejection rate and signaling overhead through simulations. The results have shown that iCAR, with the proposed distributed signaling protocols, can effectively reduce the request rejection rate with a limited increase in signaling overhead when compared with a conventional cellular system.

This work represents a first step in evolving cellular systems to support connection-oriented data traffic. It is to be followed by further work, which will evaluate the connection latency and the soft capacity improvement in the CDMA systems, and enhance the signaling and routing protocol by allowing the ARS to maintain and utilize the relaying bandwidth of not only the neighboring ARSs but also those multihop away, and taking advantage of the assistance of PSC. In addition, the ARS mobility will be studied and supported by the protocol.

REFERENCES

[1] Qiao, C., H. Wu, and O. Tonguz, "iCAR: An Integrated Cellular and Ad-Hoc Relay System," *IEEE International Conference on Computer Communications and Networks (IC3N'00)*, Las Vegas, NV, October 2000, pp. 154–161.

[2] Wu, H., et al., "An Integrated Cellular and Ad-Hoc Relaying System: iCAR," *IEEE Journal on Selected Areas in Communications*, Vol. 19, No. 10, October 2001, pp. 2105–2115.

[3] Wu, H., and C. Qiao, "Modeling iCAR System Via Multi-Dimensional Markov Chain," to appear in *Mobile Networks & Applications Journal (MONET)*, 2003.

[4] Das, S., et al., "Comparative Performance Evaluation of Routing Protocols for Mobile, Ad Hoc Networks," *7th Int. Conf. on Computer Communications and Networks (IC3N)*, 1998, pp. 153–161.

[5] Perkins, C., and E. Royer, "Ad Hoc On-Demand Distance Vector Routing," *Proceedings of IEEE WMCSA'99*, 1999, pp. 90–100.

[6] Johnson, D., and D. Maltz, "Dynamic Source Routing in Ad Hoc Wireless Networks," *Mobile Computing*, Vol. 5, 1996, pp. 153–181.

[7] Rappaport, T., *Wireless Communications Principle and Practice*, Englewood Cliffs, NJ: Prentice Hall, 1996.

[8] Garg, V., and J. Wilkes, *Wireless and Personal Communications Systems*, Englewood Cliffs, NJ: Prentice Hall, 1996.

[9] Bagrodia, R., et al., "Parsec: A Parallel Simulation Environment for Complex Systems," *Computer*, Vol. 31, No. 10, October 1998, pp. 77–85.

[10] Zeng, X., R. Bagrodia, and M. Gerla, "GloMoSim: A Library for Parallel Simulation of Large-Scale Wireless Networks," *Proc. Workshop on Parallel and Distributed Simulation*, 1998, pp. 154–161.

CHAPTER 21

Reducing Link and Signaling Costs in Mobile IP*

Ian F. Akyildiz and Young J. Lee

21.1 Introduction

IP mobility support is becoming very important since the Internet is growing quickly and wireless communications technology is advancing. The basic Mobile IP [1] was proposed to provide IP mobility support. It introduces three new functional entities: *mobile node* (MN), HA, and FA.

Although the basic Mobile IP proposes a simple and elegant mechanism to provide IP mobility support, there is a major drawback, where each packet destined to the MN must be routed through the HA along an indirect path. This is known as the triangle routing problem.

The so-called *Route Optimization Protocol* (RO) [2] was proposed by the IETF to solve the triangle routing problem. When packets are sent from a *correspondent node* (CN) to an MN, if the CN has a binding cache entry for the MN, they can be directly tunneled without the help of the HA to the COA indicated in the binding cache. In this scheme, route optimization is achieved by sending binding update messages from the HA to the CN. In Mobile IPv6 [3] binding update messages are sent from the MN to the CN. Moreover, the FA smooth handoff scheme [2] allows packets in flight or sent based on the out-of-date binding cache to be forwarded directly to the MN's new COA.

The major drawback of the IETF RO [2] is that there are additional control messages such as binding warning and binding update, which cause communication overhead and introduce high signaling and processing load on the network and on certain nodes. Some mechanisms such as the local anchoring scheme [4], regional registration [5], and the hierarchical

* This work is supported by the NSF under Grant No. CCR-99-88532.

management scheme [6] have recently been proposed to reduce signaling costs and communication overhead.

This chapter is motivated by the question: "Does route optimization need to be performed whenever an MN hands off, and a previous FA receives packets destined to the MN? What if route optimization is performed only when certain conditions are satisfied by doing that?" If the route optimization is not performed as often as it is in the IETF RO, signaling and processing load will be reduced. This question naturally leads to two issues:

1. How to guarantee that the packets destined to the MN are routed temporarily along a suboptimal path without performing route optimization;

2. When to perform route optimization.

For the first issue, the FA smooth handoff scheme [2] gives an answer. By keeping the previous FAs serving as forwarding pointers until route optimization is performed, it can be guaranteed that IP datagrams are routed along a suboptimal path. This mechanism is named a *route extension* because it simply extends the routing path from the previous FAs to the current FA. For example, in Figure 21.1, FA1 forwards packets to FA2, FA2 forwards them to FA3, and finally the packets are delivered to the MN through the FA3.

This chapter focuses on the second issue. Although the route optimization increases the network utilization by allowing packets to be routed along an optimal path from the CN to the MN, it will also increase the signaling load of the network and the processing load of certain nodes. One knows from this fact that there is a trade-off between the network resources consumed by the routing path and the signaling and processing load incurred by

FIGURE 21.1
Route extension.

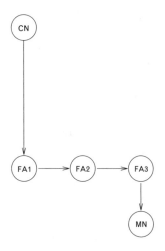

the route optimization. The decision of when to perform route optimization needs to be considered based on the following:

1. The network resources consumed by the routing path;

2. The signaling and the processing load;

3. The QoS requirements.

In IETF RO, when an FA receives a tunneled packet, and if it has the binding cache entry for the MN and does not have the visitor list entry for this MN at that point, the previous FA then sends a binding warning message to the MN's HA advising it to send a binding update message to the CN (Figure 21.2).

Regarding this FA-initiated route optimization, it is proposed that the previous FA should not send the binding warning message to the HA. In the new scheme, it is proposed that route optimization should be initiated by the current FA.

A mathematical model is developed to determine when to perform route optimization. Link cost and signaling cost functions are introduced to capture the trade-off. The objective of this chapter is to find a cost-efficient scheme for route optimization that minimizes the total cost function defined as the sum of the link and signaling cost functions. The simulation results show that the proposed scheme significantly reduces the signaling costs caused by IETF RO and provides the lowest total costs.

The rest of this chapter is organized as follows. In Section 21.2, the mathematical model is described and the decision model is provided. In Section 21.3, the performance evaluation is presented. Finally, Section 21.4 concludes the chapter.

FIGURE 21.2
*Route extension
and RO.*

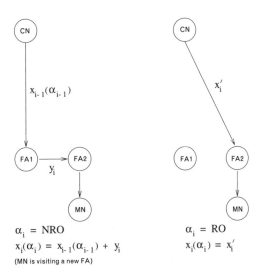

$$\alpha_i = \text{NRO} \qquad\qquad \alpha_i = \text{RO}$$
$$x_i(\alpha_i) = x_{i-1}(\alpha_{i-1}) + y_i \qquad x_i(\alpha_i) = x_i'$$
(MN is visiting a new FA)

21.2   The New Cost-Efficient Scheme

21.2.1   The Mathematical Model

The decision must be made in the time interval $[t_i, t_{i+1}]$ between the current handoff at t_i and the next handoff at t_{i+1} whether to perform route optimization or not.

As stated before, there is a trade-off between the network resources utilized by the routing path and the signaling and processing load incurred by route optimization. Two cost functions are introduced to capture the trade-off: the link and signaling cost functions. The link cost function is denoted by $g(x_i(\alpha_i))$ where $x_i(\alpha_i)$ is the number of links in the routing path between the CN and the FA during the ith period and $\alpha_i \in \{RO, NRO\}$ where RO is the action that performs route optimization, and NRO is the action without route optimization. The signaling cost function is denoted by $h(y_i, \alpha_i)$ where y_i is the number of links in the shortest path between the current FA and the previous FA during the ith period.

The total cost function is then defined as the sum of these two cost functions:

$$f(x_i(\alpha_i), y_i, \alpha_i) = g(x_i(\alpha_i)) + h(y_i, \alpha_i) \tag{21.1}$$

Let $\pi(i)$ denote a sequence of actions, $(\alpha_1, \alpha_2, ..., \alpha_i)$, which are taken sequentially during the occurrence of i handoffs. This sequence $\pi(i)$ is called a *route optimization sequence*. Let $G_i^{\pi(i)}$ denote the accumulative link cost, $H_i^{\pi(i)}$ the accumulative signaling cost, and $F_i^{\pi(i)}$ the total accumulative cost under the route optimization sequence $\pi(i)$, respectively. Then,

$$G_i^{\pi(i)} = \sum_{j=1}^{i} g(x_j(\alpha_j)) \tag{21.2}$$

$$H_i^{\pi(i)} = \sum_{j=1}^{i} h(y_j(\alpha_j)) \tag{21.3}$$

$$F_i^{\pi(i)} = \sum_{j=1}^{i} f(x_j(\alpha_j), y_j, \alpha_j) = \sum_{j=1}^{i} \{g(x_j(\alpha_j)) + h(y_j, \alpha_j)\} \tag{21.4}$$

where $\pi(i) = (\alpha_1, ..., \alpha_i)$. Here $x_j(\alpha_j)$ and y_j are network parameters.

When a route optimization is performed under the sequence $\pi(i)$ during the ith period, a signaling cost, $h(y_i, \alpha_i)$, is incurred. In this case the signaling cost, $h(y_i, \alpha_i)$, can be decomposed as follows:

$$h(\gamma_i, \alpha_i) = v(\gamma_i) + k_i(\alpha_i) \tag{21.5}$$

where $v(\gamma_i)$ is a variable signaling cost function that is independent of α_i, and $k_i(\alpha_i)$ is a portion of signaling cost that depends on α_i; that is,

$$k_i(\alpha_i) = \begin{cases} K & \text{if } \alpha_i = RO \\ 0 & \text{if } \alpha_i = NRO \end{cases} \tag{21.6}$$

In (21.5) the two terms reflect the cost of sending a binding update message from the current FA to the previous FA, sending a binding warning message from the current FA to the HA, and sending a binding update message from the HA to the CN. Here, it is assumed that every cost function is linear. Then, the link cost function, $g(x_i(\alpha_i))$, during the ith period becomes

$$g(x_i(\alpha_i)) = A \cdot T_i \cdot x_i(\alpha_i) \tag{21.7}$$

where A represents the average link cost per link that captures the bandwidth consumed by the routing path of length $x_i(\alpha_i)$, and T_i represents the sojourn time of the MN from the ith handoff to the next handoff.

The variable signaling cost function v during the ith period becomes

$$v(\gamma_i) = B \cdot \gamma_i \tag{21.8}$$

where B represents the average signaling cost per link in the path of length γ_i.

Thus, (21.9) is obtained from (21.4), (21.5), (21.7), and (21.8).

$$F_i^{\pi(i)} = \sum_{j=1}^{i} (A \cdot T_j \cdot x_j(\alpha_j) + B \cdot \gamma_j + k_j(\alpha_j)) \tag{21.9}$$

where $\pi(i) = (\alpha_1, \ldots, \alpha_i)$.

21.2.2 Optimal Solution

The objective of this section is to find the optimal sequence which is denoted as $\pi_{opt}(i)$, which minimizes the expected value of total cost $F_i^{\pi(i)}$ in (21.9).

$$E[F_i^{\pi_{opt}(i)}] = \min_{\pi(i) \in \Pi} E[F_i^{\pi(i)}] = \min_{\pi(i) \in \Pi} \sum_{j=1}^{i} (A \cdot x_j(\alpha_j) \cdot E[T_j] + B \cdot \gamma_j + k_j(\alpha_j)) \tag{21.10}$$

where $\pi(i) = (\alpha_1, \ldots, \alpha_i)$, and Π is the set of all possible sequences of $\pi(i)$.

If a route optimization is performed during the ith period, the shortest path between the CN and the current FA will be selected as the routing path. Thus, the length of the routing path will be x_i' in this case, where x_i' is the number of links in the shortest path between the CN and the current FA during the ith period. If the route optimization is not performed during the ith period, then an extended path will be the routing path, and the length of the routing path during the period will be $x_{i-1}(\alpha_{i-1}) + z_i$ where

$$z_i = \begin{cases} y_i & \text{if MN visits a new FA} \\ -y_i' & \text{otherwise} \end{cases} \qquad (21.11)$$

where y_i' is the number of links that will be reduced after the ith handoff. For example, in Figure 21.2, y_{i+1}' will be y_i if MN handoffs back to FA1 at time t_{i+1}.

This situation is detailed in Figure 21.2 and can be summarized as follows:

$$x_i(\alpha_i) = \begin{cases} x_{i-1}(\alpha_{i-1}) + z_i & \text{if } \alpha_i = NRO \\ x_i' & \text{if } \alpha_i = NRO \end{cases} \qquad (21.12)$$

where $x_i' = x_{i-1}(\alpha_{i-1}) + z_i$.

In general, the source routing is not being adopted in the Internet. Even though it is being used, network parameters cannot be known completely as networks grow bigger and become more complex [7, 8]. Thus, $x_i(\alpha_i)$ and x_i' are not available in every node. Without knowledge of these parameters, (21.10) cannot be solved. However, it can be easily solved if the model is restricted within intradomain (intrasubnet) handoff where $y_i = y_j$ for $i \neq j$, and a reasonable assumption is made—when handoffs occur in the same domain (subnet), the length of the shortest path between the CN and any FA is the same (i.e., $x_i' = x_j'$ for $i \neq j$). This assumption is reasonable because the shortest path between the CN and any FA in the same domain will pass through the main router of the domain.

Let i-stage denote the decision stage when the decision whether to perform route optimization or not is made during the ith period. In the i-stage, it can be thought that the routing path has been extended n times without performing route optimization after the last one was performed, where n is an integer. Thus, in the i-stage,

$$x_{i-1}(\alpha_{i-1}) = x_{i-n-1}' + \sum_{j=1}^{n} Z_{i-j} \qquad (21.13)$$

But $x'_{i-n-1} = x_i'$ by the above assumption. Hence, (21.13) becomes

$$x_{i-1}(\alpha_{i-1}) = x_i' + \sum_{j=1}^{n} Z_{i-j} \qquad (21.14)$$

Finally, (21.12) becomes

$$x_i(\alpha_i) = \begin{cases} x_i + Z_{i,n} & \text{if } \alpha_i = NRO \\ x_i' & \text{if } \alpha_i = NRO \end{cases} \qquad (21.15)$$

where $Z_{i,n} = \sum_{j=0}^{n} Z_{i-j}$.

Note that $Z_{i,n}$ is the length of the path between the COA and the current FA that is known within a domain. The expected value $E[F_i^{\pi(i)}]$ in (21.10) will be sequentially minimized with fixed $\alpha_1, \alpha_2, ..., \alpha_{i-1}$ by solving

$$E[F_i^{\pi_{opt}(i)}] = \min_{\alpha(i)\in\pi(i)} \sum_{j=1}^{i} (A \cdot x_j(\alpha_j) \cdot E[T_j] + B \cdot y_j + k_j(\alpha_j)) \qquad (21.16)$$

To calculate the total accumulative cost $F_i^{\pi(i)}$ in (21.4) and make an appropriate decision in the i-stage, all that is needed to know are the current length $x_{i-1}(\alpha_{i-1})$ of the routing path, the length y_i of the path between two adjacent FAs, and the length x_i' of the shortest path between the CN and the current FA. Let $s_{i-1} = (x_{i-1}(\alpha_{i-1}), x_i', y_i)$ denote the current state vector in the i-stage. The decision can be made only based on the current state vector s_{i-1}, and the next state vector s_i will be determined based on s_{i-1} and the action α_i taken in the i-stage. From this fact, it can be known that the decision model is Markovian, that is, memoryless.

The decision in each stage must constitute the optimal sequence $\pi_{opt}(i)$. $f(x_i(NRO), y_i, NRO)$ is the total cost that will be incurred during the ith period when the action NRO is taken, and $f(x_i(RO), y_i, RO)$ is the one that will be incurred during the ith period when the action RO is taken. Then the expected value of total accumulative cost, $E[F_i^{\pi opt(i)}]$, becomes

$$E[F_i^{\pi_{opt}(i)}] = E[F_{i-1}^{\pi_{opt}(i-1)}] + \min(f(x_i(NRO), y_i, NRO), f(x_i(RO), y_i, RO))] \qquad (21.17)$$

From (21.17) a decision rule can be found.

Decision Rule:

if $E[f(x_i(NRO), y_i, NRO)] \geq E[f(x_i(RO), y_i, RO)]$
then $\alpha_i = RO$
else $\alpha_i = NRO$
end if

where $E[f(x_i(NRO), y_i, NRO)] = A \cdot E[T_i] \cdot (x_i(NRO) + B \cdot (y_i + 0 = AE[T_i] (x_i' + Z_{i,n}) + B \cdot y_i$, and $E[f(x_i(RO), y_i, RO)] = A \cdot E[T_i]x_i(RO) + B \cdot y_i + K = A \cdot E[T_i] \cdot x_i' + B \cdot y_i + K$, which can be derived from (21.15). The condition $E[f(x_i(NRO), y_i, NRO)] < E[f(x_i(RO), y_i, RO)]$ means that $A \cdot E[T_i] \cdot Z_{i,n} < K$; that is, $Z_{i,n} < K / (A \cdot E[T_i])$.

Thus, the optimal sequence, $\pi_{opt}(i)$, can be obtained by following the above decision rule in each decision stage.

21.3   Performance Evaluation

In this section, the performance of the proposed scheme, $\pi_{opt}(N)$, is evaluated for route optimization, and it is compared with other schemes that are explained below:

- Scheme 1: The proposed optimal sequence, $\pi_{opt}(N)$;

- Scheme 2: Always perform route optimization. $\pi_{ARO}(N) = (\alpha_1, ..., \alpha_N)$ where $\alpha_i = RO$ for $i = 1, ..., N$;

- Scheme 3: Never perform route optimization. $\pi_{NRO}(N) = (\alpha_1, ..., \alpha_N)$ where $\alpha_i = NRO$ for $i = 1, ..., N$.

The sequence $\pi_{ARO}(N)$ represents the IETF RO [2] while $\pi_{NRO}(N)$ is a heuristic scheme. To obtain numerical results, it is assumed that N intradomain handoffs occur during a session and that B is equal to A. The performance metrics are the total cost per session $F_N^{\pi(N)}$ (21.9), the signaling cost per session $H_N^{\pi(N)}$ (21.3), and the number of route optimizations per session. In the simulation model, the number of handoffs within a domain N is assumed to be uniform random variable whose average value M is assigned during a session, and the sojourn time of an MN within a subnet is assumed to be exponentially distributed.

21.3.1   The Total Cost

The total cost per session is the sum of link cost and signaling cost per session. Equation (21.9) is used to compute the total cost per session for sequences $\pi_{opt}(N)$, $\pi_{ARO}(N)$, and $\pi_{NRO}(N)$.

In (21.9) the first term reflects the network resources utilized by the routing path during a session, while the others explain the signaling load incurred by the route optimization. The sequence $\pi_{opt}(N)$ is the one that can be found by following the decision rule of the previous section in each decision stage.

Figure 21.3 shows the total cost versus average link cost per link A during a session. The sequence $\pi_{opt}(N)$ shows the lowest total cost among the given sequences. The numerical result shows that the total cost under the sequence $\pi_{opt}(N)$ is 20.0 % lower than that under the sequence $\pi_{ARO}(N)$ on the average. For each sequence the total cost increases as A does. When the average link cost per link A is low, no difference can be observed between the results of $\pi_{opt}(N)$ and $\pi_{NRO}(N)$ because there is no advantage of performing route optimization. Under the sequence $\pi_{ARO}(N)$, however, route optimization is performed regardless of A causing the additional signaling cost. As A increases, the frequency of route optimization becomes higher, and thus, it can be observed that the results of $\pi_{opt}(N)$ and $\pi_{ARO}(N)$ converge.

Figure 21.4 shows the total cost versus average number of intradomain handoffs M during a session. The sequence $\pi_{opt}(N)$ shows the lowest total cost among the given sequences. When the frequency of intradomain handoff is low, there is only a slight difference among the results of the given sequences. However, the results of the sequences diverge, and thus the gap between the results of $\pi_{opt}(N)$ and $\pi_{ARO}(N)$ becomes bigger as the frequency increases.

FIGURE 21.3
Total cost versus average link cost per link
A ($B = A$, $K = 3$,
$M = 20$).

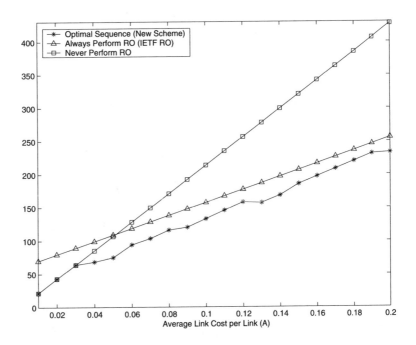

FIGURE 21.4
*Total cost versus aver-
age number of intrado-
main handoffs M (A
= B = 0.1, K = 3).*

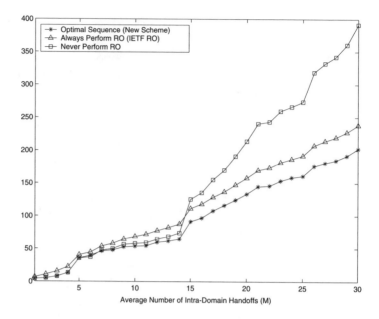

21.3.2   The Signaling Cost

The signaling cost per session is incurred by performing route optimizations during the session. Equation (21.3) is used to compute the signaling cost per session for sequences $\pi_{opt}(N)$, $\pi_{ARO}(N)$, and $\pi_{NRO}(N)$.

Whenever each handoff occurs, the decision must be made whether to perform route optimization or not. If route optimization is determined to be performed after a handoff, it causes the additional signaling cost and (21.3) captures it.

Figure 21.5 shows the signaling cost versus average link cost per link A during a session. As it can be seen in the figure, the signaling cost of each sequence increases as A does. When A is low, no difference is seen between the results of the sequences $\pi_{opt}(N)$ and $\pi_{NRO}(N)$ because no route optimization is performed in that period. It is observed that the signaling cost of $\pi_{opt}(N)$ is significantly reduced compared with that of $\pi_{ARO}(N)$. The numerical result shows that the signaling cost under the sequence $\pi_{opt}(N)$ is 83.0% lower than that under the sequence $\pi_{ARO}(N)$ on the average.

Figure 21.6 shows the signaling cost versus average number of intradomain handoffs M during a session. In this figure it can be also observed that the signaling cost of $\pi_{opt}(N)$ is significantly reduced compared with that of $\pi_{ARO}(N)$.

Under the sequence $\pi_{ARO}(N)$, the signaling cost grows linearly as M increases because the more frequently intradomain handoffs occur the more route optimization is performed. Under the sequence $\pi_{opt}(N)$, however, the signaling cost grows slightly as A increases because route optimization is not

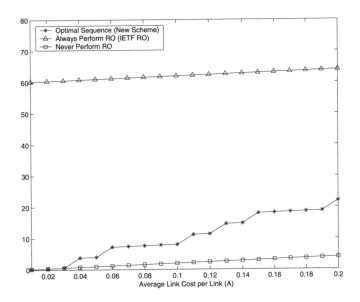

FIGURE 21.5
Signaling cost versus average link cost per link A (B = A, K = 3, M = 20).

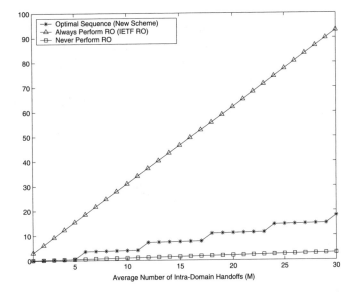

FIGURE 21.6
Signaling cost versus average number of intradomain handoffs M (A = B = 0.1, K = 3).

always performed whenever intradomain handoff occurs, which reduces the signaling cost significantly.

21.3.3 The Number of Route Optimizations

Figure 21.7 shows the number of route optimizations versus average link cost per link A. As A increases, the number of route optimizations under $\pi_{opt}(N)$ grows slowly. When A is low, no route optimization is performed

FIGURE 21.7
*Number of route op-
timizations versus av-
erage link cost per link
A (B = A, K = 3,
M = 20).*

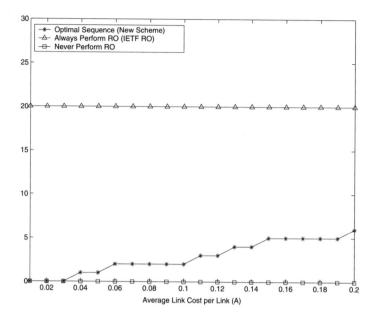

under the sequence $\pi_{opt}(N)$ because there is no advantage in performing route optimization. The number of route optimizations increases as A does because it is more profitable to perform route optimization from a cost point of view.

Figure 21.8 shows the number of route optimizations versus number of intradomain handoffs during a session. In this figure it can be observed that the number of route optimizations under the sequence $\pi_{opt}(N)$ increases

FIGURE 21.8
*Number of route
optimizations versus
average number of
intradomain handoffs
M (A = B = 0.1,
K = 3).*

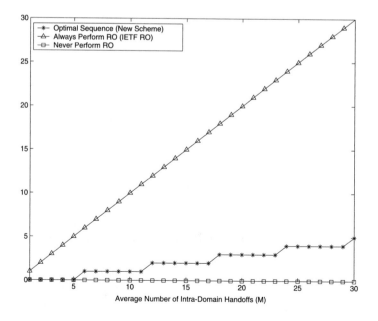

slowly as the frequency of intradomain handoff increases, which results in the reduction of the signaling cost and the total cost.

In the simulation the sequence $\pi_{opt}(N)$ shows the best performance compared with other sequences. The optimal sequence $\pi_{opt}(N)$ reduces the signaling cost caused by route optimization and provides the lowest total cost.

21.4 Summary

In this chapter, a cost-efficient scheme for route optimization was proposed to reduce the signaling cost caused by the route optimization. Link cost function represents the network resources utilized by the routing path; while signaling cost reflects the signaling and processing load incurred by route optimization. A Markovian decision model was presented to find an optimal sequence for route optimization. The model was restricted to intradomain handoff to simplify the decision process. A decision rule is derived from this model. The optimal sequence π_{opt} is obtained by following the decision rule in each decision stage.

The performance of the optimal sequence π_{opt} is compared with the other sequences π_{ARO} and π_{NRO}. The simulation results show that the optimal sequence π_{opt} provides the lowest total costs among the given sequences.

REFERENCES

[1] Perkins, C., "IP Mobility Support for IPv4," RFC 3344, August 2002.

[2] Perkins, C., and D. Johnson, "Route Optimization in Mobile IP," Internet draft, IETF, September 2001, work in progress.

[3] Johnson, D., C. Perkins, and J. Arrko, "Mobility Support in IPv6," Internet draft, IETF, June 2002, work in progress.

[4] Ho, J., and I. Akyildiz, "Local Anchor Scheme for Reducing Signaling Costs in Personal Communications Networks," *IEEE/ACM Trans. Networking*, Vol. 4, No. 5, October 1996, pp. 709–725.

[5] Gustafsson, E., A. Jonsson, and C. Perkins, "Mobile IPv4 Regional Registration," Internet draft, IETF, March 2002, work in progress.

[6] Soliman, H., et al., "Hierarchical MIPv6 Mobility Management," Internet draft, IETF, July 2002, work in progress.

[7] Guerin, R., and A. Orda, "QoS Routing in Networks with Inaccurate Information: Theory and Algorithms," *IEEE/ACM Trans. Networking*, Vol. 7, No. 3, June 1999, pp. 350–364.

[8] Orda, A., "Routing with End-to-End QoS Guarantees in Broadband Networks," *IEEE/ACM Trans. Networking*, Vol. 7, No. 3, June 1999, pp. 365–374.

CHAPTER 22

Enabling WAP Handoffs Between GSM and IEEE 802.11b Bearers with Mobile IP

Daniel Prötel, Nikolaus A. Fikouras, Stefan Aust, and Carmelita Görg

22.1 Introduction

Mobile Internet access accounts for only a small proportion of Internet users. However, wireless Internet access devices are becoming increasingly popular and are expected to overtake the desktop computer as the preferred Internet access medium [1]. Projections predict that the penetration of Internet-ready wireless phones will sustain a double and triple digit growth through 2004 [1]. By that time a wide range of wireless access technologies (bearers) with different properties but capable of bearing Internet traffic will be available. This will lead to a demand for roaming services between homogeneous and heterogeneous bearers without further customization or interruption of active communications.

WAP [2] is the de facto world standard for the presentation and delivery of wireless information and telephony services on mobile phones based on wide manufacturer commitment representing 90% of the world market [3]. WAP was created as a global wireless protocol standard with enough capacity to be deployed across all wireless bearers. The original specification, however, did not include any considerations for enabling WAP mobile clients and other architectural components (server, gateway) to roam while communicating. This is a significant disadvantage because it forces users to remain under the influence of a single provider or bearer technology despite utilizing mobile devices. It is also forecasted that mobile users might wish to host their own architectural components such as WAP servers or WAP gateways in order to provide WAP services to themselves or those around them (e.g., a scenario that includes ad hoc networks). Such a vision is not supported in the present standardization of WAP.

A solution to this problem can be provided through Mobile IP [4]. This is an extension of the IP that introduces mobility management facilities into the basic design. Despite the incompatibilities between the WAP and TCP/IP protocol stacks, certain configurations of WAP rely on a TCP/IP infrastructure for packetization (datagram construction) and routing. More specifically, the connection-oriented mode of the *WAP Wireless Session Protocol* (WSP) [5] operates on top of the *Wireless Transaction Protocol* (WTP) [6], which requires a datagram service such as UDP. In turn, UDP datagrams form the payload of IP packets that traverse the Internet fabric with respect to IP routing mechanisms. Consequently, if IP routing mechanisms can be brought through Mobile IP to support mobility management, then WAP communications can seamlessly benefit from this.

IP was originally designed to interconnect heterogeneous wireline networks. This integrating property of IP is inherited by its Mobile IP extension that focuses on mobile environments. For IP/Mobile IP the individual properties of the underlying network access technology that delivers the IP traffic is transparent. As such, the problem of integrating heterogeneous bearers is reduced into a matter of Internet routing [7]. In this manner WAP handoffs between various bearers can be realized.

The WAP specification defines the default size of the WSP *service data unit* (SDU) at 1,400 bytes [5]. It is within the WSP capabilities to negotiate a larger size. Moreover, the WTP layer is able to handle greater SDU sizes by using *segmentation and reassembly* (SAR). However, the most popular WAP gateway implementations do not support increased SDU sizes [8] and therefore restrict the total amount of data transported to a mobile client during a communication to the default size. This poses a significant problem when determining the performance of the WAP protocol stack over high-bandwidth WLANs such as IEEE 802.11b [9]. In order to overcome the aforementioned problem, a client-server software (*wapbench*) was developed that emulates a WAP communication between a mobile client and a WAP gateway. The *wapbench* allowed for the realization of extended communications that enabled the evaluation of WAP performance over GSM and IEEE 802.11b as well as during Mobile IP handoffs between them.

The investigated scenarios involve WAP communications based on the second WTP transaction service class that provides a reliable one-way request communication [6] similar to the Stop & Wait ARQ protocol. For Mobile IP handoffs, the SunLabs Mobile IP implementation [10] was extended to support the *Hinted Cell Switching* (HCS) Move Detection algorithm [11] for enhanced performance.

This study is restricted to the investigation of WAP handoffs with Mobile IP between GSM and IEEE 802.11b. The study describes the mobility support, based on Mobile IP, for all architectural components used by WAP. All conclusions drawn, however, should be applicable for mobility between any other Internet-capable wireless and wireline bearer. Moreover, the Mobile IP provides mobility support, which can be used for all IP-based network components (clients, gateways, and servers).

22.2   Reinventing in the Name of Wireless

WAP is a protocol stack with a focus on wireless environments developed by the WAP Forum. To best address the needs of the widest possible population of end users, WAP was designed to work optimally with all air interfaces (bearers). Based on the exceptional character of wireless environments and handheld devices, WAP defined a new standard for the IP. Meanwhile, through wide manufacturer commitment representing more than 90% of the world market, WAP managed to become the de facto standard for the presentation and delivery of wireless data on mobile phones [3].

To enable access to Internet resources, the WAP architecture introduced the WAP gateway, a proxy responsible for performing protocol and content translation between TCP/IP and WAP stacks [2]. Figure 22.1 illustrates the protocol profile of a WAP communication over GSM *circuit switched data* (CSD) where a mobile terminal is accessing Internet resources residing at the content provider. It can be seen that the WAP gateway intervenes in the communication between the mobile terminal and its communication peer.

The introduction of a portal between the global Internet and any WAP network provided two alternative service model options. The WAP Forum was confronted with the fundamental design question of either enabling a WAP terminal to access any public Internet resource or restricting it to the service provider's custom applications. Deciding that handheld Internet access is not about Web surfing, the WAP Forum chose for custom provider

FIGURE 22.1
The protocol profile of WAP over GSM-CSD.

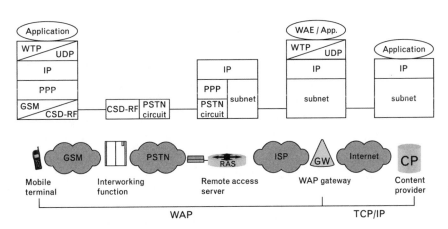

services, one of which would provide limited Internet access [12]. The most important disadvantage of WAP, however, in the scope of this study, is its lack of considerations for supporting user roaming between providers or bearers in spite of being designed especially for mobile and heterogeneous environments.

The purpose of this study is to show how WAP roaming between homogeneous and heterogeneous bearers can be realized based on Internet protocol extensions for mobility support. In this manner, a WAP terminal may switch providers or even bearer service while communicating. Even though the investigated scenario involves roaming between GSM and IEEE 802.11b WLAN bearers, the same considerations should be applicable for other Internet-capable bearers.

22.3   Converging in the Name of Progress

Certain configurations of WAP rely on the UDP and IP protocols for packetization and delivery. More specifically, for the provision of WAP services over GSM-CSD (currently the most popular WAP bearer), the presence of the UDP, IP, and PPP protocols is required (see Figure 22.1). Given this architecture, the mobile terminal uses PPP to establish, configure, and test the data link connection via GSM to the dial-in access facilities of a *remote access server* (RAS). The *interworking function* (IWF) provides the modem functionality along with access to the fixed telephone network (PSTN) [13]. Eventually, WAP traffic is delivered to the WAP gateway within UDP packets that are routed beyond the RAS server with respect to the information contained within the IP header and according to Internet hop-by-hop routing. As such, if Internet routing can be brought to support mobility management, then WAP can seamlessly benefit from this.

22.3.1   The Mobile IP

The IP was originally designed without any mobility support. The routing mechanisms of IP assume that a terminal maintains a point of attachment to the Internet, indicated by its IP address. Should a mobile terminal change its point of attachment and move to a new location incompatible with its IP address, it would be unable to send or receive traffic.

A solution to the problem of Internet mobility can be provided through the Mobile IP [4], an extension of IP. Mobile IP introduces three new network entities: the home agent, the foreign agent, and the mobile node. Alternative Mobile IP configurations may omit the FAs by distributing their functionality amongst the mobile terminal and the network infrastructure.

Every mobile terminal is permanently allocated an IP address in its home network, where the HA also resides. Every time that a mobile terminal moves, it is required to register its current point of attachment to the Internet with the HA. In that case, a mobile terminal is said to have performed a Mobile IP handoff.

The most basic functionality of Mobile IP resembles the post-office forwarding service. For every registered mobile terminal the HA is required to act as a proxy in the home network, intercept all incoming traffic from any CN, and redirect it to the mobile terminal's most recently registered location. As a result, the mobile terminal can change its location while remaining reachable on a permanently allocated IP address and without having to interrupt active communications.

The fundamental OSI assumption of layer independence dictates that layers should be unaware of other layers above and below them. Consequently, as long as a bearer can provide connectivity for the delivery of Internet traffic to the mobile terminal, then its existence is transparent to Mobile IP. In this manner the problem of integrating heterogeneous bearers can be reduced into a matter of Internet routing [7]. Similarly, the way by which IP/Mobile IP routes packets between two communication peers is kept transparent from higher layers. As such, WAP traffic can be seamlessly transported over Mobile IP in order to provide roaming services to WAP users.

In Figure 22.1 the RAS server and the WAP gateway are connected through the intranet of an ISP. In fact, the WAP gateway and the respective access server could be placed further apart in the Internet. Furthermore, by integrating the HA functionality into the WAP gateway it is possible for the mobile terminal to change its access server (even bearer service) while still remaining reachable. This is the easiest way to integrate Mobile IP with WAP. The HA must be placed inside the WAP network to intercept all traffic destined for the home addresses of the mobile nodes that it is serving.

Figure 22.2 illustrates the case when the HA has been integrated with the WAP gateway and the mobile terminal is roaming between two remote networks A and B. Every location switch between these networks is also associated with a change of the bearer service as network A provides network access over GSM while network B provides access over IEEE 802.11b.

The capacity of a WAP terminal to roam beyond the network of a WAP gateway does not reduce the importance of the gateway. Apart from providing custom provider applications, a WAP gateway is still required to perform content and protocol translation in both the receive and return communication paths when accessing Internet resources. Bypassing the WAP gateway to avoid triangle routing in both directions involves either adopting a pure Internet stack or making communication peers WAP aware. Considering the large amount of WAP-enabled devices today while

FIGURE 22.2
Experimental setup.

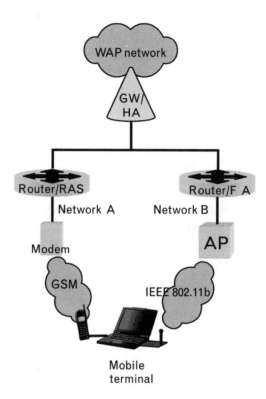

also recognizing its limitations, this study aims to converge these two worlds.

For the purposes of this study, the scenario illustrated in Figure 22.2 has been experimentally investigated. Due to the limitations of existing implementations of WAP, a client server application that emulates a WAP communication has been developed. A detailed presentation of the emulator is presented in the follow section.

22.3.2   The WAP Transport Protocols

In the WAP protocol stack transport is managed by the pair of WTP and WDP protocols [6]. For those bearer services with support for the IP (e.g., GPRS, GSM-CSD, and WLAN) WDP is replaced by UDP and IP [14]. WTP is a protocol based on top of a datagram service (WDP or UDP) that provides three different classes of transaction service to the upper layers: unreliable one-way request (class 0), reliable one-way request (class 1), and reliable two-way request (class 2). The transaction class is set by the initiator and indicated in the invoke message sent to the responder. Transaction classes cannot be negotiated [6]. WTP is optimized to adapt to the small bandwidth of the radio interface, trying to reduce the total amount of replayed transactions between the client and server.

The transport protocol has a fixed header structure, which defines the size of an invoke PDU (32 bits), a result PDU (24 bits), and an acknowledgment PDU (24 bits). A PDU type field indicates what type of WTP PDU the PDU is (invoke, ack, and so on). This provides information to the receiving WTP provider as to how the PDU data should be interpreted and what action is required [6].

The class of the transaction service is defined by the WSP, which is the application layer of the WAP stack. The WSP is also responsible for determining the SDU, which represents the total length of the compiled byte stream that a mobile terminal may download during a session.

The WAP specification defines the default size of the WSP SDU at 1,400 bytes [5]. It is, however, within the WSP capabilities to negotiate a larger size. Similarly, the WTP layer is able to transport extended SDU sizes by using SAR. After experimentation with WAP-enabled mobile phones, it was found that extended SDU size support is optional, and in practice each device has its own maximum size [15]. This was considered as a significant problem because it limits the duration of a WAP communication and prohibits the investigation of WAP performance during MIP handoffs. For this, it was decided to implement a client-server application, termed *wapbench*, that emulates a WAP communication of extended duration. This enabled the study of WAP performance during MIP handoffs as well as determining the throughput and efficiency of the second WTP transaction service class in high-bandwidth networks such as the IEEE 802.11b. The conformance of *wapbench* to the WAP specification was highly dependent on the successful emulation of the WTP. The reason for this is because WTP is the lowest layer the WAP micro-browser absolutely requires and the protocol that manipulates the datagram service to manage a communication [12].

In order to identify the correct operation of WTP, apart from studying the WAP specification, the communication between a WAP-enabled mobile phone and a popular WAP gateway was monitored with the *ethereal* [16] network analyzer. It was identified that WTP does not support any type of flow control for any of the transaction service classes [12]. Furthermore, it was determined that the second WTP transaction service class resembles the Stop & Wait ARQ protocol.

The ARQ protocol provides error correction to check the received frame for possible transmission errors and returns a short control message either to acknowledge its correct receipt or to request that another copy of the frame be sent. For the Stop & Wait scheme, the sender must wait for an acknowledgment after sending a frame [17]. This functionality was implemented in *wapbench*.

In Section 22.5, the experimental setup is presented along with the tests undertaken with *wapbench*.

22.4   Experimental Setup

The experimental test bed is illustrated in Figure 22.2. Its purpose is to provide two Internet administrative domains, each providing network access through a different bearer service. Specifically, the router of network A provides dial-in access through a standard modem, while the router of network B provides FA services through a IEEE 802.11b–compliant Cisco Aironet 340 access point. A third machine performs the combined WAP gateway and HA functionality and runs the *wapbench* server. The mobile terminal is a laptop computer with two access interfaces that provide network connectivity in each of the Internet administrative domains. Specifically, the laptop is attached to a Siemens S-35 mobile phone that contains an integrated modem. In addition, the laptop maintains a PCMCIA Cisco Aironet 340 client adapter that provides access to the WLAN managed by the access point. On the mobile terminal runs the *wapbench* client that initiates a WAP communication with the WAP gateway. Even though the mobile has been allocated an IP address from within the WAP network, it may through MIP remain reachable outside this network. Mobile IP software on the mobile terminal performs periodic Mobile IP handoffs between the networks A and B during which the mobile terminal registers its location with the WAP gateway/HA. Subsequently, all traffic for the mobile terminal is encapsulated by the WAP gateway/HA and redirected to the terminal's registered location.

The two routers and the WAP gateway/HA are AMD Thunderbird PCs at 800 MHz; while the mobile terminal is a Samsung laptop with an Intel Pentium III at 600 MHz. Where applicable, wireline network connectivity has been provided with Fast Ethernet. All machines have been installed with the Linux Mandrake 7.2 distribution and the Linux Kernel 2.2.17. The Mobile IP functionality has been provided with the Sun Microsystems Mobile IP implementation [10] that was extended to sup-port the HCS move detection algorithm for accelerated Mobile IP handoffs [11].

The investigated scenario involves a WAP communication with the second WTP transaction service class between the mobile terminal and the WAP gateway. During the communication a single Mobile IP handoff occurs. For every state, before and after the Mobile IP handoff, the properties of the communication (round-trip time, throughput, and efficiency) were measured. The experimental results are presented in the following section.

22.5   Experimental Results

In the beginning of each experimental trial, the mobile terminal used the point of attachment made available by its dial-up connection to exchange WAP traffic over a 9.6-Kbps GSM-CSD link. At time 20 seconds the mobile terminal performed a single MIP handoff that caused the communication to proceed through the point of attachment made available by the 11-Mbps IEEE 802.11b WLAN. Figure 22.3 illustrates the packet trace of the aforementioned scenario. It can be observed that the packet transmission rate becomes significantly faster following the Mobile IP handoff, which swiftly completes without any packet loss. This is derived from the fact that no packet retransmissions are observed after the handoff.

The lack of any flow control or windowing considerations in the second WTP transaction service class has an impact on the transmission rate before and after the Mobile IP handoff. However, due to log scaling of the y-axis, this can not be observed in Figure 22.3. For this purpose the RTT measurements of the communication are illustrated in Figure 22.4. That is, for every WTP data packet transmitted by the WAP gateway, the mobile terminal was forced to respond with an immediate WTP acknowledgment. The time interval between the transmission of the data packet and the receipt of the corresponding acknowledgment was identified as the RTT. It can be seen that the RTT remains constant over GSM-CSD with an average of 2.438 seconds, while the corresponding value over IEEE 802.11b has been found to be 0.006 second with sporadic deterioration due to internal processing. In this case, the acceleration of the RTT justifies the increase in the transmission rate.

FIGURE 22.3
Packet trace of a WAP communication during a Mobile IP handoff between GSM-CSD and IEEE 802.11b.

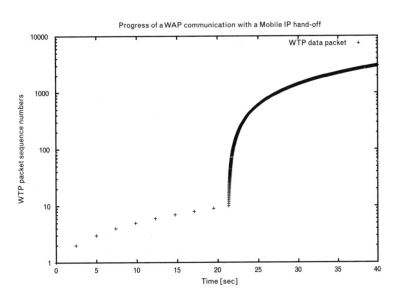

FIGURE 22.4
*RTT of a WAP
communication during
a Mobile IP handoff
between GSM-CSD
and IEEE 802.11b.*

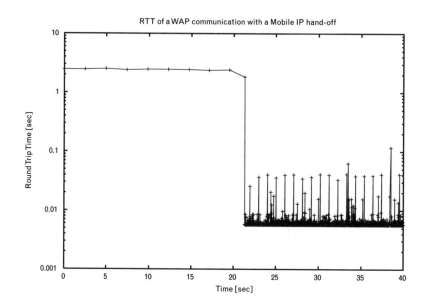

An additional point of interest is the RTT of the first WTP transmission following the Mobile IP handoff (at time 20 seconds). It is significantly shorter than the average RTT over GSM-CSD but still greater than that of IEEE 802.11b. This signifies that this packet exchange occurred during the MIP handoff. That is, the mobile terminal received a WTP data packet through its GSM-CSD connection but sent its acknowledgment through IEEE 802.11b.

In order to identify the individual delays that cause WTP to suffer extended RTTs over GSM, the *ping* application was used to determine the GSM-CSD traversal delay between the mobile terminal and the WAP gateway. The results of this investigation are presented in Figure 22.5. The *ping* has been found to demonstrate a constant RTT with an average of 821 ms. Considering that at 9.6 Kbps GSM-CSD requires 107 ms for the transmission of the *ping* request and reply messages, the rest of 714 ms is considered to be the traversal delay of the GSM network. This interval is roughly 100 ms larger than the corresponding delay measured in other investigations [13]. However, the traversal delay of GSM-CSD has been found to be network provider specific [18].

For the transmission of a full WTP data packet and its corresponding acknowledgment, GSM-CSD requires 1.2 seconds and 0.025 second, respectively. Considering an additional 714 ms for the GSM-CSD traversal delay, then from the 2.438-second average RTT there is an outstanding amount of almost 0.5 second. This can be explained by the fact that the GSM-CSD traversal delay has been found to vary during bulk data transfer and take on values as high as 1.2 seconds [18].

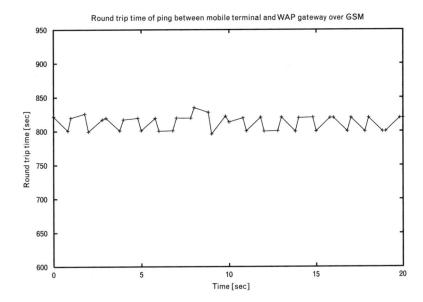

FIGURE 22.5
Round-trip time of ping between the mobile terminal and the WAP gateway over GSM-CSD.

The Stop & Wait character of the second WTP transaction service class dictates that a single WTP data packet may be transmitted per RTT. Consequently, the throughput of the WAP communication is directly dependent on its value. The throughput of the WAP communication is illustrated in Figure 22.6. As expected, it increases significantly beyond the MIP handoff, where the mobile terminal resumes access over the IEEE 802.11b WLAN. When determining the efficiency of the communication, however, the shortcomings of the second WTP transaction service class start to become apparent.

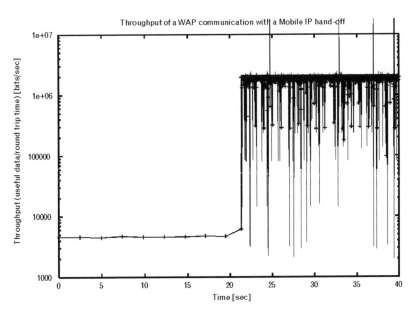

FIGURE 22.6
Throughput of a WAP communication during a MIP handoff between GSM-CSD and IEEE 802.11b.

Communication efficiency is defined as the throughput of the communication over the available bandwidth. Considering that over IEEE 802.11b, the WAP gateway may transmit a single full WTP data packet every 6 ms, then the theoretical maximum throughput is about 1.78 Mbps. However, the IEEE 802.11b WLAN provides connectivity at 11 Mbps. From this it is determined that the lack of flow control in WTP does not enable it to utilize high-bandwidth rates when they are available. This is also verified by results presented in Figure 22.7. Similarly, in links with small bit rates but long RTTs (e.g., GSM-CSD) WTP still manages a low efficiency. It can be seen that even before the Mobile IP handoff, WTP manages to provide an efficiency short of 50%.

22.6   Mobile WAP Network Components

The standardization of WAP defines the three main architectural components, the WAP client, the WAP gateway, and the WAP server. The WAP communication model is depicted in Figure 22.8. It is envisioned that in the future, mobile users may wish to host their own architectural components such as WAP server and WAP gateway in order to provide WAP services to themselves or those around them. For this scenario the current standardization of WAP has not foreseen mobility support for mobile network components.

FIGURE 22.7
Efficiency of a WAP communication during a MIP handoff between GSM-CSD and IEEE 802.11b.

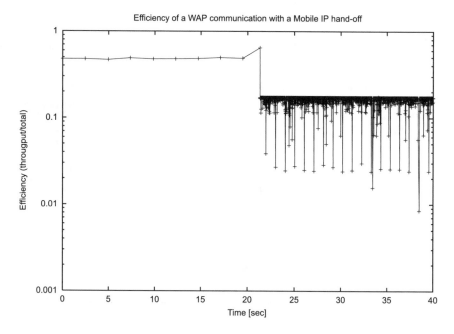

FIGURE 22.8
*AP communication
model.*

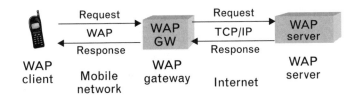

Mobile IP provides mobility support, which allows a simple integration of macro-mobility that can be used by network components. With Mobile IP, stationary WAP gateways and WAP server systems can be realized as mobile network entities, which can roam between different network domains. WAP applications can be downloaded while the WAP gateway or the WAP server is roaming. The handoff between the access networks will be managed by Mobile IP.

The investigation of Mobile IP handoffs allows an efficient design of mobile network components based on a Mobile IP infrastructure. Handoffs between current bearer systems such as GSM and WLAN are the base for new scenarios for mobile network components that include roaming features and can be installed everywhere inside an IP-based communication network.

22.7 Summary

In this study WAP handoffs between the GSM and IEEE 802.11b bearers with Mobile IP were investigated. Through an experimental approach it was identified that the proposed solution is feasible. Due to the limitations of existing implementations of WAP that restrict its communication length, a client-server application that emulates a WAP communication has been developed. It was determined that the second WTP transaction service class resembles the Stop & Wait ARQ protocol. Its lack of any flow control or windowing considerations provides low efficiency rates over both investigated bearers.

The study has also described that application of Mobile IP can be extended also for mobile architectural components. WAP network components such as WAP gateways or WAP servers can be supported by Mobile IP to provide roaming features.

REFERENCES

[1]   Cahners In-Stat Group, "Wireless Phones, Pagers and Modems to Surpass PCs as Most Popular Internet Access Devices," http://www.instat.com.

[2]   WAP Forum, "WAP Architecture," http://www.wapforum.org.

[3]   WAP Forum, "WAP White Paper," http://www.wapforum.org.

[4]   Perkins, C. E., "RFC 2002: IP Mobility Support," October 1996.

[5]   WAP Forum, "Wireless Session Protocol," http://www.wapforum.org.

[6]   WAP Forum, "Wireless Transaction Protocol," http://www.wapforum.org.

[7]   Fikouras, N. A., C. Görg, and I. Fikouras, "Achieving Integrated Network Platforms through IP," *Proceedings of the Wireless World Research Forum Kick-Off Meeting 2001 (WWRF)*, Munich, Germany, March 2001.

[8]   "Kannel: Open Source WAP and SMS Gateway," http://www.kannel.org.

[9]   IEEE Std 802.11-1997, "Wireless LAN Medium Access Control (MAC) and Physical Layer (PHY) Specification," 1997.

[10] Networking and Security Center Sun Microsystems Laboratories, "Mobile IP," http://playground.sun.com.

[11] Fikouras, N. A., and C. Görg, "Performance Comparison of Hinted and Advertisement Based Movement Detection Methods for Mobile IP Handoffs," Elsevier Science, Computer Networks, Vol. 37, Iss. 1, September 2001, pp. 55–62.

[12] Khare, R., "W★ Effect Considered Harmful," *IEEE Internet Computing*, Vol. 3, No. 4, July/August 1999, pp. 89–92.

[13] Ludwig, R., and B. Rathonyi, "Link Layer Enhancements for TCP/IP over GSM," *Proceedings of the Conference on Computer Communications (IEEE Infocom)*, New York, March 1999.

[14] WAP Forum, "WAP Wireless Datagram Protocol," http://www.wapforum.org.

[15] Banahan, M., "Impressions of Using WAP/WML," GBdirect, http://www.gbdirect.co.uk.

[16] Combs, G., "Ethereal - (version 0.8.18)," http://www.ethereal.com.

[17] Halsall, F., *Data Communications, Computer Networks and Open Systems*, Reading, MA: Addison-Wesley, 1996.

[18] Xylomenos, G., et al., "TCP Performance Issues over Wireless Links," *IEEE Communications Magazine*, April 2001.

CHAPTER 23

Interworking and Handover Mechanisms Between WLAN and UMTS

Mahbubul Alam

23.1 Introduction

Current telecommunications and computer networks are on the verge of providing mobile multimedia connectivity, where nomadic users would have ubiquitous access to remote information storage and computing services. As an evolutionary step toward 4G mobile communication, mobility in heterogeneous IP networks with both UMTS and IEEE 802.11 WLAN systems is seen as one of the central issues in the realization of 4G telecommunications networks and systems.

In the future, mobile access to the Internet will be through a collection of different wireless services, often with overlapping areas of coverage. No one technology or service can provide ubiquitous coverage, and it will be necessary for a mobile terminal to have various points of attachment to maintain connectivity to the network at all times. The most attractive solution for such consideration is to utilize high bandwidth data networks such as IEEE 802.11a/b/g WLAN whenever they are available and switch to an overlay public network such as UMTS with lower bandwidth when there is no WLAN coverage. Think of a scenario where users wish to be connected to WLAN at a low cost and with high bandwidth in the home, airport, hotel, or shopping mall, and also want to connect to cellular technologies (e.g., GPRS or UMTS) from the same terminal. In particular, the users in this scenario require support for vertical handover (handover between heterogeneous technologies) between WLAN and UMTS.

This chapter describes five possible network layer level architectures for interworking and handover between WLAN and UMTS without making any major changes to existing networks and technologies, especially at the lower layers such as MAC and PHY layers [1, 2]. This will ensure that existing networks will continue to function as before without requiring current users to change to the new approach. The implementation involves incorporating new entities like emulators and protocols that operate at the network or higher layers to enable interworking and intertechnology roaming that will be transparent to the mobile user to the extent possible.

Section 23.2 describes the general approach to an interconnection philosophy and the essential aspects of making interconnection between both technologies. Strategy and consequences behind the five interconnection approaches are described in Sections 23.3 through 23.7. Sections 23.8 and 23.9 describe the handover mechanism and procedure between WLAN and UMTS beased on Mobile IP. Finally, Section 23.10 presents the conclusions and future directions.

23.2   Interworking System Architectures

By connecting the IEEE 802.11 WLAN to a UMTS core network as a complementary radio access network, a second form of mobile packet data services are provided by this heterogeneous IP-based system.

Figure 23.1 represents the five interconnection points between WLAN and UMTS. These interconnection architectures involve minimum changes to the existing standards and technologies and especially for MAC and PHY layer to ensure that existing standards and networks continue to function as before. The first two interconnections in Figure 23.1 will always have interaction between WLAN *access point* (AP) and the *packet switched* (PS) part of the UMTS *core network* (CN) [3]. This means that the gateway to the IEEE 802.11 WLAN network is attached to the PS domain. This interconnection is possible through the 3G-SGSN entity and GGSN entity, which are the elements of the UMTS PS CN. In both cases the WLAN network appears to be a UMTS cell or *routing area* (RA), respectively. The UMTS network will be a master network and the IEEE 802.11 WLAN network will be a slave network. This means that the mobility management and security will be handled by the UMTS network, and the WLAN network will be seen as one of its own cells or RAs. This may require dual-mode PCMCIA cards to access two different physical layers. In addition, all traffic will first reach the UMTS 3G-SGSN or 3G-GGSN before reaching its final destinations even if the final destination is in the WLAN network.

In the third interconnection the *virtual access point* (VAP) reverses the roles played by the UMTS and WLAN as in the first two interconnect architectures. This is called a tight coupling because there is always interaction between both networks. Here, the IEEE 802.11 WLAN is a master

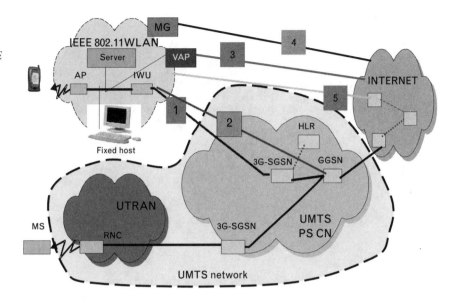

network and the UMTS is the slave network. Mobility management is according to the WLAN, and the *Inter-Access Point Protocol* (IAPP) is the protocol that is specified for this management.

In the fourth interconnection architecture a mobility gateway/mobile proxy (MG) is employed in between the UMTS and IEEE 802.11 WLAN networks. Here they are both peer-to-peer networks. The MG is a proxy that is implemented on either the UMTS or the WLAN sides and will handle the mobility and routing.

The fifth interconnection architecture is based on Mobile IP protocol. This is called "no-coupling" and both networks are peers. Mobile IP handles the mobility management. A HA/FA entity is involved with this architecture, which makes the IP layer aware of the agent advertisements of the mobility agent (HA/FA) and does a binding update periodically [4].

23.3 Interconnection Between 3G-SGSN and WLAN AP by Emulating RNC

With this interconnection the IEEE 802.11 WLAN network is connected to the UMTS CN via the Iu-ps interface. Figure 23.2 shows this heterogeneous network architecture.

The IEEE 802.11 WLAN-based RAN is connected via an *interworking unit* (IWU) as shown in Figure 23.2, which is an RNC emulator. This is needed to exchange the packets between the IEEE 802.11 WLAN network and UMTS. The function of the IWU is similar to an RNC in the

FIGURE 23.2
*Interconnection
between IEEE
802.11 WLAN
AP and 3G-SGSN
through an IWU that
emulates RNC
controller.*

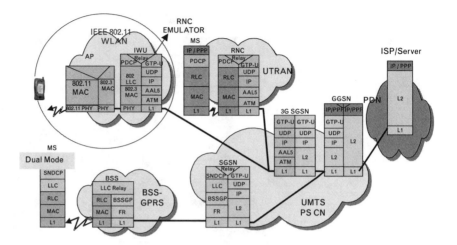

UTRAN. It has to relay the Iu bearer service on the core network side to the distribution network (IEEE 802.3 LAN) bearer service on the other side. The adapted UMTS bearer concept includes an appropriate location and mobility management for the terminals in the IEEE 802.11 WLAN coverage area. Due to the very small cell size of IEEE 802.11 WLAN systems, the access points are not directly connected to the UMTS core network in order to reduce the signaling load caused by mobility and location management. A distribution network connects the WLAN APs and enlarges the coverage area of this radio access form. The IEEE 802.11 WLAN is treated as an RA associated with the 3G-SGSN. Thus, the WLAN looks like an RNC to the UMTS network. A user, whether he or she is connected to the UMTS network or the WLAN, will always be treated as a UMTS user. The UMTS mobility management will have to maintain information about the user even when it is connected to a WLAN network. The IWU entity is the RNC emulator, which is presented in Figure 23.3.

The RNC emulator could be an LAN entity or an UMTS entity implemented in the networks. The LAN entity avoids encapsulation for routing to the UMTS network. For this type interconnection, a dual IEEE 802.11 WLAN/UMTS mode MS is required to use both networks as shown in Figure 23.4.

The intertechnology roaming arises when the user is connected to the WLAN network. For this interconnection, the users have to interface to the UMTS network through the RNC emulator. UMTS-specific protocol such as PDCP is on top of the IEEE 802.11 MAC and the PHY layers implemented. UMTS-related signaling protocols are carried out between the protocols in the MS and the RNC emulator. The RNC emulator is a black box that hides WLAN-specific features from the UMTS network. IP is used to transfer packet switched data over the Iu-interface as well as in the

FIGURE 23.3
*Protocol stack of the
RNC emulator being
a UMTS and
WLAN entity.*

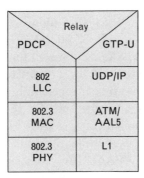

IWU as a RNC
emulator being a
UMTS entity

IWU as a RNC
emulator being a
WLAN entity

FIGURE 23.4
*IEEE 802.11
WLAN/UMTS
dual-mode protocol
stacks architecture of
the MS.*

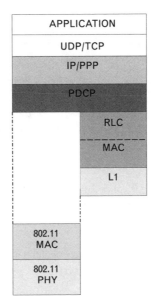

CN. The *GPRS Tunneling Protocol for UMTS* (GTP-U) on the top of this transport IP layer provides a tunneling service through the CN until the access network encapsulates the user data. Hence, if IP packets are transmitted on the user level, two IP layers exist in the packet switched architecture.

The IEEE 802.11 WLAN coverage area is represented as one routing area for the CN. If the mobile node leaves or enters a routing area, an update message is sent to the core network of UMTS. Hence, the 3G-SGSN can simply distinguish the different radio access networks via the routing areas. Running IP sessions are not interrupted because the IP address of a terminal is not changed. The procedure is completely transparent to the user. However, if a mobile leaves the IEEE 802.11 WLAN coverage area, the service quality will degrade, especially for those sessions that made use of the high throughput capabilities of IEEE 802.11 WLAN system.

The current UMTS QoS approach foresees that within the CN, Diff-Serv are used on transport IP level to differentiate between different traffic classes. This approach can be mapped quite easily on both the IEEE 802.11 WLAN distribution network and the IEEE 802.11 WLAN bearer. If switched Ethernet implements the IEEE 802.11 WLAN distribution network, the DiffServ classes can be mapped onto IEEE 802.1p priorities and then to IEEE 802.11 WLAN MAC connections, and vice versa. Figure 23.5 shows the UMTS bearer concept [5] with IEEE 802.11 WLAN access integrated. The UMTS bearer is not changed in respect to the different radio interfaces. The RAB must be adapted to the new, underlying distribution network (DN) bearer and the IEEE 802.11 WLAN bearer.

Pros and Cons

The main advantage of this interconnection, together with the dual-mode WLAN/UMTS protocol stack on MS as shown in Figure 23.5, is that the mobility management, roaming, billing, and location related issues are taken care of by the UMTS network. *Subscriber Identity Module* (SIM) and *UMTS SIM* (USIM)-based authentication of a subscriber for WLAN offers a 2G/2.5G and 3G operator the following benefits:

- The WLAN subscriber credentials are of identical format to 2G/2.5G or 3G and therefore easier to integrate subscriber into the current HLR. Therefore, all existing roaming capabilities and settlements are inherited from GSM.

- The security level offered by WLAN will be identical to that of GSM/GPRS/UMTS. GSM SIM-based security is based on a

FIGURE 23.5
Adapted UMTS bearer concept using IEEE 802.11 WLAN bearer.

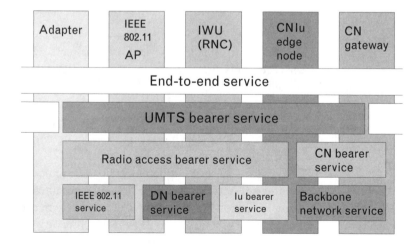

challenge-response mechanism and it offers better tamper resistance because SIM runs an operator-specific confidential algorithm that takes a 128-bit *random number* (RAND) and a secret key Ki stored on the SIM as an input to produce a 32-bit response (SRES) and a 64-bit data encryption keys Kc(n) as an output. Kc(n) are never sent over the air nor are they used in calculation for message authentication code for RAND and SRES. Kc(n) together with IMS⁎ and Ki are used by the network and the client to calculate independently the key K, that will be used for encryption of data over the air interface. So the only data exposed over the air interface are the random numbers K(n).

Strong security provided in the UMTS network and quality of services for real-time services may now be provided over WLAN as shown in Figure 23.6, thereby resolving the drawbacks of current IEEE 802.11a/b/g WLAN threats. Minimum changes are required in the UMTS network, and this will create a master-slave relationship between UMTS and WLAN as discussed in Section 23.2, which is not optimal. Using UMTS PDCP frames over WLAN may create bottlenecks. In this scenario the UMTS backbone may be a bottleneck for the WLAN traffic. WLAN data rate, currently 11 Mbps with IEEE 802.11b and 54 Mbps with IEEE 802.11a, would be degraded to the speed of the UMTS (2 Mbps). This type of interconnection requires modifications to standard WLAN terminals, which in turn would make them more expensive. The two most attractive WLAN components (i.e., speed and price) would be lost in this type of connection.

23.4   Interconnection Between GGSN and WLAN AP by Emulating 3G-SGSN

As an alternative to the interconnection of WLAN to 3G-SGSN, it can also directly be connected to the GGSN of the UMTS network, as shown in Figure 23.6.

The architecture in Figure 23.6 is a modification to that of Figure 23.2 in that the interface between WLAN and UMTS is now via a 3G-SGSN-like device, which is called a 3G-SGSN emulator. The protocol stack is very similar to the one depicted in Figure 23.3. In this interconnection the Iu-interface and the protocols between 3G-SGSN and 3G-GGSN are not used, and hence the functions supported by those protocols are not available. It is possible to bypass some of the RNC-related functionalities by using a 3G-SGSN emulator, and mobility management is again handled by UMTS.

FIGURE 23.6
Interconnection be-
tween IEEE 802.11
WLAN AP and
GGSN through an
IWU by emulating
3G-SGSN.

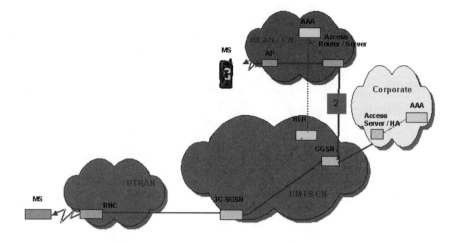

Pros and Cons

The disadvantage of the RNC emulator–based interconnection is the master-slave situation; bottlenecks and inefficient routing also exist in such interconnection type architecture. The advantage of this interconnection architecture is that also some overhead caused by available but not needed functionality is avoided. In this architecture the IWU requires the adaptation between WLAN and UMTS packet formats. If the 3G-SGSN emulator does this adaptation, the GGSN could remain unaltered. In this interconnection architecture the GGSN throughput might become a problem if the GGSN capacity is not designed to fulfill the growth of traffic. If the adaptation is done by the GGSN, taking into account the increased need for bandwidth, the speed of WLAN could be exploited in full.

23.5   Interconnection Between UMTS and WLAN Through VAP

The VAP reverses the roles played by the UMTS and WLAN in the first two interconnection architectures. Here, the WLAN is the master network and UMTS is the slave network. Figure 23.7 depicts the architecture of this interconnection type.

The difference in this interconnection compared to the other architectures is the existence of a VAP instead of RNC/3G-SGSN emulators. This is the dual mode of the RNC/3G-SGSN emulators. Mobility is managed according to the IEEE 802.11 WLAN and IAPP specifications by the WLAN standard. The intertechnology roaming that the WLAN observes is between different APs in the extended service set and the VAP that appears as yet another AP to the IEEE 802.11 WLAN. From the WLAN point of view, the entire UMTS network appears as a basic service set or a picocell

FIGURE 23.7
VAP-based intercon-
nection to the UMTS.

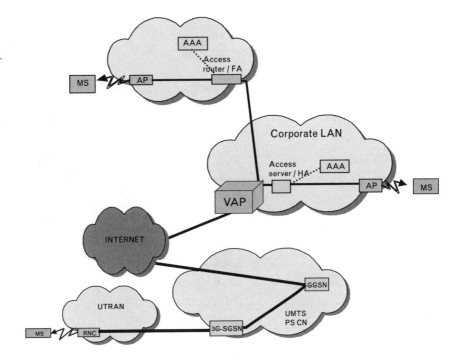

associated with another AP (in this case VAP). The function of the VAP is to communication with mobile stations connected through UMTS, deencapsulate their packets, and transmit them on the LAN. After this is done, the packets will reach the final destination through the router attached to the LAN. The protocol stacks architecture is a modified version of that in Figure 23.2. Only the VAP entity protocol stacks are placed after the GGSN protocol stacks. The protocol stack of the VAP entity is presented in Figure 23.8(a). The VAP-based interconnection also requires some modification on the MS protocol stacks, which is presented in Figure 23.8(b).

In this interconnection the VAP become an adopted unit in the user plane protocol of the UMTS architecture. For this reason the IEEE 802.11 MAC protocol is implemented on top of the protocols of the UMTS GGSN part. This is done for both the MS and the VAP entity. From the GGSN part of the UMTS network all protocols up to GTP-U level will be mapped onto IEEE 802.3 MAC so that the WLAN network sees the VAP as an AP. On the VAP side the UDP/TCP is on the top of the stack. The 802.11 MAC protocol that is implemented in the MS is a level below the UDP/TCP. Hence, the protocols below the 802.11 MAC protocols including it are mapped to the 802 LLC. The 802.3 MAC protocol in the GGSN part is mapped onto VAP below the IP/PPP in the VAP protocol stack.

FIGURE 23.8
(a) Protocol stacks architecture associated with the VAP based interconnection. (b) Protocol stacks architecture for the MS associated with the VAP-based interconnection.

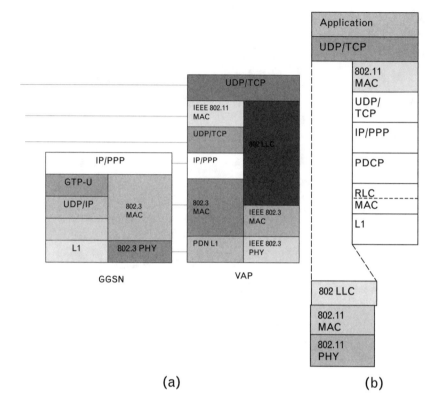

(a)   (b)

Pros and Cons

This type of interconnection does not have significant advantages over the other four types of interconnection. It is not clear how the VAP will operate with the regular 802.11 WLAN APs. From Figure 23.8, it is clear that over the UMTS air interface, each packet will have twice the UDP/TCP and twice the IP/PPP headers. In addition, there will be an IEEE 802.11 MAC header along with the UMTS-related headers. The overhead of packets on the low bandwidth air interface is quite large, which makes this interconnection inefficient.

23.6   Interconnection Between UMTS and WLAN Through Mobility Gateway

The interconnection architecture is presented in Figure 23.9. An intermediate server (mobile proxy) is placed on either the UMTS or the IEEE 802.11 WLAN sides, and the *mobility gateway* (MG) will handle the routing and mobility issues.

FIGURE 23.9
*MG-based intercon-
nection between IEEE
802.11 WLAN and
UMTS.*

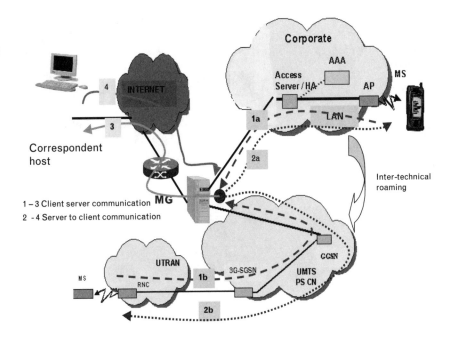

Referring to the figure, when an MS is attached to an AP, the communication path between the MS and a CH on the Internet will be 1a–3. The CH-MS communication path will be 4–2a. When the MS is on the UMTS network, this path will be 1b–3 and reverse path is then 4–2b. It should be observed that segments 3 and 4 in both paths do not change regardless of where the MS is located. Only links 1 and 2 will be continually changing depending on the movement of the MS. Clearly, only the communication between MS and the proxy server alone is subject to change, while maintaining the proxy-CH connection unchanged supports mobility. Nothing changes on protocol stacks architecture of the WLAN network. The protocol architecture of the UMTS network is a modification in which the MG is placed next the GGSN part of the protocol stacks of the UMTS. On the MS there will be some protocols adopted if the user wants roaming between both networks when he or she is still in one of the networks. So this functionality does require a dual-mode stack implementation on the MS. Figure 23.10 shows the protocol stack architecture for both the MG entity and the MS.

Pros and Cons

There are several advantages to employing proxy architecture for intertechnology roaming. The proxy architecture is scaleable. There is the possibility of further minimizing the encapsulation and routing inefficiencies with Mobile IP. However, the real reduction in overhead may not be very

FIGURE 23.10
Protocol stacks corre-
sponding with the MS
and the MG.

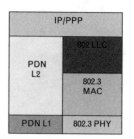

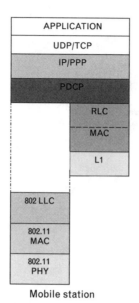

significant due to the need for additional control protocols. If the proxy is under the control of the same organization that owns the MHs/MS, it is possible to configure the proxy to support the peculiar needs of its population of mobile hosts. An optimized protocol may be run between the mobile and proxy depending on the link being employed. The proxy can manage resource-poor connections more efficiently. For example, it can drop structured data such as e-mail headers and frames in a MPEG stream selectively or drop unstructured data.

Proxies are already in place in many organizations as firewalls or Web caching servers. These may be reused for mobility management and inter-technology roaming. Proxies can be used for logging the characteristics of a connection, details of which may be usefully employed in various applications including accounting and management.

The main disadvantages of the proxy architecture are as follows. First, the architecture is not standardized and therefore requires proprietary protocols for intertechnology roaming. Second, the performance of a proxy is poor since a significant latency is added to the client-server communication path. Potentially, the end-to-end semantics of the transport protocol may also be violated. If a single proxy is employed and it fails, this may result in the failure of the entire network, so there is a need to have some fault tolerance. Third, there are still some open issues in the development of protocols for mobility management with the proxy architecture. The placement and number of proxies to be employed may depend on the situation. It is preferable to have the proxy connected to the last links of each services that the MS may use so that it can gather information about quality of each of the last links. However, the ownership of such a proxy will be contentious. The number of proxies that have to be placed for optimum performance is also subject to network conditions.

23.7 Interconnection Between UMTS and WLAN Based on Mobile IP

The interconnection architecture related to Mobile IP is presented in Figure 23.11. Mobile IP is employed to restructure connections when a mobile station roams from one data network to another. Outside of its home network, the MS is identified by a COA associated with its point of attachment, and a colocated foreign agent that manages deencapsulation and delivery of packets.

The MS registers its COA with an HA. The HA resides in the home network of the MS and is responsible for intercepting datagrams addressed to the MS's home address as well as encapsulating them to the associated COA. The datagrams to an MS are always routed through the HA. Datagrams from the MS are relayed along an optimal path by the Internet routing system, though it is possible to employ reverse tunneling through the HA.

The required dual-mode MS protocol stack is the same as in Figure 23.10. It is clear that both networks are peer networks and that the functionality of the HA/FA exists at the IP layer.

FIGURE 23.11
Interconnection architecture between WLAN and UMTS based on Mobile IP.

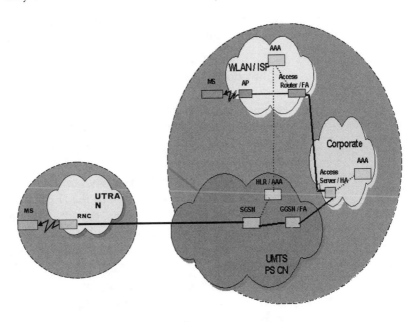

Pros and Cons

The advantage of this interconnection is that it is based upon the Mobile IP, which makes the IP address mobile. The same IP address is used, which solves the multiple address problems. To solve the packet duplications due to the lifetime of the routers, some conventions on both IEEE 802.11 WLAN and UMTS networks are needed. The databases of both networks may need to communicate to overcome packet duplication.

The main disadvantage of Mobile IPv4 is the triangle routing. This could be overcome with Mobile IP with optimized routing. This is important for real-time applications like video and audio transmission.

23.8   Handover Between IEEE 802.11 and UMTS

The motivation for intertechnology (vertical handover) in the hybrid mobile data networks arises from the fact that no one technology or service can provide ubiquitous coverage, and it will be necessary for a mobile terminal to employ various points of attachment to maintain connectivity to the network at all times. There is a clear distension between the two types of handover: horizontal and vertical handover. Horizontal handover refers to handover between node Bs or APs that are using the same kind of network interface. Vertical handover refers to handover between a node B and an AP or vice-versa that are employing different wireless technologies. In the case of a vertical handover, two diffrences are seen:

1. Upward vertical handover, which occurs from IEEE 802.11 WLAN AP with small coverage to an UMTS Node B with wider coverage;

2. A downward vertical handover, which occurs in the reverse direction.

A downward vertical handover has to take place when coverage of a service with a smaller coverage (as in WLAN service) becomes available when the user still has connection to the service with the UMTS coverage. An upward vertical handover takes place when MS moves out of the IEEE 802.11 WLAN coverage to UMTS services when it becomes available even though the user still has connection to the IEEE 802.11 WLAN coverage. In the case of the vertical handovers, the MS/MH decides that the current network is not reachable and performs handover to the higher overlay UMTS network when several beacons from the serving WLAN service are not available. It instructs the WLAN to stop forwarding packets and routes this request via Mobile IP registration procedure through the UMTS core network. When it is connected to the UMTS network, MS listens to the lower layer WLAN AP, and if several beacons are received successfully, it will switch to the IEEE 802.11 WLAN network via the Mobile IP registration process. Thus, the vertical handover decisions are made on the basis of the presence or absence of beacon packets

23.9   Handover Aspects Between IEEE 802.11 WLAN and UMTS Based on Mobile IP

Handover is the mechanism by which ongoing connection between MS and CH is transferred from one point of access to another point while maintaining the connectivity. When an MS moves away from an AP or from a node B, the signal level degrades and there is a need to switch communications to another point of attachment that gives access to the existing IEEE 802.11 WLAN network or UMTS network. Handover mechanism in an overlay UMTS and underlay WLAN network could be performed such that the users attached to the UMTS just occasionally checks for the availability of the underlay WLAN network. A good handover algorithm is needed to make the decision when to make handover in order to avoid unnecessary handover (i.e., the ping-pong effect). handover algorithm is beyond the scope of this chapter. This section discusses the handover procedure and the mechanism from WLAN to UMTS and vise-versa based on the *received signal strength* (RSS) metrics. Which means that the handover initiation or the handover triggering is sensitive to these signals. Figure 23.12 shows the handover procedure from one network to another.

FIGURE 23.12
*Handover procedure
between WLAN and
UMTS.*

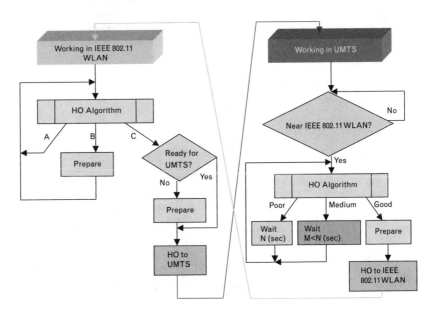

An MS moving from the WLAN network coverage may suddenly experience severe degradation of service and will have to perform handover very quickly to maintain the higher-layer connection. The following stages occur while an MS moves away from the coverage of WLAN within the UMTS coverage (Figure 23.13):

1. The signal received from the AP in the WLAN network is initially strong and the MS is connected to the WLAN network, which is also the home network of the MS and also the HA in this network.

2. The signals from the AP become weaker when the mobile moves away. The mobile scans the air for another AP. If no AP is available, or if the signal strengths from the available AP is not strong enough, the handover algorithm uses this information along with other possible information to make a decision on handing over to the higher-overlay UMTS network. Connection procedures are initiated to activate the UMTS PCMCIA card.

3. The handover algorithm in the MS decides to dissociate from the WLAN and associate with the UMTS network.

4. The FA is activated, used by the MS dual PCMCIA card and the Mobile IP, and the MS gets a COA for visiting the UMTS network as a foreign network.

FIGURE 23.13
Messages and signaling of the handover procedure from IEEE 802.11 WLAN to UMTS.

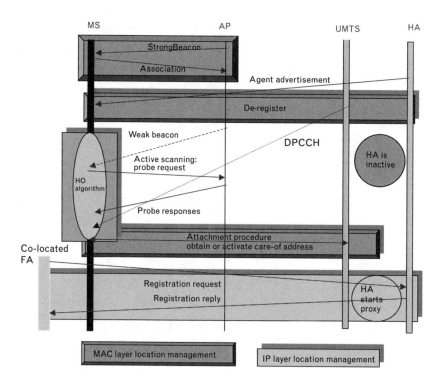

5. The HA in the WLAN is informed about the new IP address through a Mobile IP registration procedure and it performs the proxy *Address Resolution Protocol* (ARP) and intercepts the datagram. The HA encapsulated datagrams tunnels any packets arriving for the MS to the FA of the UMTS networks. At the end of the delivery the MS will deencapsulate and get the datagrams.

In this case, the handover algorithm determines that there is no local coverage available via WLAN, and handover must be performed to the UMTS network, assuming that a UMTS service is always available to the MS.

Once the MS is attached to the UMTS, it constantly monitors the air at repeated intervals to see whether or not a high data rate WLAN service is available. As soon as such a service becomes available, the handover algorithm should initiate an association procedure to the newly discovered AP.

The procedure for this reverse handover from UMTS to WLAN is as follows:

1. The signal from the WLAN AP is initially not detected.

2. The MS then detects a beacon, which indicates that the underlay WLAN network is available.

3. The handover algorithm decides on making an handover from UMTS to the WLAN network.

4. The FA in the UMTS network is deactivated and updated by Mobile IP, and the home IP address is used.

5. The HA in the WLAN network is instructed by the MS to no longer do a proxy ARP on its behalf through the Mobile IP.

23.10 Conclusions

This chapter presented five interconnections and seamless intertechnology handovers dealing with interworking aspects between IEEE 802.11 WLAN and UMTS. The concepts focused on the network layer and the link layer in order to minimize existing networks and technologies, especially at the lower layers such as MAC and PHY layer.

Based on the arguments presented for the given interworking approaches, Mobile IP–based interconnection architecture is selected as the most suitable solution. IP layer mobility management provides an efficient way to interconnect heterogeneous packet-oriented networks. Interworking cannot be handled within a proprietary protocol of one network technology, and it has to be handled either in the existing layers above or a new layer has to be added solely for the purpose of handling intertechnology roaming. In either case, there is a need for modification of existing protocols at least between the MS and a network entity that handles the mobility.

Finally, intersystem handover is a topic that will become more and more important in the evolutionary path towards UMTS and 4G wireless infrastructures. Mobile IP–based intertechnology roaming is an obvious step on this path. In particular, fast handover required for real-time services needs to be studied further within a platform.

REFERENCES

[1]   Prasad, N., and A. Prasad, *WLAN Systems and Wireless IP for Next Generation Communications*, Norwood, MA: Artech House, 2002.

[2]   IEEE, *IEEE standard for WLAN MAC & PHY specifications*, IEEE, 1997, p. 445.

[3]   Prasad, N. R., "An Overview of General Packet Radio Services (GPRS)," *Wireless Personal Multimedia Communications (WPMC'98)*, Yokosuka, Japan, November 4–6, 1998.

[4]   IP Mobility Support for IPv4, http://www.ietf.org/rfc/rfc3220.txt.

[5]   "Quality of Services Concept," UMTS 23.07 V0.4.0, http://www.3gpp.org.

CHAPTER 24

Location-Based Push Architectures for the Mobile Internet

Günther Pospischil, Johannes Stadler, and Igor Miladinovic

24.1 Introduction

Push functionality for the mobile Internet has gained much attention recently. It allows the establishment of a connection to a user terminal without user intervention and the delivery of content as soon as it is available. In particular, the "always on-line" paradigm and the limited air interface bandwidth make push an important component of mobile Internet services/applications. Using the telecommunications nomenclature, the term "application" is used to describe the implementation of dedicated function blocks (e.g., a call control application or a specific push application). If a general, abstract functionality is considered, the term "service" is used.

Several standardization groups are currently working on architectures for push services in mobile networks. In today's GSM, services are usually based on WAP. In the most recent specifications, the WAP Forum has defined an architecture for WAP Push [1]. For UMTS, standardization is done by 3GPP, which specifies the UMTS service architecture and an advanced messaging concept called *Multimedia Messaging Service* (MMS) [2]. Packet switched UMTS sessions will be handled via SIP [3], adopted by the IETF. With some extensions, SIP can also be used for messaging and asynchronous notifications [4, 5].

The convergence of IP networks (Internet) and SS7-based mobile networks towards a common mobile Internet makes a unified service architecture for push scenarios possible. Before developing such an architecture, a global view of push mechanisms is given that considers the possible use cases and their technological implications. This chapter starts with a section defining various push scenarios and deduces the key aspects that a push architecture should provide. In Section 24.3 possible technical solutions based on

SMS, WAP, and SIP are discussed. Section 24.4 provides an assessment of location-based push in two major use cases.

24.2 Push for Mobile Networks

Push services use the limited bandwidth of mobile networks efficiently because communication only occurs if there is information available. In pull-based communications, the user performs periodic checks for information—even if no new content is available. This requires the establishment of a data channel, which takes considerable time and consumes resources in the network (air interface bandwidth, buffers, and IP addresses) and terminal (transmit power, memory).

Additionally, push services are suitable for the notification of asynchronous events. This may happen frequently in mobile networks, because users can select relevant information a priori (e.g., subscribe to traffic information channels) and receive information directly ("always connected" paradigm). Especially for mobile networks, notifications are used for location-based events (see Section 24.3).

Mobile networks are early adopters of push features. It started with the SMS of GSM. Currently, an extension called MMS [2] is being standardized. In MMS, which uses WAP or Mobile Execution Environment Applications (MexE; i.e., JAVA+TCP/IP), a message may consist of text, pictures, audio, and video. SIP integrates all these types of content and offers possibilities for real-time sessions, like streaming or VoIP. Additionally, SIP *user agents* (UAs) will be available in all UMTS multimedia terminals, supporting bearer establishment directly. As there are also SIP UAs for conventional PCs, "service portability" is automatically given (i.e., a user may access a service with any kind of terminal).

24.2.1 Types of Push

There are many types of push, as follows:

- *Confirmed/unconfirmed:* Confirmed push uses an acknowledgment mechanism to indicate successful data delivery. Unconfirmed push does not provide any delivery guarantees. For important data or specific billing models (pay-per message), confirmed push should be used. If data is not so important (e.g., because it is valid only a short time), unconfirmed push can be used. An example of important and time-critical data is stock news, while location information data may be less important because it is replaced with a new location estimate after a few minutes.

- *Visible/hidden:* Visible push includes user notification, while hidden push delivers new information to a specific application (application-related push). Examples are multimedia messages or text/HTML-based advertisements (visible push) and location updates within a navigation service (hidden push). For the application-related push, an application addressing scheme is necessary.

- *Unicast/multicast/broadcast:* Unicast data is typically related to specific applications (hidden push) (e.g., delivery of location estimates to a navigation application or a user-to-user data transfer) like MMS or SMS. Multicast and broadcast data can either be application related (e.g., delivery of traffic information to all users of a navigation application) or visible (e.g., advertising to all users who have subscribed to the "advertising channel").

- *Global/regional (location based):* Current messaging systems offer global service (i.e., content is not related to the user's location). UMTS provides the additional potential to create location-based push services (e.g., location-based advertising or traffic information).

- *Connection oriented/connectionless [1]:* Connection-oriented push is just a network-initiated session establishment, which allows the transfer of any kind of information afterwards. Connectionless push is the delivery of a single piece of information without establishing a connection for subsequent data transfers.

24.2.2 Push Scenarios

Three kinds of push services are considered in this chapter: location/navigation, information distribution/advertising, and extended Internet applications.

24.2.2.1 Location/Navigation Push Service

Location-based services will often need updates on the user's position. As user movements are not deterministic, a pull-based approach is inefficient; push communication is more adequate. Depending on the service, there may be an active connection between the mobile terminal and the network, or not. If the connection is available, any remote communication concept can be used—for example, JAVA *remote method invocation* (RMI), *Common Object Request Broker Architecture* (CORBA), or socket connections. If no connection is established, an architecture that supports bearer establishment is needed. SIP provides all required functionality for this unconfirmed, hidden, unicast push.

24.2.2.2   Broadcasting Push Service

This group of push services consists of information distribution (e.g., traffic news, stock exchange data, and advertising). Optionally it may incorporate a location-based component (i.e., the service contacts only users that are in a specified region). This allows only potentially affected users to be informed about a traffic jam or be presented special offers in a user's vicinity. The communication pattern is unconfirmed (or confirmed), visible, multicast, connectionless, regional push.

This type of services offers new revenue models for operators and subscribers, including sponsoring of airtime. This scenario is already popular in the Internet, where many services are free of charge for the user because they are financed via advertising.

24.2.2.3   Extended Internet Applications

Basically, these are electronic commerce applications as already available in the Internet. They can be extended for mobile commerce (e.g., by using mobile payment solutions or SIM-based security mechanisms). These services use push for messaging/notification (e.g., deliver new mail without polling or notify a user about transaction status). All types of push may occur, depending on the application.

24.2.3   Security for Push

This topic is addressed only briefly in this chapter. There are three basic security concerns: protecting data traffic, ensuring authenticity of sender and receiver, and maintaining privacy. The privacy issue is discussed in Section 24.4, where a special focus is placed on the location information of a user. The other aspects (protecting data traffic and authenticity) are not specific to push, rather, they are relevant for all SIP-based communications.

The simplest solution is to use a trusted network that is state of the art for mobile networks. In the Internet world, however, services run on top of a network without any guarantees. Therefore, concepts like *Pretty Good Privacy* (PGP) [6] have been developed for authentication and protection of content and/or signaling.

SIP is usually based on IP networks, which is also assumed for SIP push mechanisms. Hence, any IP layer security concept, like the *Secure Internet Protocol* (IPsec) [7] can be used for SIP sessions. Additional SIM-based security mechanisms are possible in UMTS mobile networks [8]. Transport security is provided via the UMTS *Authentication and Key Agreement* (AKA) protocol [9].

A specific problem of application-related push is that a push initiator could probe which applications a user has installed. This problem can be solved with SIP quite easily as presented in Section 24.3.3.7.

24.3 Technical Aspects

Push services rely on a concept for bearer establishment and a well-defined data distribution architecture. Consequently, bearer establishment is discussed in the next paragraph, followed by a detailed presentation of different push architectures. This section ends with a paragraph on push data management.

24.3.1 Bearer Establishment

Setting up an IP bearer in the packet switched UMTS domain is called PDP context activation. The GGSN starts a network-initiated PDP context activation as soon as it receives a PDP PDU.

Current GSM/GPRS networks do not support network-initiated PDP context activation; therefore, real push is not possible with GPRS. Workarounds are the use of GSM circuit switched data connections or SMS in conjunction with a "call back" application in the terminal (which "pulls" the information after a SMS push notification). This limits the applicability of SIP push in current networks because (1) a supporting call back application is needed in the terminal, and (2) additional bearer level messages are exchanged, causing a fairly large protocol overhead (data and delay).

It should be noted, however, that the described problems for bearer establishment are inherent to the existing GPRS networks—they are not SIP specific. Standardization regarding network-enabled PDP context activation is almost finished—it will be possible to use this feature in GPRS networks soon. UMTS networks using SIP to establish VoIP sessions and other multimedia sessions avoid these problems. In the *IP Multimedia Subsystem* (IMS) of UMTS, a combined bearer and session establishment with SIP will be possible [10].

24.3.2 Application Layer Signaling

There are several application layer signaling protocols in mobile networks and the Internet. Most of them are tailored to specific session types (e.g., SMS for short text "sessions" or WAP for simple text-based mobile Internet content). The remainder of this section gives a brief overview on SIP, SMS, WAP, and H.323 protocols.

24.3.2.1 SIP

SIP is an application layer control protocol developed in the MMUSIC working group of the IETF [3]. SIP can establish, modify, and terminate all types of sessions. With some extensions SIP can also be used for presence and instant messaging applications. SIP is a text-based, HTTP-like protocol, with two types of messages: requests and responses. A SIP message can transport any MIME content. For VoIP applications and multimedia conferencing applications the *Session Description Protocol* (SDP) [11] is commonly used. Multimedia transmission usually adopts the *Real Time Control Protocol* (RTCP) and the RTP.

SIP uses e-mail–like addressing, consisting of a user identifier and a network identifier (e.g., somebody@somehost.somedomain). There are two types of entities in SIP: SIP user agents and SIP network servers. Figure 24.1 shows two terminals with SIP UAs (U_1 and U_2) and three SIP network servers (S_1, S_2, S_3). SIP network servers are used to forward SIP session requests within the network until they reach the called SIP UA. There are three types of SIP network servers: SIP proxy server, SIP redirect server, and SIP register server. A proxy server forwards SIP requests towards the called user (i.e., to another proxy server on the path or directly to the called SIP UA). A redirect server simply responds with the next server on the path. In a second step, the caller has to establish the connection to the next server. Obviously, SIP network servers need some knowledge about the route to the called SIP UA. For this purpose, SIP reuses existing mechanisms, like finger, conventional packet routing, or multicast on a local network. Additionally, SIP provides a registration service to maintain a database with mappings between SIP address and physical address. A user has to send a register message, including his or her current network address (e.g., IP address), to the corresponding register server every time he or she wants to be reachable. Validity of a register message is always time limited. A user can also unregister by sending a register message with a time limit of zero. Registration functionality is usually included in a SIP proxy.

In Figure 24.1, the SIP UA of U_1 forwards all SIP requests to SIP proxy S_1 (step 1). S_1 tries to contact U_2 via the register server S_2 in U_2's domain (step 2). S_2 uses non–SIP communication to obtain the current registration data from the register database server S_{2a} (steps 2a, 2b). S_2 responds with U_2's registered address, which is S_3 in this scenario (step 3). Now S_1 forwards the request to S_3 (step 4). But S_3 acts as SIP proxy again and forwards the request to terminal T_2 of user U_2 (step 5).

Now user U_2 may choose to accept the request and create a positive response (200 OK), which traverses the network via S_3 (step 6) and S_1 (step 7) until it finally reaches the caller (step 8).

As mentioned above, there are SIP extensions for presence and instant messaging [4, 5]. Three new request methods are proposed: SUBSCRIBE

FIGURE 24.1
SIP architecture.

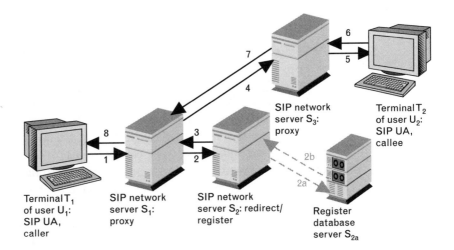

and NOTIFY for presence and MESSAGE for instant messaging. These methods can be used for push services, too. For instance, a user subscribes for a push channel (e.g., traffic alerting) with the SUBSCRIBE method. All events are pushed to subscribed users with the NOTIFY method. This replaces the periodic queries for new information (pull scenario) with a single subscribe message. The user can also unsubscribe a channel if hen or she is not interested anymore. A detailed description of the SIP push architecture is given in Section 24.3.3.

The benefits of using SIP for push services include the following:

- SIP UAs are available for Internet terminals and future mobile multimedia terminals.

- SIP architecture is decentralized and scalable, based on an open Internet standard.

- SIP uses an easy-to-implement, text-based protocol. Still, it supports flexible MIME content and description/control of real-time sessions, using SDP and RTCP.

There are also some disadvantages if SIP is used for push data transfers:

- SIP is developed for session control, not for messaging. Therefore, the mentioned extensions are necessary.

- SIP is a relatively new protocol that currently does not enjoy widespread use. However, the number of SIP implementations is growing fast, and 3GPP has decided to include SIP in UMTS.

24.3.2.2 SMS

SMS is an integrated element of GSM and UMTS. It is suitable for push delivery of short text messages. The SMS architecture consists of a receiving SMS UA (integrated into the mobile phone) and several network elements that store and forward the messages (Figure 24.2). Originators of short messages may be mobile terminals with SMS UAs or a fixed network entity. All short messages are forwarded to the *service center* (SC), which is responsible for storing and forwarding the messages (step 1). The SC itself is not part of the mobile network, but it may be integrated into the MSC of the mobile network. An SC forwards the short message to the SMS-Gateway MSC (SMS-GMSC, step 2). The SMS-GMSC contacts the HLR, which is the user database of the mobile network. The HLR responds with the current user status and location (steps 2a, 2b). Now the SMS-GMSC connects to the current serving network for the desired receiver (step 3). If the receiver is currently connected to a circuit switched GSM network, the message is forwarded to the S-MSC. In packet switched networks (GPRS or UMTS), the short message is delivered via the SGSN. The SGSN stores the current address of the terminal internally, and the S-MSC connects to the VLR to obtain the required information (steps 3a, 3b). Finally, the message is delivered to the recipient (step 4), and a confirmation is sent back to the SC automatically (steps 5, 6, 7). Optionally, a confirmation can also be sent to the originator of the short message (step 8).

The advantages of the SMS push concept are as follows:

- The system is implemented and proven in all GSM networks.

- SMS scales well.

- Roaming is possible.

But GSM/UMTS short messages have major disadvantages, too:

Figure 24.2
SMS architecture.

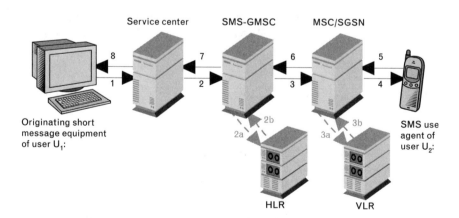

- Interworking with the Internet is only possible in one direction: Internet hosts may send but not receive short messages.

- SMS content is limited to 160 characters of text. Ringtones or animations can be transferred via the *Enhanced Messaging Service* (EMS).

- Application addressing is not possible, although there are dedicated short messages for specific features of the mobile terminal (e.g., to configure the WAP browser).

- The initiator of a short message has no control of the message delivery—there are no hard delivery guarantees.

- Automatic message forwarding to other terminals is not possible —multipoint communication is not available.

24.3.2.3   WAP

The WAP push architecture (Figure 24.3) is a candidate for the UMTS MMS. It is quite similar to the SIP architecture, including a WAP push user agent (similar to the SIP UA) in the terminal and a WAP *push proxy gateway* (PPG). The WAP PPG is similar to the SIP proxy. It accepts push requests from a push initiator in the Internet (step 1), using the *Push Access Protocol* (PAP). Before establishing a connection to the terminal, the PPG may connect to an authentication and preference server (steps 1a, 1b). The connection to the terminal typically includes two phases. In the first phase, a data connection to the terminal has to be established. This can be done by sending a specific short message to the *session initiation application* (SIA) residing in the terminal (step 2). The SIA now establishes a data connection to the PPG (step 3); optionally it responds with some user preferences (e.g., if the

FIGURE 24.3
*WAP push
architecture.*

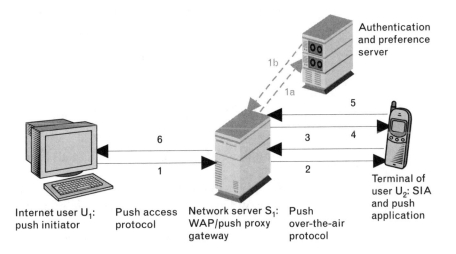

Internet user U₁:
push initiator

Push access
protocol

Network server S₁:
WAP/push proxy
gateway

Push
over-the-air
protocol

Terminal of
user U₂: SIA
and push
application

requested receiver application is available). Now phase 2 takes place, which delivers the push data to the receiver application (step 4). This task is performed by the PPG using the push *Over the Air Protocol* (OTA). The OTA is based on the WSP (for connectionless push) or the Internet HTTP (for connection-oriented/confirmed push). Finally, an acknowledgment is generated by the receiver and transferred to the push initiator via the PPG (steps 5 and 6).

The PAP is primarily used to deliver the push content to the PPG. It may also control the PPG (e.g., cancel a pending push request, inquire the status of a push request, or obtain the client capabilities). PAP content is given in the *Extensible Markup Language* (XML) and MIME format.

Client capabilities are identified according to the WAP user agent profile (UAProf) format, which is based on the composite capability/preference profiles (CC/PP) [12], specified by the *World Wide Web Consortium* (W3C). Based on these capabilities, the PPG may perform content translation [e.g., the *Wireless Markup Language* (WML) into the *Wireless Binary XML* (WBXML)].

WAP push addressing consists of client addressing and application addressing. The client address is an arbitrary identifier that is understood by the PPG (e.g., mail address, phone number, or IP address)—the application address is basically an "application name," given as *uniform resource identifier* (URI).

WAP push addresses many of the goals that are considered relevant for advanced push architectures:

- Integrated security mechanisms are present.

- Application addressing is possible.

- Connection-oriented/confirmed push and connectionless/unconfirmed push are supported.

- Limited multimedia support, based on non-real-time MIME data transfer, is available.

There are also several drawbacks of WAP push:

- Streaming and other real-time multimedia features are not supported.

- WAP lacks Internet integration (i.e., usually there are no WAP browsers or WAP applications in Internet terminals).

- Currently there is no support for advanced push concepts like location-based push or data forwarding.

As stated above, the lack of network-initiated bearer establishment is also a problem for WAP push. In the case of WAP, the solution is based on SMS and an SIA. The WAP push architecture is asymmetric, needing a specific WAP push initiator application. The proposed SIP push architecture is symmetric, where (in principle) every SIP UA may issue a push request.

24.3.2.4 H.323

H.323 is a series of recommendations for multimedia communications issued by the ITU. It has been designed for networks without QoS guarantees, like IP-based LANs and the Internet. H.323 consists of a group of mandatory protocols for control (H.245), connection establishment (H.255.0), and audio (G.7xx). Additional media types and services, like video (H.26x), data (T.120), or interworking with circuit-switched services (G.246), are optional.

All H.323 terminals support audio, based on RTP/RTCP and G.711 codec, and the control protocols H.245 and H.225. H.245 is used for negotiating channel usage and capabilities; call signaling and setup is defined in H.225. A H.323 terminal is similar to the SIP UA.

H.323 gateways provide interworking functions between IP networks and circuit switched networks, like ISDN. Interworking needs signaling translations and codec conversions. In an LAN, H.323 terminals can communicate directly with each other; gateways are optional.

Gatekeepers are used for call control, especially address translations (symbolic name to IP address), bandwidth management, admission control, and equipment registration (H.323 zone management). Gatekeepers are optional; they are comparable to SIP redirect/register servers. An H.323 system may additionally include *multipoint control units* (MCUs) for large multiparty sessions.

H.323 is specifically designed to support interworking between different networks (e.g., the IP-based Internet and ISDN). Figure 24.4 shows such a scenario. User U_1 is connected to an LAN and wants to establish a session to user U_2, using an ISDN terminal. The caller contacts the gatekeeper (step 1), which obtains the settings of user U_2 from the registration database (steps 1a, 1b). The registration data is returned to U_1 (step 2). This feature of H.323 is quite similar to an SIP redirect/register server. Now the actual session can be established via the gateway S_2 (steps 3, 4). User U_2 accepts the connection, and a connection response is sent back to user U_1 (steps 5, 6).

For transport of push data (e.g., an HTML file), the T.120 recommendation can be used. T.120 is a part of H.323 and provides optional data capabilities in H.323 (e.g., application sharing, file transfer, and instant messaging). Since H.323 uses standardized services and lacks

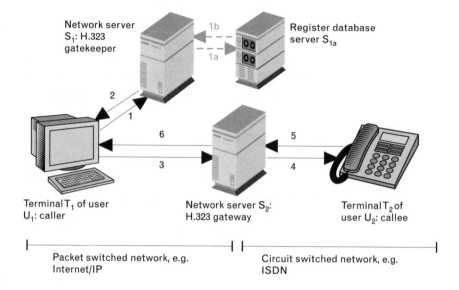

FIGURE 24.4
H.323 architecture.

SUBSCRIBE/NOTIFY functionality, integration of push concepts into H.323 is not straightforward.

The advantages of H.323 are as follows:

- Standardized codecs for audio and video transmission;

- System and network independence;

- Several implementations are already available;

- Integrated bandwidth management based on the gatekeeper;

- Multipoint support.

The disadvantages of using H.323 are as follows:

- Difficult system administration and integration as well as dynamic network ports make H.323 incompatible with usual packet filtering firewalls.

- H.323 uses binary protocols, which are bandwidth efficient but difficult to implement and debug.

- H.323 is a very complex system, based on several hundred pages of standards. Implementation into "weak" terminals, like mobile phones, is difficult.

- There is no instant messaging support. H.323 uses detailed standards, making extensions (e.g., for instant messaging) very difficult, even if the basic capabilities (e.g., T.120 data recommendation) are available.

24.3.3 Architecture Components for SIP-Based Push

This section provides a detailed description of an SIP architecture for push services in mobile networks. Typical usage scenarios are shown in Section 24.4. The SIP concept was chosen because it fits nicely into the standardized UMTS service environment. The extendible, text-based, lightweight concept of SIP makes it suitable for implementation into virtually all kinds of terminals. This is the prerequisite for true service portability.

24.3.3.1 Overview

In Figure 24.1, the basic SIP architecture, consisting of SIP UAs and several SIP network servers was shown. All these entities belong to the network provider domain (i.e., the transport infrastructure). In Figure 24.5 this architecture is extended with the service provider domain and the PARLAY/OSA [13] interface that connects both domains. By using these extensions, it is possible to offer advanced push services, like location-based push or content provider–initiated group push (to several users simultaneously).

The network provider runs a SIP-enabled communication network (e.g., a UMTS network). Call management is performed via the SIP proxy, which is called CSCF in UMTS. It is used to forward SIP messages to their destinations and to run some applications. It may need user preferences and location information. In UMTS networks, this can be achieved by contacting the HSS, the central user database, and the *gateway mobile location center* (GMLC), used to determine a user's location, using MAP/Cx and MAP/Le interfaces.

The signaling in the terminals is done by SIP UAs, making no difference if the end user is a content provider or not. SIP is designed for call control, but it may transport arbitrary MIME content in the body of a message. Even Web pages using HTML can be transferred in this way. This

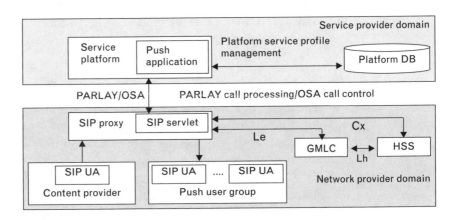

FIGURE 24.5
SIP push architecture.

circumstance offers the possibility of HTML pushing behavior instead of the usual HTTP/HTML pull concept.

The service provider operates a platform for the creation, operation, and maintenance of applications. These applications can be offered to one or more network providers, using their infrastructure for data transport between various clients and servers. A standardized interface, like PARLAY, between service provider and network provider helps to ensure service portability and to reduce implementation effort and time to market.

A detailed description of the architecture components is given in the following sections.

24.3.3.2 Push Application

The push service consists of two parts. The data delivery part runs in the SIP proxy and is described in the next section. The second part resides on the service platform of the service provider. It can be seen as business logic, being responsible for the management of push subscriptions and interaction with other services. End users may sign in for push services either explicitly (e.g., on a Web form or with a SIP SUBSCRIBE message) or implicitly. An example of the latter is a subscription to an advertising tariff, where push advertising is mandatory. All users who have signed in for a certain push service (push channel) are included in a user group that has a specific SIP address.

The platform database (platform DB in Figure 24.5) holds the list of subscribers for a push channel—that is, it performs the mapping between the channel address (SIP user address) and the SIP addresses of its members.

The second task is the communication with the SIP proxy of the network provider via the PARLAY API. This is necessary to receive push requests and return the list of subscribers after accessing the platform database.

If users subscribe to a push channel during a pending push request (i.e., before push request expiration), they should be added dynamically to the list of receivers. In this case the application has to contact the SIP proxy again to initiate the data transfer to the additional users.

24.3.3.3 Push Gateway

The push gateway consists of an SIP proxy with an additional SIP push servlet. If an SIP message (request or response) is received, the appropriate servlet is activated. The message is forwarded to this servlet, which executes some code to handle the request.

Consider a push servlet, getting a request that includes push-related information. The push information may consist of an HTML push content, the location information for location-based scenarios, and the expiration time. The latter indicates how long the push is valid in time. It is possible to use the expire header of the SIP request for this purpose. The push content and the location information are transported as multipart message in the SIP body. The location information is given in some user and system readable format, preferably XML as proposed in [14].

All this information has to be identified and stored until the push request is completed. After receiving the list of push subscribers from the push application, the servlet determines the push settings and location of every subscriber. The servlet contacts the HSS (see Section 24.3.3.4) to identify the current push settings of the desired user; the coordinates are supplied by the GMLC (see Section 24.3.3.5). If the user is in the desired region and has push enabled, the servlet forwards the push request to the user's SIP UA.

A detailed description of this behavior is presented in Section 24.4.1.

24.3.3.4 HSS/HLR

The HSS/HLR is the central user database of a UMTS/GSM mobile network. It is used to store service-specific data and general subscription information (e.g., the user's IMSI number) or tariff model. In the push scenario, two parameters are relevant: push settings and localization settings.

Via the push settings of the HSS, a user may quickly disable or enable all push services he/she is subscribed to. This avoids the situation where a user has to contact all service platforms and modify the subscription profile there. Optionally the HSS may contain lists of content providers or push applications that are allowed to perform push; it is also possible to block certain push originators.

This concept has been developed in analogy to the standardized localization security exception settings of the HSS [15]. A user may decide who is allowed to know his/her location and enter appropriate settings in the HSS. Again it is possible to enter some global settings (i.e., allow or block all localization requests) and to enter individual settings, specifying all allowed (or blocked) requestors separately.

For obvious reasons, a convenient access to the HSS via the mobile terminal is needed. Therefore, an HTML/WML interface to the HSS should be provided by the mobile operator. A special challenge in this context is authorization, but the SIM-based security mechanisms can be used for this purpose, just as it is currently done for accessing the mobile voice mailbox.

24.3.3.5 GMLC

The GMLC is used to determine the user's location for location-based push requests. There are three possible situations: (1) the user has blocked localization, (2) a user is in the region, and (3) a user is outside the push region.

If a user has blocked localization, no location-based push data will be delivered; it is assumed that the user is permanently outside the target region. If localization indicates that a user is in the push region, the push content is immediately forwarded to the user's terminal. In the latter case, a trigger is set within the GMLC. It includes the desired push region and the expiration time of the push request. If the user enters the push region before the timer expires, the trigger fires (i.e., it notifies the SIP push servlet, which forwards the push request to the user upon receiving the notification).

24.3.3.6 Content Provider

If the content provider wants to initiate a push message, it generates an SIP request including the HTML content and, in the case of location-based push, the location parameters. This feature has to be supported by the SIP UA of the content provider (i.e., it must be possible to attach MIME content to the INVITE/NOTIFY message). Since SIP UAs are available for virtually all Internet terminals, any Internet user may originate a push message. The SIP proxy and the push application may decide to process the request or discard it if the originator is not allowed to initiate push requests.

24.3.3.7 Terminal Application/User

The SIP UA of the end user receives the push message. It extracts the HTML page and presents it to the user. Similarly, any other MIME content can be attached; it may be presented via appropriate external programs. This enables application-oriented push—the only requirement is that the application is registered with the related MIME type in the SIP UA. There is no feedback if an application for a specific MIME type is available or not; therefore, there is also no privacy problem regarding application probing.

24.4   Realization

24.4.1   Content Provider Push and Location-Based Push

As mentioned above, two different scenarios of push are considered: (generic) content provider push and location-based push. Both are feasible with the architecture proposed in Section 24.3.

The following section describes in detail a location-based push and outlines the differences to the content provider push. Figure 24.6 gives an overview of the scenario. The marker /n indicates that an operation is done for every user. Without this marker, the operation is performed once (i.e., it is representative for the whole group of push subscribers).

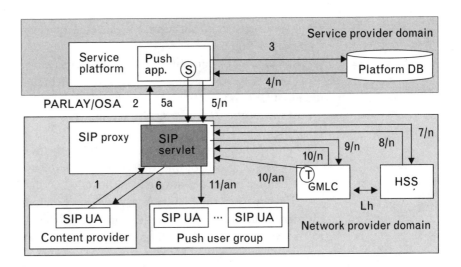

FIGURE 24.6
*Location-based push
scenario.*

The first step is to send an SIP INVITE request to the SIP proxy (1). The body of the request contains the location parameters and the push content. The content may be an HTML page, any MIME content, or multimedia session information (described with SDP). The header of the SIP request contains the expiration time as well as the channel identifier (push address). The servlet engine of the proxy activates the corresponding push servlet and forwards the request.

The push servlet has to maintain the state of the push process and store important information, like the push content, the location parameters, and the expiration value. The servlet forwards the push request, including the channel identifier, to the push application on the service platform over the open PARLAY interface (2).

Now the application retrieves the list of subscribers for the given push channel from the platform database (3, 4). Afterwards, the application returns the list of push subscribers to the push servlet (5). In parallel, it initiates a status report to the content provider (6). If the push channel is valid and contains at least one subscriber, a positive SIP response is sent to the content provider. Otherwise, or if any other problem occurs, an error report is sent.

A user may subscribe to the push channel while the push request is pending (i.e., before its expiration time). This situation is shown with the encircled S in Figure 24.6. In this case, the push application adds the new user to the channel and notifies the push servlet (5a).

Now it is checked for every user if push is allowed. This information and the current address of the user are retrieved from the HSS (7, 8). The servlet now sets up a list of users who will receive the push content. This is necessary for managing the push delivery (e.g., performing retries if a user cannot be reached immediately). It is also used for billing purposes (i.e., to report the number of successfully contacted users to the push application after completing the push request).

In the case of a content provider push, data delivery is performed at this point in time by forwarding the SIP request, including the push content, to the identified end users (11).

In the location-based case, the servlet has to contact the GMLC to find out if the requested user is present in the specified geographical area (9). The GMLC responds with the user location (10). If the user is not in the area, the servlet sets a trigger T for this user. It fires when the user enters the push area (10a). Upon a positive response from the GMLC, the push servlet forwards the SIP request to the SIP stack of the proxy, including the address of the user terminal. In the following, the proxy tries to deliver the push message to the SIP UA of the specified terminal (11). If it succeeds, the servlet updates its internal user list accordingly. After expiration time of the push request, all triggers are disabled in the GMLC. Then the push servlet sends the delivery statistics (i.e., the number of reached subscribers and billing data) to the push application.

Note that this architecture ensures a clear separation between service providers, who provide applications, and network providers, who are responsible for the transport infrastructure. Furthermore, dealing with security issues is simple and effective in this architecture. Authentication is only required between pairs of entities (e.g., content provider and push application, or push application and SIP servlet). This can be done with state-of-the-art methods like PGP [6] and IPsec [7]. Privacy of end users is automatically ensured because only statistical data is reported to the push application and the content provider. No individual user data (preferences, location) are sent outside the network provider domain. This also helps to minimize traffic over the PARLAY interface.

24.4.2 Application-Related Push for Local Location Assistant (LoL@)

The LoL@ is a prototype of a UMTS location-based service that is being developed at Forschungszentrum Telekommunikation Wien (FTW; http://www.ftw.at). It provides a tourist guide for Vienna. LoL@ includes navigation features (i.e., the user is guided from his or her current position to a selected destination). This feature uses maps where the current position and the next part of the route are shown. To update the map display (i.e., to show a new user location and/or route segment), an application-related push concept is used. In this case the architecture is much simpler than in the previous example.

The content provider requests periodic updates of the user location information from the GMLC, using the PARLAY Mobility Service. If the user has moved, the content provider initiates a push data transfer to the terminal, including the new coordinates to be shown in the map. This example shows an unconfirmed, connectionless, unicast push request that is not location based, because the user location is the content of the request.

The control flow (Figure 24.7) is as follows. The user logs into LoL@, which activates the periodic location requests (1). If a new location estimate is available, the GMLC notifies the LoL@ application (2). The application determines if the movement was big enough to send a notification to the LoL@ client running in the user terminal. In this case an SIP push request is issued (3). The SIP proxy identifies the request and activates the push servlet. The servlet now checks the push settings in the HSS (4, 5). If push is allowed, the push request is sent to the SIP UA running in the terminal (6). The UA is responsible to forward the request to the local LoL@ applet. The applet uses the received information to update the display (i.e., to show the current user location in a map). At first glance, this situation looks like a standard SIP session activation, but there are some differences: (1) checking the push preferences in the HSS, (2) the SIP session is automatically accepted and closed after data delivery, and (3) session data is directly sent to the appropriate application.

24.5 Summary

This chapter started by showing the framework around push functionality, followed by a discussion of push scenarios in Section 24.2. A comparison of different push architectures, based on Section 24.3, is given in Table 24.1. Because of its advantages, especially considering flexibility and system interworking, SIP has been selected as the architecture for advanced push functionality. Examples of advanced push scenarios were discussed in Section 24.4.

FIGURE 24.7
LoL@ push.

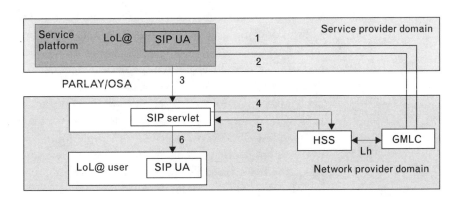

TABLE 24.1   PUSH ARCHITECTURE COMPARISON

	SIP	SMS	WAP	H.323
Internet clients	Yes	No	No	Yes
Mobile clients	Yes (UMTS Rel. 5)	Yes	Yes	No
Protocol type	Text	Binary	Binary	Binary
Protocol efficiency	Medium	High	High	High
Flexibility	High	Low	Medium	Low
Multimedia	Yes	No	No	Yes
Application addressing	Yes	No	Yes	No
Network-initiated sessions	Yes, network initiated PDP context required	Yes	Yes, network initiated PDP context required	Yes, network initiated PDP context required
Standardized content codec	No	Yes	Yes	Yes
Complexity	Medium	Low	Medium	High
Comments	Interworking between PC and MT Symmetric Roaming Delivery guarantees	Symmetric No delivery guarantees	Asymmetric Delivery guarantees	Interworking between PC and ISDN Symmetric

24.6   Conclusions and Future Directions

An SIP-based push architecture with special considerations for location-based push services in UMTS has been discussed. A comparison with other push concepts has been presented. The key advantages of SIP are its inter-working capabilities between UMTS terminals and Internet PCs, the extensibility, and multimedia features. The text-based protocol provides high flexibility and medium complexity; minor drawbacks are the message size and the lack of mandatory, standardized codecs. In H.323, a quite complex binary protocol is used together with a well-defined set of codecs. The problems are its resource consumption and the rigid protocol structure—both issues limit the applicability of H.323 for mobile smart phones. SMS and WAP are both suitable for simple mobile push applications, like notifications. Both concepts are not adequate for complex multimedia services, accessed via smart phones or Internet terminals.

Based on two examples, it was shown that a function split between network and service provider is useful for UMTS push services. It ensures a clear assignment of tasks while minimizing data traffic over the OSA/PARLAY interface.

In the future, interworking between SIP-based mobile networks and other networks, for instance ISDN, will become an interesting issue. Interworking concepts are already considered in H.323 by means of the gateway; they include signaling translation and codec conversion. Codec conversion is a challenging task for real-time sessions because of the resource and QoS requirements. For non-real-time content, conversion is usually not required. It can be assumed, for example, that an HTML push content can be displayed by all push terminals. Interworking of SIP with other signaling systems (e.g., SS7 of GSM or ISDN) is currently not possible. It can be expected, however, that UMTS will increase efforts made in this direction. Basically it is a question of defining and implementing appropriate media and signaling gateways. An SIP proxy with some kind of servlet can also be seen as a simple signaling gateway, converting SIP signaling into PARLAY function calls. Note, however, that the SIP proxy is usually not involved in the content transfer, thus it cannot perform codec conversions easily.

After finishing the implementation of LoL@ and the basic SIP push architecture, the location-based push concepts will be studied in detail. The location information will be obtained from a mobile positioning test bed, which is being developed at FTW. By now, the test bed uses the GPS, and extensions for cell-based methods are currently being implemented.

REFERENCES

[1] WAP-165, "WAP Push Architectural Overview," WAP Forum, http://www.wap-forum.org, November 1999.

[2] 3G TSG Terminals, "Multimedia Messaging Service; Functional Descr., Stage 2 (Release 4)," 3GPP, January 2001.

[3] Handley, M., et al., "SIP: Session Initiation Protocol," RFC 2543, IETF, http://www.ietf.org, March 1999.

[4] Rosenberg, J., et al., "SIP Extensions for Presence," Internet draft, IETF, http://www.ietf.org, March 2001.

[5] Rosenberg, J., et al., "SIP Extension for Instant Messaging," Internet draft, IETF, http://www.ietf.org, April 2001.

[6] Elkins, M., et al., "MIME Security with PGP," RFC 2015, IETF, http://www.ietf.org, October 1996.

[7] Kent, S., et al., "Security Architecture for IP," RFC 2401, IETF, http://www.ietf.org, November 1998.

[8] Kroeselberg, D., "SIP Security Requirements from 3G Wireless Networks," Internet draft, IETF, January 2001.

[9] 3G TSG Services and System Aspects, "TS 33.102: 3G Security; Security Architecture (Release 1999)," 3GPP, http://www.3gpp.org, December 2000.

[10] 3G TSG Services and System Aspects, "TS 23.228: IP Multimedia (IM) Subsystem – Stage 2 (Rel. 5)," 3GPP, September 2001.

[11]   Handley, M., and V. Jacobson, "SDP: Session Description Prot.," RFC 2327, IETF, http://www.ietf.org, April 1998.

[12]   Klyne, G., et al., "Composite Capability/Preference Profiles (CC/PP): Structure and Vocabularies," World Wide Web Consortium (W3C), http://www.w3.org/TR/CCPP-struct-vocab.

[13]   Davis, S., "The Parlay Concept and Overview," The Parlay Group, http://www.parlay.org, June 2000.

[14]   Korkea, M., et al., "A Common Spatial Loc. Data Set," Internet Draft, IETF, http://www.ietf.org, May 2001.

[15]   3G TSG Services and System Aspects, "TS 23.271: Funct. Stage 2 Descr. of LCS (Rel. 4)," 3GPP, January 2001.

CHAPTER 25

Signaling Network Architecture in Wireless IP Overlay Networks

Masahiro Kuroda, Takashi Sakakura, Tatsuji Munaka, Gang Wu, and Mitsuhiko Mizuno

25.1 Introduction

New wireless communications are coming up in the market. Meanwhile, the popularization of cellular communications has aroused a need for Internet access from mobile devices. There have been discussions on IP routing, QoS, and so on, for offering users the seamless data access and voice communication in wireless IP networks.

One approach for accessing the Internet is to augment a cellular system that provides the Internet access to cell phones. The i-mode in Japan is a successful example of this approach. The cellular network, however, is not the only wireless network. Many emerging wireless communications are being introduced that satisfy user requirements, such as high speed and lower communication cost. New wireless communication systems, such as *High Altitude Platform Station* (HAPS) [1], are currently under research and are expected to come to market in the near future. Each wireless communication system has its own advantages and disadvantages for providing IP connection. Besides cellular networks, Bluetooth [2], *Dedicated Short Range Communications* (DSRC) of *Intelligent Transportation Systems* (ITS) [3], *Multimedia Mobile Access Communication* (MMAC) [4], and HAPS are the candidates for IP wireless communications; and the combinations of these wireless networks satisfy various type of uses.

Users can select a communication path implicitly or explicitly in wireless overlay networks [5] when interfaces to wireless networks fit into a common interface in a mobile device. The wireless overlay networks are a future wireless solution to offer services such as multimedia data streaming and data servicing in congested areas where one wireless communication system cannot offer enough bandwidth to users. It is assumed that all the services will be merged to IP networks, and the integration of wireless networks is expected. Efforts are being made to support terminal mobility in IP that originates in the Mobile IP [6] to support seamless mobility [7]. Discussions are also being held on protocols that support micro-mobility in the IP layer [8–10]. Regarding devices, users always carry and use them at any location. The devices need to have low power consumption and connections to any type of wireless networks.

This chapter focuses on a dedicated signaling network architecture in wireless overlay networks and discusses their basic design issues. Actually, "signaling" is a telephony term that refers to sending call information across a telephone connection. The signaling information is transmitted via many techniques. In the plain telephony system, the following series of operations is called signaling: (1) opening and closing a loop to start and stop the flow of *direct current* (dc) loop current, which is used to indicate on-hook and off-hook states and to transmit dial-pulsing of digits; (2) sending of ringing voltage to alert the other side of an incoming call, sending digit information in the form of Dial Tone Multi Frequency, or sending call state information on a Digital Signal 0 time-slot by using robbed-bits. The signaling system in wired-telephony networks is augmented with a help of computer technology to support various services such as 1-800 toll free line, call forwarding, or roaming service to cellular systems. The standard signaling technology that enables these services is known as SS7. In the case of a mobile wireless network, typically a cellular phone system, its own signaling system for sending call information on the wireless network is deployed. Looking at the spectrum of frequencies ranging from several tens of megahertz to several tens of gigahertz, we find that there are dozens of (digital) communication systems. These ubiquitous systems are independently designed, implemented, and operated to meet different requirements on mobility, data rates, services, and so on. Some (if not all) of these systems can simultaneously provide services at a specific geographic location, creating a heterogeneous wireless environment for users in overlaid service areas. A signaling system in mobile wireless networks would be concerned with the location of wireless terminals to localize paging operation, a power efficient way to page the terminals in the scheme of wireless access policy, along with an implementation of a terminal standby mode. A wide-coverage signaling system can accommodate various high-speed wireless data access networks.

In order to bring a signaling system to wireless overlay networks, requirements special to the wireless overlay networks should be considered. These requirements include a wireless system discovery and selection, mobility management, and energy efficient paging [11]. The wireless device in the architecture equips multiple radio interfaces both for signaling and service. The device inquires its signaling network for the available data access networks and configures an IP for the selected data network. The device can access data and can communicate with the destination.

25.2   Network Model

The network model is flexible in order to accommodate various types of wireless communications, from short range to long range as shown in Figure 25.1.

Various wireless communication systems can be categorized as follows (we choose a typical implementation from each category as a member of wireless overlay networks):

- *Macro cell (up to 50 km):* Implemented at several frequency bands, providing link-based communication at intermediate to high speed and having its own handoff mechanism and a sparse coverage in the service area;

- *Micro cell (from 100m to 1 km):* Implemented at several frequency bands, providing link-based communication at intermediate to high speed.

- *Pico cell (from 10m to 100m):* Implemented at several frequency bands from 2.4 GHz (ISM) to several tens of gigahertz. (CSMA/CA-based IEEE 802.11b and 802.11a WLANs and Bluetooth are typical systems in this category);

FIGURE 25.1
The network model.

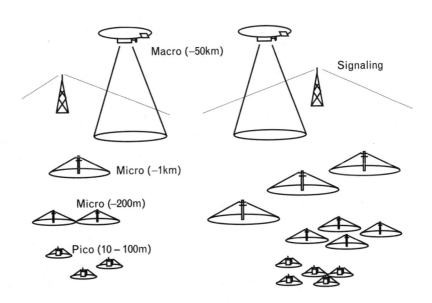

• *Signaling network:* Implemented at pager frequency range, providing two-way communications at low speed but with a wide coverage area.

25.2.1   Dedicated Signaling Network Architecture

The architecture consists of the following four network components: a *basic access network* (BAN), RANs, a *common core network* (CCN), and an external IPv6 backbone network (Figure 25.2) [12]. The BAN is used for signaling, and various types of RANs are targeted for data communications. These networks are connected to the CCN, which provides a common platform through which all terminals communicate with *correspondent nodes* (CNs) in the external IPv6 network.

A region network, which manages a wide geographical area—say, 50 Km in diameter—consists of a CCN, a BAN, and several RANs. The region network is connected to the external IPv6 backbone network via a *gateway router* (GR). The GR provides routing and seamless handover instruction among region networks using a terminal's regional location information identified by its region identification (R-id), which is kept in the *region register* (RR). A HA or a CN can communicate with a terminal in the region network by inquiring its current location directly from the RR.

The BAN is used as the dedicated signaling network to support mobility. The signaling function provides paging, location registration, and information about available RANs for a terminal. A paging message is delivered to a terminal that is not able to communicate with CNs for a certain period of time, when IP packets arrive at the GR to which the terminal is connected.

FIGURE 25.2
The dedicated signaling network architecture and its components.

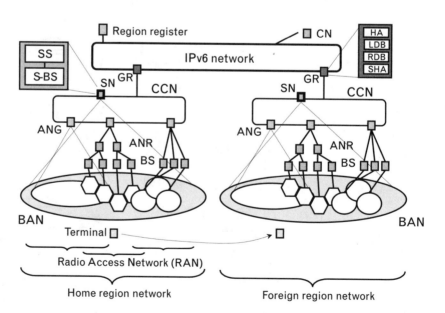

When a terminal changes its location to a new region, it needs to send a location registration message including its *terminal identification* (T-id) and the R-id to the GR in the region network through the BAN. The GR registers the terminal's location indicated by the R-id into the RR and issues the COA and its home IP address for the terminal from the GR of the terminal's home region.

After the terminal receives a reply message including available RANs in the region network, it selects an RAN from the RAN's list and configures its air interface to access to the RAN. At that time, the terminal's COA, which is similar to the IPv6 stateless address auto configuration scheme, is generated by the R-id and its MAC address.

When a terminal moves from one location to another within the same region, it does not need to reregister its current location to the RR. Mobility management within a region is handled by a micro-mobility management scheme such as Cellular IP [8]. This micro-mobility protocol has the only function to maintain a route map from a terminal to the GR in the region. The route map, which is the chain of route caches on routers from an access point of the RAN to the GR, is created when an IP packet is sent to the GR from the terminal via the access point.

On the other hand, when a terminal moves to another region, it updates the current regional location managed by the RR. If the CN fails to inquire the current terminal's location to the RR, it sends an IP packet to the terminal's home region and the packet is forwarded to the current GR by home region's GR. If the CN makes an inquiry request to the RR, it receives the COA of the destination terminal and sends IP packets to the terminal directly. As a result, it reduces the handover latency and traffic overhead on the network by changing the route map from the CN to the GR in the terminal's current region.

25.2.2 Architecture Components

The region network is a primary component in the architecture, as shown in Figure 25.2. It consists of the CCN, the BAN, and RANs. A RAN is connected to the CCN through an *access network gateway* (ANG), which is a top node router of the RAN. Each ANG has an access *gateway interface* (GRI) that handles all communication messages to the GR in the region network. The GRI enables a new wireless access network to be plugged in to the GR in the region network with low cost. The ANG also handles local authentication and authorization, which are dependent on each wireless access service, and also serves as an *access network router* (ANR) to downward nodes. The ANR is an IP router that has route caches for active terminals. A BS is an ANR disposed at the edge of the RAN that provides a layer 2 wireless interface. Separating a paging network from data access networks enables a new wireless service to be added into the network easily. An additional reason for defining the BAN as a dedicated signaling network comes from the requirement to support a wide coverage area.

25.2.2.1   RR

The RR holds a regional location entry denoted by terminal id, its region id for each terminal in all regions. The entry is updated by the GR when it receives a location registration message from a terminal that has changed its location from one region to another. Since one-to-one mapping in service coverage between a BAN and a CCN is adopted in this study, location management for paging localization is not needed. Therefore, the RR only provides an inquiry service that informs the current location of a specified terminal to CNs or the HA of each terminal so far. The RR allows a flexible configuration of BAN to keep mapping between BANs and CCNs for paging localization, and has a functionality of the CCN selection.

25.2.2.2   Gateway Router

The RR and GRs are the key components providing routings and seamless handovers among region networks. A GR is designated as an IP router of a region network and consists of the following functional entities.

25.2.3   Signaling Home Agent

The *signaling home agent* (SHA) is responsible for the terminal management of locations and mobility for within the same region network and between two different region networks. It also provides resource management in a region network for visiting terminals. An explicit signaling mechanism provided by the BAN allows for a simpler design for mobility management. The SHA and the terminal itself keep IP routes to the terminal including configuration for packet forwarding between regions by triggering each other. The SHA acts as an HA and an FA as well in packet forwarding in the Mobile IP manner, and also acts as a GR without paging cache management in the Cellular IP manner.

25.2.4   Location Database

Geographical location data including latitude, longitude, and altitude of terminals in a region network are registered in the *location database* (LDB). An entry of a terminal in the LDB is created when the terminal moves into the region for the first time and is updated when the terminal makes a registration update message through the BAN. This geographical data is used to enumerate available RANs for the terminal invoking an access network discovery.

25.2.5 Resource Database

The *resource database* (RDB) keeps resource information of the RAN in a region network, such as maps of access points and status of respective RANs.

25.2.6 HA

The HA forwards IP packets to the GR according to the cached routes kept in an SHA.

25.2.6.1 Signaling Node

The *signaling node* (SN) consists of a *signaling server* (SS) and a *signaling BS* (S–BS). When the SN receives a location update message from a terminal, it passes the message to the SHA. Signaling messages from the SHA are passed to the SS and the SS sends them to terminals via the S–BS.

In this architecture, a terminal is an IP mobile node enabling the user to get services via a selected access network. Each terminal has a *basic access component* (BAC) to communicate with the BAN, and it also has one or more radio modules to access the CCN [13]. The terminal is in either of the following states:

- *Active state:* A terminal is sending (receiving) IP packets to (from) a correspondent node.

- *Idle state:* A terminal is not sending (receiving) IP packets for a certain period of time.

When a terminal in the idle state receives a paging message through the BAN, it changes its state to active after selecting an RAN and configuring IP in the RAN.

25.3 Functions of Signaling Network

25.3.1 Terminal Addressing

A fixed home IP address is only valid within its home region network. A terminal can be identified by its fixed home IP address or by a temporary IP address (COA) in a visiting region network. In this architecture, a terminal address is defined as follows:

IP address = R–id[+QoS class] + MAC address of the terminal

This addressing scheme is similar to the IPv6 stateless address auto-configuration. In the IPv6 autoconfiguration, the prefix address is delivered to a terminal via a router advertisement. In this architecture, the prefix (an R-id) is notified to a terminal by a reply message of a location registration message through the BAN. The IP address can optionally include the QoS class that is set by applications explicitly or set by routers within the CCN. A QoS class is expressed as an alias address of a terminal within an IP subnet that does not affect existing IP QoS mechanism that uses the IP header field.

25.4 Signaling and Location Management

When a GR receives an IP packet destined for a terminal, it searches the corresponding entry in its route cache first. If the GR cannot find a valid route cache entry for the terminal, it makes an explicit paging to the region network through a BAN. As the terminal in the idle state receives the paging message specified by the terminal identification, it goes into the active state by selecting a suitable RAN, configuring the RAN, and generating a COA in the region network.

When the terminal moves into a foreign region network, a location registration message is passed to the GR through the BAN. The GR enrolls the geographical location data in the message into the LDB and updates the R-id, which is kept in the RR.

A GR, then, receives a location registration message from the terminal. It selects the candidates of RANs by available service information showing each RAN coverage and current states of RANs such as available bandwidth. The list of available RANs indicated by <R-id, RAN-id, and BS-id> is delivered to the terminal as the reply message for the registration message.

Further clarification of the paging and route cache entry setup procedure is shown in Figure 25.3. In the procedure, it is assumed that a terminal is in the idle state while staying in the home region, and the regional location of the terminal is already registered into the RR by the implicit location update messages periodically processed by a layer 2 function driver.

1. A correspondent node sends an IP packet to a terminal that is in its home region.

2. The GR managing the home region network searches the route cache entry of the destined terminal first when it receives the IP packet. If a valid entry is found in the cache, the GR forwards the packet to the current GR in the foreign region network in which the terminal exists.

FIGURE 25.3
*The paging route
create procedure.*

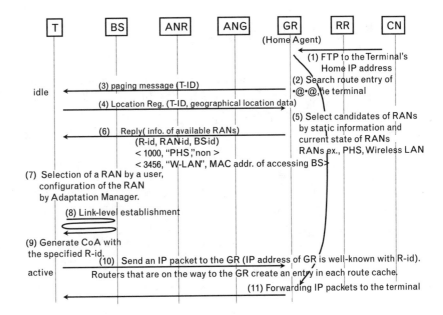

3. If the GR cannot find a valid entry, it sends a paging message to the terminal in the home region network through the BAN.

4. The terminal receiving the paging message sends a location registration message to the GR. This message plays a role of an inquiry request for retrieving the available RANs at the terminal's current location.

5. The GR selects the candidates of RANs based on the information that indicates those RANs covering the region. The current states of RANs are observed dynamically.

6. The GR sends a reply message including the list of available RANs.

7. The most suitable RAN is selected according to the user's requirement, and the terminal is reconfigured to adjust to the RAN.

8. Link-level communication is enabled.

9. The terminal generates a tentative IP address (COA) by the R–id and its MAC address.

10. The terminal sends the IP packet to the GR. The GR and routers on the way to the GR create or update the route cache entries corresponding to the terminal.

11. After the GR successfully receives the IP packet from the terminal through the configured RAN, it starts forwarding IP packets coming from the CN to the terminal.

25.5 Mobility Management

In general, the handover is defined as a process involved when an active terminal changes its point of attachment to the network, or when such change is attempted.

In the architecture, the following handovers are defined, as shown in Figure 25.4:

- *Intraregion handover:* A handover occurs when a terminal changes BSs connected to the same region network.

- *Interregion handover:* A handover occurs when a terminal changes BSs connected to the different region network.

- *Horizontal handover:* A handover occurs when a terminal changes BSs in the same RAN.

- *Vertical handover:* A handover occurs when a terminal changes BSs between different RANs.

In general, micro-mobility refers to mobility over a small area and macro-mobility refers to mobility over a large area. In the case of wireless overlay networks, intraregion handover and interregion handover are defined to make the difference of mobility clearer.

The handover procedure when a terminal moves from its home region network to a foreign region network is shown in Figure 25.5.

FIGURE 25.4
The scope of handover.

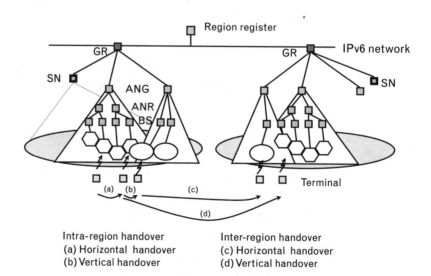

FIGURE 25.5
The interregion
handover procedure.

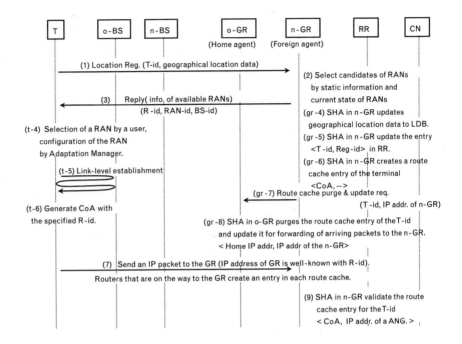

1. A terminal sends a location registration message to a GR in a newly visited foreign region network.

2. The new GR knows that the new terminal moves into the region since it cannot find the corresponding entry in its route cache. In order to establish a data communication pass to the terminal, the GR chooses the candidates of RANs by information indicating which RANs cover the region and current location of the terminal.

3. The GR sends a reply message including the list of available RANs.

The terminal and the region networks execute the following procedures in parallel. The (t–x)'s explain the steps taken by the terminal, whereas the (gr–x)'s show the steps of region networks:

• (t–4) The suitable RAN satisfying the user's requirements is selected, and the corresponding air interface is reconfigured for access to the RAN.

• (t–5) Enable link-level communication.

• (t–6) The terminal generates a tentative IP address in the foreign region network.

• (gr–4) The SHA in the new GR updates the geographical location data in the LDB.

- (gr-5) The SHA makes an update request of the regional location of the terminal denoted by <T-id, R-id> to the RR.

- (gr-6) The SHA creates a route cache entry for the terminal shown by <COA, non>. This entry shows that the route to the terminal in the CCN is not validated yet. When the GR receives an IP packet denoted to the terminal on this condition, the paging procedure is invoked by the GR.

- (gr-7) The SHA sends a route update message including <T-id, IP address of the new GR> to a SHA in the former GR.

- (gr-8) The SHA in the former GR updates the entry into <the fixed Home IP address of the terminal, an IP address of the new GR>. After updating the entry, packets arriving at the previous GR destined to the terminal are forwarded to the new GR.

- (9) The terminal sends an IP packet to the new GR. The packet is listened to by routers on the way to the GR, and they add route cache entries for the terminal.

- (10) The SHA in the new GR updates the cache entry of the terminal as follows: <COA, IP address of an ANG in the region>.

If IP packets arrive at the new GR destined to the terminal from step (gr-7) to the step (10), a paging message is sent to the terminal and succeeded paging procedures shown in Figure 25.3 are handled in the terminal.

The IP address of a terminal is defined as follows:

- If (current region = home region of a terminal), then IP address of a terminal = Home IP address of the terminal = COA in the region.

- Else then IP address of a terminal = COA in the current region.

25.5.1   Resource Management

Resources are defined as the availability of communication channels used by the RAN and functional abilities of a terminal while focusing on the availability of network communication channels. The final goal is to design an efficient communication system by taking advantages of features of the network architecture combining different types of RANs from the point of view of both users and network operators. All the resources will be defined in the near future. Information such as the map of access points is treated as a static resource. Resources such as bandwidth, memories for data buffering, and so on are dynamic resources monitored and controlled by the operating system. This resource information is stored in an RDB in each region.

25.6 Basic Issues

25.6.1 Paging and Power Consumption of Idle Terminal

To save the terminal's battery life, two approaches are considered. One is that the wireless device gets into a low-power state periodically for a certain amount of time while the terminal does not send or receive IP packets. The wireless device wakes up from the low-power state for a short time only for checking the arrival of the paging messages. Bluetooth adopts this approach in its paging scheme. In this approach, no particular interface between the IP layer and the link layer is required for implementing a paging mechanism. Because the wireless device is basically active to the IP layer, the IP layer can control the paging process by sending a paging message as an IP packet. However, the terminal may miss the paging messages while the wireless device is sleeping, such that the paging message traffic on the network increases as a result.

Another approach is to define an explicit low-power state, which only allows the wireless device to listen to the paging messages. In this approach, the wireless device is in the low-power state until receiving the paging message. The wireless device notifies the higher communication layer when it receives the paging message. The cellular system is an example of this approach. Some dedicated layer 2 functions and the higher layer functions are needed additionally for paging; however, this approach enables a better paging response, a lower paging traffic, and low power consumption.

The system discussed here uses the dedicated network, the BAN, for a paging and signaling purpose. The radio module for the BAN, BAC, is installed into the terminal independently from other radio modules for the RANs used for communications. Only the BAC can be powered on in the idle state.

25.6.2 Generic Extension Interface for Adding a New Type of Wireless Network as an RAN

Two other issues should be considered for adding a new type of wireless service. One is an issue regarding an attachment of a radio module to a terminal, and the other is regarding the connection of an RAN to a CCN.

If a newly added RAN can communicate with the terminal by an existing RF module, only firmware of base-band logic has to be added into the terminal. The terminal will equip with some of the RF modules that covers 100 MHz to several tens of gigahertz, so most of the wireless networks can be supported on the terminal RF platform. By defining the firmware interface and link control interface, a network operator who wants to join the system can develop the module for plug-in.

If IP communications are already enabled on the wireless network, then connecting the wireless network is simple. The wireless network is connected to the GR of a CCN via the ANG, which reconciles the difference of security policy between the CCN and the wireless network. Otherwise, an interworking gateway is needed on top of the link control function of the wireless network. The network operator should prepare the IP gateway in this case.

25.6.3   Capacity of the Networks as a Communication System

One purpose of designing the networks is to achieve good utilization of radio resources. The system aims to support as many wireless communications as possible at an allocated frequency band. The resource management function of the system should arrange total traffic balance in the system and improve the communication availability.

The cell layout design of the respective RANs is also an issue, especially in the case of PAN type of RAN. In the case the RANs are provided by cellular systems, the cell layout is carefully designed and power control functions are installed in the CDMA systems. In the case of PAN, a frequency band is highly reusable owing to its low transmit power. It is obvious that much deployment of the PAN improves the communication capacity of the system. Moreover, if the PAN, such as Bluetooth, adopts the WLAN medium, then the access network can easily be expanded in an ad hoc network manner.

25.6.4   Geographical Coverage of Service and BAN Capacity

The entire service area of the system is covered by the BAN. The 300–MHz frequency band is allocated to the two-way pager technology including the BAN. This is good for the geographical coverage of a medium powered base station, but the bandwidth is not good enough for the data communication. A small-scale simulation evaluation suggests that the current specification of BAN is not broad enough if the number of handoff operation exceeds three times per second. Therefore, the BAN is not suitable for data communication. The service areas should be covered entirely by some RANs for enabling the "anytime, anywhere" service.

25.6.5   Handover Latency

According to the IETF Internet draft [10], handover is defined as the process involved when an active MH changes its point of attachment to the network, or when such a change is attempted. Handover latency is the time difference between when a wireless terminal is able to send and/or receive an

IP packet by the way of the previous ANR and when the terminal is able to send and/or receive an IP packet through the new ANR.

Table 25.1 shows the handover-related processes for each kind of handover specified in Figure 20.5 in the wireless IP overlay networks.

In the case of the horizontal handover of intraregion handover, the handover latency depends on geographical location update and route cache update. Although the geographical location update process is only to update an entry in an LDB, the performance of the location management should not depend on the number of wireless terminals within an area. Route cache update is the creation of the route map from the visiting BS to the ANR that has the route entry of the wireless terminal. Cellular IP is a candidate of the micro-mobility protocols updating the route map and will be required to improve the performance according to the characteristics of the RAN.

The handover latency of the vertical handover mainly depends on I/O configuration in the terminal and resource reallocation. It is expected that the process of I/O configuration in terminal will be improved by newly introduced technology such as software radio technology. The resource reallocation process is related to the QoS mechanism in this network model and needs particular attention within the CCN. This issue is one of the most important items in the future works.

TABLE 25.1   THE HANDOVER-RELATED PROCESSES

	INTRAREGION HANDOVER (MICRO-MOBILITY)		INTERREGION HANDOVER (MACRO-MOBILITY)	
	HORIZONTAL	VERTICAL	HORIZONTAL	VERTICAL
Region-level location update to RR	Not required	Not required	Required	Required
Geographical location update to GR	Required	Required	Required	Required
Bind update from GR to HA	Not Required	Not Required	Required	Required
Resource reallocation	Not Required	Required	Required	Required
Reassignment of COA	Not Required	Not Required	Required	Required
I/O configuration in terminal	Not Required	Required	Required	Required
Route cache update	Routers up to ANR or ANG in RAN	Routers in new RAN on the way to GR	Routers in new RAN on the way to GR	Routers in new RAN on the way to GR

In the case of interregion handover, almost all the wireless terminals will be required to get an available IP address in the visiting network. The reassignment of COA process basically follows IPv6 stateless address autoconfiguration, and the binding update from a GR to the HA process follows the Mobile IPv6. However, the optimizations of these processes and region-level location update to RR are indispensable to investigate under the conditions such as the number of wireless terminals within an area and traffic load caused by terminals in addition to the discussion in the standardization activities.

This architecture assumes that all the cell categories have the same handover feature, but the Bluetooth pico cell network is small in coverage and can establish an ah hoc connectivity. There is no need for intraregion handover based on geographical location update (shown in Table 25.1) because it is managed by the ad hoc network.

25.7   Concluding Remarks

In this architecture a newly added wireless network does not need to take care of the signaling procedure because a wide-coverage signaling network, BAN, provides a common signaling mechanism covering different types of wireless networks, and mobility management among different wireless access networks is taken care of by CCN.

Further investigations and evaluations on the paging and power consumption for a terminal, the network interface to accommodate different types of wireless communications, capacity of the networks assuming a large number of users, and the trade-off between geographical coverage areas and BAN capacity are made. Capabilities to minimize the handover latency and to provide QoS at vertical handover in the wireless IP overlay networks are also topics in this research area. A system bottleneck at BAN, however, is anticipated in this network architecture, and optimization to decrease the signaling traffic on BAN, such as IP paging partially in the data networks, is a considerable issue.

Authentication, Authorization, Accounting (AAA), which is not discussed in this chapter, is indispensable for providing seamless communication services to users. It needs to be investigate how to eliminate unnecessary AAA interactions to keep the handover latency.

REFERENCES

[1]   Wu, G., R. Miura, and Y. Hase, "A Broadband Wireless Access System Using Stratospheric Platforms," *IEEE Globecom 2000*, November 2000.

[2]   Shimotsuma, Y., et. al., "Percolating Data Delivery on Cellular Ad Hoc Integrated Network," *IEICE Transactions on Communications*, Vol. E84-B No. 4, April 2001.

[3] "Dedicated Short Range Communication (DSRC) for Transport Information and Control Systems (TICS)," ARIB Standard Version 1.0, http://www.arib.or.jp/.

[4] Hase, Y., M. Inoue, and G. Wu, "Development of a Millimeter-Waveband High-Speed Multimedia Wireless LAN Prototype," *Journal of Signal Processing*, Vol. 3, No. 2, March 1999, pp. 117–126.

[5] Stemm, M., and R. H. Kats, "Vertical Handoffs in Wireless Overlay Networks," *ACM Mobile Networks and Applications*, Vol. 3, Summer 1998.

[6] Perkins, C., "IP Mobility Support," RFC2002, October 1996.

[7] Loughney, J., et. al., "SeaMoby Micro Mobility Problem Statement," Internet draft, February 2001.

[8] Campbell, A. T., et al., "Design, Implementation and Evaluation of Cellular IP," *IEEE Personal Communications*, Special Issue on IP-Based Mobile Telecommunications Networks, June/July 2000.

[9] Ramjee, R., et al., "IP Micro-Mobility Support Using HAWAII," Internet draft, June 1999.

[10] Manner, J., et al., "Mobility Related Terminology," Internet draft, November 2001.

[11] Wu, G., P. Havinga, and M.Mizuno, "MIRAI Architecture for Heterogeneous Network," *IEEE Communications Magazine*, February 2002, pp. 126–134.

[12] Mahmud, K., et al., "On the Required Features and System Capacity of Basic Access Network in the Multimedia Integrated Networks by Radio Access Innovation (MIRAI)," *Fourth Int. Symposium Wireless Personal Multimedia Communication (WMPC'01)*, Denmark, September 2001, pp. 1199–1204.

[13] Atwal, K., and R. Akers, "Transmission of IP Packets over Bluetooth Networks," Internet draft, May 2001.

Part V:
Services and Applications

CHAPTER 26

Mobile Content Distribution for Wireless IP Networks

Tao Wu and Sudhir Dixit[*]

26.1 Introduction

The last several years have seen explosive growth of the Internet and especially the WWW. Today, with more people accessing the Web and more applications becoming Web-enabled every day, WWW is quickly becoming an indispensable channel for information exchange and communications. However, the vision of accessing content and information at anytime from anywhere using a variety of devices, will not be fully realized until the wide deployment of emerging wireless IP networking technologies such as GPRS, 3G, WLAN, and Bluetooth. Future wireless IP networks will empower a multitude of data-centric services including many existing wireline Web applications through their IP infrastructure. More importantly, they will add another dimension of mobility to the Web and will enable novel context-aware services such as location-based services and m-commerce.

For wireless IP networks to deliver these promises, however, they must provide a high quality of service to the end user. This is the lesson learned from the wireline Web, which almost became the victim of its own success. During its short history, there have been numerous serious service interruptions because the infrastructure was incapable of handling exponential traffic growth and flash crowds. For example, many of these service interruptions occurred when widely expected Web content became available and was requested by more people than the infrastructure could process. As a result, origin servers often crashed due to overload, and networks were congested with repetitive requests for the same content. More critically, user

*   The authors thank Sadhna Ahuja for her comments on reliable multicast.

experience degraded dramatically during the interruption as users endured excessive Web latency only to learn that the request had timed out. Because wireless IP networks will provide more critical and interactive services such as financial transactions to millions of terminals, they need to avoid service outage by building scalability and application layer QoS into the network infrastructure.

Fortunately, novel content distribution technologies have considerably improved wireline Web user experience mostly by scaling the infrastructure; wireless IP networks can apply similar technologies to achieve scalability and accelerated user experience. The basic idea of content distribution is to distribute content to the edge of the network, closer to the end user. When a user requests Web content, the network attempts to serve the content from the edge rather than from the origin server. Content distribution offers a number of benefits, including offloading servers, reducing backbone traffic utilization, and significantly improving user experience and application layer QoS. Essential content distribution technologies include Web caching, content routing and switching, load balancing, streaming media delivery, and content management. These application layer technologies, together with advances in hardware/software design in lower layers, have transformed the wireline Web from a simple client-server model for information access to a scalable, high-performance, content-rich infrastructure and a service-enabling platform.

Content distribution in mobile wireless IP networks, however, faces new and unique challenges. First, to achieve accelerated user experience, mobile content distribution must be highly optimized on the wireless bandwidth usage. Content distribution for the wireline Web has been focusing on distributing content to the edge, since infrastructure scalability is key to wireline application layer QoS. In wireless networks, however, the wireless link is crucial to user experience because it is usually slow, expensive, and error-prone. Bandwidth of wireless links is affected by many factors, and connectivity may be lost at any time. Thus, edge distribution alone is not adequate for wireless networks, and the network must be able to intelligently provide optimal content in all connectivity situations. Second, most mobile devices used in wireless networks are unlikely to have the resources and capabilities of desktop PCs. Consequently, mobile content distribution needs to provide novel functionality of content and service adaptation to many terminals with limited and different processing, display, input and storage capabilities, and stringent power consumption requirements. Third, wireless IP networks such as GPRS and UMTS need to support terminal mobility, so current wireline content distribution techniques need to be enhanced in order to be effective in mobile networks.

Because a good understanding of content distribution is essential for analyzing and designing high-performance and efficient wireless networks, especially from the perspective of service and application development, the purpose of this chapter is to review content distribution technologies and identify opportunities and challenges in developing content distribution for wireless IP networks. Specifically, we develop a general taxonomy for content distribution techniques. In this approach, a content distribution technique must make improvements in one or more of the following three areas: network scaling, endpoint (server and client) acceleration, and protocol and content optimization. A key finding of this chapter is that while wireline content distribution focuses on network scaling and server acceleration, wireless content distribution will likely benefit most from protocol and content optimization and client acceleration with particular emphasis on the limitations of the wireless medium. In addition, recent developments in wireline content distribution such as edge services have the potential to be widely applied in mobile content distribution.

We try to include both academic and industrial contributions to the field, as this is a very active market sector and many innovations take place in an industrial R&D environment. We also note that content distribution is a fast-developing technology, and this chapter is not intended to be an exhaustive survey; rather, we feel it is more important to summarize the key techniques in content distribution and their implications on wireless IP networks.

The remainder of the chapter is organized as follows. In Section 26.2, we develop the taxonomy for content distribution techniques and use it to review existing and emerging content delivery and distribution technologies in the wireline Web. In Section 26.3, we describe wireless IP network characteristics and existing content distribution approaches addressing these characteristics. We discuss opportunities for mobile content distribution in Section 26.4 and conclude in Section 26.5.

26.2 Content Distribution in Wireline Web

26.2.1 A Taxonomy of Content Distribution

The Internet infrastructure's ability to maintain acceptable user experience has been constantly challenged in the last few years as WWW enjoys unprecedented success, popularity, and growth. During the course, a set of technologies has been developed to expedite the delivery of content from server to client. These techniques vary widely, and to the best of our knowledge, there is no formal taxonomy that gives both the visibility of content distribution as a whole and the resolution of identifying the strength and applications of each technique. Here, we propose the following categorization of content acceleration techniques: network scaling, which distributes the content in the network; endpoint acceleration, which speeds up the content delivery at content source and destination; and protocol and content optimization, which improves content transmission efficiency. This taxonomy allows us to treat different aspects on content distribution separately and at the same time maintain an integral and consistent view of the entire spectrum of content distribution.

First, technologies such as edge Web caching and *content distribution networking* (CDN) aim to scale the network and distribute content to the network edge. They significantly improve user experience by serving Web content locally. Other benefits of network scaling are server offloading and backbone traffic reduction. Recent progress in this area will enable Web services (such as dynamic content generation and personalization) to be provided at network edge, further enhancing user experience. Second, both server and client can accelerate content delivery to the end user. On the server acceleration side, load balancing at the application layer and Web switching, together with reverse proxy, were developed to scale the server farms. On the client side, caching has been widely used to enhance user perceived performance. Third, the HTTP/TCP/IP stack is the standard Web transport vehicle, and optimizations have been made to improve their efficiency in certain situations. For example, the three-way TCP connection setup often introduces excessive overhead in Web servers, and dedicated TCP multiplexers have been introduced to offload the TCP connection management from the server. Furthermore, packet-level FEC coding can be used to disseminate content to multiple recipients, eliminating the scalability limitations of TCP. Beyond this level, techniques such as HTML macros, cache-based compaction, and duplicate suppression have been proposed to transport and store content efficiently.

In the remainder of the section, we discuss these techniques within the framework of content distribution taxonomy.

26.2.2 Network Scaling

26.2.2.1 Web Caching

Web caching is one of the earliest Web content acceleration technologies, and it is a key building block for many content distribution techniques. As illustrated in Figure 26.1, a Web cache stores frequently requested content at network edge so that it can be served locally when a user requests it. Web caching can considerably improve user experience by eliminating much of the Web latency. Furthermore, it can significantly lower the backbone bandwidth usage since it eliminates repetitive transmission of popular Web objects, indirectly improving QoS of other applications. Another benefit of Web caching is that it offloads the origin server by processing requests in a distributed manner. Also, it mitigates the disastrous consequences of a single point of failure at the origin server by caching content at the edge. A Web

FIGURE 26.1
Edge Web caching.

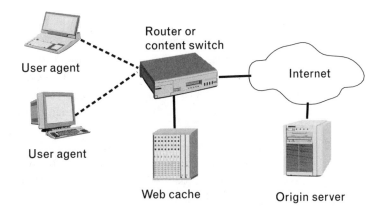

cache deployed at edge needs either user agents to be configured so that they can send requests to the cache, or router support to route the Web requests to the caches.

Web caching resembles caching in the computer systems memory hierarchy [1], and the key component of Web caching is the replacement algorithm, which decides what cached object to delete when the cache is full and room must be made to accommodate an incoming object. Beyond this, Web caching faces new problems that do not exist in memory hierarchy caching. First, in memory caching, the cost of each memory access is usually identical; thus, hit rate is directly linked to the average access time. In Web caching, however, cost of fetching a Web object can vary widely, and there is no single universal target a cache can optimize upon (although many performance measures are often positively correlated). Second, even if hit rate is selected as the optimization target, Web cache needs to deal with Web objects of different sizes. It can be shown that the Web cache replacement problem is NP-complete. Third, Web objects can be modified at the origin server and this may make the cached object stale. Although HTTP/1.1 provides mechanisms such as validation and expiration to check freshness of cached objects, it is still a weak consistency model. Fourth, Web access marginal distribution is known to be Zipf-like [2]:

$$P_N(i) = \frac{\Omega}{i^\alpha}, 1 \le i \le N$$

where N is the number of distinct documents, $P_N(i)$ is the probability of access for the ith most frequently accessed document, Ω is a normalizing constant, and $0 \le \alpha \le 1$. This divergent distribution makes the "10/90" rule in program execution invalid for Web caching. In reality, Web cache hit rates seldom exceed 50% to 60%. In addition, [3, 4] demonstrated the existence of short-term temporal correlations in Web request pattern, although simulations in [4] showed that reasonably large caches could absorb such fluctuations.

A number of Web cache replacement policies have been proposed and studied, including LRU, Hyper-G, LFU-DA, SLRU, GD, GD-Size, and GD★ [2, 5–7]. However, today's Web caches are becoming so large that content expiration occurs more often than replacement.

As the network traffic grows, Web cache itself faces scalability issues. Hierarchical caching [8] addresses this issue by constructing caching hierarchies. To facilitate inter-cache queries and messaging, ICP protocol was developed [9]. Load balancing between Web caches becomes a more important problem in cache array, too [10], since it is desirable to have the same cache serve the requests for the same content to fully utilize the request locality. On the other hand, a particular cache should not be kept overloaded to affect QoS and cache availability.

26.2.2.2   CDNs

Although Web caching solves some scalability and performance problems in the Internet, content providers rarely have control of cached objects and the QoS end user receives. CDNs (e.g., Akamai [11] and Digital Island [12], to name just two) have been developed to address these issues. Conceptually, CDNs can be thought of as a network of caches that stores and distributes managed content. By using CDNs, content providers are able to not only have global reach, but also get guaranteed and measurable QoS and control over content. New or changed content can be proactively distributed within a CDN, while stale content can be located and removed. From the user point of view, however, the best experience can be achieved when different CDNs can exchange content so that the user perceived latency is minimized [13].

26.2.2.3   Edge Services

As the Web evolves, more and more content becomes dynamically generated or personalized. These types of content are either time-sensitive (e.g., stock quotes, scoreboard) or tailored to individual users (e.g., personalized home pages made possible by the use of cookies). Web caches and CDNs currently do not support these types of content. As a result, it is not uncommon that the entire page has to be generated dynamically at the origin server every time a user sends a request, even when most of the page is static and remains the same for all users. This situation limits the usefulness of Web caching and CDNs. There is strong interest among CDNs to address this issue to enable dynamic content generation and delivery at the edge [14]. Furthermore, IETF's Open Pluggable Edge Service working group [15] is working on standardizing Web intermediary application interfaces to enable value-added edge services, such as transcoding and virus scanning. Envisioned edge services will make the Web an even more distributed computing platform, with improved versatility, performance, and scalability.

26.2.3 Endpoint Acceleration

In wireline Web, endpoint acceleration mostly focuses on the server side since this is often the bottleneck of scalability and user perceived performance. To handle heavy traffic and achieve high availability, today's large Web sites use server farms where load is distributed among multiple servers. To further enhance performance, Web caches are also used to serve static content, as shown in Figure 26.2. In this deployment, Web caches are also called reverse proxy because of the configuration.

One natural question regarding server farms is how to choose a Web server for a given request. Content switches or load balancers are usually used for this purpose, as illustrated in Figure 26.2. Content switches are specialized routers that forward requests based on application layer information rather than destination IP address, such as URLs and HTTP headers. Load balancers perform similar tasks, though they traditionally try to balance the load of Web servers. The load-balancing algorithm can be as simple as round robin or as sophisticated as measurement based [16]. In addition, similar to cache load balancing discussed at the end of Section 26.2.2.1, server load balancing needs to provide both load protection and request locality [17, 18].

In addition, in order to address the dynamic content distribution problem discussed in Section 26.2.2.3, several commercial products accelerate the origin servers by caching dynamic content at Web servers, such as SpiderCache [19] and XCache [20].

FIGURE 26.2
A simplified server farm.

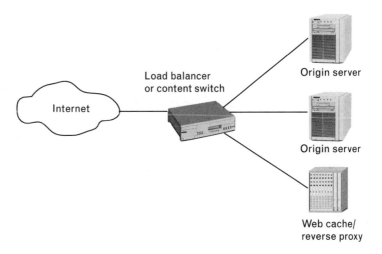

26.2.4 Protocol and Content Optimization

HTTP/TCP/IP is the predominant protocol stack used in WWW. HTTP/1.1 has optimized Web resources in many ways such as cache control and persistent connections. However, there still are situations where significant improvements can be made.

First, TCP uses three-way handshake to establish a connection. This can result in considerable resource waste in busy Web servers where a great deal of load is generated by TCP connection setup and teardown. To address this issue, a dedicated TCP multiplexer can be deployed in front of the servers to terminate the TCP connections with the clients. There is a persistent TCP connection between the server and the multiplexer, eliminating the TCP connection setup/teardown overhead [21].

Second, reliable multicast is often desirable when a server serves popular content to many clients simultaneously. However, some reliable multicast schemes that rely on the receiver to send acknowledgments have serious scalability limitations (i.e., acknowledgments from different receivers can quickly congest the network links to the sender). Channel coding techniques such as FEC can be used at the packet level to eliminate the need for acknowledgments. The encoded packets have the property that as long as the receiver obtains enough unique packets, it can perform decoding and recover the original packets [22]. This feedback-free technique is particularly suitable for applications that need to support one-to-many or many-to-one [23] reliable simultaneous content distribution. The coding and decoding overhead of this approach may be high, however, and a trade-off between coding efficiency and complexity may be necessary [22].

Third, there exists dependence among certain Web objects, which can be exploited to reduce content transport and storage cost. For example, queries about weather forecast in two cities typically generate two different dynamic replies. But if the replies are generated at the same Web site, they are usually formatted in the same way, and a significant portion of the page is static and can be reused in composing similar Web pages. Based on this observation, [24] proposed to use HTML macros and templates to separate static data from dynamic data. While dynamic data is transmitted by the server every time a request is received, the template (static data) can be cached by the client. Cache-based compaction [25] adopts a similar approach, but the requested object is constructed at the client side using objects that are cached by client as a base. Because there may be many cached objects, heuristics is needed to avoid prohibitive computation cost. Evidently, these two techniques can be viewed as applications of source coding in the way they reduce redundancy, although they work at application/object level instead of physical/bit level. A third method, duplicate suppression [26], improves content delivery efficiency at an even higher level. The motivation for duplicate suppression is that a Web object may be repetitively stored and transmitted under different URLs. Conventional Web caching is not effective for this type of objects since it matches the URL, not the object itself. Mogul [26] proposed an approach to discover such duplicates so that retransmission of the identical object can be avoided.

Fourth, prefetching can enhance performance on dedicated access links. In general, one major side effect of prefetching is that it increases bandwidth usage. In dedicated links such as dial-up modem and DSL, however, prefetching during idle periods only enhances the performance since other users could not have shared the idle bandwidth anyway. Fireclick's products are based on this idea [27].

Additionally, it has been found that prefetching the information about the connection can reduce Web latency [28]. This approach takes advantage of the locality of DNS, router and origin server memory access, and reduction of TCP connection setup.

26.3 Content Distribution in Wireless IP Networks

Contrary to the wireline Web content distribution, where the emphasis is on network and server scalability, current wireless Web research focuses on protocol and content optimization, particularly over the air interface. As discussed in this chapter's introduction, the air interface is a shared and expensive resource in wireless networks. More significantly, connectivity is not guaranteed as wireless coverage may be limited and may change over time; in fact, connectivity should not even be assumed by applications [29]. These factors have resulted in special protocol designs over the air interface, and client caching is often used to reduce the impact of expensive and unreliable connectivity. Moreover, the limited terminal resources and capability necessitate more network support to achieve user experience comparable to that in the wireline Web. Much work has been done to transform the content to match the terminal and available bandwidth. Since terminal mobility must be supported, such proxy services may also need to handoff with the terminal handoff.

26.3.1 Air Interface Protocol Optimization

The error-prone air interface can considerably degrade protocol performance that is designed for wireline networks. For example, bit errors in the air interface often cause packet corruptions, which in turn result in packet losses. TCP treats packet loss as a sign of network congestion, since irrecoverable bit errors are rare in wireline networks. Hence, TCP invokes the congestion control and connection throughput is severely affected. The best solution to this problem is to develop link layer packet level reliability rather than relying on TCP retransmission. However, this new link layer approach inevitably introduces additional delay during retransmission, which may affect the TCP RTT estimate and cause timeout. High load in cells only exacerbates the problem as the bit error rate increases. Andersson [29] reported that in an emulated GPRS environment, the average PING RTT can increase from 425 to 772 ms as the number of users in the cell increases from 4 to more than 40, compared to fixed Internet's PING RTT of 33 ms. Furthermore, both HTTP and TCP require the client and the server to exchange messages frequently. This causes their performance in wireless to degrade even further since every message incurs much longer and unpredictable delay. Finally, mobile users may move to an area with no or limited coverage, and connectivity may be lost or significantly affected. Serious functionality, security, and performance consequences could result if applications are developed with the assumption of "always-on" connectivity.

Because of the unique challenges in the radio link, many researchers specifically optimize radio transmission by using a split proxy approach and use proxy service to provide a transparent (or nearly transparent) interface to applications, as shown in Figure 26.3 [30–32]. This model effectively decouples the interactions between the wireless link and applications. The split proxy can use different physical and link layer techniques without each application having to be rewritten to be "wireless-aware." It provides a systematic way to experiment application performance on different air interface techniques. As an example, the WTP in the WAP suite uses this approach and provides the reliability and simplicity between UDP and TCP. Indeed, it should be possible for protocols such as WTP to achieve reliability comparable to that of TCP without introducing as much complexity, because this protocol tightly integrates specific link layer conditions, compared to TCP's end-to-end approach. To save additional bandwidth, wireless protocols such as WAP try to minimize message exchanges and use

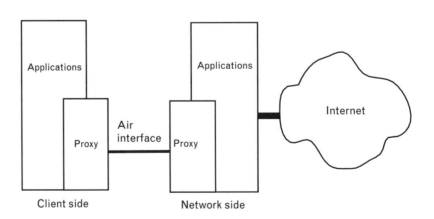

FIGURE 26.3
*Split proxy for
wireless Web.*

binary-formatted messages, as opposed to HTTP's frequent, text-formatted message exchanges.

Finally, client cache is of key importance to alleviate the impact of connectivity unreliability. By using a proxy at the client side as shown in Figure 26.3, the client proxy can cache or prefetch [33] the content whenever possible, dramatically improving user experience and reducing the impact of connectivity instability to application performance.

26.3.2 Content Adaptation and Proxy Services for Wireless Terminals

Because wireless terminals often have small and grayscale screens, limited processing capacity, and stringent power consumption requirements, it is desirable that the network transform the content so that it can be mostly efficiently transmitted over the air and rendered on the terminal. Transcoding has received a lot of attention in recent years, where a network server transforms the image or streaming media content to different formats, including different window sizes, resolutions, grayscales, and bandwidths [32, 34–36].

Terminal mobility adds a new dimension to the problem of proxy services in wireless networks. As the terminal moves within the network, it may be necessary for the proxy to hand off the proxy service such as transcoding or caching to a new proxy [37, 38]. Mobility network layer support such as context transfer [39] might be necessary so that proxies can perform proxy/service state transfer. In addition to this, proxy placement within the wireless network has also been studied [40].

26.4 Discussions

Content delivery and distribution for the wireline Web has been concentrating on network infrastructure scalability, and this will continue to be an important issue as wireless IP networks scale to millions of heterogeneous terminals. Protocol optimization used in wireline Web could be applied to wireless, too. However, the key challenge for accelerating user experience in wireless networks will likely be developing novel techniques that optimize radio bandwidth usage and overcome terminal capability and resource limitations. For example, advanced wireless IP proxy services could fully utilize the capabilities offered by emerging edge service described in Section 26.2.2.3 to provide content- and context-aware services. Such services will be intelligent and resilient to mobility, terminal limitations, connectivity irregularity, and content mismatch, and will provide transparent interface to applications to empower seamless and ubiquitous content services.

....

26.5   Concluding Remarks

Emerging wireless IP networks will provide many new services and will rely on advanced content distribution techniques to achieve adequate user experience. In this chapter, we developed taxonomy of content distribution that consists of network scaling, endpoint acceleration, and protocol and content optimization. Using this framework, we reviewed existing wireline content distribution techniques that will be useful in wireless networks. We identified key challenges for content distribution in wireless IP networks, including radio connectivity, limitations on terminal resources and capability, mobility, and discussed their implications on mobile content distribution. We found that content and protocol optimization and new service models such as edge services could significantly accelerate the user experience in wireless IP networks.

REFERENCES

[1]   Hennessy, J., and D. Patterson, *Computer Organization and Design: The Hardware/Software Interface*, San Francisco, CA: Morgan Kaufmann, 1994.

[2]   Breslau, L., et al., "Web Caching and Zipf-Life Distributions: Evidence and Implications," *Proceedings of IEEE INFOCOM '99*, New York, March 1999, pp. 126–134.

[3]   Jin, S., and A. Bestavros, "Temporal Locality in Web Request Streams: Sources, Characteristics, and Caching Implications (extended abstract)," *Proceedings of SIGMETRICS 2000*, Santa Clara, CA, June 2000.

[4]   Mahanti, A., D. Eager, and C. Williamson, "Temporal Locality and Its Impact on Web Proxy Cache Performance," *Performance Evaluation*, Vol. 42, 2000, pp. 187–203.

[5]   Jin, S., and A. Bestavros, "Greedydual★ Web Caching Algorithm: Exploiting the Two Sources of Temporal Locality in Web Request Streams," *Proceedings of the 5th International Web Caching and Content Delivery Workshop*, Lisbon, Portugal, May 2000.

[6]   Cao, P., and S. Irani, "Cost-Aware WWW Proxy Caching Algorithms," *Proceedings of the 1997 USENIX Symposium on Internet Technology and Systems*, December 1997, pp. 193–206.

[7]   Williams, S., et al., "Removal Policies in Network Caches for World Wide Web Documents," *Proceedings of ACM SIGCOMM '96*, Palo Alto, CA, August 1996, pp. 293–305.

[8]   Chankhunthod, A., et al., *A Hierarchical Internet Object Cache*, Tech. Rep. 95-611, Computer Science Department, University of Southern California, 1995.

[9]   Wessels, D., and K. Claffy, "ICP and the Squid Web Cache," *IEEE Journal on Selected Areas in Communications*, Vol. 16, No. 3, March 1998, pp. 345–357.

[10] Wu, K., and P. Yu, *Load Balancing and Hot Spot Relief for Hash Routing Among a Collection of Proxy Caches*, Tech. Rep. RC 21420 (96708), Computer Science/Mathematics Group, IBM Research, March 1999.

[11] http://www.akamai.com.

[12] http://www.digitalisland.com.

[13]   Billiris, A., et al., "CDN Brokering," *Proceedings of Sixth International Workshop on Web Caching and Content Distribution*, Boston, MA, June 2001.

[14]   http://www.edge-delivery.org/overview.html.

[15]   http://www.extproxy.org/.

[16]   Kanodia, V., and E. Knightly, "Multi-Class Latency-Bounded Web Services," *IEEE/IFIP IWQoS 2000*, Pittsburgh, PA, June 2000.

[17]   Doyle, R., et al., "The Trickle-Down Effect: Web Caching and Server Request Distribution," *Proceedings of Sixth International Workshop on Web Caching and Content Distribution*, Boston, MA, June 2001.

[18]   Aron, M., et al., "Scalable Content-Aware Request Distribution in Cluster-Based Network Servers," *Proceedings of the 2000 Annual USENIX Technical Conference*, San Diego, CA, June 2000.

[19]   http://www.spidercache.com.

[20]   http://www.xcache.com.

[21]   Thomas, S., *HTTP Essentials*, New York: Wiley, 2001.

[22]   Byers, H., et al., "A Digital Fountain Approach to Reliable Distribution of Bulk Data," *Proceedings of ACM SIGCOMM'98*, Vancouver, BC, 1998.

[23]   Byers, J., and M. Luby, *Accessing Multiple Mirror Sites in Parallel: Using Tornado Codes to Speed Up Downloads*, Tech. Rep. TR-98-021, International Computer Science Institute, 1998.

[24]   Douglis, F., A. Haro, and M. Rabinovich, "HPP: HTML Macro-Preprocessing to Support Dynamic Document Caching," *USENIX Symposium on Internetworking Technologies and Systems*, 1997.

[25]   Chan, M. C., and T. Woo, "Cache-Based Compaction: A New Technique for Optimizing Web Transfer," *IEEE INFOCOM 1999*, New York, 1999.

[26]   Mogul, J., "Squeezing More Bits Out of HTTP Caches," *IEEE Network*, Vol. 14, No. 3, 2000.

[27]   http://www.fireclick.com.

[28]   Cohen, E., and H. Kaplan, "Prefetching the Means for Document Transfer: A New Approach for Reducing Web Latency," *Proceedings of INFOCOM '00*, Tel Aviv, Israel, 2000.

[29]   Andersson, C., *GPRS and 3G Wireless Applications*, New York: Wiley, 2001.

[30]   Housel, B. C., G. Samaras, and D. B. Lindquist, "WebExpress: A Client/Intercept Based System for Optimizing Web Browsing in a Wireless Environment," *Mobile Networks and Applications*, Vol. 3, No. 4, 1998, pp. 419–431.

[31]   Liljeberg, M., et al., "Optimizing World Wide Web for Weakly Connected Mobile Workstations: An Indirect Approach," *Proceedings of 2nd International Workshop on Services in Distributed and Networks Environments (SDNE'95)*, Whistler, Canada, June 1995.

[32]   Margaritidis, M., and G. Polyzos, "MobiWeb: Enabling Adaptive Continuous Media Applications over Wireless Links," *IEEE Personal Communications*, Vol. 7, No. 6, December 2000, pp. 36–40.

[33]   Jiang, Z., and L. Kleinrock, "Web Prefetching in a Mobile Environment," *IEEE Personal Communications*, October 1998.

[34]   Fox, A., et al., "Adapting to Network and Client Variability Via On-Demand Dynamic Distillation," *Proceedings of Seventh International Conference on Architectural Support for Programming Languages and Operating Systems*, Cambridge, MA, 1996.

[35] Chandra, S., C. Ellis, and A. Vahdat, "Application Level Differentiated Multimedia Web Services Using Quality Aware Transcoding," *IEEE Journal on Selected Areas in Communications*, Vol. 18, No. 12, December 2000, pp. 2544–2564.

[36] Warabino, T., et al., "Video Transcoding Proxy for 3G Wireless Mobile Internet Access," *IEEE Communications Magazine*, October 2000.

[37] Hadjiefthymiades, S., and L. Merakos, "Using Proxy Cache Relocation to Accelerate Web Browsing in Wireless/Mobile Communications," *Proceedings of WWW10*, Hong Kong, 2001.

[38] Joshi, A., S. Weerawarana, and E. Houstis, "On Disconnected Browsing of Distributed Onformation," *Proceedings of Seventh International Workshop on Research Issues in Data Engineering*, 1997.

[39] http://www.ietf.org/html.charters/seamoby-charter.html.

[40] Jiang, Z., et al., "Incorporating Proxy Services into Wide Area Cellular IP Networks," *Proceedings of 2000 IEEE Wireless Communications and Networking Conference*, Chicago, IL, 2000.

CHAPTER 27

Perceptual QoS for Wireless and IP Networks

Anand R. Prasad

27.1 Introduction

Several access technologies are evolving and emerging [1–8]. The second-generation GSM is evolving via GPRS, HSCSD, and EDGE towards UMTS/IMT-2000. In addition, WLAN type systems like HIPERLAN/2 and IEEE 802.11 are becoming available.

Together with these, IP-based networks have evolved tremendously. The benefit of packet switching and the success of IP-based networks together with the success of wireless technology have brought about convergence of the two technologies.

In addition to the demand for wireless networking, multimedia is playing an increasing role in today's society, creating new challenges to those working in the development of telecommunications systems. Thus, the pressure for telecommunications systems to cope with increasing QoS demands is enormous. As we move towards the convergence of networks, which will have its own technical challenges, it is important to understand this demand for QoS.

QoS means providing consistent, predicted delivery of a service. In simple terms, QoS satisfies customer requirements. Today, QoS in any network (Internet, mobile, or WLAN) is measured mostly in terms of *signal-to-interference ratio* (SIR), PER, or latency. In practice, however, a user does not measure QoS based on these parameters. A user perceives QoS. This brings us to the key word of providing QoS: *perceptual QoS* (PqoS). Thus, we require PQoS measurement in telecommunications systems.

Conventional PQoS is based on subjective measurements [8, 9], which are very expensive and time consuming. The solution is to use cheaper objective PQoS measurements with high correlation with subjective measurement.

Before we move towards a converged network, we have to understand the PQoS-based performance of current networks. In this chapter we will present results for PQoS measurement results for WLANs and mobile networks (GSM and UMTS).

27.2 QoS

One can perform subjective or objective quality measurement. Usually subjective measurements are performed to get PQoS measurement. The subjective method requires several people to listen to or see a voice, video, or audio clip in a given environment. This is obviously a very time-consuming and expensive process. A faster method used by telecommunications systems is objective measurement. The goal of which is to provide consistent perceived quality. This obviously will lead to higher customer satisfaction and an increase in overall capacity.

There are standards that discuss PQoS measurement methods with reasonable correlation to the subjective measurements, [8–14]. In this section we discuss the definition of the PQoS measurement method for voice and video as described in standards.

27.2.1 Voice

Quality of voice is usually represented using a value called *mean opinion score* (MOS), which lies between 1 and 5. ITU-T standard P.861 [10] defines the method for objective measurement of voice quality. This objective measurement is represented by *perceptual speech quality measure* (PSQM).

Within PSQM, the physical signals constituting the source and coded speech are mapped onto psychophysical representations that match the internal representations of the speech signals (the representations inside our heads) as closely as possible. These internal representations make use of the psychophysical equivalents of frequency (critical band rates) and intensity (compressed sone). Masking is modeled in a simple way: only when two time-frequency components coincide in both the time and frequency domains is masking taken into account.

Within the PSQM approach, the quality of the coded speech is judged on the basis of differences in the internal representation. This difference is used for the calculation of the noise disturbance as a function of time and frequency. In PSQM, the average noise disturbance is directly related to the quality of coded speech.

The transformation from the physical (external) domain to the psycho-physical (internal) domain is performed by three operations:

1. Time-frequency mapping;

2. Frequency warping;

3. Intensity warping (compression).

Besides perceptual modeling, the PSQM method also uses cognitive modeling in order to get high correlations between subjective and objective measurements.

PSQM values can be converted to MOS values by deriving transformation functions. *Perceptual evaluation of speech quality* PESQ [11] is a new voice quality measurement standard not used for the study in this chapter.

27.2.2   Video

Video quality is often measured using *peak signal-to-noise ratio* (PSNR). PSNR is defined as the ratio between peak signal and rms noise observed between the reference video and the processed video.

PSNR, however, does not take into account human vision and thus cannot be a reliable predictor of perceived visual quality. Human observers will perceive different kinds of distortions in digital video, like jerkiness, blockiness, blurriness, and noise. These cannot be measured by PSNR.

ANSI T1.801.03-1996 standard [13, 14] defines a number of features and objective parameters related to the above-mentioned video distortions. These include the following:

- *Spatial information* (SI) is computed from the image gradient. It is an indicator of the amount of edges in the image.

- Edge energy is derived from SI. The difference in edge energy between reference and processed frames is an indicator of blurring (resulting in a loss of edge energy), blockiness, or noise (resulting in an increase of edge energy).

- The difference in the ratios of *horizontal/vertical* (HV) edge energy to non-HV edge energy quantifies the amount of horizontal and vertical edges (especially blocks) in the frame.

- *Temporal information* (TI) is computed from the pixel-wise difference between successive frames. It is an indicator of the amount of motion in the video. Repeated frames become apparent as zero TI, and their percentage can be determined for the sequence.

- Motion energy is derived from TI. The difference in motion energy between reference and processed video is an indicator of jerkiness (resulting in a loss of motion energy), blockiness, or noise (resulting in an increase of motion energy).

Different ANSI parameters can be combined to get perceptual video quality measurement (i.e., jerkiness, blurriness, and blockiness). Blockiness is a perceptual measure of the block structure that is common to all DCT-based image compression techniques. Blurriness is a perceptual measure of the loss of fine detail and the smearing of edges in the video. It is typically caused by a high-frequency attenuation at some stage of the recording or encoding process. Jerkiness is a perceptual measure of motion that does not look smooth. A primary cause of perceived jerkiness is the dropping of frames, but other factors contribute to the perception jerkiness as well.

In the real world we will have video together with audio. Depending on the content, the requirement will be high for one of them (e.g., music users will prefer high audio quality as compared to video). In our current work we have not considered perceptual quality for combined audio and video services.

27.3   Simulation Model

For the purpose of simulation we use a number of tools for quality measurement and link level simulation. In this section, first the tools are explained after which the simulation model used is explained.

27.3.1   Quality Measurement Tools

We use two tools for quality measurement in this paper.

The first tool is for voice quality measurement [8]. This tool takes as input the original speech and modified, or in our case received speech, and measures the PSQM value by comparing the two. It then calculates a MOS value based on a predefined transformation method.

A video QoS tool was used to get video quality results. This tool measures the difference between reference and processed video streams for various ANSI and perceptual parameters [8]. For this paper we have considered only a few of the parameters.

27.3.2   Link Level Simulation Tools

Link level simulation is performed by using a Matlab-based tool for 5-GHz systems (HIPERLAN Type 2 or IEEE 802.11a) [8]. These tools are implemented as per the standard for receiver and transmitter. The channel used is

Rayleigh fading and AWGN channel defined as per the standard requirement for different test conditions. Similarly, we also used a Matlab simulator for UMTS with AMR [15].

27.3.3 Simulation Model

In Figure 27.1 the simulation model used for the study is given. For all systems we have the transmitter, realistic channel, and receiver. The input data can be voice, audio, or video, which is encoded before transmitting using the required codec at the receiver. Decoding takes place at the receiver after which the transmitted and received signals are compared. This gives the PQoS value.

27.4 Video over WLAN

27.4.1 Video over 5-GHz WLAN

In this section simulation results are presented for video over 5-GHz WLAN. The simulation was done for HIPERLAN/2, and video quality

FIGURE 27.1
Simulation model.

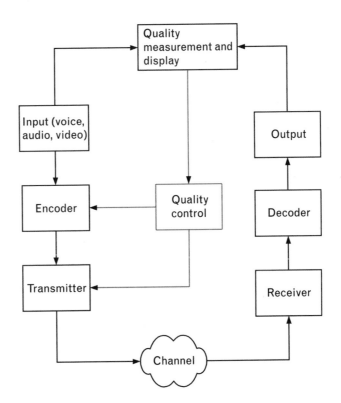

was measured using video QoS tool. The results are shown in Figures 27.2, 27.3, and 27.4. For this simulation only one user was considered. An indoor office 18 path Rayleigh fading channel as defined by the standard was used. The video used was MPEG1, 300 frames of 10 seconds length, in a studio environment. The results were generated for 12 Mbps and 18 Mbps with 20 LCH, f_d (Doppler frequency) of 1 Hz, and SNR of 10, 14, and 16 dB. Although HIPERLAN/2 was used, one can also consider these results valid for IEEE 802.11a with packet size of 1,080 bytes. As there is only one-way communication without any buffering at the receiver, one can consider this as real-time communication [8].

FIGURE 27.2
Percentage jerkiness
for different data rates
and SNR.

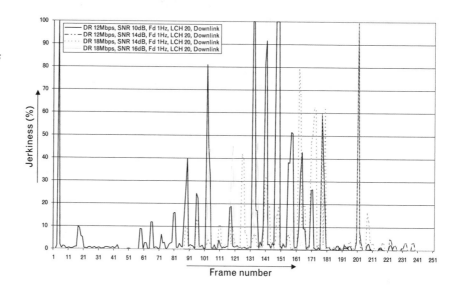

FIGURE 27.3
Percentage blockiness
for different data rates
and SNR.

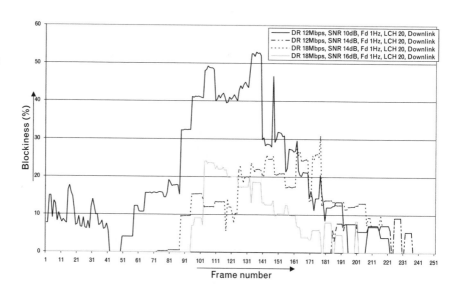

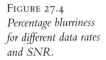

FIGURE 27.4
*Percentage blurriness
for different data rates
and SNR.*

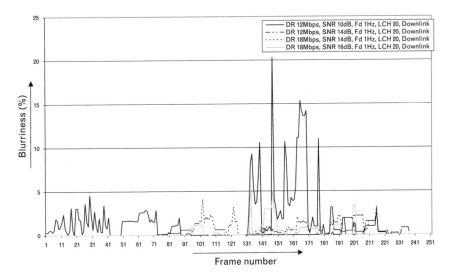

As expected, we see that the performance of the lower data rate is better than the higher data rate, and that higher SNR for a given data rate gives better performance.

27.4.2   Video over IEEE 802.11b

Streaming video quality over IEEE 802.11b WLAN with backbone IP network was measured. The method was to measure the perceptual quality at the access point and at the wireless station. The video stream used was a music video with varying background at 373 Kbps of 4 minutes and 8 seconds length. For both wireless and wired stations only jerkiness was observed; the results are given in Figure 27.5.

In the case of a wired network the reason for jerkiness is easily understood; probability of error in wired networks is negligible, on the other hand congestion can occur thus causing jerkiness. The wireless network, in our case, used CSMA/CA-based MAC protocol and there was no interference. Thus, similar to the wired network only jerkiness is observed. Of course, as the number of users increases, jerkiness will also increase.

In the case of 11-Mbps IEEE 802.11b WLAN, one observes a net throughput of about 5 Mbps [7]. Thus, the jerkiness observed is almost negligible.

27.5   UMTS Speech Quality

The AMR speech codec has been adopted as the standard codec for GSM phase 2+ [15] and UMTS 3G [15] standards. This codec supports eight

FIGURE 27.5
*Jerkiness at wireless
station using IEEE
802.11b at 11 Mbps.*

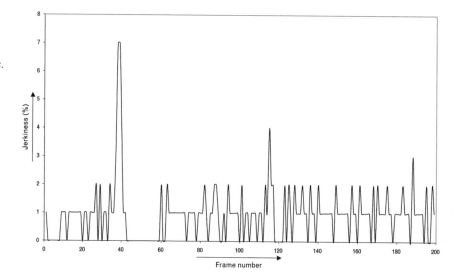

rates, which can be dynamically selected as often as every 20 ms during a conversation. The performance of the AMR codec has been characterized in the case of GSM physical layer. However, the corresponding results have not been available until recently for the case of UMTS physical layer and only then for very limited conditions [15]. In this section, the performance of AMR is characterized based on objective voice quality measurement. The performance curves presented here represent the objective voice quality in terms of *frame erasure rate* (FER).

27.5.1 Simulation Model

The simulation was done using original speech as input to the AMR codec over the UMTS physical layer. On the received side the decoded speech was stored. The input and output speech were used as input to the voice quality measurement tool to get MOS results [15].

The speech files have been obtained from the ITU database for voice quality measurement tests. The database includes 176 files, which contain clean speech. Each file contains prerecorded sentences of 8 seconds duration in PCM format. The sentences contain approximately 50% speech and 50% silence intervals. Voices from two male and two female speakers have been used in the simulations.

The AMR speech codec is based on the *Algebraic Code Excited Linear Prediction* (ACELP) technique [15]. Encoded speech samples are transmitted in 20-ms frames. The number of bits in each frame can vary from one frame to the next depending on the codec mode. There are a maximum of eight modes possible for the codec: 4.75, 5.15, 5.90, 6.70, 7.40, 7.95, 10.2, and 12.2 Kbps. The encoded speech is divided into three classes: class A, B, and

C in the decreasing order of their perceptual importance in reconstruction of speech. The frame erasure rate in AMR is defined as the average rate of speech frames for which the CRC indicates decoding errors in class A bits. These frames are marked as "bad" frames, and their occurrence invokes the error concealment procedure [15]. Additional details of AMR can be found in 3GPP standards and in [15].

The major parameters used for physical layer simulation are chip rate of 3.84 Mcps, spreading factor of 128, channel bit rate of 60 Kbps, TTI of 20 ms. Also, DTX was used for downlink and both inner and outer loop power control was used. The vehicular-A channel impulse response model [15] has been used in the simulation.

27.5.2 Results

In this section, the AMR characterization test results are presented as a function of FER and RBER. They provide an indication of the quality degradation to be expected for the implementation of the AMR speech codec in UMTS networks.

The objective speech quality results for the simulation are shown in Figure 27.6. The quality degradation is expressed in *differential MOS* (DMOS).

Figure 27.6 shows the following trends in speech quality degradation:

- There is very small difference in speech quality for all AMR rates at low error rates.

FIGURE 27.6
DMOS results as a function of FER.

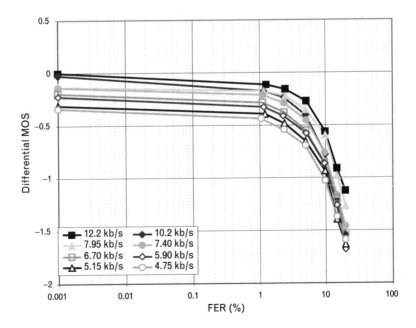

• The voice quality is degraded quite rapidly at high error rates.

In addition, the voice quality is gradually degraded for all codec rates as the FER is increased up to 1%. In the event of occasional bad frames, the error concealment procedure compensates for the lost frames.

As the FER is increased beyond 1%, for each codec rate a critical FER value is reached beyond which the voice quality is degraded exponentially with FER. It can be observed from Figure 27.6 that the speech quality for AMR rate 12.2 Kbps at FER = 7% is essentially the same as the quality for AMR rate 4.75 Kbps at FER = 1%.

27.6   Concluding Remarks

In this chapter, PQoS definition, application, and results are presented for different types of systems and networks. These solutions make use of reference (original) and output signal to give PQoS result.

For WLAN, simulation results are given for video over HIPERLAN/2. Since simulation for WLAN is done considering one user, the results are also valid for IEEE 802.11a. Since the above-mentioned results give PQoS measurement, these results can also be used for quality control. Measurement results for streaming video over IEEE 802.11b is also given. Results show that only jerkiness is observed.

A full set of AMR performance characteristic curves has also been obtained in this chapter. The results show a common trend in the AMR codec voice quality for various rates. For each codec rate, the voice quality remains practically the same with FER until the critical FER value for that codec rate is reached. Results show that the robustness of AMR codec to errors increases at higher rates as evident from the critical FER values. In particular, results show that the speech quality for AMR rate 12.2 Kbps at FER = 7% is practically the same as the quality for AMR rate 4.75 Kbps at FER = 1%.

REFERENCES

[1]   ETSI BRAN, "HIPERLAN Type 2 Functional Specification Part 1—Physical Layer," DTS/BRAN030003-1, June 1999.

[2]   Halls, G., "HIPERLAN Radio Channel Models and Simulation Results," RES10TTG 93/58.

[3]   IEEE, "802.11, Wireless LAN Medium Access Control (MAC) and Physical Layer (PHY) specifications," November 1997.

[4]   IEEE, "Supplement to Standard for Telecommunications and Information Exchange Between Systems—LAN/MAN Specific Requirements—Part 11: Wireless MAC

and PHY Specifications: High Speed Physical Layer in the 5-GHz Band," P802.11a/D7.0, July 1999.

[5] van Nee, R. D. J., and R. Prasad, *OFDM for Wireless Multimedia Communications,* Norwood, MA: Artech House, 2000.

[6] Ojanpera, T., and R. Prasad, *Wideband CDMA for Third Generation Mobile Communications,* Norwood, MA: Artech House, 2000.

[7] Prasad, N. R., and A. R. Prasad, (Eds.), *WLAN Systems and Wireless IP for Next Generation Communications,* Norwood, MA: Artech House, 2002.

[8] Prasad, A. R., et al., "Perceptual Quality Measurement and Control: Definition, Application and Performance," *WPMC 2001,* Aalborg, Denmark, September 9–12, 2001.

[9] ITU-T, "Objective Quality Measurement of Telephone-Band (300–3400 Hz) Speech Codecs," Series P: Telephone Transmission Quality, Telephone Installations, Local Line Networks, P.861, February 1998.

[10] ITU-T Recommendation P.862, "Perceptual Evaluation of Speech Quality (PESQ): An Objective Method for End-to-End Speech Quality Assessment of Narrowband Telephone Networks and Speech Codecs," February 2001.

[11] ITU-T, "Subjective Performance Assessment of Telephone Band and Wide Band Digital Codecs," *Telephone Transmission Quality: Methods for Objective and Subjective Measurement of Quality,* P.830, February 1996.

[12] ITU-R, "Methods for Objective Measurements of Perceived Audio Quality," BS.1387.

[13] ANSI, "Digital Transport of One-Way Video Signals—Parameters of Objective Performance Assessment," ANSI T1.801.03-1996, February 5, 1996.

[14] Video Quality Research, http://www.its.bldrdoc.gov/n3/video/Default.htm.

[15] Rohani, B., and B. Rohani, "Objective Characterization of AMR Speech Codec on UMTS Physical Layer," WPMC 2001, Aalborg, Denmark, September 9–12, 2001.

CHAPTER 28

Transcoding for the Mobile Internet: The Case of Video Transcoding*

Shinji Ota, Akio Yoneyama, Takayuki Warabino, Daisuke Morikawa, and Masayoshi Ohashi,

28.1 Introduction

With the advent of broadband wireless links and high-performance mobile devices, mobile users can handle multimedia data in their wireless Internet accesses. At present, mobile phones can also process video data and mobile users can enjoy video content on mobile phones. In current services for video content distribution to mobile phones, users can playback video content after the video data is downloaded completely to their mobile phones. In these services, the video content data size is limited according to the mobile phone's memory for processing video data and that of the total download time of the video data.

A video streaming technique, in which video content is played back while the video data is being downloaded from an origin server, is useful for the distribution of higher quality and longer video content in a mobile environment. With this technique, mobile devices can process video content with a shorter processing memory, and mobile users need not wait as long for playback of the video content.

In the mobile environment, however, constant end-to-end QoS is not guaranteed. Therefore, to meet the performance requirement for viewing multimedia content, adaptations for a variety of mobile devices and for QoS fluctuations over a wireless link are required. In this chapter, an overview of video transcoding for the mobile Internet is provided.

* This work is supported by the Telecommunications Advancement Organization of Japan (TAO).

28.2 Adaptation Techniques for Multimedia Content Distribution

28.2.1 Adaptation Technique for Mobile Web Accesses

In recent years, several transcoding proxies for mobile Web access have been used in order to adapt mobile device capabilities. Most mobile operators may install these proxies on the border between their mobile networks and the Internet/intranet. The reasons are that all content providers may not support adaptation proxies, and they may not prepare all contents accommodated for mobile users. As shown in Figure 28.1, transcoding proxies receive Web content such as HTML files and image data from Web servers and relay them to mobile devices after transcoding. There are two types of transcoding techniques depending on the aim of adaptation.

28.2.1.1 Format Conversions

Format conversions convert one data format into another format that the device can support. WML is a Web content markup language for mobile phones. To process Web content distributed for Internet-connected PCs on mobile phones, a format conversion from HTML to WML is needed.

Current mobile phones mainly support image data formats such as GIF, PNG, and JPEG. Mobile phones, however, usually cannot deal with high-quality image data transferred from the Internet because the processing power of mobile phones is not sufficient. Conversions among image data formats are useful in enabling mobile users to view as much image data in Web pages as possible.

FIGURE 28.1

The basic architecture of transcoding systems.

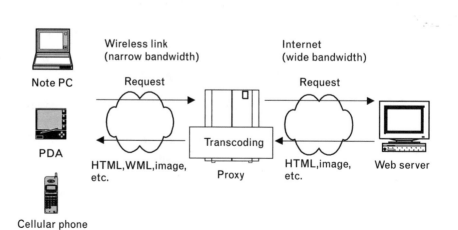

28.2.1.2   Web Page Rearrangement

Web page rearrangement refers to techniques that change Web page layout in order to fit the display size of a mobile device. Page size reduction is useful for PDAs, where a Web page layout is reduced by changing the size properties of Web page objects (e.g., frames, tables, images) while maintaining the same proportions. With this technique, mobile users can view a Web page with the look and feel of a PC.

However, the technique is not effective for mobile phones because it is hard for mobile users to view Web page objects that are too small. Furthermore, given the small size of mobile phone displays, it is also hard to reproduce Web page overview. Therefore, rearrangement of Web page objects is used for distributing Web content to mobile phones.

A rearrangement technique, in which HTML frames and tables tags are deleted from a HTML file and the other objects in the HTML page are rearranged lengthways, is proposed and used [1]. In this rearrangement, Web page objects are scaled to fit into the lateral size of mobile phone displays. Mobile phone users can view a rearranged Web page only with up and down key operations.

28.2.2   Issues of Viewing a Video Stream in a Mobile Environment

The demand of mobile video streaming is being accelerated by the new-generation wireless systems. There are some restrictions, however, in viewing a video stream in a mobile environment:

- *Lack of mobile device's capability:* Receiving video streaming data that exceeds a mobile device's capabilities will cause a lack of video frames and deterioration of the picture quality. Mobile devices have not been able to support video formats encoded by higher bit-rate compression methods (e.g., MPEG-2) as used in the Internet/intranet. The video formats used in a mobile environment must be encoded by a lower bit-rate compression method (e.g., MPEG-4), and the encoded bit rate must be adjusted to the mobile device's capability.

- *Insufficient and unstable quality of wireless links:* Insufficient quality of wireless links would cause a lack/delay of video frames and deterioration of the picture quality. The bandwidth of wireless links is usually narrower than that of the Internet link. Therefore, video-streaming data that exceeds the capabilities of a wireless link is kept waiting or discarded. Unstable wireless link QoS is also a restriction. The packet losses and delays fluctuate depending on the conditions of the wireless link (e.g., signal to noise ratio), and handoffs and initiations/terminations of other streamings in the same link. Video streaming is influenced by these restrictions because the video data is transmitting for a long time over the wireless link. These bandwidth discrepancy issues may be resolved by the QoS signaling in the future if the signaling and its control mechanism are well established over wireless links.

28.2.3   Adaptation Techniques for Video Streaming

In this section, several adaptation techniques for video streaming over the Internet are introduced.

28.2.3.1   Adaptation Approaches

Examples of adaptation approaches for best-effort networks (e.g., the Internet) are simulcast, layered coding, and transcoding. The objective of these approaches is to multicast or Webcast efficiently over the Internet to devices and networks of a broad range of capabilities.

In a simulcast [2], a sender of video data prepares a small number of video streams carrying the same video data but targeted at receivers with different capabilities. These video streams are sent to different multicast groups or different unicast sessions; a receiver can receive a suitable video stream by selecting a multicast group or a unicast session port for the video stream.

With layered coding [3], video data is decomposed into multiple video streams. Using this approach, video data is encoded into a number of layered streams $(S_1, S_2, ..., S_n)$ that can be incrementally combined by the decoder. These layered streams are sent to different multicast groups or unicast sessions $(G_1, G_2, ..., G_n)$, respectively, and a receiver can receive a suitable video stream by selectively joining multicast groups or receiving unicast sessions based on monitoring the received quality of the streams.

In real-time transcoding [4], a transcoder converts a high-rate and high-quality video stream to another low-rate and low-quality stream depending on changes in the reception quality in real time. The transcoder converts video formats corresponding to receiver capabilities.

These approaches have some advantages and disadvantages. Simulcast wastes relay networks' bandwidth if the number of supported video streams becomes large. Layered coding is more efficient than simulcast; however, it does not always provide the best quality in terms of coding efficiency because the bit rate for each layer is usually fixed. Of these options, transcoding can provide the best possible video quality to each receiver. In addition, the transcoding approach can easily deploy new functions for a new format. This approach, however, involves scalability issues, where the transcoder must perform conversions requested from multiple receivers in real time.

28.2.3.2 Types of Transcoders

The following section focuses on video transcoding techniques. Major techniques are frame dropping, requantization, and video format conversion [5].

Frame dropping [6, 7] reduces the output data rate by dropping video frames. A shortage of receiver device processing power and the lack of sufficient processing memory may cause unnecessary dropping of frames on the device. The transcoder has information about frame types and their importance, and it drops frames according to that importance in order to adjust the output frame rate of the stream.

Requantization [8] reduces the output data rate by reducing the number of bits used to encode pixels. In MPEG video encoding, if the frequency coefficients become coarser, the output rate of the encoder can be reduced at the cost of picture quality.

The video format conversion is used for swapping between different video formats. In the technique, parameters of the output stream, such as frame size, frame rate, quantizer parameter, can be adjusted to fit the receiver capabilities and the available bandwidth.

28.2.3.3 Adaptation Mechanisms

For the adaptation schemes described in Sections 28.2.3.1 and 28.2.3.2, several adaptation algorithms have been proposed. In the simulcast and the layered coding approach, receivers control the adaptation strategy [3]. A receiver monitors the quality of received video, such as frame loss and jitter, and decides whether to change multicast group, and to which multicast group the receiver should move, depending on the monitored quality.

In the transcoding approach, transcoding nodes (e.g., servers or proxies) control the adaptation strategy. Some transcoding systems have feedback mechanisms to collect information about reception quality from receivers, and operate transcoding depending on packet losses or frame losses reported by receivers [4, 6]. In this case, adaptation capability is limited by feedback frequency; that is, the system does not fully adapt to rapid changes in transmission quality.

The other approach is to estimate transmission QoS in the adaptation node instead of using a feedback mechanism. For example, transmission delays are estimated from the state of the source buffer, and transcoding is performed depending on the amount of data in the buffer [7, 8]. In this case, flow-control mechanisms for video data distribution are mandatory.

28.3 Related Works

In the TranSend project [9], a proxy for transcoding Web contents was proposed. The proxy supports "distillation" relating to images, texts, and videos. The proxy adapts according to device capabilities. The Browse-it server provided by Pumatech is based on this work. Other similar commercial products include Spyglass Prism and AvantGo.

The Vosaic [6] extended the vanilla NCSA Mosaic architecture to support the dynamic, real-time video and audio information. To cope with insufficient CPU power, the Vosaic player continuously measures the rate of frames dropped by the receiver and reports the frame loss rate to the Vosaic server. The player also measures and reports packet loss rate to the server to cope with network congestion. The server adjusts data rate to the receiver by dropping frames based on the reports from the player.

MobiWeb [7] enhances the performance of adaptive real-time streams over wireless Internet links. In the system, two symmetric proxy layers are added to both a mobile device and a BS. The proxies provide an admission control to reserve wireless resources for each stream, as well as a dynamic prioritization scheme to guarantee the stable transmission of high-priority streams. In addition, an adaptation scheme for reducing the data rate of low-priority streams depending on the wireless link quality is also provided. It is presumed that the information used in the controls, such as throughput, error rate of the link transmission, received signal power and handoff notifications, is provided from the link layer.

Another proxy-based system is proposed in [10]. Under this system, the *mobile proxy server* (MPS) is located at the boundary of the wired and wireless network (e.g., a WLAN access point), and the MPS transcodes an MPEG stream from a server into a format suitable to the mobile client. The purpose of the transcoding is to support a wide range of mobile devices; however, no adaptation is provided against QoS fluctuations on the wireless link.

PacketVideo provides MPEG-4 multimedia distribution in a wireless environment. The player, PVPlayer, supports data rates ranging from 9.6 Kbps to 384 Kbps. The FrameTrack technology used in the system applies dynamic rate adaptation against bandwidth fluctuations during video streaming. During the adaptation, the server regulates the frame rate based on PVPlayer reports.

28.4 Standardization Issues

Protocols for streaming over the Internet have been standardized by the IETF. RTP [11] is a protocol that provides end-to-end data delivery services with real-time characteristics, such as interactive audio and video.

A receiver of RTP data observes the QoS, and the QoS information [referred to as *receiver reports* (RRs)] is conveyed in RTP control protocol packets. The highest sequence number received, the number of packets lost, the estimated packet interval jitter, and timestamps for computing end-to-end delays are included in the RR. Furthermore, *Real Time Streaming Protocol* (RTSP) [12] for streaming session setup and control has been standardized.

In MPEG-4 standard, *binary format for scene description* (BIFS) is adopted for providing facilities to compose a set of multimedia objects into a scene. MPEG-4 also supports a framework for representing MPEG-4 scene description using a textual syntax. The framework provides interoperability with *synchronized multimedia integration language* (SMIL) [13], which is an XML-based textual language for enabling simple authoring of interactive audiovisual presentations.

The CC/PP [14] framework will specify how client devices express their capabilities and preferences to servers that originate content. The framework is supported by WAP and 3GPP, and it is expected that it will also be supported by various types of devices for Internet access.

Support of streaming services in the 3G wireless environment is also being discussed in the 3G specification groups such as 3GPP and 3GPP2. In the specifications, it is recommended that Internet standard/de facto standard protocols and languages for streaming should be supported in the mobile environment. QoS requirements and features of video streaming service in the 3G wireless environments are provided in [15]. In the specification, it is suggested that a streaming service should support dynamic adjustment of streaming parameters; however, definite approaches for the adjustment will be one of the discussion items in the near future.

28.5 Video Transcoding System Architecture and Control Mechanism

28.5.1 Basic Concept of the System

Mobile packet services have been commercialized for efficient Internet accesses, and these services will become major in future mobile systems. In the packet services, the throughput fluctuations occur because resources will be allocated to each user in a best-effort manner. To achieve sufficient video streaming quality, a *video transcoding proxy* (VTP) for IMT-2000 packet service is proposed [16]. VTP is based on the following concepts.

1. *Adaptations for long-term bandwidth fluctuations:* The quality fluctuations of the wireless link can be classified into two categories. Short-

term fluctuations are caused by the transmission characteristics of the air interface such as fading, interference, and so on. Handoffs of mobile devices and initiations/terminations of streaming cause long-term fluctuations. In IMT-2000 systems based on CDMA, the outer-loop control of a transmitting power control operates to keep the frame error rate of the mobile device constant. Consequently, it is considered that no short-term fluctuation should ideally occur in the CDMA systems. From this point of view, VTP is mainly looking at adaptations for the long-term fluctuations.

2. *Adjust an output data rate to throughputs:* In order to control the output rate of the stream to an arbitrary level, a transcoding technique is used. VTP tries to adjust the output data rate for a mobile device to the throughput reported from the mobile device.

3. *Coping with capability variation in mobile devices:* There are capability gaps between Notebook PCs and mobile phones (e.g., display size, CPU power, and battery life), even in future mobile communication systems. Therefore, adaptations for the various mobile devices are also required.

4. *Reducing the cost of implementation:* It is supposed that two symmetric proxies are added to both a mobile device [a *client proxy* (CP)] and a mobile network (a VTP). The VTP collects information about reception quality from the CP. To minimize the feedback delay, the VTP should be placed at the nearest node in the mobile communications network to the CP (e.g., the BS). However, it makes control mechanisms of the VTP complicated and costly. For example, a mobile device must seamlessly switch from one VTP on the old BS to another VTP on the new BS during its handoff processes. By placing the VTP functions at the gateway nodes as shown in Figure 28.2, the impact of such enhancement will be kept to a minimum.

28.5.2   System Architecture

The architecture of the video transcoding system is illustrated in Figure 28.3. The system consists of a CP located on a mobile device and a VTP located on a gateway.

The CP collects and manages the profile of the mobile device (device profile) and user preferences relating to VTP operations (user profile).

The device profile includes display size, processing power, and video format type supported by the mobile device. The user profile includes the maximum bandwidth allowed to the user (e.g., subscribed bandwidth). These profiles are transmitted to the VTP before the beginning of video streaming. The CP also periodically measures download throughput (transmission profile) and passes it to the VTP during video streaming.

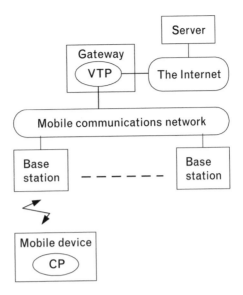

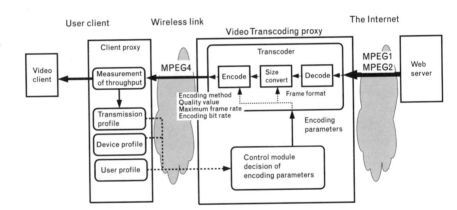

The VTP has two components, a transcoder and a control module. The transcoder is a video stream conversion engine. It decodes an input video stream and then performs size conversion and reencoding in real time. The control module decides appropriate encoding parameters based on the profiles from the CP and periodically passes them to the transcoder. The control module also observes the transmission profile and increases the encoding bit rate if no rate reduction request is received over a period of time.

28.5.3 Control Mechanism

The control module decides the following encoding parameters by evaluating the profiles reported from the CP.

1. *Frame format:* The control module determines the frame format for the video stream to be fitted to the display size of the mobile device. For instance, from the relationship between frame format and size shown in Table 28.1, the control module chooses the largest frame format for which both width and height are within the display size in the device profile.

2. *Maximum frame rate:* The control module determines the maximum frame rate (F_{max}) for the video stream based on the mobile device's processing capability. The following equations show the method of F_{max} calculation (F and $F_{max} \leq 30$; A: constant value):

$$P_d = A \cdot F \cdot S \leq P_t \tag{28.1}$$

$$F_{max} = \frac{P_t}{A} \cdot \frac{1}{S} \tag{28.2}$$

where P_d and P_t indicate the amount of processing that is required for decoding in real time and which the client device can perform, respectively. F is the frame rate of the video stream, while S is the frame size of the video stream and is derived by multiplying width and height. Equation (28.1) shows P_d is proportional to F and S, and for decoding in real-time, P_d must be less than P_t. Then, (28.2) is derived from (28.1) and F_{max} can be calculated. In calculating, as P_t depends on the capability of the CPU and video codec LSI present on the client device, P_t is determined based on the processing power information in the device profile (Figure 28.4). The transcoder always keeps the frame rate of output video stream less than F_{max} so that the mobile device can playback the video stream successfully.

3. *Encoding bit rate:* The control module determines the encoding bit rate of a video stream from the VTP to the CP based on the transmission profile. When the throughput from the VTP to the CP becomes low, the encoding bit rate of the output video stream will be reduced according to the transmission profile. However, the control module cannot detect the increase in bandwidth because the exact available bandwidth cannot be known before use. So the control module

TABLE 28.1   THE RELATIONSHIP BETWEEN FRAME FORMAT AND SIZE

FRAME FORMAT	CIF	QCIF	SQCIF
Width [pixels]	352	176	128
Height [pixels]	288	144	96

environment. With the conversion, the output rate from the trans-coder is controlled within the range suitable for a mobile device (MPEG-4: 5 Kbps ~ 5 Mbps), and the error resilience feature can also be enhanced (target bit error rate is from 10^{-3} to 10^{-4}).

For the development of the transcoder, the above three factors were combined. The transcoder adapts frame rate and picture quality according to the setting of the target bit rate and picture size. Under this system, the input MPEG-2 stream is first decoded to a picture image and stored in the frame memory. At the same time, motion vector information is extracted from the original MPEG video stream. Then a picture-resizing operation is applied to the decoded pictures before reencoding. Generally, MPEG video encoding requires over twice as much computation power as MPEG video decoding. The most computationally expensive processing is motion esti-mation processing and it takes up about half of the whole encoding process. In the developed transcoder, the extracted motion vector information from the original MPEG-2 video stream is used for the motion estimation opera-tion at the MPEG-4 video reencoding stage so that computational com-plexity is significantly reduced.

28.6   Features of the Video Transcoding System

28.6.1   Experimental Video Transcoding System

An experimental video transcoding system was built (Figure 28.5). The CP and a *video client* (VC) are located at the same *client emulator* (CE). The VTP is located at the *mobile WWW gateway* (MWG). The *mobile network simulator* (MNS) can shape traffic flows from the MWG to the CE, and the peak-rate of the wireless link is emulated. The *dummy traffic generator* (DTG) can gen-erate dummy traffic and send them to the CE with the requested data rate. By using MNS and DTG, the bandwidth fluctuations can be emulated.

FIGURE 28.5
Structure of experi-mental system.

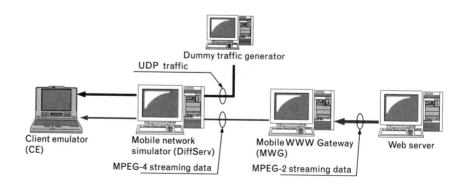

MWG is built on a PC with Pentium III CPU (800 MHz) and 786–MB memory. The system parameters are shown in Table 28.2. The control period is the time interval, with which the control module calculates encoding parameters. The increase cycle of the data rate provides the timing at which the control module tries to raise the output data rate; and the increase ratio provides the percentage with which the bandwidth is increased. The frame rate of the output stream for the CE is limited to 15 fps (frames per second) by the maximum frame rate. In these experiments, MPEG-2 content shown in Table 28.3 was used as the input. The length of the input stream is 180 seconds.

28.6.2 Characteristics of the Experimental System

Figure 28.6 shows the characteristics of this transcoding system when 150-Kbps dummy traffic is periodically inserted and removed.

If the bandwidth adaptation is not performed, the packets to the CP are delayed by a TCP flow control between the CP and the VTP. In the no-adaptation case, the playback duration is 220 seconds, and it is longer than that of the input stream (180 seconds). When the VTP bandwidth adaptation is used, the playback duration is 180 seconds, and the video streaming is completed in real time without transmission delays.

TABLE 28.2 SYSTEM PARAMETERS FOR EXPERIMENTS

Maximum rate	350 Kbps
Control period	5 seconds
Increase cycle	2
Increase ratio	20%
Maximum frame rate	15 fps

TABLE 28.3 SPECIFICATION OF THE MPEG-2 INPUT STREAM

Stream type	MPEG-2 PS
Video	MPEG-2 Video
Bit rate	1.5 Mbps
Picture size	320 pixel × 240 pixel
Frame rate	30 fps
Audio	MPEG-1 layer II
Bit rate	224 Kbps

FIGURE 28.6
*The characteristics of
the transcoding system
when 150-Kbps
dummy traffic is peri-
odically inserted and
removed.*

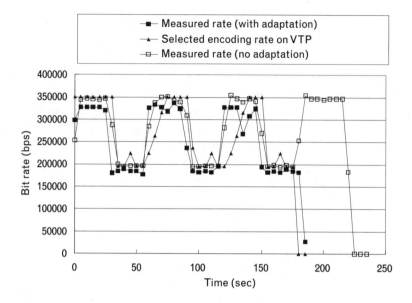

In the no adaptation case, frequent interruptions of the audio and video outputs were observed when the bandwidth was narrowed. When adaptations were used, the number of interruptions was reduced. In the system, interruptions were observed when adaptations were performed to fit the output rate into the narrowed bandwidth and when bouncings of the selected encoding rate are observed. Bouncings are caused by our control algorithm, in which the control module gradually increases the encoding bit rate if it observes no rate reduction over a period of time.

28.7   Concluding Remarks

In this chapter, an overview of video transcoding for the mobile Internet is provided. A real-time video transcoding system to overcome the issue of QoS fluctuations in mobile environments is introduced.

In the near future mobile communications systems (which have advanced concepts from original IMT-2000 systems) such as cdma2000 1x EV-DO and 4G systems will emerge. New enhanced transcoding mechanisms will be necessary.

Continuing studies toward further improvement of the transcoding system performance must be carried out. Furthermore, development of various adaptation techniques by examining transcoder control methods to overcome the short-term fluctuations is needed.

REFERENCES

[1]   Morikawa, D., et al., "A Proposal of Web Page Transcoding Method for Mobile Computing Environment," *6th International Workshop on Mobile Multimedia Communications*, March 2000 (in Japanese).

[2]   Cheung, S. Y., M. H. Ammar, and X. Li, "On the Use of Destination Set Grouping to Improve Fairness in Multicast Video Distribution," *Proceedings of INFOCOM '96*, IEEE, March 1996.

[3]   McCanne, S., and V. Jacobson, "Receiver-Driven Layered Multicast," *ACM SIGCOMM'96*, Stanford, CA, August 1996.

[4]   Amir, E., S. McCanne, and H. Zhang, "An Application Level Video Gateway," *ACM Multimedia Conference*, San Francisco, CA, November 1995.

[5]   Yeadon, N., et al., "Filters: QoS Support Mechanisms for Multipeer Communications," *IEEE Journal on Selected Areas in Communications,* Vol. 14, No. 7, September 1996, pp. 1245–1262.

[6]   Chen, Z., et al., "Real Time Video and Audio in the World Wide Web," *Proceedings of Fourth International World Wide Web Conference*, Boston, MA, December 1995.

[7]   Margaritidis, M., and G. C. Polyzos, "MOBIWEB: Enabling Adaptive Continuous Media Applications over Wireless Links," *IEEE International Conference on Third Generation Wireless Communications*, San Francisco, CA, June 2000.

[8]   Duffield, N. G., K. K. Ramakrishnan, and A. R. Reibman, "SAVE: An Algorithm for Smoothed Adaptive Video over Explicit Rate Networks, *IEEE/ACM Trans. Networking*, Vol. 6, No. 6, December 1998.

[9]   Fox, A., et al., "Adapting to Network and Client Variation Using Infrastructural Proxies: Lessons and Perspectives," *IEEE Personal Communications*, August 1998.

[10] Vass, J., et al., "Efficient Mobile Video Access in Wireless Environment," *IEEE Wireless Communications and Networking Conference*, New Orleans, LA, and Los Angeles, CA, September 1999.

[11] Schulzrinne, H., et al., "RTP: A Transport Protocol for Real-Time Applications," IETF RFC 1889, January 1996.

[12] Schulzrinne, H., A. Rao, and R. Lanphier, "Real Time Streaming Protocol (RTSP)," IETF RFC 2326, April 1998.

[13] W3C Recommendation, "Synchronized Multimedia Integration Language (SMIL 2.0)," http://www.w3.org/TR/2001/REC-smil20-20010807, August 2001.

[14] W3C Working Draft, "Composite Capabilities/Preference Profiles Requirements and Architecture," http://www.w3c.org/TR/2000/WD-CCPP-ra-20000721, July 2000.

[15] 3GPP2 S.R0021, "Video Streaming Sevices-Stage1," July 2000.

[16] Warabino, T., et al., "Video Transcoding Proxy for 3G Wireless Mobile Internet Access," *IEEE Communications Magazine*, October 2000.

CHAPTER 29

On Security in Wireless and IP Networks

Senthil Sengodan

29.1   Introduction

The security services/mechanisms/algorithms that are present in today's wireless networks are very different from those that are present in today's fixed IP networks. With the convergence of these networks, differences between these networks needs to be investigated and the consequent effect on end-to-end security needs to be determined. In this chapter, we provide an overview of security technologies used within various wireless and IP networks. This paves the way to understand implications of end-to-end security when wireless and IP networks converge.

Figure 29.1 illustrates two broad scenarios that are applicable in this context. Figure 29.1(a) depicts the case where the wireless network has been designed with the application in mind. For instance, second-generation cellular networks, such as Telecommunications Industry Association/Electronics Industries Association-95 (TIA/EIA-95), TIA/EIA-136, and GSM, fall under this category. Such networks have been designed primarily to carry voice. Interface to fixed IP networks, in this case, usually takes place via gateways.

In this scenario, it is important to understand the security features/capabilities of each network. Such an understanding permits an understanding of the end-to-end security when the communicating endpoints are on different networks. The security of the entire systems is only as strong as the weakest link.

Figure 29.1(b) depicts the case where the wireless network is a bearer network. In other words, the wireless network carries payload that could be an IP datagram. Link layer wireless technologies such as Bluetooth, Infrared Data Association (IrDA), and WLANs fall under this category. The 2G

FIGURE 29.1
Two scenarios of wire-less IP convergence: (a) the case in which the wireless network has been designed from the application stand-point, and (b) the case in which the wireless network is a bearer network.

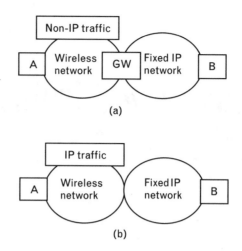

cellular networks may also be used as bearer networks when they are used to transport data—for instance, by connecting one's laptop to one's GSM phone. Generally speaking, however, 2G cellular networks fall under the previous category since such networks have been designed to carry voice.

In this scenario, an understanding of the security features/capabilities of wireless and IP networks would permit the understanding of layers in the protocol stack in which the security functionality may either be turned on or off. As seen in the figure, IP traffic rides atop the wireless network. Consequently, depending on the particular system, it may not be necessary to utilize all security features of each layer in the protocols stack (e.g., network and wireless link layer). Instead, one may decide to turn off the security functionality in the wireless link layer if such functionality is redundant.

The rest of the chapter is organized as follows. Section 29.2 gives an overview of the security features of several cellular networks: AMPS, TIA/EIA 41, TIA/EIA 136, IS-95, GSM, GPRS, and UMTS. Section 29.3 describes the security features within two kinds of wireless link layer technologies: Bluetooth and IEEE 802.11 LANs. Section 29.4 describes the security protocols that are used within TCP/IP networks—specifically, IPsec, *Transport Layer Security* (TLS), and *Internet Key Exchange* (IKE). Section 29.5 describes several commonly used encryption and hash algorithms. Section 29.6 describes issues such as private addressing [e.g., *network address translators* (NAT) and *Realm Specific IP* (RSIP)], firewalls, and *distributed denial of service* (DDoS) attacks. Finally, we conclude in Section 29.7.

29.2   Cellular Networks Security

The cellular systems discussed here include AMPS, IS-136, IS-95, GSM, GPRS, and UMTS.

29.2.1 AMPS

The AMPS specification only deals with the air interface (unlike 2G systems like TIA/EIA-136 and GSM that deal with the entire system).

The original AMPS system only provided weak authentication, and it did not provide any encryption or integrity. The authentication was based on the use of the correct *electronic serial number* (ESN) corresponding to the *mobile identification number* (MIN). With a table lookup at the *mobile telephone switching office* (MTSO), authentication either succeeded or failed. The MIN and ESN were sent on the reverse control channel, thereby making it easy for eavesdroppers to "pick-up" ESNs corresponding to MINs. These were then used in "cloned" terminals. Later implementations used a *personal identification number* (PIN) on the voice channel in order to authenticate the user. Since the voice channel is dynamically assigned, detection of the PIN by eavesdroppers is more difficult (see Figure 29.2).

Newer AMPS phones (or dual-mode phones) use strong authentication and encryption; these techniques are similar to those used in digital systems.

29.2.2 TIA/EIA-41

TIA/EIA-41 was defined in order to standardize roaming between different operators.

As in the case of AMPS, the MS transmits the ESN and MIN in a registration message over the air interface. As seen in Figure 29.3, upon reception, the MSC sends a REGISTRATION_NOTIFICATION_INVOKE message to the VLR (message 1); and the VLR sends a QUALIFICATION_REQUEST_INVOKE (or REGISTRATION_NOTIFICATION_INVOKE) message to the HLR (message 2). Messages 1 and 2 each carry the ESN and MIN supplied by the MS. Upon receiving message 2, the HLR verifies the ESN corresponding to the MIN. The HLR then either accepts or rejects the request by sending a QUALIFICATION_REQUEST_RESULT (or REGISTRATION_NOTIFICATION_RESULT) message (message 3); which prompts the

FIGURE 29.2
AMPS system.

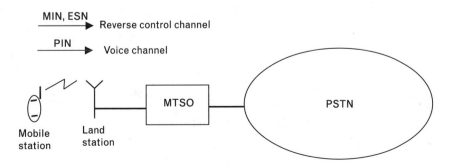

VLR to send the REGISTRATION_NOTIFICATION_RESULT message (message 4) to the MSC.

In addition, the HLR and the previous VLR, as well as the previous VLR and its MSC (not shown Figure 29.3), exchange a REGISTRATION_CANCELLATION_REQUEST and REGISTRATION_CANCELLATION_RESULT between them. Similarly, PROFILE_REQUEST_INVOKE and PROFILE_REQUEST_ RESULT messages are exchanged between the VLR and the HLR.

29.2.3 TIA/EIA-136 (NA-TDMA)

TIA/EIA-136 (referred to as IS-136) also known as North American Time Division Multiple Access (NA-TDMA) is a second-generation cellular system that has been standardized by TIA.

IS-136 introduces several new features that provide enhanced security. A unique 64-bit key, known as the A-key, is stored securely in each mobile terminal and its corresponding *authentication center* (AuC). The *cellular authentication and voice encryption* (CAVE) algorithm is used to authenticate the mobile as well as to derive a *shared secret data* (SSD) that is used for encryption over the air interface.

Figure 29.4(a) illustrates the case of generating an SSD that is used for encryption, while Figure 29.4(b) illustrates the case of generating an authenticator that is used for confirming secure SSD generation as well as for authentication. Figure 29.5 illustrates the message exchanges during SSD computation, while the authentication itself takes place using a challenge-response mechanism (as described in Section 29.2.3.2).

FIGURE 29.3
Authentication during registration in IS-41.

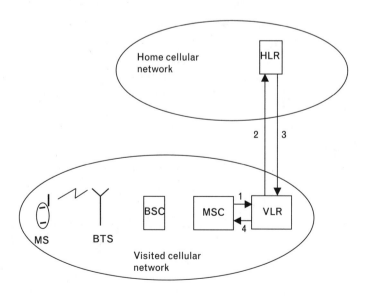

FIGURE 29.4
*Using CAVE to
compute (a) SSD and
(b) AUTHBS.*

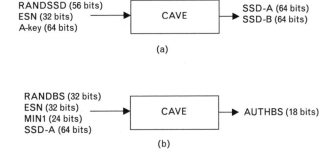

FIGURE 29.5
*Message exchanges
during SSD
computation.*

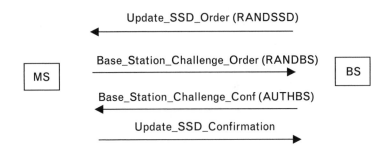

29.2.3.1   SSD Generation

The AuC generates a 56-bit random number (RANDSSD) and using the CAVE algorithm computes SSD (SSD-A and SSD-B). The RANDSSD and computed SSD are sent by the AuC to the BS. The BS sends an UPDATE_SSD_ORDER message containing the RANDSSD to the mobile. Using the CAVE algorithm, the mobile computes the SSD. It then generates a random number (RANDBS), with which AUTHBS is computed. The mobile includes RANDBS in the BASE_STATION_ CHALLENGE_ORDER message sent to the BS. The BS then computes AUTHBS and includes it in the BASE_STATION_ CHALLENGE_ CONFIRMATION message sent back to the mobile. If the AUTHBS computed by the mobile matches that sent by the BS, the mobile responds with an UPDATE_SSD_CONFIRMATION message. This ensures that the mobile and the BS both share the same SSD. SSD-A is then used for authentication, while SSD-B is used for voice encryption.

29.2.3.2   Authentication

In a two-way challenge-response handshake mechanism, the BS sends the Unique_Challenge_Order message containing RANDBS to the MS, which then computes AUTH as in Figure 29.4(b) and returns it back to the BS in the Unique_Challenge_Order_Conf message.

An additional authentication mechanism uses a counter increment scheme. Here, the BS increments the value of an 8-bit counter in the MS by the use of the Parameter_Update command. In order to authenticate itself, the MS would then have to produce the correct value of the counter.

29.2.4 IS-95 (CDMA)

IS-95 is based on CDMA. The authentication and privacy mechanisms employed in IS-95 are identical to those employed in IS-136. IS-95 uses an additional privacy mechanism based on a private long code mask that is securely stored within the MS and its corresponding AuC. This is initiated either by the BS or by the MS by transmission of a Long_Code_Transition_Order message.

29.2.5 GSM

GSM provides several security mechanisms. These include anonymity, authentication, and confidentiality. Anonymity is provided by the assignment of a temporary identity, referred to as the TMSI, to the MS. The TMSI only has local significance within the MSC/VLR area that the MS is visiting.

In GSM, a unique authentication key K_i is stored securely within the SIM card and the AuC corresponding to the subscriber. In order to authenticate a user, the AuC generates a random number (between 0 and $2^{128} - 1$) called RAND. Using RAND and K_i as inputs to algorithm A3, the AuC generates SRES. RAND is also transmitted to the MS as a challenge, and if the MS returns an SRES that is the same as the one computed by the AuC, the MS is authenticated.

Algorithm A8 is used to generate the 64-bit encryption key K_c used for voice stream encryption. A8 takes as input RAND and K_i and outputs K_c. Figure 29.6 illustrates algorithms A3 and A8. Algorithm A5 is the encryption algorithm itself.

GSM separates the *mobile equipment* (ME) from the *mobile user* (MU). The ME is identified by the *mobile equipment identity* (MEI), while the MU is identified by the IMSI. In order to detect stolen MEs, the EIR stores a list of stolen MEIs. If the MEI of the ME is listed, then the EIR knows that the ME has been stolen.

29.2.6 GPRS

The GPRS is an enhancement to GSM-based networks in order to make the networks data friendly. Figure 29.7 depicts the logical architecture of GPRS networks, where the two GPRS entities, namely, the (SGSN and GGSN) are shown. The Gp interface that exists between two different *public*

FIGURE 29.6
Authentication (A3)
and encryption key
generation (A8)
algorithms in GSM.

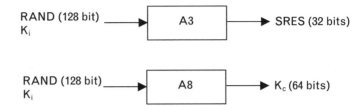

FIGURE 29.7
GPRS logical
architecture. (Source:
ETSI.)

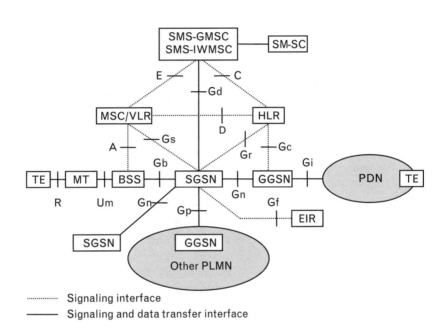

and mobile networks (PLMNs) provides security functionality that is similar to a VPN. This is operator specific.

The security mechanism that GPRS uses is similar to those that are used within GSM, the distinction being that the SGSN performs the security and access control functions that the MSC would perform in GSM networks.

29.2.6.1 Authentication of Subscriber and Ciphering

Figure 29.8 illustrates the authentication procedure within GPRS networks. Such authentication procedures may be invoked, for instance, when an MS performs a GPRS attach with an SGSN, in order to establish a mobility context. The SGSN needs to have at least one authentication triplet before it can pass on a challenge to the MS. This authentication triplet is obtained by the SGSN from the HLR. The SGSN sends a "Send Authentication info" message to the HLR with the IMSI of the MS as a parameter, and one or more authentication triplets are returned in the "Send Authentication Info Ack" message. Each authentication triplet contains (RAND,

FIGURE 29.8
*Subscriber authentica-
tion. (Source: ETSI.)*

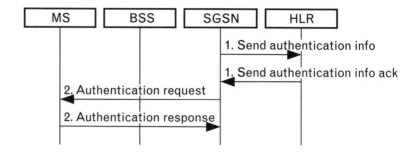

SRES, K_c), where RAND is the random number (challenge), SRES is the response to the challenge, and K_c is the encryption key.

In GSM networks, the ciphering algorithm is fixed (algorithm A5), whereas in GPRS networks, the ciphering algorithm can be determined by the HLR. The choice of the ciphering algorithm is included in the "Authentication Request" message that the SGSN sends to the MS. In addition to sending the ciphering algorithm, the SGSN includes the RAND as a challenge for the MS. The MS then computes the response SRES and returns it to the SGSN in the "Authentication Response" message.

Using RAND and the secret key K_i, the MS also determines the encryption key K_c. Encryption then takes place using the "ciphering algorithm" specified by the SGSN and the encryption key K_c. The MS starts encryption immediately after sending the "Authentication Request" message, while the SGSN starts encryption after receipt of the "Authentication Response" message. It may be noted that while in GSM, the ciphering takes place between the MS and the BTS; in the case of GPRS, the ciphering occurs between the MS and the SGSN.

29.2.7 UMTS

The UMTS specifies the *authentication and key agreement* (AKA) algorithm. One of the distinguishing factors of AKA (compared to GPRS or GSM) is that AKA allows for mutual authentication (i.e., the MS authenticates the network as well). Figure 29.9 illustrates the AKA procedure—note that the message sequence is identical to that in Figure 29.8 for GPRS networks. The specific parameters and algorithms are slightly different, though.

As in the case of GPRS, the SGSN sends an "Authentication Data Request" message to the HLR with the IMSI as a parameter. The HLR then returns a set of *authentication vectors* (AV) in the "Authentication Data Response" message, which are equivalent to the authentication triplets in the case of GPRS. Included in an AV are (RAND, XRES, CK, IK).

When the SGSN wishes to authenticate an MS, it sends a "User Authentication Request" message to the MS. This message contains a RAND, which is used as a challenge, as well as AUTN, which is used by the

FIGURE 29.9
*Illustrating the AKA
procedure of UMTS.*

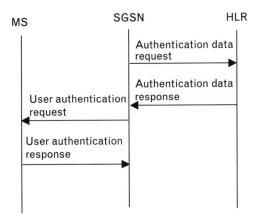

MS to authenticate the SGSN. After the MS authenticates the SGSN using the AUTN, the RAND is used by the MS to compute the response RES to the challenge. This response is then returned in the "User Authentication Response" message. The *cipher key* (CK) and the *integrity key* (IK) are computed by the MS as well, and these keys are known to the SGSN as they are sent by the HLR in the AV.

29.3   Link Layer Security

In this section, we will give an overview of some of the security mechanisms that are present in wireless link layer technologies—specifically, Bluetooth and IEEE 802.11 WLANs.

29.3.1   Bluetooth Security

Bluetooth provides device authentication as well as message confidentiality. In order to achieve this, two kinds of keys are defined:

1. *Link key:* This 128-bit key, also known as the authentication key, is used for device authentication as well as for generating the encryption key. The link key can be one of four types:

   • *Unit key:* This is a semipermanent key that is generated within a *Bluetooth device* (BD) without interacting with other BDs. A unit key may be used as the link key between a pair of BDs if the memory available in the BD does not allow creation/storage of a combination key.

- *Initialization key:* This is a temporary key that has significance for a pair of BDs. The initialization key is deleted after creation of the combination key.

- *Combination key:* This is a semi-permanent key that is created for every pair of interacting BDs.

- *Master key:* This is a key that is shared among several BDs in a piconet, and is useful in a multicast configuration.

2. *Encryption key:* This key, derived from the current link key, is used for message encryption and can be between 8 and 128 (in octet multiples) bits long.

A semipermanent key is one that is stored in nonvolatile memory and is valid even after session termination. A session is defined as the duration that a BD is connected to a piconet. A piconet is the set of BDs that use the same channel. A temporary key, on the other hand, is only valid for the session duration. Unit and combination keys are semipermanent, while initialization and master keys are temporary.

29.3.1.1   Key Generation Techniques

Unit Key Generation
The unit key is generated when the BD is being operated for the first time. The BD generates a 128-bit random number RAND, which is input to algorithm E2 along with the 48-bit BD_ADDR of the device to produce the 128-bit unit key (see Figure 29.10).

Initialization Key Generation
The initialization procedure is needed when a pair of BDs (say, BD_A and BD_B) do not share a link key or when the link key is lost, and communication between the two is desired. Prior to initialization, a PIN is entered securely into each of the BDs—either manually or by automated means. This PIN (of length L octets) is augmented with BD_ADDR_B (assuming that BD_B is the claimant) to produce PIN' (of length 16 octets) as follows:

FIGURE 29.10
Illustrating link key generation for unit/combination keys (mode 1) and master/initialization keys (mode 2).

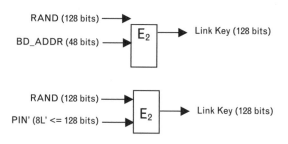

PIN' = PIN [0 ... L–1] U BD_ADDR$_B$ [0 ... min{5, 15–L}]
if L < 16

PIN' = PIN
if L = 16

The verifier, BD$_A$, generates a 128-bit random number RAND, which along with PIN' is input to algorithm E2, producing the 128-bit initialization key as output. RAND is transmitted by the verifier to the claimant, which then computes the key as well.

Combination Key Generation

This key is produced after generation of the initialization key when a pair of BDs need to communicate. Once the combination key has been generated, the initialization key is discarded. The combination key may also be regenerated between a pair of BDs at a later time. To generate a combination key, BD$_A$ and BD$_B$ first generate a temporary key (say, KEY$_A$ and KEY$_B$) using a procedure identical to that used to generate a unit key. Hence, RAND$_A$ and BD_ADDR$_A$ are input to algorithm E2 by BD$_A$ to produce KEY$_A$, while RAND$_B$ and BD_ADDR$_B$ are input to algorithm E2 by BD$_B$ to produce KEY$_B$. BD$_A$ then XORs RAND$_A$ with the current link key (which is the initialization key soon after initialization) and transmits the result to BD$_B$; similarly, BD$_B$ XORs RAND$_B$ with the current link key and transmits the result to BD$_A$. Each BD regenerates the RAND used by the other BD, and using the BD_ADDR of the other BD computes the KEY generated by the other BD. Knowing KEY$_A$ as well as KEY$_B$, the two are XORed by each BD to compute the combination key.

Master Key Generation

This key is generated when the same key needs to be used for several BDs (one master and two or more slaves). The master generates two random numbers RAND$_1$ and RAND$_2$, and passes them through algorithm E2 to produce the master key K$_{master}$. A third random number RAND$_3$ is generated by the master and passed on to each of the slaves. (Different slaves may receive different random numbers, if so desired; but, this is not necessary.) The slave inputs RAND$_3$ and the current link key with the master through algorithm E2 to produce a temp key K$_{temp}$. The master does the same to produce K$_{temp}$ as well. The master XORs K$_{temp}$ with K$_{master}$ and transmits the result to the slave. The slave computes K$_{master}$ by XORing the received value with K$_{temp}$.

Encryption Key Generation

Figure 29.11 illustrates the generation of the encryption key using the current link key. The 96-bit *ciphering offset number* (COF) is chosen as follows.

FIGURE 29.11
Encryption key
generation.

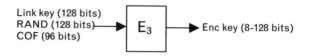

Link key (128 bits)
RAND (128 bits)
COF (96 bits) → E_3 → Enc key (8-128 bits)

Authentication and Encryption

Device authentication in Bluetooth is based on a challenge-response mechanism as illustrated in Figure 29.12. The verifier BD_A generates a random number $RAND_A$ and sends it to the claimant BD_B. The verifier and the claimant generate the authenticator SRES based on the E_1 algorithm The claimant transmits the authenticator (SRES) to the verifier. If this matches the authenticator generated at the verifier, then the claimant has been authenticated. Encryption and decryption in Bluetooth is as shown in Figure 29.13.

29.3.2   IEEE 802.11 WLAN Security

The security mechanism for IEEE 802.11 wireless LANs is termed *Wired Equivalent Privacy* (WEP). WEP does not specify any key management techniques, nor does it provide message authentication. The primary security service provided by WEP is message confidentiality based on a 40-bit key, 24-bit *initialization vector* (IV) RC4 algorithm. The short key length (40 bits) and the short IV (24 bits) that WEP uses have largely contributed to its criticism as being insufficient for today's security needs.

The security drawbacks of the current WEP version are being addressed in the next version of WEP. Three new security services are being proposed: (1) key management and distribution, (2) data authentication, and (3) replay attack prevention. In addition to the mandatory support for the RC4 algorithm, two optional algorithms are being introduced: the *advanced*

FIGURE 29.12
Authentication in
Bluetooth.

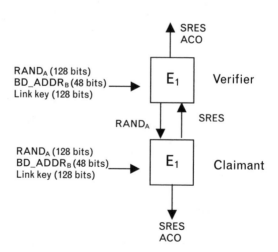

SRES
ACO

$RAND_A$ (128 bits)
BD_ADDR_B (48 bits)
Link key (128 bits) → E_1 Verifier

$RAND_A$ SRES

$RAND_A$ (128 bits)
BD_ADDR_B (48 bits)
Link key (128 bits) → E_1 Claimant

SRES
ACO

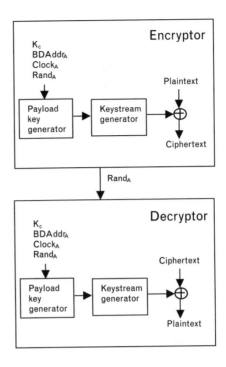

encryption standard (AES) and another algorithm currently termed WEP2. In addition to data authentication at the link layer, an upper layer authentication mechanism is also being proposed. With these enhancements to WEP, most of the security concerns regarding WEP will be addressed.

29.4 TCP/IPsec

The security protocols that are discussed here are IPsec, TLS, and IKE. IPSec is a protocol that is used to obtain certain security services such as authentication and encryption at the IP layer. In addition to providing such functionality at the transport layer, TLS also includes a handshake protocol for establishing the security context between the communicating endpoints. IKE is a protocol that is used for secure key exchange. While being a general purpose key exchange protocol, its applicability is largely towards establishing the security context for IPsec endpoints.

29.4.1 IPsec

IPsec is a protocol standardized within the IETF, and which provides security services at the network layer (i.e., the IP layer) for IP-based communications. IPsec offers two security protocols—*authentication header* (AH) and *encapsulating security payload* (ESP)—each of which provides certain security services. The security services provided by AH are connectionless integrity, data origin authentication, optional anti-replay service and access control. The security services provided by ESP are message confidentiality, and optionally connectionless integrity, data origin authentication, anti-replay service, and access control. While the security services offered by ESP include all those that are offered by AH, the set of fields over which the services are applicable varies in the two cases. AH and ESP are defined as separate headers that are carried within the IP datagram.

29.4.1.1   AH

IPSec AH provides connectionless integrity, data origin authentication, optional anti-replay service, and access control. With IPsec AH, an authentication header is inserted between the IP header and the UDP or TCP header. The format of this header is shown in Figure 29.14(a). The actual authenticator, such as the MAC or the digital signature, goes in the last field labeled "authentication data." The sequence number field is provided to combat replay attacks. The *security parameters index* (SPI), which is an arbitrary 32-bit value assigned by the destination, in conjunction with the destination IP address and the security protocol (which is AH in this case) uniquely identifies the *security association* (SA).

IPsec AH can exist in two modes: transport mode and tunnel mode. These are illustrated in Figure 29.14(b, c), respectively. Transport mode is used when the SA exists between the two communicating EPs. Tunnel mode, which uses IP in IP encapsulation, is used when at least one of the terminating points of the SA is at a security gateway instead of the communicating EP. In either case, the entire IP datagram is authenticated.

29.4.1.2   ESP

IPsec ESP provides message confidentiality, and optionally connection-less integrity, data origin authentication, anti-replay service, and access control. With IPsec ESP, an ESP header is inserted between the IP header and the UDP/TCP header. In addition, an ESP trailer is inserted after the UDP/TCP payload. The format of the ESP header and trailer is shown in Figure 29.15(a). As in the case of AH, the 3-tuple of (destination IP address, ESP, SPI) is used to identify the ESP SA. Similarly, the sequence number field is used to combat replay attacks. Since several encryption algorithms are block algorithms (i.e., they operate on blocks of fixed size), the UDP/TCP payload may have to be padded to an integral multiple of the block size. As in the case of AH, ESP may be used either in transport or tunnel mode. ESP also provides optional authentication, and the authenticator (MAC or digital signature) is carried in the "Authentication data" field. Comparing Figures 29.14 and 29.15, however, it is seen that while IPsec

FIGURE 29.14
(a) AH, (b) AH in transport mode, and (c) AH in tunnel mode.

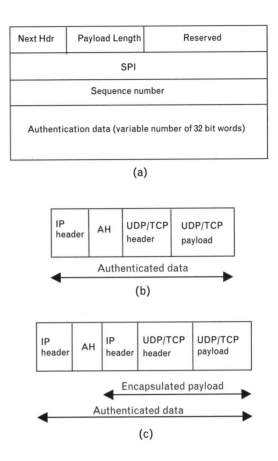

(a)

(b)

(c)

AH authenticates the entire IP datagram, authentication within IPsec ESP does not cover the IP header.

29.4.1.3 Outbound and Inbound Processing

Consider the case where packets are sent between two communicating endpoints, and IPSec ESP is used to encrypt the traffic using the *data encryption standard* (DES) algorithm. It is assumed that an ESP SA has already been established between the two endpoints (say, using IKE). Let the IP address of the transmitting endpoint (say, A) be 192.100.70.10, that of the receiving endpoint (say, B) be 185.25.35.6, the source port at A be 1355 and the destination port at B be 1656. Let the SPI (which is chosen by B during SA establishment) associated with this SA be 35. Figure 29.16 illustrates the outbound security processing that occurs at A, while Figure 29.17 illustrates the inbound security processing that occurs at B.

As seen in Figure 29.16, each IP datagram at A is sent to an IPsec control unit, which consults a *security policy database* (SPD) to determine what action,

FIGURE 29.15
*(a) ESP, (b) ESP in
transport mode, and
(c) ESP in tunnel
mode.*

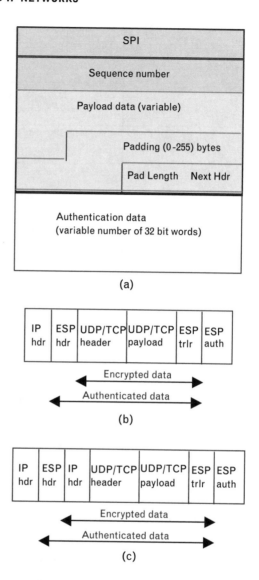

if any, needs to be applied to the packet. Upon consulting the SPD, it is seen that packets with source IP address 192.100.70.10, destination IP address 185.25.35.6, source port number 1355, destination port number 1656, and protocol field of UDP need to be processed using a certain SA. The SA that is used to process the packet is described in the *security association database* (SAD), and a pointer to the entry in the SAD exists in the SPD. The packet is suitably processed using this SA (which happens to be DES with Key 2), and sent out the outbound interface.

As seen in Figure 29.17, upon receiving the packet, B looks up the 3-tuple of (destination IP address, security protocol, SPI), based on which the SA is ascertained using the SAD. In this example, the 3-tuple of

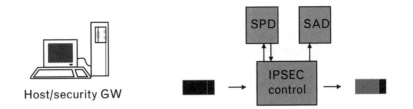

FIGURE 29.16
*IPsec outbound
processing.*

Host/security GW

IPSec databases

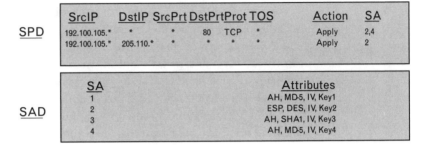

	SrcIP	DstIP	SrcPrt	DstPrt	Prot	TOS	Action	SA
SPD	192.100.105.*	*	*	80	TCP	*	Apply	2,4
	192.100.105.*	205.110.*	*	*	*	*	Apply	2

	SA	Attributes
	1	AH, MD-5, IV, Key1
SAD	2	ESP, DES, IV, Key2
	3	AH, SHA1, IV, Key3
	4	AH, MD-5, IV, Key4

FIGURE 29.17
*IPsec inbound
processing.*

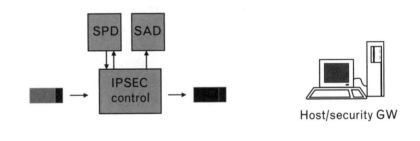

Host/security GW

	DestIP	SecProt	SPI	Attributes
				AH, MD-5, IV, Key1
SAD				ESP, DES, IV, Key2
				AH, SHA1, IV, Key3
				AH, MD-5, IV, Key4

(185.25.35.6, ESP, 35) points to the appropriate SA in the SAD at B. The attributes of this SA (which includes the key) are used to suitably decrypt the packet.

29.4.2 TLS

The most commonly used security protocol on IP networks (specifically, the Internet) today is probably the *Secure Socket Layer* (SSL) protocol. SSL is a proprietary protocol that has not been standardized by any organization (but is an industry de facto standard). The TLS protocol is based on SSL, and has been standardized by the IETF. TLS has a mode where it can fall back to SSL. TLS contains two main layers:

1. A handshake layer that negotiates the secure connection;

2. A record layer that provides the security services using the security association that is created by the handshake layer.

One-way or two-way authentication using public key mechanisms are used during the handshake layer. The security services provided by the record layer are primarily (1) confidentiality using symmetric key encryption (DES, RC4 and so on), and (2) message integrity using keyed MAC mechanisms such as keyed *secure hash algorithm* (SHA) and keyed *message digest 5* (MD5). The TLS record protocol requires a reliable transport layer, such as TCP, for its operation. When UDP is used for transport, as is the case with real-time media streams, TLS cannot be used.

Two modes of handshake exist:

1. Full handshake is illustrated in Figure 29.18. The ClientHello and ServerHello messages are used to exchange security capabilities of the TLS client and server, and at the end of the exchange the following attributes are established: protocol version, session ID, cipher suite, compression method, and random numbers. The messages that follow are used for key exchange purposes. A ChangeCipherSpec and a finished message are sent to indicate that the security attributes that were established as part of this handshake will be used to secure the message exchanges. The asterisks in the figure imply that the message is an optional one.

FIGURE 29.18
TLS full handshake.

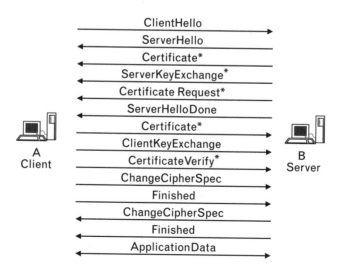

2. Abbreviated handshake may be used when the client wishes to resume or duplicate an existing session that is identified by the session ID.

29.4.3 ISAKMP and IKE

We now give a brief description of ISAKMP and IKE.

29.4.3.1 ISAKMP

ISAKMP provides a framework for authentication and key management. Several key exchanges may be used within the ISAKMP framework. OAKLEY describes several key exchanges (or modes), and SKEME describes one key exchange that may be used within ISAKMP. Depending on the nature of the key exchange protocol, one or more of certain security services is provided for the SA.

ISAKMP provides several payload types: SA, proposal, transform, key exchange, identification, certificate, certificate request, hash, signature, nonce, notification, delete, and vendor ID. Figure 29.19 shows the structure of some of these payload types, along with the generic ISAKMP message header. One or more payload types may be present in an ISAKMP message, and the "Exchange type" field within the ISAKMP header

FIGURE 29.19
ISAKMP payload types.

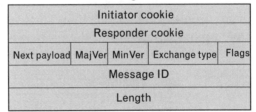

determines the presence and ordering of the payloads. The exchange types that have been defined within ISAKMP include: base, identity protection, authentication only, aggressive, and informational. Protocols using the ISAKMP framework, such as IKE, may define additional key exchanges for their purpose.

Figure 29.20 provides an example of using of these payload types for the purpose of establishing an SA. The ISAKMP message shown on the left is sent by endpoint A to endpoint B, offering a set of security capabilities. The ISAKMP message shown on the right, is subsequently sent by B to A conveying the security context that is chosen by B. The ISAKMP message sent by A to B contains two proposals, denoted as Proposal #1 and Proposal #2. Proposal #1 is a proposal for IPsec ESP and IPsec AH. With IPsec ESP, one of two security transforms may be used: the DES encryption algorithm with specified attributes, and the 3-DES algorithm with specified attributes. With IPsec AH, only one security transform is offered, that using hybrid message authentication code-MD5 (HMAC-MD5) with certain specified attributes. Proposal #2, on the other hand, is a proposal for IPsec AH only. This IPSec AH proposal offers only one security transform, that using HMAC-SHA with certain specified attributes.

Upon receiving these two proposals, B chooses between them (assuming that at least one of the two is supported by B). In the example, B chooses Proposal #1, which implies that IPsec ESP along with IPSec AH is being chosen. Since two transforms were offered with IPsec ESP, B needs to choose one of the two, and in this case, the 3-DES encryption algorithm is chosen rather than the DES algorithm.

It may be noted that a complete exchange is not presented here in Figure 29.20, rather only an illustrative one (which is incomplete) is provided.

FIGURE 29.20
SA establishment example.

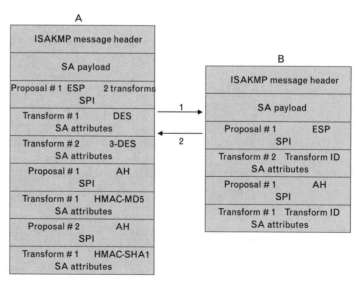

29.4.3.2 IKE

While ISAKMP provides a framework for authentication and key exchange, the IKE protocol makes use of this framework. IKE utilizes parts of the OAKLEY and SKEME key exchange mechanisms within the ISAKMP framework in order to define a robust key exchange mechanism.

IKE operates in two phases: phase 1 establishes the general IKE SA, while phase 2 establishes specific protocol SAs such as IPsec AH SA, IPsec ESP SA, TLS SA, and so on. The advantage of a two-phase approach is that a common IKE SA may be used to create multiple phase 2 SAs, thereby presenting a faster and more efficient mechanism to create SAs. Certain modes have been specified within IKE, and the security services vary depending on the specific mode used. For instance, the establishment of a phase 1 SA (i.e., a general IKE SA) may use one of two modes: main mode or aggressive mode. While the main mode is more robust, the aggressive mode is achieved with fewer message exchanges at the expense of compromising identity protection. Main mode is an instantiation of the ISAKMP identity protect exchange, while the aggressive mode is an instantiation of the ISAKMP aggressive exchange. Similarly, a quick mode has been defined for use in creating phase 2 SAs. Quick mode is not an instantiation of any of the predefined exchanges within ISAKMP, and has been specified only within IKE.

29.5 Security Algorithms

29.5.1 Encryption Algorithms

Encryption algorithms are employed within protocols such as IPsec or TLS in order to provide confidentiality service. Such algorithms are one of two kinds: block ciphers and stream ciphers. Block ciphers—which include DES, 3-DES, Blowfish, Rijndael, RC2, and RC5—operate on input blocks of certain sizes to produce ciphertext (encrypted text) of the same size. The size of the block depends on the encryption algorithm (i.e., cipher) used. Block sizes of 64, 128, 196, 256 bits are typical. Stream ciphers —which include RC4—on the other hand, operate on a single bit or a single byte at a time.

29.5.1.1 Operation Modes

Since block ciphers operate one block at a time, it is possible to chain or feed information performed in one block ciphering operation to that in the subsequent block ciphering information. Depending on whether and how this is done, four modes of operation are commonly defined: *electronic code book*

(ECB), *cipher block chaining* (CBC), *cipher feedback* (CFB), and *output feedback* (OFB).

In ECB mode, each block of data is encrypted independently of any other block of data, as shown in Figure 29.21(a). In the figure, the input blocks are represented as P_i for plaintext, while the cipher output blocks are represented as C_i for ciphertext. In CBC), the ciphertext of one cycle is XORed with the plaintext of the next before feeding it to the cipher algorithm. For the first cycle, an IV is used. This is shown in Figure 29.21(b).

CFB and OFB modes are a way of approximating the behavior of stream ciphers using block ciphers. J-bit CFB or OFB mode operates on j-bits (where j does not exceed the block size) of plaintext in each cycle to produce j-bits of ciphertext. In CFB, for the first cycle, the leading j-bits of the output of the ciphering operation is XORed with the j-bit plaintext (P1) to produce the j-bit ciphertext (C1). C1 is also fed to rightmost j-bits of a left-shift register, which is initialized to IV. The contents of this shift register along with the key K constitute the input to the next ciphering operation. The process continues, as illustrated in Figure 29.21(c). OFB is similar to CFB, with the difference being that in OFB, the j-bit output from the ciphering operation in one cycle is fed into the next cycle, instead of the j-bit ciphertext that is fed in the case of CFB.

29.5.1.2 DES

The DES is the most widely used encryption algorithm to date. The algorithm was standardized by the United States government's *National Institute of Standards and Technology* (NIST) in 1977 and has enjoyed widespread

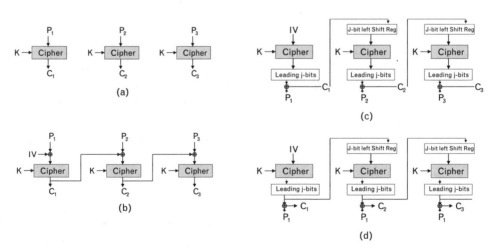

FIGURE 29.21
Illustrating (a) ECB, (b) CBC, (c) j-bit CFB, and (d) j-bit OFB modes for block ciphers.

popularity since. DES is a symmetric block cipher, which implies that the same key is used for encryption and decryption. Generally speaking, symmetric ciphers are considerably faster than asymmetric ciphers (which use a different key for encryption and decryption), and hence are widely used for media stream and other bulk encryption. DES takes a block input of 64-bit length, and using a 56-bit key, outputs a block of 64-bits. The key length specified is actually 64 bits, although every eighth bit is used for parity purposes, thereby making the actual key length 56 bits. Triple-DES, because of its increased security over DES, is currently being used increasingly widely. Double-key triple-DES uses two different keys each 56 bits long, while triple-key triple-DES uses three keys each 56 bits long.

29.5.1.3 AES

In order to replace DES with a stronger algorithm, NIST chose the Rijndael cipher as its AES in October 2000. AES is a symmetric block cipher that can operate on blocks of size 128, 196, or 256 bits. Independent of the block size, the key size may also be either 128, 196, or 256 bits long.

29.5.1.4 RSA Algorithm

Rivest, Shamir, Adleman (RSA) algorithm, which has been named after its three inventors, is the most popular asymmetric cipher. Asymmetric ciphers, also known as public key algorithms, use the public key of the recipient for encryption, and the recipient uses his private key for decryption. The patent for the RSA algorithm expired in October 2000, and it is likely to see an increase in use because of this. Since RSA is a public key algorithm, it is considerably slower than symmetric algorithms like DES, 3-DES, and more. Consequently, it is not used for media stream encryption, but is usually used for operations such as key exchange. The algorithm is straightforward:

$C = P^e \bmod n$ is the encryption algorithm

$P = C^d \bmod n$ is the decryption algorithm

where P is the binary representation of the plaintext, and C is the corresponding ciphertext. The private key is the 2-tuple (d,n), and the public key is the 2-tuple (e,n). The strength of the algorithm lies in the difficulty in factoring a product of two large prime numbers.

29.5.2   Hash and Signature Algorithms

One-way hash functions (commonly referred to as simply hash functions) are the core of authentication and integrity services. A hash function takes an input of arbitrary size and creates an output of fixed size called the message digest. This output, which is quite small, acts as a fingerprint of the input. In other words, if two different inputs are provided to a hash function, the likelihood of getting the same output should be small. This desirable functionality of hash functions is referred to as collision avoidance. The one-way nature of hash functions implies that it is extremely unlikely to determine the input, given the output of the hash function.

Hash functions may be used in several ways to provide authentication/integrity services. For instance, in conjunction with a secret key, they are used to generate MACs. Two popular mechanisms exist for this purpose: keyed MAC and *hashed MAC* (HMAC). When used in conjunction with a private key, they are used to generate digital signatures. Two popular digital signature algorithms are the RSA algorithm and the *digital signature algorithm* (DSA).

When the output of a hash function along with a user's private key is input to a suitable signing algorithm, a digital signature is obtained. This digital signature may be used for authentication, integrity, and nonrepudiation purposes. Authentication is provided because only the holder of the private key could have produced the digital signature. Nonrepudiation is provided for a similar reason, and because the recipient does not possess the private key. Message integrity is provided because any alteration of the message in transit would result in a mismatch between the included and computed message digest at the receiver.

29.5.2.1   MD5

MD5 is probably the most widely used hash function. MD5 takes an input of arbitrary size and produces a message digest of 128 bits. The MD5 algorithm internally operates with inputs of size 512 bits, but the algorithm itself (or this may also be viewed as a wrapper to the core algorithm) has functionality to break an arbitrary input into 512-bit blocks. Padding and length indication functionality is also included within the algorithm specification. Recently, certain security attacks (brute force and other forms of cryptanalysis) have been mounted successfully upon the MD5 algorithm, causing its use in certain applications is being replaced by the SHA-1 hash function.

29.5.2.2   Secure Hash Algorithm 1 (SHA-1)

The SHA-1 was standardized by the NIST in 1995. SHA-1 takes inputs of arbitrary length and creates a message digest of 160 bits. As with MD5,

SHA-1 internally takes inputs of block 512 bits. As with MD5, SHA-1 itself has functionality to break an arbitrary sized input to blocks of 512 bits to append any padding and length fields. Compared to MD5, SHA-1 is more resistant to brute force and other forms of cryptanalysis.

29.5.2.3 RSA Algorithm

The RSA algorithm may be used to compute a digital signature. The hash of the original message is signed using the sender's RSA private key to create the digital signature. The digital signature is concatenated with the original message and sent to the recipient. The recipient recomputes the hash from the message, and compares this with the output that is obtained when the digital signature is processed using the sender's RSA public key. If these two match, then the integrity of the message is intact.

29.5.2.4 DSA

The DSA was standardized by NIST within the *Digital Signature Standard* (DSS) in 1991, with revisions in 1993 and 1996. The hash of the original message, a random number, the sender's private key, and a set of global parameters is input to a signature algorithm, which outputs two parameters s and r, which are concatenated with the original message and sent to the receiver. The receiver computes the hash of the message, and sends this hash along with the parameters s, r, the sender's public key, and the set of global parameters to a verifier, which should output r if the integrity of the message is not compromised.

29.6 DDoS, Firewalls, and Private Addressing

In this section, we discuss issues dealing with firewalls, DDoS attacks and private addressing.

29.6.1 Always-On and DDoS

Both GPRS and 3G systems are always on with respect to the data path. In addition, wireline connectivity technologies such as DSL and cable-modem are also always on. The always-on nature of these technologies, while a convenience to subscribers, is an added security threat. Once an attacker is able to penetrate the user's always-on device, he or she could potentially do one of two things:

- *Access confidential information belonging to the subscriber:* This directly hurts the subscriber since sensitive information such as credit card numbers can be stolen.

- *Utilize subscriber system resources to run "zombie" programs and launch attacks such as DDoS attacks:* The victims in this case are both the subscriber whose device resources are being utilized as well as the server on which the DDoS attack is launched. In addition to his or her resources being utilized, the subscriber may also be held liable for damages to the server if he or she has not taken suitable measures to combat such an attack.

The use of personal firewalls decreases the chances of an attacker successfully attacking a user's device. In addition, some recent efforts within the IETF that facilitate in countering DDoS attacks include the following:

- Efforts are ongoing within the itrace working group in the area of ICMP traceback. The idea here is that routers, in addition to forwarding packets, would also generate ICMP packets with a small probability. These ICMP packets are destined to the same destination as the other packets. Using such ICMP traceback packets, the destination can determine the path that the packets traversed from the source. Such forward path information is useful to combat source address spoofing that is commonly used in DDoS and other attacks. While the generation of such ICMP traceback messages has clear advantages, some of the major challenges that such an effort faces include dealing with user privacy issues, ISPs wishing to keep their network topology confidential, and message authentication.

- While *intrusion detection systems* (IDS) are available for detecting network, host and application intrusion, the *Intrusion Detection Exchange Format Working Group* (IDWG) aims to specify a common message format. This would facilitate better coordination among various IDSs, thereby reducing intrusion instances by attackers.

29.6.2   Private Addressing

Private IPv4 addresses may be assigned to mobile/wireless devices in order to cope with limited IPv4 address space. Since assignment of private IP addresses comes with no added cost to the user, this is attractive from a pricing perspective as well. However, when a device within an addressing domain (usually a PLMN) needs to communicate with a device that is in a different addressing domain, one of two possible approaches is resorted to:

- NATs residing at the boundary of the addressing domains replace the private IP address with a public address when the packet leaves the PLMN, and vice-versa.

- RSIP is a mechanism whereby a suitable IP address (private or public) is assigned to the device depending on whether the remote communication endpoint resides within the same addressing domain or outside it.

29.6.2.1 NAT

Several flavors of NATs exist, and their functionality depends on the particular flavor in use. Some of the NAT flavors are traditional NAT [which include basic NAT and *network address port translator* (NAPT)], two-way NAT or bidirectional NAT, twice NAT, and multihomed NAT. Figure 29.22 illustrates the operation of a basic NAT.

One of the security-related features that NAT provides is the privacy of the endpoint within the PLMN. The remote endpoint outside the PLMN is not aware of the IP address or topology of the local endpoint within the PLMN. When a local and a remote endpoint communicate with each other at two different instances, the NAT feature prevents the remote endpoint from detecting that it is the same local endpoint, thereby facilitating privacy.

Although NATs provide privacy to some extent, they suffer from two serious drawbacks:

1. Certain applications (such as FTP and H.323) that embed network addresses within APDUs require the use of an *application level gateway*

FIGURE 29.22
Illustrating basic NAT operation.

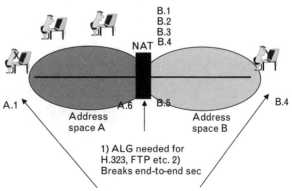

WIRELESS IP AND BUILDING THE MOBILE INTERNET

(ALG) in order to work with NATs. Such ALGs, which are usually collocated with the NAT functionality, replace the private IP address in the APDU with a public IP address.

2. NATs suffer from a serious security drawback in that they break the end-to-end IPsec security model. When a local endpoint within a PLMN establishes an IPsec AH or ESP SA for packet authentication purposes, any modification in the IP address and/or port number by the NAT would result in the packet being discarded at the remote endpoint.

29.6.2.2   RSIP

RSIP uses a different approach compared to NATs while dealing with private addressing. The approach with RSIP is that when an endpoint A (EP A), which is assigned a private address, needs to communicate with an endpoint B (EP B), which lies outside this private addressing domain, then EP A leases a public address from an RSIP server for the purpose of this communication. Since this public IP address is then used by the application, the problem faced with NATs is not encountered here.

Figure 29.23 illustrates the operation of RSIP. As seen in the figure, EP A, which is located within a private addressing domain (address space A), initiates communication with EP B, which is located in a public addressing domain (address space B). Certain IP addresses are denoted in the figure, and for illustration let us assume that IP address A.x represents the private address 10.0.0.x, while IP address B.x represents the public address 195.0.0.x. It is seen that EP A has a private address of 10.0.0.1 assigned to it, while EP B has a public address of 195.0.0.20 assigned to it. The RSIP server that resides at the boundary of the private and public domains has two interfaces: that to

FIGURE 29.23
*Illustrating RSIP
operation.*

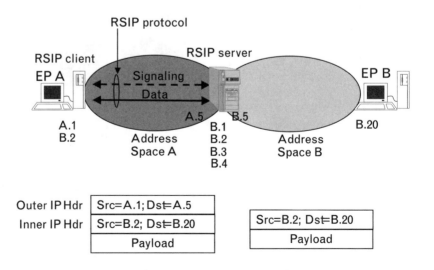

the private domain has an IP address 10.0.0.5, while that to the public domain has an IP address 195.0.0.5. The RSIP server also has a pool of four public IP addresses—195.0.0.1 through 195.0.0.4—that it may lease to its clients within the private domain.

Since EP A wishes to communicate with EP B, which is a host in the public addressing domain, EP A (which is an RSIP client) leases the public IP address 195.0.0.2 from the RSIP server using the RSIP. Certain criteria may be associated with this lease—such as the duration of the lease and whether the leased IP address may be used for communication only with EP B or with other EPs as well. Now, the IP address that the application at EP A is aware of is 195.0.0.2, and not 10.0.0.1. Any APDUs that are sent by EP A to EP B contain a network address value of 195.0.0.2. Similarly, the source IP address of the IP datagram created at EP A also contains 195.0.0.2. In order to transport this IP datagram from EP A to EP B, the datagram is tunneled through another IP datagram from EP A to the RSIP server. This outer IP datagram, which has a source IP address of 10.0.0.1 and a destination IP address of 10.0.0.5, is needed because the inner IP datagram cannot be routed within the private domain. Upon reaching the RSIP server, the outer IP header is stripped, and the original IP datagram is routed through the public domain to EP B.

Since the only address that EP B sees for EP A is 195.0.0.2, this is the address that is used for EP A. Messages sent by EP B to EP A are subsequently tunneled to EP A when they reach the RSIP server. Thus, it is seen that with the use of RSIP, one does not require an ALG in the presence of private addressing. With NATs, on the other hand, an ALG is needed.

Using RSIP, a user in a PLMN (with IPv4 addressing) can obtain end-to-end IPsec authentication while coping with the problem of limited IPv4 addresses. Since a public transport address is assigned a local device communicating with a remote endpoint outside the PLMN, no address/port translation is needed, thereby facilitating interworking with IPsec authentication.

29.6.3 Firewalls

Firewalls have traditionally been used at the periphery of corporate networks in order to protect these networks from outside attacks. In addition, they also act as a policy enforcement point whereby different policies may be enforced for different users/hosts within the protected domain. With the advent of IP within wireless networks, the resulting open architectures warrant firewalls.

Typically, firewalls have come in three categories:

1. *Packet filters:* These are of two kinds: static packet filters and statefull packet filters. Static packet filters have a static configuration based on

which packets are either granted or denied access. They do not maintain any state about the session, and they handle packets on a per-packet basis. Statefull packet filters, on the other hand, maintain information on session state, and based on this information, packets may dynamically be granted or denied access. The static nature of static packet filters makes them unsuitable for PLMNs due to the typical dynamic nature of port assignment for several PLMN sessions. Statefull packet filters are neither adequate nor economically feasible, and the performance requirements are too high for a low-end single-point-of-access gateway.

2. *Circuit gateways:* The most common circuit gateway is SOCKS. Here, a SOCKS server typically residing at the edge of the PLMN establishes a TCP connection (or circuit) to each of the communicating endpoints, and the session data is relayed by the SOCKS server. The SOCKS client within the PLMN authenticates the establishment of such a virtual circuit. Such a mechanism is also complex and not very suitable for UDP traversal.

3. *Application level gateways:* Application level gateways (also known as proxies) operate by being aware of the application. While being very secure, they are not scalable since such a solution would require that each application require a proxy at the edge of the PLMN. In addition, this is an expensive solution.

In light of the unsuitability of some of the traditional firewall solutions for PLMNs, newer solutions are being considered. One technique that holds promise is based on a firewall control interface between a device within the PLMN and the firewall. Using such an open interface, all application level functionality can be moved towards the end devices, and the end device controls the firewall to allow legitimate traffic to pass through.

29.7   Concluding Remarks

In this chapter, we discussed the security features/mechanisms present within various network technologies. Security features within several types of cellular networks (AMPS, EIA/TIA-41, EIA/TIA-136, IS-95, GSM, GPRS, and UMTS) were discussed. The security mechanisms employed by wireless link layer technologies (Bluetooth and IEEE 802.11 WLANs) were discussed. We also gave an overview of the security mechanisms employed within TCP/IP networks (IPsec, TLS, ISAKMP, and IKE). Finally, we discussed aspects dealing with firewalls, private addressing (NAT, and RSIP), and DDoS attacks.

We see that the security mechanisms and algorithms employed within wireless networks have traditionally been different from those employed within traditional wired TCP/IP networks. This has been due to a combination of factors: differing threat models, choice of mechanisms/algorithms suited to the particular environment, as well as varying standards bodies responsible for standardizing these. Although converged mechanisms for security within wireless and wireline systems are beginning to emerge and their significance is being increasingly recognized, deployments based on current disparate security technologies will have to be considered for quite some time in system design. In such an environment, an understanding of the security features employed within various network technologies is critical to understand implications to end-to-end security.

Lately, with IP technology emerging as a glue across networks, common IP level solutions are beginning to be seen. For instance, although earlier versions of WAP used *wireless TLS* (WTLS) for security, versions that are currently being developed specify the use of TLS. Such converged security mechanisms have been made possible with increased bandwidth availability, increased processing power of wireless devices, and greater coordination among standardization bodies. Significant work is currently ongoing in various standardization bodies to improve several aspects dealing with security, while taking into account this convergence aspect. These include the IETF (next-generation IPsec, AAA, Mobile IP), W3C (XML Signatures, XML Encryption), 3GPP (AKA), IEEE 802.11 (WEP), OASIS (SAML, XACML, Web services Security) and Liberty Alliance. We expect continued efforts in improving overall system security as the convergence of wireless and IP networks continues.

SELECTED BIBLIOGRAPHY

Bluetooth SIG, "Specification of the Bluetooth System: Core," July 1999.

Cheswick, W. R., and S. M. Bellovin, *Firewalls and Internet Security*, Reading, MA: Addison-Wesley, 1994.

Daemen, J., and V. Rijmen, "The Rijndael Block Cipher," 2000, http://csrc.nist.gov/encryption/aes/.

Dierks, T., and C. Allen, "The TLS Protocol Version 1.0," RFC-2246, IETF, January 1999.

ETSI, "Digital Cellular Telecommunications System (Phase 2+); General Packet Radio Service; Service description; Stage 2," EN 301 344, ETSI, July 1998.

Goodman, D. J., *Wireless Personal Communication Systems*, Reading, MA: Addison-Wesley, 1997.

Gopal, R., et al., "DOS Detection, Prevention and IDS System for Wireless Networks," *RSA Conference 2001*, Amsterdam, October 15–18, 2001.

Harkins, D., and D. Carrel, "The Internet Key Exchange (IKE)," RFC-2409, IETF, November 1998.

IEEE 802.11, http://grouper.ieee.org/groups/802/11/index.html.

Kent, S., and R. Atkinson, "IP Authentication Header," RFC-2402, Internet Engineering Task Force (IETF), November 1998.

Kent, S., and R. Atkinson, "IP Encapsulating Security Payload," RFC-2406, Internet Engineering Task Force (IETF), November 1998.

Kent, S., and R. Atkinson, "Security Architecture for the Internet Protocol," RFC-2401, November 1998.

Krawczyk, H., M. Bellare, and R. Canetti, "HMAC: Keyed-Hashing for Message Authentication," RFC-2104, February 1997.

Kumar, V., M. Korpi, and S. Sengodan, *IP Telephony with H.323: Architectures for Unified Networks and Integrated Services*, New York: Wiley, 2001.

Leech, M., et al., "SOCKS Protocol Version 5," RFC-1928, IETF, March 1996.

Liberty Alliance, http://www.projectliberty.org.

Maughan, D., et al., "Internet Security Association and Key Management Potocol (ISAKMP)," RFC-2408, IETF, November 1998.

Mehrotra, A., *GSM System Engineering*, Norwood, MA: Artech House, 1997.

NIST, "DES Modes of Operation," FIPS PUB 81, December 1980, http://www.itl.nist.gov/div897/pubs/fip81.htm.

NIST, "Secure Hash Standard," FIPS PUB 180-1, April 1995, http://csrc.nist.gov/fips/fip180-1.ps.

Rivest, R., "A Description of the RC2 Encryption Algorithm," RFC-2268, March 1998.

Rivest, R., "MD5 Digest Algorithm," RFC-1321, April 1992.

Schneier, B., *Applied Cryptography*, Second Edition, New York: Wiley, 1995.

Srisuresh, P., and M. Holdredge, "IP Network Address Translator (NAT) Terminology and Considerations," RFC-2663, IETF, August 1999.

Stallings, W., *Cryptography and Network Security*, Second Edition, Englewood Cliffs, NJ: Prentice Hall, 1999.

Thayer, R., N. Doraswamy, and R. Glenn, "IP Security Document Roadmap," RFC-2411, November 1998.

About the Editors

Sudhir Dixit received a B.E. from Maulana Azad College of Technology (MACT), Bhopal, India, an M.E. from Birla Institute of Technology and Science (BITS), Pilani, India, and a Ph.D. from the University of Strathclyde in Glasgow, Scotland, all in electrical engineering. He also received an M.B.A. from the Florida Institute of Technology in Melbourne, Florida. Dr. Dixit is currently a senior R&D manager and a site manager at Nokia Research Center in Burlington, Massachusetts. His main areas of interest are mobile/wireless Internet, optical networks, and content delivery networks. From 1991 to 1996 he was a broadband network architect at NYNEX Science and Technology (now Verizon Communications). Prior to that, he held various engineering and management positions at other major companies such as GTE, Motorola, Wang, Harris, and STL—now Nortel Europe Labs. He has published or presented more than 150 papers and has almost 30 patents either granted or pending. He has edited *IP over DWDM: Building the Next Generation Internet* (John Wiley, 2003). He has been a technical cochair and a general chair of the IEEE International Conference on Computer, Communications, and Networks, in 1999 and 2000, respectively; a technical cochair of the SPIE Conference Terabit Optical Networking from 1999 to 2001; a general chair of the Broadband Networking in the Next Millennium Conference in 2001; and a general cochair of the Opti-Comm 2002 Conference. He has also been an ATM Forum Ambassador since 1996. He has served as a guest editor for *IEEE Network, IEEE Communications Magazine,* and *Optical Networks Magazine.* From 1999–2002 Dr. Dixit was a Lightwave Series editor for the *IEEE Communications Magazine.* He is now associate editor of the *IEEE Optical Communications Magazine.* He is also on the editorial board of the *Wireless Personal Communications Journal,* the *International Journal on Wireless and Optical Communications,* and the *Journal of Communications and Networks.* His e-mail address is sudhir.dixit@ieee.org.

Ramjee Prasad received a B.Sc. (eng.) from the Bihar Institute of Technology in Sindri, India, in 1968, and an M.Sc. (eng.) and a Ph.D. from the Birla Institute of Technology (BIT), Ranchi, India, in 1970 and 1979, respectively.

He joined BIT as a senior research fellow in 1970 and became an associate professor in 1980. While he was with BIT, he supervised a number of research projects in the area of microwave and plasma engineering. From 1983 to 1988, he was with the University of Dar es Salaam (UDSM), Tanzania, where he became a professor of telecommunications in the Department of Electrical Engineering in 1986. At UDSM, he was responsible for the collaborative project Satellite Communications for Rural Zones with Eindhoven University of Technology, Netherlands. Dr. Prasad has previously worked with the Telecommunications and Traffic Control Systems Group at DUT, where he was actively involved in the area of wireless personal and multimedia communications (WPMC). He was the founding head and program director of the Center for Wireless and Personal Communications (CEWPC) of International Research Center for Telecommunications—Transmission and Radar (IRCTR). Since 1999, Dr. Prasad has been with Aalborg University, as the codirector of the Center for PersonKommunikation (CPK), and he holds the chair of wireless information and multimedia communications. He was involved in the European ACTS project FRAMES (Future Radio Wideband Multiple Access Systems) as a DUT project leader. He is a project leader of several international, industrially funded projects. He has published more than 300 technical papers, contributed to several books, and has authored, coauthored, and edited 10 books: *CDMA for Wireless Personal Communications*, *Universal Wireless Personal Communications*, *Wideband CDMA for Third Generation Mobile Communications*, *OFDM for Wireless Multimedia Communications*, *Third Generation Mobile Communication Systems*, *WCDMA: Towards IP Mobility and Mobile Internet*, *Towards a Global 3G System: Advanced Mobile Communications in Europe, Volumes 1 & 2*, *IP/ATM Mobile Satellite Networks*, and *Simulation and Software Radio for Mobile Communications*, all published by Artech House. His current research interests lie in wireless networks, packet communications, multiple-access protocols, advanced radio techniques, and multimedia communications.

Dr. Prasad has served as a member of the advisory and program committees of several IEEE international conferences. He has also presented keynote speeches and delivered papers and tutorials on WPMC at various universities, technical institutions, and IEEE conferences. He was also a member of the European Cooperation in the Scientific and Technical Research (COST-231) project dealing with the evolution of land mobile radio (including personal) communications as an expert for the Netherlands, and he was a member of the COST-259 project. He was the founder and chairman of the IEEE Vehicular Technology/Communications Society Joint Chapter, Benelux Section, and is now the honorary chairman. In addition, Dr. Prasad is the founder of the IEEE Symposium on Communications and Vehicular Technology (SCVT) in the Benelux, and he was the symposium chairman of SCVT'93.

Dr. Prasad is the coordinating editor and editor-in-chief of the *Kluwer International Journal on Wireless Personal Communications* and a member of the editorial board of other international journals, including the *IEEE Communications Magazine* and *IEE Electronics Communication Engineering Journal*. He was the technical program chairman of the PIMRC'94 International Symposium held in The Hague, Netherlands, September 19–23, 1994, and also of the Third Communication Theory Mini-Conference in Conjunction with GLOBECOM'94, held in San Francisco, California, November 27–30, 1994. He was the conference chairman of the 50th IEEE Vehicular Technology Conference and the steering committee chairman of the second International Symposium WPMC, both held in Amsterdam, the Netherlands, September 19–23, 1999. He was the general chairman of WPMC'01, which was held in Aalborg, Denmark, September 9–12, 2001.

Dr. Prasad is also the founding chairman of the European Center of Excellence in Telecommunications, known as HERMES. He is a fellow of IEE, a fellow of IETE, a senior member of IEEE, a member of the Netherlands Electronics and Radio Society (NERG), and a member of IDA (Engineering Society in Denmark).

Authentication header (AH), 599–600
  defined, 600
  illustrated, 601
  in transport mode, 601
  in tunnel mode, 601
  See also IPsec
Automatic repeat requests (ARQ), 188
  EMBSD of, 199
  error correction, 477
  FULL, 196, 198, 199
  relative number of retransmissions for, 199
  send-and-wait, 196
  SPB, 196, 198, 199
  SPB-MARK, 195
Automatic retransmission/fragmentation
    (ARQ/F), 285–86, 299, 300

B

Backbone networks, 9
Bandwidth management, 83
Base Station Subsystem GPRS Protocol (BSSGP), 33
Basic access network (BAN), 528, 529
  capacity, 538
  system bottleneck at, 540
Bearer service manager (BSM), 138, 139
  capability, 139
  defined, 138
Belief function (BF) calculus, 327–30
  abstract example, 327–30
  defined, 327
  example frames and compatibility relations, 329
  future directions, 330
  motivating factors, 327
  overview, 327
Billing system, ad hoc networks, 110
Binary format for scene description (BIFS), 577
Binary Runtime Environment for Wireless
    (BREW), 25
Bit error rate (BER), 153, 155, 198
  residual, 155
  tolerated, 153
Bluetooth, 525, 595–98
  authentication in, 598
  combination key, 596, 597
  defined, 595
  encryption/decryption in, 598, 599
  encryption key, 596, 597–98

initialization key, 596–97
key generation techniques, 596–98
link key, 595–96
master key, 596, 597
unit key, 595, 596
See also Security
Broadband wireless access (BWA), 279
  network components, 280
  next-generation (2G-BWA), 280, 284–89
Building wide experiment, 410–12
  results, 410–12
  setup, 410
  summary of results, 411
  See also Handoff initiation
Burst tolerance (BT), 84

C

Caches
  HBR entries, 368
  paging, 120
  routing, 120
  See also Web caching
Call admission control (CAC), 84, 228, 231–48
  analytical method, 240–42
  defined, 231
  dynamic systems, 232
  evaluation metrics, 232–33
  MdCAC, 233, 242
  MsCAC, 233, 242–43
  overview, 232–33
  performance comparison, 244–48
  performance evaluation, 233–48
  performance measures, 240
  policies, 232
  QoS differentiation paradigms, 237–39
  soft decision (SCAC), 228, 233, 243–44
  static systems, 232
  system modeling, 233–44
  system parameter summary, 245–46
  system principles, 233–37
  traffic model, 239
  use, 231
Care-of-address (COA), 118–19
  example, 125–26
  link-local address, 124
  Mobile IP, 368
  obtaining, 123–24

Recent Titles in the Artech House Mobile Communications Series

John Walker, Series Editor

FIGURE 28.4
F_{max} and restriction of processing power.

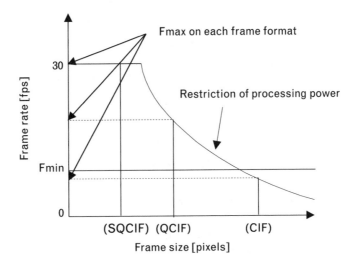

gradually increases the encoding bit rate if it observes no rate reduction over a period of time.

28.5.4 MPEG Transcoder

Generally, transcoders decode the original stream and then reencode the decoded media using a different encoding setting. In this case, it is possible to change the bit rate dynamically according to the mobile device's condition. For the bit rate downconversion of MPEG video streams, there are three main factors:

1. *Frame rate conversion:* It reduces the frame rate compared to the original stream. In MPEG video encoding, there are three picture coding types: I-pic (INTRA-coded pictures), P-pic (INTER-frame predictive coded pictures), and B-pic (bidirectionally predicted pictures). Since B-pics are not used as reference pictures by other frames, simple frame rate reduction can be achieved by eliminating the B-pic data from the original stream. This method, however, is not applicable for the elimination of I-pics or P-pics since it temporarily degrades picture quality. Therefore, full decoding of the original stream is required to realize the conversion to an arbitrary frame rate.

2. *Image size conversion:* It reduces the size of the image from the original stream. It also requires full decoding of the original stream to realize the conversion to arbitrary image size. If the H/V reduction ratio of the image size is a power of two, the resizing of the images can be done only by a simple subsampling with a two-tap filter.

3. *Format conversion:* It converts the coding format from MPEG-2 to MPEG-4, which will be commonly used in the Internet and mobile

For further information on these and other Artech House titles,
including previously considered out-of-print books now available through our In-Print-Forever® (IPF®) program, contact:

Artech House
685 Canton Street
Norwood, MA 02062
Phone: 781-769-9750
Fax: 781-769-6334
e-mail: artech@artechhouse.com

Artech House
46 Gillingham Street
London SW1V 1AH UK
Phone: +44 (0)20 7596-8750
Fax: +44 (0)20 7630-0166
e-mail: artech-uk@artechhouse.com

Find us on the World Wide Web at:
www.artechhouse.com